Advances in Experimental Medicine and Biology

Volume 735

For further volumes:
http://www.springer.com/series/5584

John D. Lambris • V. Michael Holers • Daniel Ricklin
Editors

Complement Therapeutics

Springer

Editors
John D. Lambris
Department of Pathology
and Laboratory of Medicine
University of Pennsylvania
Philadelphia, PA, USA

V. Michael Holers
Departments of Medicine
and Immunology
University of Colorado
School of Medicine
Aurora, CO, USA

Daniel Ricklin
Department of Pathology
and Laboratory of Medicine
University of Pennsylvania
Philadelphia, PA, USA

ISSN 0065-2598
ISBN 978-1-4614-4117-5 ISBN 978-1-4614-4118-2 (eBook)
DOI 10.1007/978-1-4614-4118-2
Springer New York Heidelberg Dordrecht London

Library of Congress Control Number: 2012945942

Printed on acid-free paper

Springer is part of Springer Science+Business Media (www.springer.com)

Preface

The past decade has witnessed a new surge in the development of complement-targeted therapeutics, fueled by the successful launch of the first complement pathway-specific inhibitors and the discovery of both orphan and highly prevalent diseases with contributions from complement. Apart from offering new hope for patients suffering from such diseases, the study of complement pathways, mutations, and deficiencies also teaches us important lessons about the role of complement in health and disease as well as allows us to refine our models and tools for applied and basic research. These exciting developments were at the center of the Fifth International Complement Therapeutics Conference (June 22–27, 2011) in Rhodes, Greece, where expert scientists from academia, clinical institutions, and the pharmaceutical industry convened to discuss current progress, challenges, trends, and emerging applications in complement therapeutics. This volume represents a collection of highlights from this conference covering topics from functional and molecular descriptions of complement targets to development of complement-directed inhibitors, drug targeting to cells, tissues, and biomaterials, emerging clinical applications (e.g., in ocular, dental, and transplantation medicine), as well as novel diagnostic approaches.

We would like to express our sincere thanks to all the contributing authors for the time and effort they have devoted to writing exceptionally informative chapters in a book that will have a significant impact on the field of complement-related drug discovery and development. We would also like to thank Rodanthi Lambris for collating the chapters and Dimitrios Lambris of Conferex LLC for managing the organization of this exceptional meeting. Finally, we also thank Samantha Lewis of Springer Publishers for her supervision in this book's production.

Philadelphia, PA, USA — John D. Lambris
Aurora, CO, USA — V. Michael Holers
Philadelphia, PA, USA — Daniel Ricklin

Contents

Chapter 1
Progress and Trends in Complement Therapeutics

Daniel Ricklin and John D. Lambris

Abstract The past few years have proven to be a highly successful and exciting period for the field of complement-directed drug discovery and development. Driven by promising experiences with the first marketed complement drugs, increased knowledge about the involvement of complement in health and disease, and improvements in structural and analytical techniques as well as animal models of disease, the field has seen a surge in creative approaches to therapeutically intervene at various stages of the cascade. An impressive panel of compounds that show promise in clinical trials is meanwhile being lined up in the pipelines of both small biotechnology and big pharmaceutical companies. Yet with this new focus on complement-targeted therapeutics, important questions concerning target selection, point and length of intervention, safety, and drug delivery emerge. In view of the diversity of the clinical disorders involving abnormal complement activity or regulation, which include both acute and chronic diseases and affect a wide range of organs, diverse yet specifically tailored therapeutic approaches may be needed to shift complement back into balance. This chapter highlights the key changes in the field that shape our current perception of complement-targeted drugs and provides a brief overview of recent strategies and emerging trends. Selected examples of complement-related diseases and inhibitor classes are highlighted to illustrate the diversity and creativity in field.

1.1 Of Dogmas, Challenges, and Opportunities: The Changing Field of Complement Research

It is very rare that complement research in general, and complement-directed drug discovery in particular, finds itself in the spotlight of media attention. Yet success stories about the off-label use of the clinical anti-C5 antibody Eculizumab (Soliris, Alexion Pharmaceuticals) in the recent outbreak of enterohemorrhagic *E. coli* in Europe (Laursen 2011; Lapeyraque et al. 2011) or the promising results with a soluble form of complement receptor 1 (sCR1, Mirococept) in transplantation medicine (Sample 2010; Sacks and Zhou 2012) sparked a general interest in the field. While this attention may not persist at such a high level, it clearly underscores a new perception of the role of complement in health and disease and highlights the promise of therapeutic intervention in the complement cascade. Its upstream positioning in inflammatory processes and modulatory involvement in many (patho)

D. Ricklin (✉) • J.D. Lambris
Department of Pathology and Laboratory Medicine, University of Pennsylvania, 401 Stellar Chance, 422 Curie Blvd, Philadelphia, PA 19104-6100, USA
e-mail: ricklin@upenn.edu; lambris@upenn.edu

J.D. Lambris et al. (eds.), *Complement Therapeutics*, Advances in Experimental Medicine and Biology 735, DOI 10.1007/978-1-4614-4118-2_1, © Springer Science+Business Media New York 2013

physiological processes indeed render complement an attractive target system. Research in recent years has unraveled some of the mysteries about complement, shaken various dogmas, and revealed fascinating new insights that are of importance for work related to complement-directed drug discovery and beyond.

The most well-known function of complement is undoubtedly its role in microbial defense, where it recognizes, tags, and helps to eliminate intruders such as bacteria, viruses, fungi, or parasites. However, the surface recognition properties of complement are not restricted to pathogen-associated molecular patterns (PAMPs) but also include danger-, damage-, or disease-related patterns of host cells/tissues, immune complexes, or other foreign surfaces such as biomaterials. The severity and outcome of complement response to these distinct triggers have to be tuned carefully and may include opsonization, clearance, elimination, and/or danger signaling to inflammatory and adaptive systems. This tuning is dependent on the context-specific interplay of some 50 different proteins encompassing pattern recognition proteins, proteases and their complement component substrates, soluble and membrane-bound regulators, and various receptors (Ricklin and Lambris 2007a; Ricklin et al. 2010). While often organized in three distinct initiation pathways, that is, the classical, lectin, and alternative pathways (CP, LP, and AP, respectively; Fig. 1.1), it becomes increasingly evident that there are several interconnectivities and bypasses of the complement activation pathways; the involvement of these pathways may therefore greatly vary depending on the trigger, as well as other factors. Independent of the initiation route, amplification of the response by the AP, via formation of C3 convertases that cleave the central component C3 into an anaphylatoxin (C3a) and an opsonin (C3b) fragment, often causes the lion's share of overall complement activation. Opsonization with C3b and its degradation fragments iC3b and C3d facilitates both phagocytosis and adaptive immune signaling via complement receptors CR1 to CR4. Deposited C3b not only fuels amplification by forming additional C3 convertases but also induces the generation of C5 convertases. Cleavage of C5 generates a highly potent anaphylatoxin (C5a) with chemotactic and proinflammatory capacities as well as C5b, which initiates the formation of the terminal complement complex (TCC) that may induce lysis of susceptible cells or participate in signaling events. On host cells, a panel of "regulators of complement activation" (RCA) and other inhibitors tame amplification and the accumulation of effector molecules (Ricklin et al. 2010; Carroll and Sim 2011). While the underlying processes within the cascade during complement activation are highly complex and diverse, this level of complexity is not only essential for an adequate response to distinct triggers but also offers a wide panel of potential targets for therapeutic intervention. Likewise, it becomes apparent that complement is by far not an isolated system but heavily intertwined with other immune, defense, and inflammatory systems (e.g., toll-like receptors, the cytokine network, adaptive immunity, and coagulation) and involved in homeostatic, developmental, regenerative, or metabolic processes (Ricklin et al. 2010). For example, some proteases involved in coagulation (e.g., thrombin) or other networks (e.g., elastase) have been shown to directly activate C3 and C5 (Fig. 1.1) (Amara et al. 2010; Oikonomopoulou et al. 2012).

At the same time, we are getting to know the key players, and thereby potential drug targets, in this intricate complement network more closely than ever before. A wealth of high-resolution structural models and sensitive functional assays have offered unprecedented insight into the shape, conversions, and ligand interaction sites of complement components and revealed fascinating molecular mechanisms that help explain the ability of complement to direct and tune its activities to the specific task and target surface. While crystallography is an essential driving force in this development, orthogonal biophysical technologies like NMR, hydrogen/deuterium exchange mass spectrometry, small-angle x-ray spectroscopy, electron microscopy, or interaction analysis by surface plasmon resonance have filled many gaps and critically contributed to fully describe the structural and functional complexity of complement proteins (Arlaud et al. 2007; Ricklin and Lambris 2007b; Schuster et al. 2007; Perkins et al. 2002). Exemplary for this development is the structural characterization of C3, which progressed from describing the C3d fragment (Nagar et al. 1998) to the publication of full native C3 (Janssen et al. 2005) and provided critical insight into the dynamic activation process and

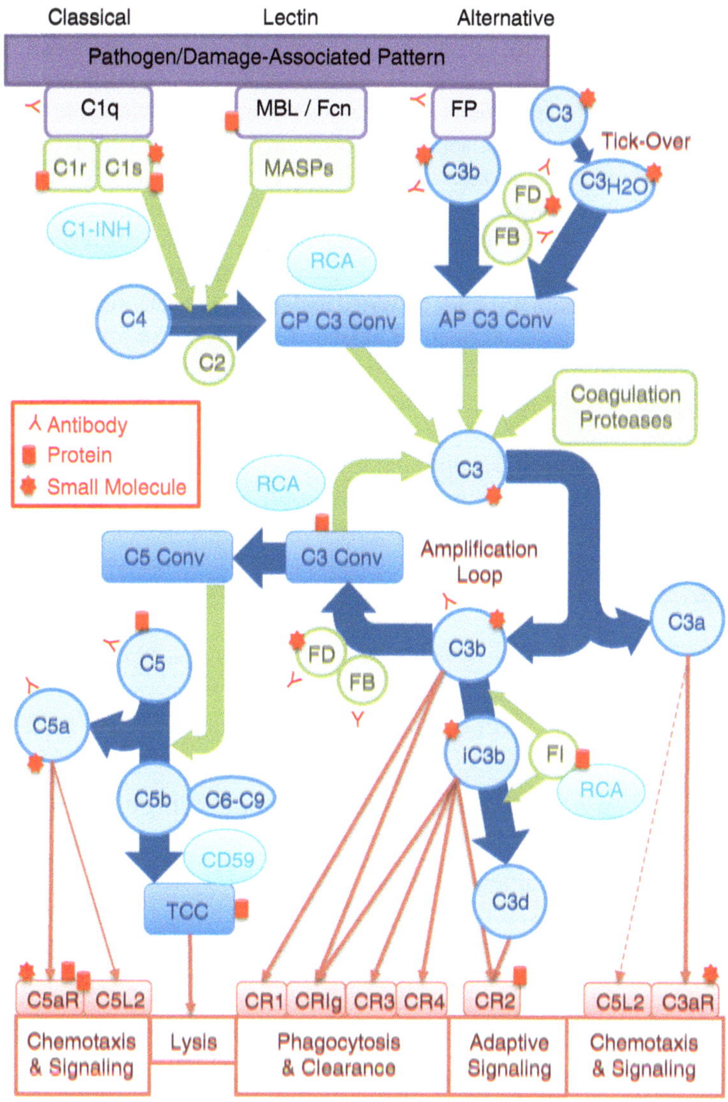

Fig. 1.1 Simplified scheme of the complement cascade and major points of therapeutic intervention. Pattern recognition molecules are colored in *purple*, proteases in *green*, complement components in *blue*, regulators in *cyan*, and receptors in *dark red*. *Red symbols* mark major therapeutic classes (small molecules, proteins, and antibodies; see legend) and are depicted next to their target protein. Abbreviations used: *C1–C9* complement components 1–9, *C1-INH* C1 esterase inhibitor, *C3aR* C3a receptor, *C5aR* C5a receptor, *Conv* convertase, *CR* complement receptor, *CRIg* complement receptor of the immunoglobulin family, *FB* factor B, *Fcn* ficolins, *FD* factor D, *FI* factor I, *FP* properdin, *MASPs* MBL-associated serine proteases, *MBL* mannose-binding lectin, *RCA* regulators of complement activation, *TCC* terminal complement complex

newly exposed binding sites in C3b (Janssen et al. 2006; Schuster et al. 2008; Wiesmann et al. 2006). Meanwhile, structural data of C3b-ligand complexes have included activators (Forneris et al. 2010; Torreira et al. 2009), regulators (Wu et al. 2009), and receptors (Wiesmann et al. 2006). More recently, the integration of structural and functional studies revealed mechanistic aspects of the C3 convertase (Forneris et al. 2010; Rooijakkers et al. 2009) and shed light into self-recognition and regulation by factor H (FH) (Morgan et al. 2011; Kajander et al. 2011), the interaction of C3d with CR2 (van den Elsen and Isenman 2011; Shaw et al. 2010), or the structure of iC3b (Alcorlo et al. 2011). Yet structural insight is hardly restricted to C3 and the AP, and important contributions have recently been made in respect to the CP (e.g., pattern recognition by C1q (Garlatti et al. 2010)), the LP (e.g., MBL-MASP1 complex (Gingras et al. 2011)), or the terminal pathway (e.g., the structures of C5 or the C5b-C6 complex (Laursen et al. 2010, 2011; Hadders et al. 2012)). Importantly, and in addition to providing mechanistic insight, such structural studies also offer a highly valuable base for understanding disease-related effects of mutations and polymorphisms in complement proteins (Rodriguez de Cordoba et al. 2011). Clearly, this progress in providing molecular details of complement proteins will largely and beneficially influence complement-related drug discovery, either by understanding underlying processes or by allowing improved rational design. And there is a good chance that many more structures covering major drug targets will be released in the future.

Finally, genome-wide association studies (GWAS) and genomic/proteomic analyses have provided a steady flow of information that confirmed or extended many disease links and put new potential disease hot spots on the map. Whereas the link between polymorphisms in FH and the risk for developing AMD likely represents the most prominent case, strong connections have also been found for other complement components in AMD, for some RCA's in kidney diseases like atypical hemolytic uremic syndrome (aHUS) and membranoproliferative glomerulonephritis type II (MGPN II), or for several members of the CP in systemic lupus erythematosus (SLE) (Degn et al. 2011). More recently, important disease associations have also been revealed between MASP-1/3 and 3MC syndrome (Sirmaci et al. 2010) and between CR1 and the development of Alzheimer's disease and depression (Hamilton et al. 2012; Lambert et al. 2009).

1.2 Finding the Achilles' Heel of Complement, One Disease at a Time

1.2.1 A Broad Range of Diseases Warrant a Broad Range of Strategies

In view of its deep involvement in many key physiological processes and its complex interplay of several dozen specialized proteins, it is not surprising that any disruption in the balance of complement activation and regulation may have pathological consequences. Indeed, the list of diseases with contribution of complement has been steadily growing in the past few decades and encompasses autoimmune, inflammatory, hematological, and neurodegenerative disorders, as well as cancer, ischemia/reperfusion (I/R) injuries, and sepsis (Ricklin and Lambris 2007a; Ricklin et al. 2010; Lachmann and Smith 2009) (Fig. 1.2). In addition, and despite progress in material sciences, the foreign surfaces present in biomaterials ranging from medical implants and hemodialysis filters to drug delivery systems such as liposomes and micro-/nanoparticles may still trigger complement to a significant degree and contribute to clinical complications (Ekdahl et al. 2011; Nilsson et al. 2010). Finally, transplantation medicine faces multiple complement-related challenges that include antibody-mediated rejection of the allo- or xenotransplant, as well as I/R effects (Asgari et al. 2010). The reported diversity in involved components and pathways, affected organs, time courses, and case numbers for different diseases renders a "one-size-fits-all" complement treatment virtually impossible and suggests that therapeutic concepts have to be tailored to specific disorders.

The direct activation of complement by danger- and disease-associated patterns places it upstream of many inflammatory reactions that are triggered by foreign or damaged surfaces, thereby supporting the concept that complement inhibition should be considered in various inflammatory diseases. The same might be true for age-related diseases that are influenced by the slow accumulation of debris or plaques that may in turn trigger complement activation; such connections have, for example, been proposed for both age-related macular degeneration (AMD) and Alzheimer's disease, yet with distinct mechanisms (Anderson et al. 2010; Fonseca et al. 2011; Veerhuis et al. 2011; Proitsi et al. 2012). Again, complement inhibition could offer an early point of intervention that could retard, prevent, or

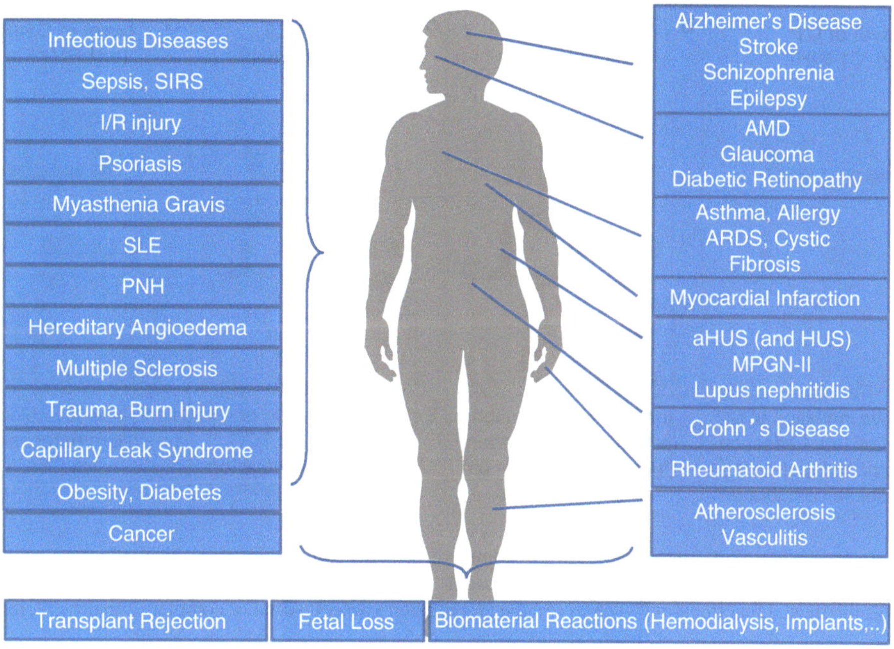

Fig. 1.2 Diseases and clinical disorders/complications with demonstrated or suspected involvement of complement. Abbreviations used: *aHUS* atypical hemolytic uremic syndrome, *AMD* age-related macular degeneration, *I/R* ischemia/reperfusion, *PNH* paroxysmal nocturnal hemoglobinuria, *SIRS* systemic inflammatory response syndrome, *SLE* systemic lupus erythematosus

even reverse disease progression. Especially in the case of such chronic and slowly accumulative diseases, we begin to recognize that the kinetics of complement turnover may be essential and that even small changes in the activity of individual complement proteins may add up to a significant effect over the progression of the disease (Heurich et al. 2011).

1.2.2 *Is There an Optimal Point of Intervention?*

In principle, the complement cascade offers points of intervention at almost any level ranging from initiation and primary activation to amplification, effector signaling, and lysis. While discussions about complement-related drug discovery occasionally center on the identification of the ideal target for "general" complement inhibition, the diversity of triggering patterns, pathomechanisms, and involved pathways and components more likely requires a careful and disease-specific selection of targets, treatment regimens, and delivery routes. The ideal compromise between sufficient blockage of disease-causing complement activation and preservation of the network's immune surveillance and homeostatic capacities therefore must be reassessed for each clinical disorder. In general, upstream intervention at the level of a specific initiation step, such as the inhibition of C1r/s by C1 esterase inhibitor (C1-INH), may effectively shut down activation and subsequent generation of effector molecules caused by an individual pathway without affecting the protective functions of the other pathways (Fig. 1.3a). However, this requires profound knowledge about the triggering mechanism and may not sufficiently work if more than one route contributes to the overall response. On the other hand, inhibiting at the level of C3 activation, either by blocking C3 directly (e.g., using compstatin) or by acting on the convertase, will efficiently block all activation, amplification, and effector routes independent of the disease mechanism (Fig. 1.3b) but may theoretically bear a higher risk of affecting

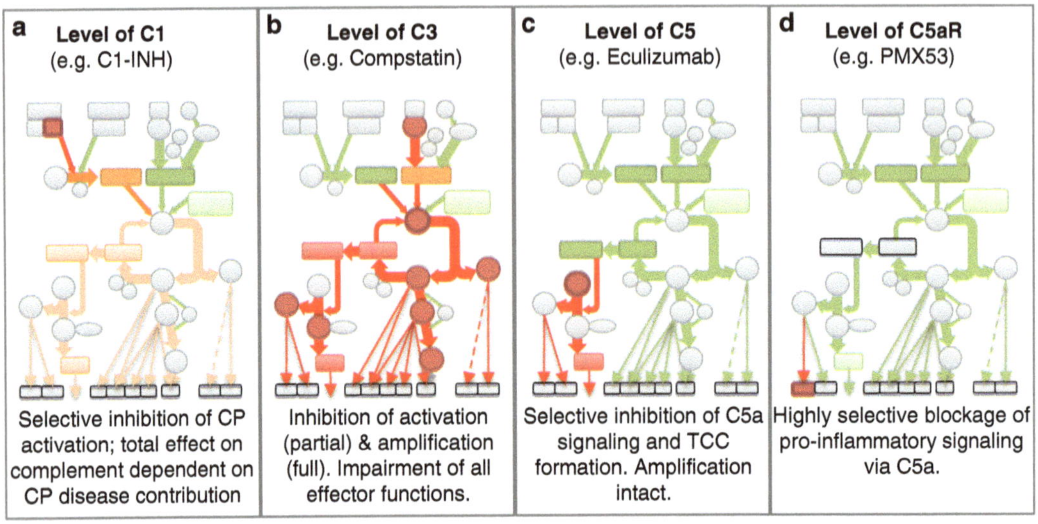

Fig. 1.3 Points of therapeutic intervention in the complement cascade and their theoretical effect on activation and effector mechanisms. The cascade organization in all panels corresponds to the one depicted in Fig. 1.1. *Green coloring* symbolizes unaffected functionality, whereas *orange* and *red* tones reflect partial or complete impairment, respectively

physiologically beneficial complement functions (see below). Finally, blockage at the terminal pathway level can often be tailored to eliminate one or several effector steps depending on the target. For example, blocking C5 (e.g., by Eculizumab) will impair both the formation of the TCC and inflammatory signaling by C5a (Fig. 1.3c), whereas only the latter, signaling function, is blocked when targeting the C5a receptor (C5aR) with antagonists like PMX53 (Fig. 1.3d). Such downstream interventions are often highly efficient in suppressing the major clinical manifestations of excessive complement activity while fully preserving key functions like opsonization. Yet, they also tend to be more "symptomatic" and do not eliminate the causative activation and amplification of the cascade. This dilemma is perhaps best illustrated in the case of paroxysmal nocturnal hemoglobinuria (PNH), where blockage of C5 by Eculizumab efficiently prevents erythrocyte lysis as the predominant clinical issue, thereby dramatically increasing the patient's quality of life. However, studies have also revealed that uninhibited amplification via the AP leads to an accumulation of C3-derived opsonins, which increases extravascular hemolysis of the resulting "ghost erythrocytes." Inhibition at the level of C3 or C3 convertases, for example, by anti-C3b antibodies, compstatin, or targeted regulators (TT30; see below), is therefore being evaluated as an alternative option (Risitano et al. 2011; Parker 2012; Luzzatto et al. 2010).

Besides the point of intervention, decisions about the type of administration (i.e., local vs. systemic) and treatment duration (i.e., acute/short term vs. chronic/continuous) are at least equally important and have to be critically evaluated for each disease (Fig. 1.4). In the case of AMD, for example, local injection of complement-targeted drugs into the eye combines several advantages concerning tissue targeting, pharmacokinetics, and safety (at least regarding preservation of systemic complement activation) but comes at the price of comparatively inconvenient, invasive, and costly intravitreal injection. As AMD has been considered to reflect a local manifestation of a more systemic complement imbalance, the potential of treating AMD by systemic administration of complement drugs has been evaluated with some promising results; however, careful evaluation of the therapeutic and economic benefits will certainly be necessary in such cases. On the other hand, the suppression of inflammatory effects and increased organ preservation by complement inhibitors in the case of severe sepsis, as recently demonstrated with compstatin treatment, will most likely be based on a short-term systemic administration in a hospital setting (e.g., via i.v. infusion) and less dependent on pharmacokinetic restrictions.

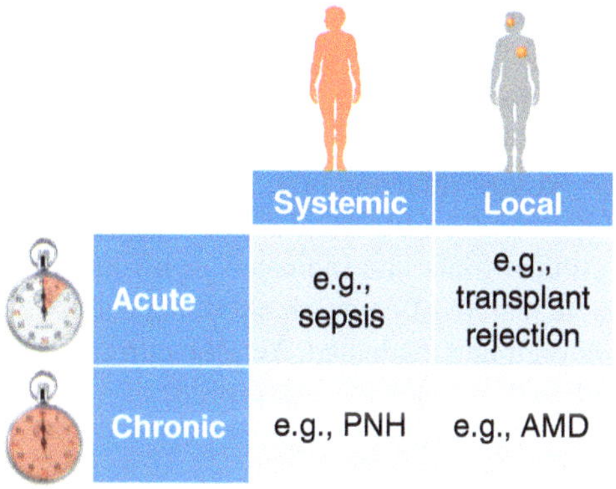

Fig. 1.4 Examples of complement-related disorders with different potential requirements concerning drug administration

1.2.3 Safe Ride or Tightrope Walk? A (Cautious) Risk Assessment

The strong involvement of complement in microbial defense and many other physiological processes intuitively raises questions about the safety and potential risks of therapeutic complement inhibition, especially in the case of systemic treatment over long periods. While long-term clinical data are still scarce, several clinical trials involving diverse complement inhibitors and the experiences of patients receiving complement-directed drugs (i.e., Eculizumab and C1-INH preparations) have largely shown beneficial safety profiles. Previous discontinuations of clinical trials with complement inhibitors were mostly caused by lower-than-expected efficacy or strategic business decisions rather than safety concerns. Importantly, in the case of Eculizumab, the long-term blockage of C5 activation in PNH patients did not only prove to be largely safe (Roth et al. 2011) but was also shown to normalize important immune parameters, such as the number of B lymphocytes, which are normally altered in patients suffering from PNH (Alfinito et al. 2011).

As evident from patients with primary deficiencies or total dysfunctions in specific complement proteins, the highest potential risk may be a higher susceptibility to bacterial infections. Whereas some individuals deficient in C5 have shown a higher rate of neisserial infections, the pathogen spectrum appears somewhat more extended in the case of C3 deficiencies (Skattum et al. 2011; Reis et al. 2006). Interestingly, though, these infections appear to be more frequent in childhood and tend to improve when deficient individuals become adults, and it has been speculated that complement-related bacterial defense becomes less important once our body gains full capacity to produce high IgG levels (Lachmann and Smith 2009). Importantly, across Eculizumab-related clinical PNH studies, the overall infection rate was not significantly increased between patients receiving the antibody drug or a placebo (three cases of *Neisseria meningitidis* were reported during these trials) (Dmytrijuk et al. 2008). In some cases, the risk or severity of potential infections can be further reduced by vaccination, as in the case of Eculizumab where treatment with a *N. meningitidis* vaccine is required prior to the start of therapy.

Of course, the question arises whether complete systemic complement blockage is actually necessary, or even achievable, in all cases. In many diseases, it is an imbalance between activation and regulation rather than a complete dysfunction that drives complement's contribution to pathological states. Even a partial or targeted inhibition strategy may therefore sufficiently help restore the balance of complement activity and potentially reverse disease symptoms. In the case of complement targets with high abundance (e.g., C3) or very rapid turnover in the body (e.g., FD), it may also prove technically difficult to achieve full complement inhibition over a long treatment period. Such partial inhibition may pose an even lower risk for infection; indeed, an assessment of patients who were positive for nephritic factor (i.e., antibodies that stabilize convertases and thereby largely deplete C3 stores) indicated that residual C3 levels far below 10% of the normal range were sufficient for fighting intruders, as these patients did not display clinically significant infection rates (Skattum et al. 1997). Finally, many of the potential applications for complement-targeted drugs involve acute and/or time-restricted treatment (e.g., during hemodialysis or cardiopulmonary bypass surgery) or local administration to tissues with comparatively low risk of infection (e.g., vitreous in the case of AMD), thereby further reducing potential risks.

Clearly, more data from clinical trials and indicated or off-label use of the marketed drugs are needed to paint a more complete picture of complement therapeutics-related risks, and future studies still need to be carefully monitored and evaluated. Yet, the currently available data so far appear to indicate a rather beneficial and well-manageable risk profile.

1.3 A Handyman's Toolbox of Complement Inhibition

The points concerning disease diversity, pathway involvement, wealth of intervention points, safety and treatment regimens discussed above strongly support the notion that a one-size-fits-all approach will likely not suffice for managing a diverse set of complement-related disorders. Ideally, we should arrive at a panel of specific inhibitors that allow for the selective inhibition of initiation, amplification, and effector steps. Luckily, the field has experienced a high level of creativity and productivity over the past few years that have already produced an impressive panel of inhibitors, some of which have been evaluated in clinical trials. Similar to drug discovery in general, complement inhibitors have seen a shift from small molecules to biopharmaceuticals such as proteins, antibodies, aptamers, and peptides, though this trend seems to be much more pronounced for complement-related drugs. The following sections quickly summarize concepts and trends in the major drug classes and name a few examples of drugs in clinical development.

1.3.1 Towards More Selective Protease Inhibitors

Owing to complement's cascade-type architecture with strong involvement of several serine proteases, and due to the high drugability of proteases, enzymes such as C1s or FB have been among the earliest targets for complement-directed drugs. While several attempts have been undertaken to arrive at small molecule inhibitors of complement proteases, poor pharmacokinetic profiles and insufficient target specificity have so far prevented such drugs from entering clinical applications (Qu et al. 2009). For example, BCX-1470 (BioCryst) was found to inhibit both FD and C1s with high activity and has been used to prevent skin edema during Arthus reaction in rats (Szalai et al. 2000); however, no clinical development has been reported.

Today, human C1 esterase inhibitor (C1-INH) is the only complement-directed protease inhibitor in the clinic. This large glycoprotein that belongs to the SERPIN family is primarily used in connection with hereditary angioedema (HAE) (Tourangeau and Zuraw 2011), a genetic disease caused by a deficiency of functional C1-INH and characterized by a swelling of subcutaneous tissue. While C1-INH preparations have been available for the treatment of HAE for several years in Europe and other regions, the introduction of such drugs into the US market happened only recently. Whereas the C1-INH drug Cinryze (ViroPharma) was approved by the FDA for prophylactic treatment of HAE in late 2008, Berinert (CSL Behring) was approved in 2009 for the treatment of acute HAE attacks (Gompels and Lock 2011; Keating 2009). Although C1 is eponymous of the inhibitor, C1-INH is not C1s-specific but blocks additional proteases of the complement (C1r, MASP2), fibrinolytic (plasmin), coagulation (thrombin, factor Xa), and contact systems (plasma kallikrein, FXIIa) (Davis et al. 2010).

In addition, C1-INH may exert immunomodulatory activities that are not related to protease inhibition (Zeerleder 2011; Thorgersen et al. 2010). In fact, it is likely that the pathogenesis of HAE, and thereby the main pharmacological activity of the C1-INH treatment for this indication, is more closely related to the kallikrein-bradykinin system than to complement.

Despite the uncertain complement connection for the current use of C1-INH in HAE treatment, the availability of approved drugs with activity towards the classical pathway may largely facilitate applications in complement-related diseases (Beinrohr et al. 2008; Kirschfink and Mollnes 2001). Indeed, C1-INH has shown promising effects in various disease models ranging from I/R injury and transplantation (Banz and Rieben 2011; Tillou et al. 2010) to sepsis (Igonin et al. 2011). Whereas changes in the C1-INH status have also been observed in AMD, the therapeutic implications have to be further explored (Gibson et al. 2012).

Current efforts to modulate the action of complement proteases appear to primarily focus on the inhibition of substrate binding via monoclonal antibodies (e.g., against FB, FD, or MASP; see next chapter). In addition, peptide mimics of the scissile loop areas have been suggested for inhibiting C2 and FB (Ruiz-Gomez et al. 2009; Halili et al. 2009). More recently, a targeted phage-display library approach starting from the sequence of sunflower trypsin inhibitor has been used to arrive at peptidic inhibitors of MASP-1 and MASP-2 (Kocsis et al. 2010). Yet, the availability of improved structural models for activated protease states as in the case of FD (Forneris et al. 2010) may well revive the design and optimization of new small molecule protease inhibitors.

1.3.2 Block It: Protein Interaction Inhibitors from Big to Small

Structural studies of several target-ligand complexes of the complement system in recent years have impressively demonstrated the exceptionally high prevalence of large protein–protein interaction interfaces in complement activation and regulation. For example, the four regulatory domains of FH occupy an area of 4,500 Å^2 on the surface of C3b despite a comparatively weak interaction ($K_D \sim 10\ \mu M$) (Wu et al. 2009). In the case of the C3b-FB complex, the formation of additional contacts upon binding and extension of the interaction footprint largely drives convertase assembly and, thereby, complement activity (Forneris et al. 2010). Competitive or allosteric blockage of such interaction interfaces therefore offers a promising strategy for developing complement inhibitors.

Whereas the development of small molecule protein–protein interaction inhibitors has been gaining traction despite the inherent challenges, monoclonal antibodies and antibody fragments are undoubtedly the most rapidly growing class of complement inhibitors. Indeed, several complement-targeted antibodies are currently on the market (anti-C5, Eculizumab, Soliris, Alexion) or in clinical development (e.g., anti-FD Fab, Genentech; anti-FB, TA106, Alexion; anti-MASP2, Omeros; anti-properdin, NovelMed). In addition, minibodies (anti-C5, Mubodina, Adienne), aptamers (anti-C5, ARC1905, Ophthotech), and spiegelmers (anti-C5a, NOX-D19, Noxxon) are being developed as alternative macromolecular blocking entities. A majority of these new inhibitors in development are currently positioned as treatment options for AMD. In the case of Eculizumab, Alexion recently received FDA approval for the treatment of atypical hemolytic uremic syndrome (aHUS) in children and adults (Alexion 2011) after it was found to reduce complement-mediated thrombotic microangiopathy and other clinical parameters in this rare but severe genetic disease (Waters and Licht 2011; Nurnberger et al. 2009; Tschumi et al. 2011). In addition to PNH and aHUS, Alexion is currently evaluating Eculizumab in several other disorders ranging from AMD and *E. coli*-induced HUS to transplantation medicine.

Even though protein–protein interactions are more challenging to inhibit with small molecules, peptides and other low size entities have nevertheless proven successful in several cases and promise advantages in administration and cost when compared to therapeutic antibodies (Wells and McClendon

2007; Mullard 2012). In the case of complement, the limitation of low cell permeability, often seen for peptidic drugs (Mullard 2012), does not have to be taken into account since all potential targets are available in circulation or extracellularly. Compstatin, a cyclic 13-residue peptide that binds to C3 and prevents complement activation and amplification by all pathways (Ricklin and Lambris 2008; Sahu et al. 1996), is currently the only member of that class in clinical development. One compstatin analog (POT-4, Potentia; AL-78898A, Alcon) has successfully passed phase I clinical trials for AMD and is currently evaluated in a phase II trial (Yehoshua et al. 2011). In primates, this compstatin analog was shown to reverse the formation of drusen, which are a hallmark of early AMD (Chi et al. 2010). At the same time, compstatin has been further optimized and has shown promising effects in other disorders such as sepsis, transplantation, PNH, hemodialysis-associated complications, severe asthma, and chronic obstructive pulmonary disease (Kourtzelis et al. 2010; Silasi-Mansat et al. 2010; Qu et al. 2011; Qu et al. 2012) (currently developed by Amyndas Biotherapeutics and Apellis Pharmaceuticals). An alternative approach based on antisense homology has been explored to generate a peptide that is complementary to a region of C5a and binds to the anaphylatoxin (C5aIP (Fujita et al. 2004)). This peptide interfered with C5a-mediated activation of neutrophils and was later shown to attenuate cross talk between complement and coagulation in a model of islet transplantation; in addition, evaluation of C5aIP in a phase II clinical trial for sepsis has previously been announced (Fujita et al. 2004; Tokodai et al. 2010).

1.3.3 Build Your Own Regulator

Our body produces a natural panel of highly effective complement inhibitors that primarily act at the level of convertases or the TCC, and which can be exploited for therapeutic purposes. Among them, the "regulator of complement activation" (RCA) protein family has attracted particular interest. These modular proteins composed of 4–30 complement control protein (CCP) domains include FH and C4b-binding protein (C4BP) in circulation as well as membrane-associated CR1 (CD35), decay accelerating factor (DAF/CD55), and membrane cofactor protein (MCP/CD46); they either accelerate the decay of the C3 convertase or mediate the degradation of C3b by FI. With the strongest activity on both CP- and AP-mediated complement activations, soluble CR1 (sCR1; TP10, Avant) has been the most extensively developed RCA and used in a variety of disease models. Despite promising data in preclinical and phase I studies for preventing complement activation during cardiac surgery (Li et al. 2006), questions regarding gender specificity and efficacy in phase II trials affected its clinical progression (Lazar et al. 2007). More recently, Celldex has picked up development of sCR1 (CDX1135) and is assessing its use in rare renal diseases and antibody-mediated transplant rejection. In addition, recombinant FH (Optherion) has been evaluated for use in AMD and renal diseases that are often affected by polymorphisms in the FH gene. A key disadvantage of these RCAs is their large size (150–300 kDa), which may render production and therapy both challenging and costly.

The modular nature of RCA proteins has inspired various efforts to extract, combine, or extend selected domain modules to arrive at tailored or targeted regulators (Fig. 1.5). Early examples in clinical development include the combination of the short four CCP-domain regulators MCP and DAF into an inhibitor with extended regulatory functions (CAB-2, Millennium) or the truncation of CR1 into shorter fragments. More recently, targeting of regulatory modules to sites of activation or tissues of interest by addition of addressing moieties/modules has regained attention and led to the development of highly promising inhibitors and therapeutic approaches (e.g., TT30, Mirococept; Fig. 1.5; see Sect. 1.4).

Besides RCAs and C1-INH (see above), CD59 is another regulator with therapeutic potential. By intercalating with components C8 and C9, membrane-bound CD59 prevents formation of the TCC and, consequently, TCC-mediated lysis or signaling. While currently not clinically developed, soluble and targeted forms of CD59 have recently been explored for the prevention of choroidal neovascularization (CNV) as it occurs in the wet form of AMD (Cashman et al. 2011; Bora et al. 2010).

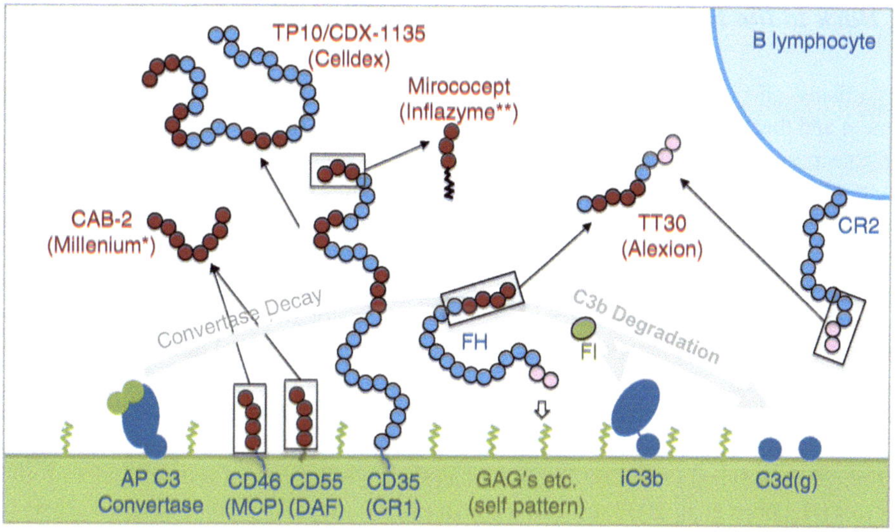

Fig. 1.5 Modular concept of complement regulators and receptors composed of CCP domains and their use for designing therapeutic complement inhibitors. *Colored circles* depict individual CCP domains, with regulatory and targeting entities marked in *red* and *pale magenta*, respectively. *CAB-2 is not currently listed in the pipeline of Millennium; **Even though Mirococept is not currently listed in the pipelines of pharmaceutical companies, it has recently been evaluated for transplant protection

1.3.4 A Friendly Reception: Prevention of Inflammatory Signaling

Anaphylatoxins (C3a, C5a) are among the strongest effectors generated by complement activation and induce chemotaxis, cell activation, and inflammatory signaling via binding to their respective G-protein-coupled receptors (GPCR), that is, C3aR and C5aR (CD88). While a third anaphylatoxin receptor (C5L2, GPR77) has been identified, its binding specificity, signaling pattern, and functional role have not yet been fully elucidated (Ricklin et al. 2010; Klos et al. 2009). In view of their strong inflammatory effects and the potential drugability as GPCRs, anaphylatoxin receptors have long been identified as attractive drug targets. However, the development of C3aR antagonists has proven to be considerably challenging, and a promising clinical candidate (SB 290157, Merck) was later shown to exert partial-agonistic activities (Mathieu et al. 2005). In contrast, the clinical development of antagonists for C5aR has been much more rewarding and has produced several low molecular weight candidates (e.g., PMX53 and PMX205, Cephalon/Teva; JPE-1375, Jerini; CCX168, ChemoCentryx; NGD-2000-1, former Neurogen; see also reviews in (Ricklin and Lambris 2007a; Qu et al. 2009; Woodruff et al. 2011; Monk et al. 2007)). Among them, the PMX family of molecules have been particularly well studied and successfully used in a variety of preclinical studies ranging from AMD and sepsis to trauma, transplantation, cancer, and Alzheimer's disease (Woodruff et al. 2011; Kohl 2006; Bosmann and Ward 2012; Fonseca et al. 2009; Recknagel et al. 2012; Lewis et al. 2008; Markiewski et al. 2008). Although both PMX53 and JPE-1375 have been evaluated in clinical trials and have shown beneficial safety profiles, none of these trials have yet been extended. In addition, C5aR agonists (e.g., EP54/EP67 (Kollessery et al. 2011)), inverse agonists (e.g., NDT 9513727, former Neurogen (Brodbeck et al. 2008)), or modified recombinant C5a with dual antagonistic activity for C5aR and C5L2 ($A8^{\Delta 71-73}$, (Otto et al. 2004)) have been explored. Clearly, anaphylatoxin receptor modulators remain a highly important class of drug candidates with high implication in inflammatory diseases.

1.3.5 Back to the Roots: Exploiting Natural Concepts

Microorganisms and parasites that enter circulation are exposed to the defensive action of complement and therefore have been required to develop powerful evasion strategies, many of which rely on the secretion of inhibitors against host complement components (Lambris et al. 2008). While they may not be used as therapeutic inhibitors directly due to immunogenicity concerns, these pathogen-derived inhibitors may nevertheless serve as templates for developing complement-targeted drugs. An example of a pathogen-derived inhibitor that is in clinical development is the chemotaxis inhibitory protein of *Staphylococcus aureus* (CHIPS), which acts as an antagonist of the C5aR. A truncated and mutated form of CHIPS (ADC-1004, Alligator Bioscience) has been developed and successfully tested in a preclinical model of I/R injury during myocardial infarction (Gustafsson et al. 2009; van der Pals et al. 2010).

Another promising agent from natural sources is the tick-derived *Ornithodoros moubata* complement inhibitor (OmCI), which binds to C5 and potently inhibits its activation to C5a and C5b (Nunn et al. 2005; Roversi et al. 2007). OmCI has shown efficacy in a rodent model of myasthenia gravis in rats (Hepburn et al. 2007) as well as in an *ex vivo* model of *E. coli*-induced sepsis in human and porcine blood (Barratt-Due et al. 2011). Orthopox viruses, on the other hand, produce several RCA mimics, such as the vaccinia virus complement control protein (VCP) and the smallpox inhibitor of complement enzymes (SPICE) (Ahmad et al. 2007); VCP has been explored in several disease studies (Ghebremariam et al. 2010; Kulkarni et al. 2011). Finally, *S. aureus* appears to be an especially rich source of complement inhibitors and immune modulators that act on a variety of targets and show a fascinating diversity of inhibitory mechanisms (Laarman et al. 2010; Chavakis et al. 2007), including allosteric modulation of C3b (Chen et al. 2010).

1.4 Aim Before You Shoot: Targeting of Complement Drugs

In the vast majority of complement-related disorders, the detrimental activation and amplification process occurs directly on surfaces that trigger the defensive actions of complement, including microbial particles, foreign biomaterials (e.g., implants or drug delivery systems), transplanted cells or organs, or diseased cells and tissues. Systemic complement inhibitors have frequently proven effective in suppressing such surface-induced inflammatory effects (Ekdahl et al. 2011; Nilsson et al. 2010), as, for example, in the case of hemodialysis filter membranes (Kourtzelis et al. 2010) or implantable devices (Sokolov et al. 2011). Yet rather than saturating circulation with soluble inhibitors, the selective blockage of complement activation directly on the triggering surface appears to be a more elegant and focused approach that may better preserve systemic complement activity and reduce drug doses and cost. Fortunately, recent years have produced a panel of creative and diverse strategies of targeted surface protection from complement attack (Fig. 1.6). Similar to systemic inhibition, the task is a delicate one and has to be tailored to the clinical situation. For example, biomaterials can be coated with inhibitors during production, and transplants may be perfused with inhibitor solutions *ex vivo*, thereby allowing the use of targeting entities with broader specificity. In contrast, the targeted inhibition on selected cells types such as erythrocytes during PNH requires a much more selective strategy.

In the case of biomaterials, coating with heparin has been explored and demonstrated to largely prevent activation of both complement and coagulation; however, the demand on production and cost of such surfaces as well as the broad binding specificity of heparin impose potential limitations. PEG coatings, on the other hand, effectively reduce protein binding and certain initiation processes but have shown ambiguous success in preventing complement activation. The use of small molecular entities to recruit physiological regulators such as C4BP or FH to surfaces by means of mimicking self-protection

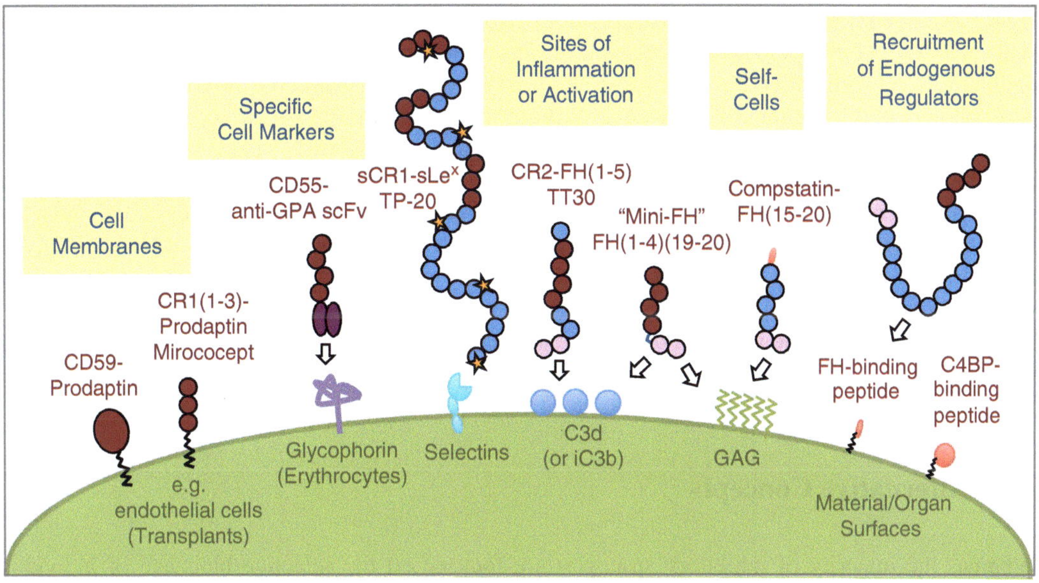

Fig. 1.6 Approaches to target complement inhibitors to various surfaces either in circulation or after *ex vivo* coating/perfusion. Regulatory/inhibitory and recognition domains are depicted in *dark red* and *pale pink*, respectively. Antibody fragments are shown in *purple*, sialyl Lewis x moieties are represented as *orange stars*, and inhibitory or recruiting peptides in *bright red*

of human cells or complement evasion strategies by certain pathogens has recently resulted in promising applications (Ekdahl et al. 2011). For example, coating of model biomaterials with small peptides that bind to the nonregulatory part of FH and thereby attracting this regulator in a functional way were shown to significantly reduce complement activation (Wu et al. 2011).

Perfusion of organ transplants with membrane-targeted inhibitors after removal from the donor has emerged as another highly promising application. Mirococept (APT070) is a particularly interesting and well-studied candidate in this context. This combination of a truncated CR1 fragment (CCP1-3) with a lipopeptide anchor that attaches to cell membranes was initially described several years ago (Smith and Smith 2001; Smith 2002) and had been shown to ameliorate I/R-injury effects in rat intestine and models of myocardial infarction and renal isograft (Banz et al. 2007; Souza et al. 2005; Patel et al. 2006). Though the compound was not clinically developed for several years, Mirococept has recently regained attention after its clinical evaluation in kidney transplantations indicated highly promising effects based on significantly increased life spans for transplants that were perfused with this targeted inhibitor (Sample 2010). A similarly targeted variant of CD59 (APT542) has also been described (Fraser et al. 2003) and was recently tested in a model of CNV (Bora et al. 2010).

Whereas targeting of a systemically administered complement inhibitor to a specific cell type imposes considerable challenges, this problem can be partially circumvented by instead targeting sites of ongoing activation/amplification. Surfaces under attack by the complement system accumulate opsonins such as C3b that are subsequently transformed into cleavage fragments (e.g., iC3b, C3d). By coupling regulatory RCA modules to moieties that bind to such opsonization fragments, one can achieve complement inhibition that is focused on activation sites. A well-described inhibitor that follows this promising strategy is TT30 (Alexion), a chimeric protein that combines regulatory domains from human FH (CCP1-5) with C3d-binding domains from CR2 (CCP1-4) and potently inhibits AP activation (Fridkis-Hareli et al. 2011). TT30 and its rodent-specific analogs have shown encouraging results in a series of AP-mediated disease models ranging from CNV and I/R injury to arthritis and PNH (Fridkis-Hareli et al. 2011; Rohrer et al. 2012; Rohrer et al. 2010; Khan et al. 2011; Banda et al. 2009;

Huang et al. 2008). Alternative approaches for *in vivo* targeting of cell surfaces include the coating of sCR1 with sialyl Lewis x to address selectins at sites of inflammation (Schmid et al. 2001), the combination of RCAs with antibodies against cellular markers (Spitzer et al. 2004, 2005), or the coupling of compstatin to the C-terminus of FH to achieve protection of (attacked) self-surfaces (Zipfel et al. 2011) (Fig. 1.6). Very recently, the new structural information concerning the N- and C-terminus of FH (Wu et al. 2009; Morgan et al. 2011; Kajander et al. 2011) was utilized to design a truncated regulator construct with full regulatory (via CCP1-4) and versatile targeting activities (to both sites of activation and self-surfaces via CCP19-20) that showed promising results in PNH models (Schmidt et al. 2012). The modularity of RCA and CR proteins, the potent activity of RCAs on the convertase level, and the increased knowledge about the molecular involvement of individual CCP domains based on mutational analysis will likely uphold regulator design and targeting as a powerful source of complement-directed therapeutics.

1.5 Alternative Concepts

Whereas the majority of inhibitory strategies are focused on the selective blockage of individual complement components, alternative approaches based on depletion, reconstitution, or gene therapy have also been pursued. Cobra venom factor (CVF) is the most prominent example of the first category; this snake protein isolated from certain cobra species shares high structural and sequence similarity with C3 and is able to form highly stable convertase complexes that rapidly activate and consume C3 and C5 in circulation (Vogel and Fritzinger 2010). CVF and a recombinant humanized form created by replacing small stretches of the C3 sequence with the stabilizing parts of CVF (HC3-1496, InCode BioPharmaceutics) have been evaluated in various disease models including myocardial I/R injury, CNV, arthritis, PNH, or transplant accommodation (Vogel and Fritzinger 2010; Fritzinger et al. 2010; Chen Song et al. 2011).

Even though administration of purified proteins is preferred for treating diseases caused by deficient, dysfunctional, or mutated complement components, such an option is not available or economically feasible in all cases. Reconstitution with functional complement components can be achieved in such cases by infusion with fresh frozen plasma. This type of therapeutic approach has been utilized for C3 dysfunction (Nilsson et al. 1994) and aHUS (Licht et al. 2005).

In most cases, and as illustrated above, the therapeutic use of complement regulators involves the administration of RCA constructs, their recruitment to surfaces, or reconstitution from plasma. However, in the case of cancer, complement regulators may undesirably protect tumor cells from complement attack, thereby warranting their therapeutic modulation. For example, CD59 inhibitors have been shown to enhance the anticancer activity of cancer-directed antibodies in certain cases (Ge et al. 2011). Similarly, genetic knock-down of membrane-associated RCAs (CD46, CD55, CD59) using small interfering RNA (siRNA) has been beneficial as an adjuvant strategy to antibody-based cancer immunotherapy (Geis et al. 2010). In view of the dual role that complement may have in different types of cancer (Markiewski and Lambris 2009), the potential of using complement inhibitors and "enhancers" has to be more closely explored.

Finally, the complex intertwinement and collaboration of complement with other defense and inflammatory pathways suggests that certain diseases may benefit from combinatorial treatment strategies. Such an approach based on simultaneous inhibition of toll-like receptor and complement pathways (via anti-CD14 mAb and compstatin, respectively) has shown promising and synergistic effects in models of bacteria-induced sepsis (Christiansen et al. 2012). Also, simultaneous inhibition of complement and coagulation targets promises advantages for preventing inflammatory responses induced by biomaterials or transplants (Ekdahl et al. 2011; Fujiwara et al. 1997). Clearly, such combinatorial therapies are just starting to be explored and may be highly potent for the treatment of diseases with complex etiologies.

1.6 Pinpointing Activation Sources: Diagnostic Strategies

The early involvement of complement in many disease processes and its upstream position in the inflammatory system potentially render it an interesting option for biomarker analysis in clinical diagnostics. In this context, approaches have been exploited that measure either the plasma levels of intact complement components, the presence of complement activation products, or the opsonization of tissue with fragments of C3 or C4.

Examples of using total complement component concentrations as markers are the measurement of C3 and C4 levels in SLE, antiphospholipid syndrome, atopic asthma, or cardiovascular disease (Mosca et al. 2011; Engstrom et al. 2007; Palikhe et al. 2007; Ramos-Casals et al. 2004). On the other hand, many activation products, such as C3a, C5a, or FB fragments, are increased in a variety of inflammatory and immune-related diseases and have been used as predictive markers for disorders like pregnancy complications and preeclampsia, heart failure, or adult respiratory distress syndrome (Gombos et al. 2012; Lynch et al. 2008, 2011; Zilow et al. 1990). Recently, several studies also investigated the systemic concentrations of complement proteins and activation products of the alternative pathway in the case of AMD and revealed significant correlations for several markers (Hecker et al. 2010; Scholl et al. 2008; Reynolds et al. 2009). Whereas biomarker analysis at the protein level has several advantages over genetic studies, including the potential assessment of posttranslational modifications and activation fragments, protein-based diagnostic approaches are often dependent on the method and difficult to multiplex, thereby complicating biomarker discovery and analysis. However, recent developments in advanced screening methods are increasingly facilitating such endeavors. For example, assay platforms based on SOMAmers (i.e., chemically optimized aptamers that form tight complexes with protein targets; SOMAscan, SomaLogic) offer multiplexing capabilities and diagnostic versatility in low sample quantities (Kraemer et al. 2011; Gold et al. 2010). Several other diagnostic platforms are also available based on mass spectrometry or antibody arrays (Sanchez-Carbayo 2011; Blonder et al. 2011).

In many cases, though, the most profound accumulation of complement activation products occurs directly on the surface of diseased organs, as in the case of various kidney diseases (Berger 1974; Brown et al. 2007) and organ/cell transplantations (Asgari et al. 2010; Kato et al. 2003), as well as in AMD where such fragments are found both in drusen and on the subretinal tissue (Anderson et al. 2010). Opsonizing complement fragments such as C3b or C4d have long been used for diagnostic purposes, but their deposition has mainly been determined through tissue biopsies and immunohistochemistry applications. More recently, noninvasive technologies have been developed that utilize the same principles of addressing sites of ongoing complement activation as described above for inhibitor targeting. For example, iron oxide nanoparticles have been coated with the C3d-binding domains of CR2 to perform magnetic resonance imaging of inflammatory kidney disease in a mouse model of lupus nephritides with promising results (Serkova et al. 2010). Similar approaches with targeted contrast agents or other diagnostic entities may therefore be developed and could allow a more dynamic and noninvasive route to monitor complement activation *in vivo*; they also might be useful in the decision-making process to determine which patients may benefit from complement-targeted treatments.

1.7 Conclusions and Outlook

With its strong involvement in many inflammatory, I/R-injury-related, immune and degenerative diseases, its 50 potential targets, and its cascade organization that allows various points of intervention, complement presents itself as an increasingly attractive target network for pharmaceutical intervention. Currently, the approved therapeutics only cover orphan diseases (PNH, aHUS, HAE), thereby influencing both cost and recognition in the field. However, additional and much bigger markets ranging from transplantation medicine to prevalent diseases like AMD are already looming on the horizon. It will be interesting to follow the changes and new opportunities in the field of complement-directed therapeutics once new drugs that target different levels of the cascade hit the market. Similarly, much may be learned when those new therapeutics, along with existing drugs such as Eculizumab and rC1-INH, have their applications gradually extended through off-label use or newly approved indications. The shifting dogmas and increased knowledge about complement in health and disease, and the high level of diversity and creativity in complement-targeted drug discovery, surely paint a bright picture and promise an exciting future for treating complement-mediated disorders.

Acknowledgments We thank Dr. Robert A. DeAngelis for critically reading the manuscript. This work was supported by National Institutes of Health grants AI068730, AI030040, AI072106, AI071028, AI097805, DE021685, GM97747, and EY020633.

Disclosure J.D.L. is the inventor of the C3 inhibitor compstatin and holds several patents about the development and clinical application of compstatin analogs. He has previously served as a member on the Scientific Advisory Board of Potentia Pharmaceuticals and is the founder of Amyndas Biotherapeutics, which perform clinical development of compstatin analogs for various indications.

References

Ahmad M, Pyaram K, Mullick J, Sahu A (2007) Viral complement regulators: the expert mimicking swindlers. Indian J Biochem Biophys 44(5):331–343

Alcorlo M, Martinez-Barricarte R, Fernandez FJ, Rodriguez-Gallego C, Round A, Vega MC et al (2011) Unique structure of iC3b resolved at a resolution of 24 A by 3D-electron microscopy. Proc Natl Acad Sci USA 108(32):13236–13240

Alexion (2011) Soliris® (eculizumab) Approved by FDA for All Patients with Atypical Hemolytic Uremic Syndrome (aHUS). Alexion Pharmaceuticals Press Release

Alfinito F, Ruggiero G, Sica M, Udhayachandran A, Rubino V, Pepa RD et al (2011) Eculizumab treatment modifies the immune profile of PNH patients. Immunobiology 217(7):698–703

Amara U, Flierl MA, Rittirsch D, Klos A, Chen H, Acker B et al (2010) Molecular intercommunication between the complement and coagulation systems. J Immunol 185(9):5628–5636

Anderson DH, Radeke MJ, Gallo NB, Chapin EA, Johnson PT, Curletti CR et al (2010) The pivotal role of the complement system in aging and age-related macular degeneration: hypothesis re-visited. Prog Retin Eye Res 29(2):95–112

Arlaud GJ, Barlow PN, Gaboriaud C, Gros P, Narayana SV (2007) Deciphering complement mechanisms: the contributions of structural biology. Mol Immunol 44(16):3809–3822

Asgari E, Zhou W, Sacks S (2010) Complement in organ transplantation. Curr Opin Organ Transplant 15(4):486–491

Banda NK, Levitt B, Glogowska MJ, Thurman JM, Takahashi K, Stahl GL et al (2009) Targeted inhibition of the complement alternative pathway with complement receptor 2 and factor H attenuates collagen antibody-induced arthritis in mice. J Immunol 183(9):5928–5937

Banz Y, Rieben R (2012) Role of complement and perspectives for intervention in ischemia-reperfusion damage. Ann Med 44(3):205–217

Banz Y, Hess OM, Robson SC, Csizmadia E, Mettler D, Meier P et al (2007) Attenuation of myocardial reperfusion injury in pigs by Mirococept, a membrane-targeted complement inhibitor derived from human CR1. Cardiovasc Res 76(3):482–493

Barratt-Due A, Thorgersen EB, Lindstad JK, Pharo A, Lissina O, Lambris JD et al (2011) *Ornithodoros moubata* complement inhibitor is an equally effective C5 inhibitor in pigs and humans. J Immunol 187(9):4913–4919

Beinrohr L, Dobo J, Zavodszky P, Gal P (2008) C1, MBL-MASPs and C1-inhibitor: novel approaches for targeting complement-mediated inflammation. Trends Mol Med 14(12):511–521

Berger J (1974) Complement deposition in the kidney. Adv Nephrol Necker Hosp 4:37–48

Blonder J, Issaq HJ, Veenstra TD (2011) Proteomic biomarker discovery: it's more than just mass spectrometry. Electrophoresis 32(13):1541–1548

Bora NS, Jha P, Lyzogubov VV, Kaliappan S, Liu J, Tytarenko RG et al (2010) Recombinant membrane-targeted form of CD59 inhibits the growth of choroidal neovascular complex in mice. J Biol Chem 285(44):33826–33833

Bosmann M, Ward PA (2012) Role of C3, C5 and anaphylatoxin receptors in acute lung injury and in sepsis. Adv Exp Med Biol 946:147–159
Brodbeck RM, Cortright DN, Kieltyka AP, Yu J, Baltazar CO, Buck ME et al (2008) Identification and characterization of NDT 9513727 [N, N-bis(1,3-benzodioxol-5-ylmethyl)-1-butyl-2,4-diphenyl-1H-imidazole-5-methanam ine], a novel, orally bioavailable C5a receptor inverse agonist. J Pharmacol Exp Ther 327(3):898–909
Brown KM, Sacks SH, Sheerin NS (2007) Mechanisms of disease: the complement system in renal injury – new ways of looking at an old foe. Nat Clin Pract Nephrol 3(5):277–286
Carroll MV, Sim RB (2011) Complement in health and disease. Adv Drug Deliv Rev 63(12):965–975
Cashman SM, Ramo K, Kumar-Singh R (2011) A non membrane-targeted human soluble CD59 attenuates choroidal neovascularization in a model of age related macular degeneration. PLoS One 6(4):e19078
Chavakis T, Preissner KT, Herrmann M (2007) The anti-inflammatory activities of *Staphylococcus aureus*. Trends Immunol 28(9):408–418
Chen Song S, Zhong S, Xiang Y, Li JH, Guo H, Wang WY et al (2011) Complement inhibition enables renal allograft accommodation and long-term engraftment in presensitized nonhuman primates. Am J Transplant 11(10):2057–2066
Chen H, Ricklin D, Hammel M, Garcia BL, McWhorter WJ, Sfyroera G et al (2010) Allosteric inhibition of complement function by a staphylococcal immune evasion protein. Proc Natl Acad Sci USA 107(41):17621–17626
Chi ZL, Yoshida T, Lambris JD, Iwata T (2010) Suppression of drusen formation by compstatin, a peptide inhibitor of complement C3 activation, on cynomolgus monkey with early-onset macular degeneration. Adv Exp Med Biol 703:127–135
Christiansen D, Brekke OL, Stenvik J, Lambris JD, Espevik T, Mollnes TE (2012) Differential effect of inhibiting MD-2 and CD14 on LPS- versus whole *E. coli* bacteria-induced cytokine responses in human blood. Adv Exp Med Biol 946:237–251
Davis AE 3rd, Lu F, Mejia P (2010) C1 inhibitor, a multi-functional serine protease inhibitor. Thromb Haemost 104(5):886–893
Degn SE, Jensenius JC, Thiel S (2011) Disease-causing mutations in genes of the complement system. Am J Hum Gen 88(6):689–705
Dmytrijuk A, Robie-Suh K, Cohen MH, Rieves D, Weiss K, Pazdur R (2008) FDA report: eculizumab (Soliris) for the treatment of patients with paroxysmal nocturnal hemoglobinuria. Oncologist 13(9):993–1000
Ekdahl KN, Lambris JD, Elwing H, Ricklin D, Nilsson PH, Teramura Y et al (2011) Innate immunity activation on biomaterial surfaces: a mechanistic model and coping strategies. Adv Drug Deliv Rev 63(12):1042–1050
Engstrom G, Hedblad B, Janzon L, Lindgarde F (2007) Complement C3 and C4 in plasma and incidence of myocardial infarction and stroke: a population-based cohort study. Eur J Cardiovasc Prev Rehabil 14(3):392–397
Fonseca MI, Ager RR, Chu SH, Yazan O, Sanderson SD, LaFerla FM et al (2009) Treatment with a C5aR antagonist decreases pathology and enhances behavioral performance in murine models of Alzheimer's disease. J Immunol 183(2):1375–1383
Fonseca MI, Chu SH, Berci AM, Benoit ME, Peters DG, Kimura Y et al (2011) Contribution of complement activation pathways to neuropathology differs among mouse models of Alzheimer's disease. J Neuroinflammation 8(1):4
Forneris F, Ricklin D, Wu J, Tzekou A, Wallace RS, Lambris JD et al (2010) Structures of C3b in complex with factors B and D give insight into complement convertase formation. Science 330(6012):1816–1820
Fraser DA, Harris CL, Williams AS, Mizuno M, Gallagher S, Smith RA et al (2003) Generation of a recombinant, membrane targeted form of the complement regulator CD59: activity in vitro and in vivo. J Biol Chem 278(49):48921–48927
Fridkis-Hareli M, Storek M, Mazsaroff I, Risitano AM, Lundberg AS, Horvath CJ et al (2011) Design and development of TT30, a novel C3d-targeted C3/C5 convertase inhibitor for treatment of human complement alternative pathway-mediated diseases. Blood 118(17):4705–4713
Fritzinger DC, Dean R, Meschter C, Wong K, Halter R, Borlak J et al (2010) Complement depletion with humanized cobra venom factor in a mouse model of age-related macular degeneration. Adv Exp Med Biol 703:151–162
Fujita E, Farkas I, Campbell W, Baranyi L, Okada H, Okada N (2004) Inactivation of C5a anaphylatoxin by a peptide that is complementary to a region of C5a. J Immunol 172(10):6382–6387
Fujiwara I, Nakajima H, Akioka K, Matsuda T, Oka T (1997) Soluble complement receptor type 1 and antithrombin-III combination therapy prolongs xenograft survival: the role of thrombin and prostacyclin in hyperacute rejection. Transplant Proc 29(1–2):935–937
Garlatti V, Chouquet A, Lunardi T, Vives R, Paidassi H, Lortat-Jacob H et al (2010) Cutting edge: C1q binds deoxyribose and heparan sulfate through neighboring sites of its recognition domain. J Immunol 185(2):808–812
Ge X, Wu L, Hu W, Fernandes S, Wang C, Li X et al (2011) rILYd4, a human CD59 inhibitor, enhances complement-dependent cytotoxicity of ofatumumab against rituximab-resistant B-cell lymphoma cells and chronic lymphocytic leukemia. Clin Cancer Res 17(21):6702–6711
Geis N, Zell S, Rutz R, Li W, Giese T, Mamidi S et al (2010) Inhibition of membrane complement inhibitor expression (CD46, CD55, CD59) by siRNA sensitizes tumor cells to complement attack in vitro. Curr Cancer Drug Targets 10(8):922–931

Ghebremariam YT, Engelbrecht G, Tyler M, Lotz Z, Govender D, Kotwal GJ et al (2010) Vaccinia virus complement control protein (VCP) improves kidney structure and function following ischemia/reperfusion injury in rats. J Surg Res 159(2):747–754
Gibson J, Hakobyan S, Cree AJ, Collins A, Harris CL, Ennis S et al (2012) Variation in complement component C1 inhibitor in age-related macular degeneration. Immunobiology 217(2):251–255
Gingras AR, Girija UV, Keeble AH, Panchal R, Mitchell DA, Moody PC et al (2011) Structural basis of mannan-binding lectin recognition by its associated serine protease MASP-1: implications for complement activation. Structure 19(11):1635–1643
Gold L, Ayers D, Bertino J, Bock C, Bock A, Brody EN et al (2010) Aptamer-based multiplexed proteomic technology for biomarker discovery. PLoS One 5(12):e15004
Gombos T, Forhecz Z, Pozsonyi Z, Szeplaki G, Kunde J, Fust G et al (2012) Complement anaphylatoxin C3a as a novel independent prognostic marker in heart failure. Clin Res Cardiol. doi:10.1007/s00392-012-0432-6
Gompels MM, Lock RJ (2011) Cinryze (C1-inhibitor) for the treatment of hereditary angioedema. Expert Rev Clin Immunol 7(5):569–573
Gustafsson E, Forsberg C, Haraldsson K, Lindman S, Ljung L, Furebring C (2009) Purification of truncated and mutated chemotaxis inhibitory protein of *Staphylococcus aureus* – an anti-inflammatory protein. Protein Expr Purif 63(2):95–101
Hadders MA, Bubeck D, Roversi P, Hakobyan S, Forneris F, Morgan BP et al (2012) Assembly and regulation of the membrane attack complex based on structures of C5b6 and sC5b9. Cell Rep 1:1–8
Halili MA, Ruiz-Gomez G, Le GT, Abbenante G, Fairlie DP (2009) Complement component C2, inhibiting a latent serine protease in the classical pathway of complement activation. Biochemistry 48(35):8466–8472
Hamilton G, Evans KL, Macintyre DJ, Deary IJ, Dominiczak A, Smith BH et al (2012) Alzheimer's disease risk factor complement receptor 1 is associated with depression. Neurosci Lett 510(1):6–9
Hecker LA, Edwards AO, Ryu E, Tosakulwong N, Baratz KH, Brown WL et al (2010) Genetic control of the alternative pathway of complement in humans and age-related macular degeneration. Hum Mol Genet 19(1):209–215
Hepburn NJ, Williams AS, Nunn MA, Chamberlain-Banoub JC, Hamer J, Morgan BP et al (2007) In vivo characterization and therapeutic efficacy of a C5-specific inhibitor from the soft tick *Ornithodoros moubata*. J Biol Chem 282(11):8292–8299
Heurich M, Martinez-Barricarte R, Francis NJ, Roberts DL, Rodriguez de Cordoba S, Morgan BP et al (2011) Common polymorphisms in C3, factor B, and factor H collaborate to determine systemic complement activity and disease risk. Proc Natl Acad Sci USA 108(21):8761–8766
Huang Y, Qiao F, Atkinson C, Holers VM, Tomlinson S (2008) A novel targeted inhibitor of the alternative pathway of complement and its therapeutic application in ischemia/reperfusion injury. J Immunol 181(11):8068–8076
Igonin AA, Protsenko DN, Galstyan GM, Vlasenko AV, Khachatryan NN, Nekhaev IV et al (2011) C1-esterase inhibitor infusion increases survival rates for patients with sepsis. Crit Care Med 40:770–777
Janssen BJ, Huizinga EG, Raaijmakers HC, Roos A, Daha MR, Nilsson-Ekdahl K et al (2005) Structures of complement component C3 provide insights into the function and evolution of immunity. Nature 437(7058):505–511
Janssen BJ, Christodoulidou A, McCarthy A, Lambris JD, Gros P (2006) Structure of C3b reveals conformational changes that underlie complement activity. Nature 444(7116):213–216
Kajander T, Lehtinen MJ, Hyvarinen S, Bhattacharjee A, Leung E, Isenman DE et al (2011) Dual interaction of factor H with C3d and glycosaminoglycans in host-nonhost discrimination by complement. Proc Natl Acad Sci USA 108(7):2897–2902
Kato M, Morozumi K, Takeuchi O, Oikawa T, Koyama K, Usami T et al (2003) Complement fragment C4d deposition in peritubular capillaries in acute humoral rejection after ABO blood group-incompatible human kidney transplantation. Transplantation 75(5):663–665
Keating GM (2009) Human C1-esterase inhibitor concentrate (Berinert). BioDrugs 23(6):399–406
Khan MA, Jiang X, Dhillon G, Beilke J, Holers VM, Atkinson C et al (2011) CD4+ T cells and complement independently mediate graft ischemia in the rejection of mouse orthotopic tracheal transplants. Circ Res 109(11):1290–1301
Kirschfink M, Mollnes TE (2001) C1-inhibitor: an anti-inflammatory reagent with therapeutic potential. Expert Opin Pharmacother 2(7):1073–1083
Klos A, Tenner AJ, Johswich KO, Ager RR, Reis ES, Kohl J (2009) The role of the anaphylatoxins in health and disease. Mol Immunol 46(14):2753–2766
Kocsis A, Kekesi KA, Szasz R, Vegh BM, Balczer J, Dobo J et al (2010) Selective inhibition of the lectin pathway of complement with phage display selected peptides against mannose-binding lectin-associated serine protease (MASP)-1 and -2: significant contribution of MASP-1 to lectin pathway activation. J Immunol 185(7):4169–4178
Kohl J (2006) Drug evaluation: the C5a receptor antagonist PMX-53. Curr Opin Mol Ther 8(6):529–538
Kollessery G, Nordgren TM, Mittal AK, Joshi SS, Sanderson SD (2011) Tumor-specific peptide-based vaccines containing the conformationally biased, response-selective C5a agonists EP54 and EP67 protect against aggressive large B cell lymphoma in a syngeneic murine model. Vaccine 29(35):5904–5910

Kourtzelis I, Markiewski MM, Doumas M, Rafail S, Kambas K, Mitroulis I et al (2010) Complement anaphylatoxin C5a contributes to hemodialysis-associated thrombosis. Blood 116(4):631–639
Kraemer S, Vaught JD, Bock C, Gold L, Katilius E, Keeney TR et al (2011) From SOMAmer-based biomarker discovery to diagnostic and clinical applications: a SOMAmer-based, streamlined multiplex proteomic assay. PLoS One 6(10):e26332
Kulkarni AP, Govender DA, Kotwal GJ, Kellaway LA (2011) Modulation of anxiety behavior by intranasally administered vaccinia virus complement control protein and curcumin in a mouse model of Alzheimer's disease. Curr Alzheimer Res 8(1):95–113
Laarman A, Milder F, van Strijp J, Rooijakkers S (2010) Complement inhibition by gram-positive pathogens: molecular mechanisms and therapeutic implications. J Mol Med (Berlin) 88(2):115–120
Lachmann PJ, Smith RA (2009) Taking complement to the clinic – has the time finally come? Scand J Immunol 69(6):471–478
Lambert JC, Heath S, Even G, Campion D, Sleegers K, Hiltunen M et al (2009) Genome-wide association study identifies variants at CLU and CR1 associated with Alzheimer's disease. Nat Genet 41(10):1094–1099
Lambris JD, Ricklin D, Geisbrecht BV (2008) Complement evasion by human pathogens. Nat Rev Microbiol 6(2):132–142
Lapeyraque AL, Malina M, Fremeaux-Bacchi V, Boppel T, Kirschfink M, Oualha M et al. (2011) Eculizumab in severe Shiga-toxin-associated HUS. N Engl J Med 364(26):2561–2563
Laursen L (2011) E. coli crisis opens door for Alexion drug trial. Nat Biotechnol 29(8):671
Laursen NS, Gordon N, Hermans S, Lorenz N, Jackson N, Wines B et al (2010) Structural basis for inhibition of complement C5 by the SSL7 protein from *Staphylococcus aureus*. Proc Natl Acad Sci USA 107(8):3681–3686
Laursen NS, Andersen KR, Braren I, Spillner E, Sottrup-Jensen L, Andersen GR (2011) Substrate recognition by complement convertases revealed in the C5-cobra venom factor complex. EMBO J 30(3):606–616
Lazar HL, Keilani T, Fitzgerald CA, Shapira OM, Hunter CT, Shemin RJ et al (2007) Beneficial effects of complement inhibition with soluble complement receptor 1 (TP10) during cardiac surgery: is there a gender difference? Circulation 116(11 Suppl):I83–I88
Lewis AG, Kohl G, Ma Q, Devarajan P, Kohl J (2008) Pharmacological targeting of C5a receptors during organ preservation improves kidney graft survival. Clin Exp Immunol 153(1):117–126
Li JS, Jaggers J, Anderson PA (2006) The use of TP10, soluble complement receptor 1, in cardiopulmonary bypass. Expert Rev Cardiovasc Ther 4(5):649–654
Licht C, Weyersberg A, Heinen S, Stapenhorst L, Devenge J, Beck B et al (2005) Successful plasma therapy for atypical hemolytic uremic syndrome caused by factor H deficiency owing to a novel mutation in the complement cofactor protein domain 15. Am J Kidney Dis 45(2):415–421
Luzzatto L, Risitano AM, Notaro R (2010) Paroxysmal nocturnal hemoglobinuria and eculizumab. Haematologica 95(4):523–526
Lynch AM, Gibbs RS, Murphy JR, Byers T, Neville MC, Giclas PC et al (2008) Complement activation fragment Bb in early pregnancy and spontaneous preterm birth. Am J Obstet Gynecol 199(4):354 e1–8
Lynch AM, Gibbs RS, Murphy JR, Giclas PC, Salmon JE, Holers VM (2011) Early elevations of the complement activation fragment C3a and adverse pregnancy outcomes. Obstet Gynecol 117(1):75–83
Markiewski MM, Lambris JD (2009) Is complement good or bad for cancer patients? A new perspective on an old dilemma. Trends Immunol 30(6):286–292
Markiewski MM, DeAngelis RA, Benencia F, Ricklin-Lichtsteiner SK, Koutoulaki A, Gerard C et al (2008) Modulation of the antitumor immune response by complement. Nat Immunol 9(11):1225–1235
Mathieu MC, Sawyer N, Greig GM, Hamel M, Kargman S, Ducharme Y et al (2005) The C3a receptor antagonist SB 290157 has agonist activity. Immunol Lett 100(2):139–145
Monk PN, Scola AM, Madala P, Fairlie DP (2007) Function, structure and therapeutic potential of complement C5a receptors. Br J Pharmacol 152(4):429–448
Morgan HP, Schmidt CQ, Guariento M, Blaum BS, Gillespie D, Herbert AP et al (2011) Structural basis for engagement by complement factor H of C3b on a self surface. Nat Struct Mol Biol 18(4):463–470
Mosca T, Menezes MC, Dionigi PC, Stirbulov R, Forte WC (2011) C3 and C4 complement system components as biomarkers in the intermittent atopic asthma diagnosis. J Pediatr (Rio de Janeiro) 87(6):512–516
Mullard A (2012) Protein–protein interaction inhibitors get into the groove. Nat Rev Drug Discov 11(3):173–175
Nagar B, Jones RG, Diefenbach RJ, Isenman DE, Rini JM (1998) X-ray crystal structure of C3d: a C3 fragment and ligand for complement receptor 2. Science 280(5367):1277–1281
Nilsson B, Nilsson UR, Karlsson-Parra A, Sjolin-Forsberg G, Hallgren R (1994) Reconstitution of the alternative pathway of complement by plasma infusions given to a patient with an SLE-like syndrome associated with a hereditary C3 dysfunction. Ann Rheum Dis 53(10):691–694
Nilsson B, Korsgren O, Lambris JD, Ekdahl KN (2010) Can cells and biomaterials in therapeutic medicine be shielded from innate immune recognition? Trends Immunol 31(1):32–38

Nunn MA, Sharma A, Paesen GC, Adamson S, Lissina O, Willis AC et al (2005) Complement inhibitor of C5 activation from the soft tick *Ornithodoros moubata*. J Immunol 174(4):2084–2091
Nurnberger J, Philipp T, Witzke O, Opazo Saez A, Vester U, Baba HA et al (2009) Eculizumab for atypical hemolytic-uremic syndrome. N Engl J Med 360(5):542–544
Oikonomopoulou K, Ricklin D, Ward PA, Lambris JD (2012) Interactions between coagulation and complement – their role in inflammation. Semin Immunopathol 34(1):151–165
Otto M, Hawlisch H, Monk PN, Muller M, Klos A, Karp CL et al (2004) C5a mutants are potent antagonists of the C5a receptor (CD88) and of C5L2: position 69 is the locus that determines agonism or antagonism. J Biol Chem 279(1):142–151
Palikhe A, Sinisalo J, Seppanen M, Haario H, Meri S, Valtonen V et al (2007) Serum complement C3/C4 ratio, a novel marker for recurrent cardiovascular events. Am J Cardiol 99(7):890–895
Parker CJ (2012) Paroxysmal nocturnal hemoglobinuria. Curr Opin Hematol 19(3):141–148
Patel H, Smith RA, Sacks SH, Zhou W (2006) Therapeutic strategy with a membrane-localizing complement regulator to increase the number of usable donor organs after prolonged cold storage. J Am Soc Nephrol 17(4):1102–1111
Perkins SJ, Gilbert HE, Aslam M, Hannan J, Holers VM, Goodship TH (2002) Solution structures of complement components by X-ray and neutron scattering and analytical ultracentrifugation. Biochem Soc Trans 30(Pt 6): 996–1001
Perkins SJ, Gilbert HE, Aslam M, Hannan J, Holers VM, Goodship TH (2002) Solution structures of complement components by X-ray and neutron scattering and analytical ultracentrifugation. Biochem Soc Trans 30(Pt 6):996 1001
Proitsi P, Lupton MK, Dudbridge F, Tsolaki M, Hamilton G, Daniilidou M et al (2012) Alzheimer's disease and age-related macular degeneration have different genetic models for complement gene variation. Neurobiol Aging 33(8):1843.e9-1843.e17
Qu H, Ricklin D, Bai H, Chen H, Reis ES, Maciejewski M, Tzekou A, DeAngelis RA, Resuello RRG, Lupu F, Barlow PN, Lambris JD (2012) New analogs of the clinical complement inhibitor compstatin with subnanomolar affinity and enhanced pharmacokinetic properties. Immunobiology (in press). doi:10.1016/j.imbio.2012.06.003
Qu H, Ricklin D, Lambris JD (2009) Recent developments in low molecular weight complement inhibitors. Mol Immunol 47(2–3):185–195
Qu H, Magotti P, Ricklin D, Wu EL, Kourtzelis I, Wu YQ et al (2011) Novel analogues of the therapeutic complement inhibitor compstatin with significantly improved affinity and potency. Mol Immunol 48(4):481–489
Ramos-Casals M, Campoamor MT, Chamorro A, Salvador G, Segura S, Botero JC et al (2004) Hypocomplementemia in systemic lupus erythematosus and primary antiphospholipid syndrome: prevalence and clinical significance in 667 patients. Lupus 13(10):777–783
Recknagel S, Bindl R, Kurz J, Wehner T, Schoengraf P, Ehrnthaller C et al (2012) C5aR-antagonist significantly reduces the deleterious effect of a blunt chest trauma on fracture healing. J Orthop Res 30(4):581–586
Reis SE, Falcao DA, Isaac L (2006) Clinical aspects and molecular basis of primary deficiencies of complement component C3 and its regulatory proteins factor I and factor H. Scand J Immunol 63(3):155–168
Reynolds R, Hartnett ME, Atkinson JP, Giclas PC, Rosner B, Seddon JM (2009) Plasma complement components and activation fragments: associations with age-related macular degeneration genotypes and phenotypes. Invest Ophthalmol Vis Sci 50(12):5818–5827
Ricklin D, Lambris JD (2007a) Complement-targeted therapeutics. Nat Biotechnol 25(11):1265–1275
Ricklin D, Lambris JD (2007b) Exploring the complement interaction network using surface plasmon resonance. Adv Exp Med Biol 598:260–278
Ricklin D, Lambris JD (2008) Compstatin: a complement inhibitor on its way to clinical application. Adv Exp Med Biol 632:273–292
Ricklin D, Hajishengallis G, Yang K, Lambris JD (2010) Complement: a key system for immune surveillance and homeostasis. Nat Immunol 11(9):785–797
Risitano AM, Perna F, Selleri C (2011) Achievements and limitations of complement inhibition by eculizumab in paroxysmal nocturnal hemoglobinuria: the role of complement component 3. Mini Rev Med Chem 11(6):528–535
Rodriguez de Cordoba S, Harris CL, Morgan BP, Llorca O (2011) Lessons from functional and structural analyses of disease-associated genetic variants in the complement alternative pathway. Biochim Biophys Acta 1812(1):12–22
Rohrer B, Long Q, Coughlin B, Renner B, Huang Y, Kunchithapautham K et al (2010) A targeted inhibitor of the complement alternative pathway reduces RPE injury and angiogenesis in models of age-related macular degeneration. Adv Exp Med Biol 703:137–149
Rohrer B, Coughlin B, Bandyopadhyay M, Holers VM (2012) Systemic human CR2-targeted complement alternative pathway inhibitor ameliorates mouse laser-induced choroidal neovascularization. J Ocul Pharmacol Ther. doi:10.1089/jop.2011.0212
Rooijakkers SH, Wu J, Ruyken M, van Domselaar R, Planken KL, Tzekou A et al (2009) Structural and functional implications of the alternative complement pathway C3 convertase stabilized by a staphylococcal inhibitor. Nat Immunol 10(7):721–727

Roth A, Hock C, Konik A, Christoph S, Duhrsen U (2011) Chronic treatment of paroxysmal nocturnal hemoglobinuria patients with eculizumab: safety, efficacy, and unexpected laboratory phenomena. Int J Hematol 93(6):704–714
Roversi P, Lissina O, Johnson S, Ahmat N, Paesen GC, Ploss K et al (2007) The structure of OMCI, a novel lipocalin inhibitor of the complement system. J Mol Biol 369(3):784–793
Ruiz-Gomez G, Lim J, Halili MA, Le GT, Madala PK, Abbenante G et al (2009) Structure-activity relationships for substrate-based inhibitors of human complement factor B. J Med Chem 52(19):6042–6052
Sacks SH, Zhou W (2012) The role of complement in the early response to transplantation. Nat Rev Immunol 12(6):431–442
Sahu A, Kay BK, Lambris JD (1996) Inhibition of human complement by a C3-binding peptide isolated from a phage-displayed random peptide library. J Immunol 157(2):884–891
Sample I (2010) Mirococept could double life of transplant organs. The Guardian 2010. http://www.guardian.co.uk/science/2010/sep/15/mirococept-transplant-organs
Sanchez-Carbayo M (2011) Antibody microarrays as tools for biomarker discovery. Methods Mol Biol 785:159–182
Schmid RA, Hillinger S, Hamacher J, Stammberger U (2001) TP20 is superior to TP10 in reducing ischemia/reperfusion injury in rat lung grafts. Transplant Proc 33(1–2):948–949
Schmidt CQ, Bai H, Risitano AM, Barlow PN, Ricklin D, Lambris JD (2012) Rational engineering of a novel complement regulator affords double targeting of host cells and controls innate immunity in disease condition (submitted)
Scholl HP, Charbel Issa P, Walier M, Janzer S, Pollok-Kopp B, Borncke F et al (2008) Systemic complement activation in age-related macular degeneration. PLoS One 3(7):e2593
Schuster MC, Chen H, Lambris JD (2007) Hydrogen/deuterium exchange mass spectrometry: potential for investigating innate immunity proteins. Adv Exp Med Biol 598:407–417
Schuster MC, Ricklin D, Papp K, Molnar KS, Coales SJ, Hamuro Y et al (2008) Dynamic structural changes during complement C3 activation analyzed by hydrogen/deuterium exchange mass spectrometry. Mol Immunol 45(11):3142–3151
Serkova NJ, Renner B, Larsen BA, Stoldt CR, Hasebroock KM, Bradshaw-Pierce EL et al (2010) Renal inflammation: targeted iron oxide nanoparticles for molecular MR imaging in mice. Radiology 255(2):517–526
Shaw CD, Storek MJ, Young KA, Kovacs JM, Thurman JM, Holers VM et al (2010) Delineation of the complement receptor type 2-C3d complex by site-directed mutagenesis and molecular docking. J Mol Biol 404(4):697–710
Silasi-Mansat R, Zhu H, Popescu NI, Peer G, Sfyroera G, Magotti P et al (2010) Complement inhibition decreases the procoagulant response and confers organ protection in a baboon model of *Escherichia coli* sepsis. Blood 116(6):1002–1010
Sirmaci A, Walsh T, Akay H, Spiliopoulos M, Sakalar YB, Hasanefendioglu-Bayrak A et al (2010) MASP1 mutations in patients with facial, umbilical, coccygeal, and auditory findings of Carnevale, Malpuech, OSA, and Michels syndromes. Am J Hum Genet 87(5):679–686
Skattum L, Martensson U, Sjoholm AG (1997) Hypocomplementaemia caused by C3 nephritic factors (C3 NeF): clinical findings and the coincidence of C3 NeF type II with anti-C1q autoantibodies. J Intern Med 242(6):455–464
Skattum L, van Deuren M, van der Poll T, Truedsson L (2011) Complement deficiency states and associated infections. Mol Immunol 48(14):1643–1655
Smith RA (2002) Targeting anticomplement agents. Biochem Soc Trans 30(Pt 6):1037–1041
Smith GP, Smith RA (2001) Membrane-targeted complement inhibitors. Mol Immunol 38(2–3):249–255
Sokolov A, Hellerud BC, Lambris JD, Johannessen EA, Mollnes TE (2011) Activation of polymorphonuclear leukocytes by candidate biomaterials for an implantable glucose sensor. J Diabetes Sci Technol 5(6):1490–1498
Souza DG, Esser D, Bradford R, Vieira AT, Teixeira MM (2005) APT070 (Mirococept), a membrane-localised complement inhibitor, inhibits inflammatory responses that follow intestinal ischaemia and reperfusion injury. Br J Pharmacol 145(8):1027–1034
Spitzer D, Unsinger J, Bessler M, Atkinson JP (2004) ScFv-mediated in vivo targeting of DAF to erythrocytes inhibits lysis by complement. Mol Immunol 40(13):911–919
Spitzer D, Unsinger J, Mao D, Wu X, Molina H, Atkinson JP (2005) In vivo correction of complement regulatory protein deficiency with an inhibitor targeting the red blood cell membrane. J Immunol 175(11):7763–7770
Szalai AJ, Digerness SB, Agrawal A, Kearney JF, Bucy RP, Niwas S et al (2000) The Arthus reaction in rodents: species-specific requirement of complement. J Immunol 164(1):463–468
Thorgersen EB, Ludviksen JK, Lambris JD, Sfyroera G, Nielsen EW, Mollnes TE (2010) Anti-inflammatory effects of C1-Inhibitor in porcine and human whole blood are independent of its protease inhibition activity. Innate Immun 16(4):254–264
Tillou X, Poirier N, Le Bas-Bernardet S, Hervouet J, Minault D, Renaudin K et al (2010) Recombinant human C1-inhibitor prevents acute antibody-mediated rejection in alloimmunized baboons. Kidney Int 78(2):152–159
Tokodai K, Goto M, Inagaki A, Nakanishi W, Ogawa N, Satoh K et al (2010) Attenuation of cross-talk between the complement and coagulation cascades by C5a blockade improves early outcomes after intraportal islet transplantation. Transplantation 90(12):1358–1365

Torreira E, Tortajada A, Montes T, Rodriguez de Cordoba S, Llorca O (2009) 3D structure of the C3bB complex provides insights into the activation and regulation of the complement alternative pathway convertase. Proc Natl Acad Sci USA 106(3):882–887
Tourangeau LM, Zuraw BL (2011) The new era of C1-esterase inhibitor deficiency therapy. Curr Allergy Asthma Rep 11(5):345–351
Tschumi S, Gugger M, Bucher BS, Riedl M, Simonetti GD (2011) Eculizumab in atypical hemolytic uremic syndrome: long-term clinical course and histological findings. Pediatr Nephrol 26(11):2085–2088
van den Elsen JM, Isenman DE (2011) A crystal structure of the complex between human complement receptor 2 and its ligand C3d. Science 332(6029):608–611
van der Pals J, Koul S, Andersson P, Gotberg M, Ubachs JF, Kanski M et al (2010) Treatment with the C5a receptor antagonist ADC-1004 reduces myocardial infarction in a porcine ischemia-reperfusion model. BMC Cardiovasc Disord 10:45
Veerhuis R, Nielsen HM, Tenner AJ (2011) Complement in the brain. Mol Immunol 48(14):1592–1603
Vogel CW, Fritzinger DC (2010) Cobra venom factor: structure, function, and humanization for therapeutic complement depletion. Toxicon 56(7):1198–1222
Waters AM, Licht C (2011) aHUS caused by complement dysregulation: new therapies on the horizon. Pediatr Nephrol 26(1):41–57
Wells JA, McClendon CL (2007) Reaching for high-hanging fruit in drug discovery at protein–protein interfaces. Nature 450(7172):1001–1009
Wiesmann C, Katschke KJ, Yin J, Helmy KY, Steffek M, Fairbrother WJ et al (2006) Structure of C3b in complex with CRIg gives insights into regulation of complement activation. Nature 444(7116):217–220
Woodruff TM, Nandakumar KS, Tedesco F (2011) Inhibiting the C5-C5a receptor axis. Mol Immunol 48(14): 1631–1642
Wu J, Wu YQ, Ricklin D, Janssen BJ, Lambris JD, Gros P (2009) Structure of complement fragment C3b-factor H and implications for host protection by complement regulators. Nat Immunol 10(7):728–733
Wu YQ, Qu H, Sfyroera G, Tzekou A, Kay BK, Nilsson B et al (2011) Protection of nonself surfaces from complement attack by factor H-binding peptides: implications for therapeutic medicine. J Immunol 186(7):4269–4277
Yehoshua Z, Rosenfeld PJ, Albini TA (2011) Current clinical trials in dry AMD and the definition of appropriate clinical outcome measures. Semin Ophthalmol 26(3):167–180
Zeerleder S (2011) C1-inhibitor: more than a serine protease inhibitor. Semin Thromb Hemost 37(4):362–374
Zilow G, Sturm JA, Rother U, Kirschfink M (1990) Complement activation and the prognostic value of C3a in patients at risk of adult respiratory distress syndrome. Clin Exp Immunol 79(2):151–157
Zipfel PF, Lauer NL, Ricklin D, Heinen SH, Dahse HMD, Tzekou A et al (2011) The new chimeric alternative complement inhibitor COMP_CFH15-20 protects surfaces from complement-mediated damage (conference abstract). Mol Immunol 48(14):1713

Chapter 2
Inhibition of the Serine Proteases of the Complement System

Péter Gál, József Dobó, László Beinrohr, Gábor Pál, and Péter Závodszky

Abstract Proteases play important roles in human physiology and pathology. The complement system is a proteolytic cascade, where serine proteases activate each other by limited proteolysis in a strictly ordered manner. Serine proteases are essential in both the initiation and the amplification of the cascade. Since uncontrolled complement activation contributes to the development of serious disease conditions, inhibition of the complement serine proteases could be an attractive therapeutic approach. In this chapter, we give a brief overview of the major types of natural serine protease inhibitors and their role in controlling the complement cascade. A special emphasis is laid on C1-inhibitor, a natural complement protease inhibitor, which is approved for clinical use in hereditary angioedema (HAE). We also examine the potential of developing artificial complement protease inhibitors. Synthetic small-molecule drugs can be very efficient serine protease inhibitors, but they usually lack sufficient specificity. A promising approach to yield more specific compounds is the alteration of natural protease inhibitors through engineering or directed evolution resulting in new variants with fine-tuned specificity and enhanced affinity.

2.1 The Key Role of Serine Proteases in Cascade Systems

Proteases (other commonly used names: peptidases, proteinases) are enzymes that are able to degrade protein or peptide substrates via catalyzing the hydrolysis of the peptide bond. Proteases are ubiquitous in all forms of life from bacteria to mammals. Typically 2–4 % of the genes in a genome encode for proteases (Puente et al. 2005). While some of the proteases have broad substrate specificity and thought to be mainly involved in the nonspecific degradation of unwanted or dietary proteins (e.g., the proteasome or trypsin), many proteases play key roles in various biological processes through mediating highly specific cleavages on certain proteins (Neurath 1984). According to their mechanisms of action, proteases can be classified into six distinct classes: serine, cysteine, threonine, aspartic, glutamic, and metalloproteases, with the name reflecting on the centrally important catalytic enzyme residue or compound (López-Otín and Bond 2008). More than one third of all known proteases belong

P. Gál (✉) • J. Dobó • L. Beinrohr • P. Závodszky
Institute of Enzymology, Research Centre for Natural Sciences,
Hungarian Academy of Sciences, Budapest, Hungary
e-mail: gal@enzim.hu

G. Pál
Department of Biochemistry, Eötvös Loránd University, Budapest, Hungary

J.D. Lambris et al. (eds.), *Complement Therapeutics*, Advances in Experimental Medicine and Biology 735,
DOI 10.1007/978-1-4614-4118-2_2, © Springer Science+Business Media New York 2013

to the class of serine proteases (Page and Di Cera 2008). The catalytic mechanism of the serine proteases is typically based on the catalytic triad of serine, histidine, and aspartate residues (Hedstrom 2002). In the first step of the proteolysis, the nucleophilic serine of the active site attacks the carbonyl moiety of the substrate peptide bond resulting in the formation of an acyl-enzyme complex through an unstable tetrahedral intermediate in parallel with the release of the C-terminal half of the substrate peptide. During the catalysis, the nucleophilicity of the serine is enhanced by the adjacent histidine, while the basic character of the histidine is enhanced by the aspartate. Deacylation (decomposition of the acyl-enzyme intermediate) occurs via a similar mechanism, but in this case a water molecule is the attacking nucleophile and an ester is the scissile bond. The biological significance of this catalytic mechanism is underlined by the fact that the same catalytic triad has been developed four times by independent evolutionary events in different organisms (convergent evolution). The archetype of the serine proteases is the chymotrypsin. Chymotrypsin-like enzymes form the largest and most studied family (S1) of serine proteases (clan PA). In the human plasma, proteases having chymotrypsin-fold and trypsin-like specificity (i.e., they cleave after positively charged side chains: Arg or Lys) form extracellular signaling cascades mediating various physiological processes such as blood clotting, fibrinolysis, angiogenesis, and inflammatory and complement responses (Krem and Di Cera 2002). In fact, serum serine proteases form a sophisticated network. Division of this network into distinct cascades is rather artificial since there are numerous cross talks between the more or less arbitrarily defined cascades (Fig. 2.1).

The driving force behind the signaling cascades is the sequential activation of serine protease zymogens. In the resting state of the cascades, the serine proteases are present as nearly inactive proenzymes (zymogens) which become fully activated by proteolytic cleavage at the N-terminus. The very first enzymatic event is usually an autoactivation step where the first proenzyme of the cascade activates itself by self-cleavage. The proteolytically activated initiator protease then activates the subsequent member of the cascade launching the cascade reaction. From this point, each successive zymogen is proteolytically activated by the previous member of the cascade. Since one protease molecule of the system can activate many subsequent zymogen protease molecules, by the end of the cascade reaction an enormous signal-amplification effect takes place. Consequently, activation of the proteolytic cascades must be tightly regulated in space and time by inhibitors to avoid self-damage. Inhibition of the initial proteases could be very efficient in regulating the whole cascades.

The complement system is a sophisticated network of about 30 proteins circulating in serum or bound to cell surface (Walport 2001; Ricklin et al. 2010). It contains nine serine proteases: eight of them (C1r, C1s, MASP-1, MASP-2, MASP-3, factor D, C2, factor B) are involved in the initiation and propagation of the cascade, while factor I is a cascade-regulating protease (Sim and Tsiftsoglou 2004). All complement proteases, except factor D, are mosaic proteins containing several noncatalytic domains preceding the catalytical serine protease domain (Fig. 2.2). These noncatalytic domains form protein–protein interactions with other complement components contributing to the narrow substrate specificity of the complement proteases. The complement proteases do not act alone, but they are usually part of larger multimolecular complexes, and consequently their substrates are also frequently found in such complexes.

There are three major ways of complement activation: the classical, the lectin, and the alternative pathway. The common characteristic of the classical and the lectin pathway is that they use soluble pattern recognition molecules ("receptors") associated with serine protease zymogens to initiate the cascade. The pattern recognition molecule of the classical pathway is C1q which associates with a heterotetramer of C1r and C1s molecules (C1s-C1r-C1r-C1s) to form the C1 complex (Nayak et al. 2011). When C1q binds to its target (IgG- or IgM-containing immune complex, C-reactive protein, apoptotic cell surface, etc.), C1r autoactivates (Gál et al. 2009). Active C1r then cleaves C1s which is the executioner protease of the C1 complex. C1s cleaves C4 and C4b-associated C2 to form the classical pathway C3 convertase C4b2a.

In the lectin pathway, we currently know five pattern recognition molecules: mannan-binding lectin (MBL) (Dommet et al. 2006; Thiel 2007), three ficolins (ficolin-1, ficolin-2, and ficolin-3)

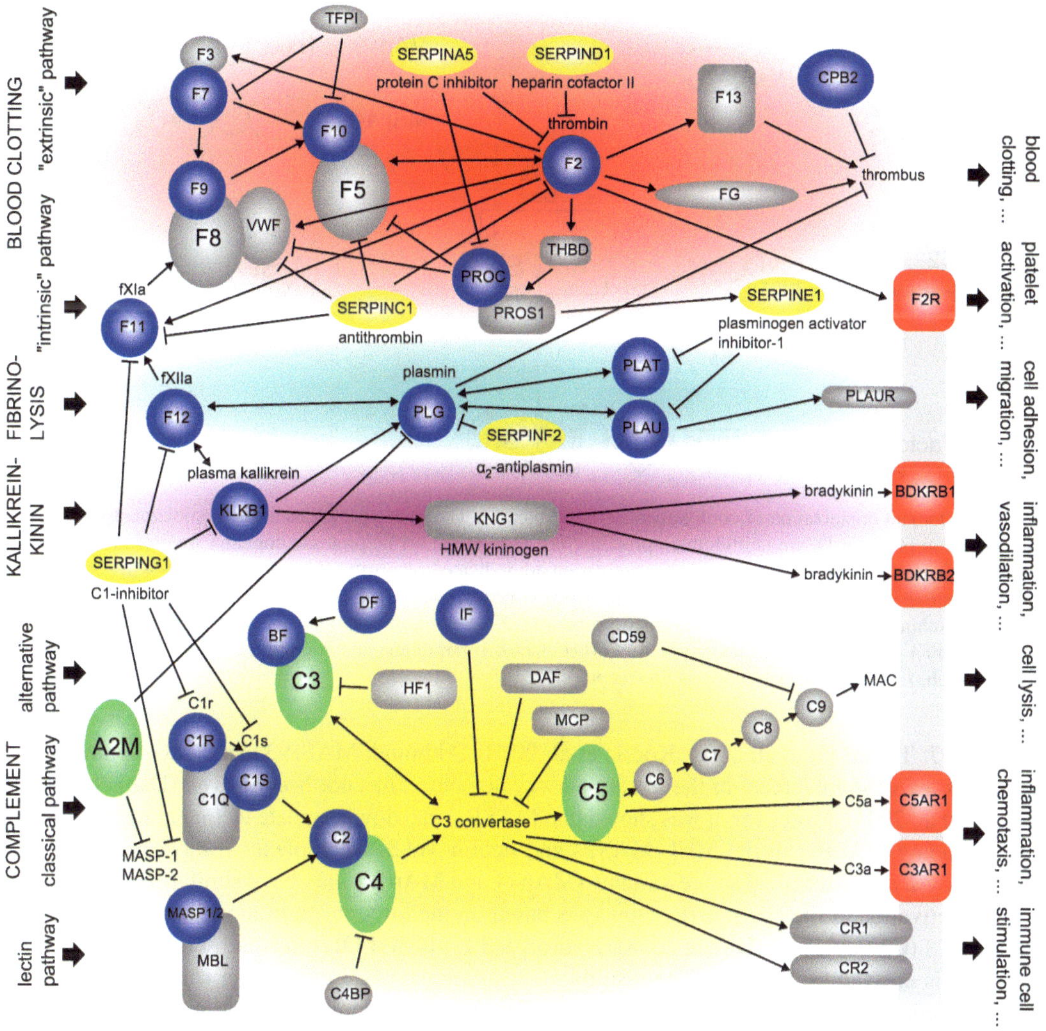

Fig. 2.1 Cascade systems in the blood. In the blood plasma, trypsin-like serine proteases form a sophisticated network, which can be divided into distinct cascade systems. The blood clotting, fibrinolytic, kallikrein-kinin, and the complement system are shown. The figure illustrates numerous cross talks between the cascades both in respect of activators and inhibitors

(Endo et al. 2011), and collectin-11 (CL-K1) (Hansen et al. 2010). There are three serine proteases and two nonenzymatic proteins associated with the pattern recognition molecules. MASP-1, MASP-3, and MAp44 (Degn et al. 2009; Skjoedt et al. 2010a) are alternative splice products of the *MASP-1* gene while MASP-2 and MAp19 (Stover et al. 1999; Takahashi et al. 1999) are that of the *MASP-2* gene. MASP-1, MASP-2, and MASP-3 contain different serine protease domains linked to five noncatalytic domains, whereas MAp44 and MAp19 consist of four and two N-terminal-interacting domains of MASP-1 and MASP-2, respectively (Fig. 2.2). The activation mechanism of the lectin pathway is not fully understood. It was shown that MASP-2 combines the activity of C1r and C1s in a sense that it autoactivates and cleaves C4 and C2 (Ambrus et al. 2003; Rossi et al. 2001; Gál et al. 2005). A MASP-2 dimer associated with an MBL molecule is able to initiate the complement cascade (Vorup-Jensen et al. 2000). However, the physiological situation is much more complex since several lines of evidence suggest that MASP-1 significantly contributes to the lectin pathway activation through cleaving C2 and facilitating the activation of MASP-2 (Chen and Wallis 2004; Moller-Kristensen

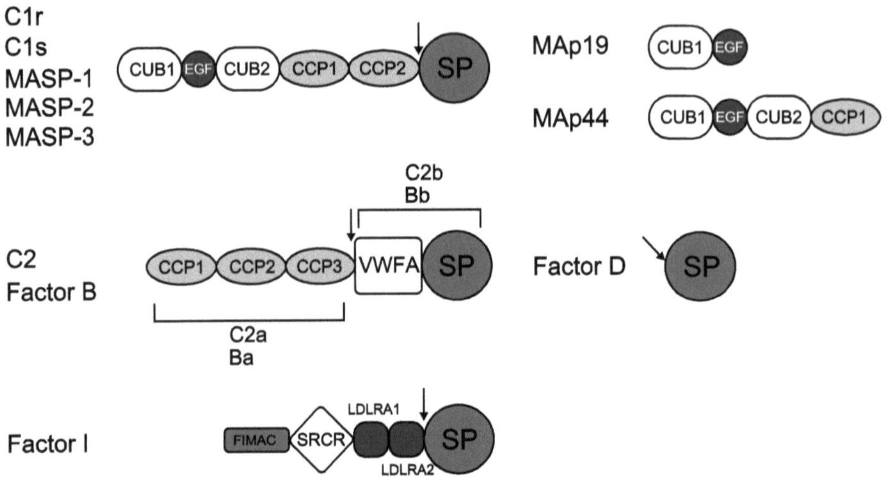

Fig. 2.2 Domain organization of complement serine proteases. The serine proteases of the complement system, with the exception of factor D, are multidomain proteins. The C-terminal serine protease domains are preceded by several noncatalytic modules. The cleavage sites for activation are indicated by arrows. MAp19 and MAp44 are the noncatalytic alternative splice products of the *MASP-2* and *MASP-1* genes, respectively. Domain name abbreviations: *CUB* C1r/C1s, sea urchin Uegf, and bone morphogenetic protein-1, *EGF* epidermal growth factor, *CCP* complement control protein, *VWFA* von Willebrand factor type A, *FIMAC* factor I membrane-attach complex, *SRCR* scavenger receptor cysteine-rich, *LDLR* low-density lipoprotein receptor

et al. 2007; Takahashi et al. 2008; Kocsis et al. 2010). Although MASP-3 can cleave synthetic substrate, its active involvement in the lectin pathway activation has not been proven (Dahl et al. 2001; Zundel et al. 2004). Moreover, it was shown that MASP-3 can downregulate lectin pathway activation by displacing MASP-2 in the MBL-MASPs and ficolin-MASPs complexes (Skjoedt et al. 2010b). The noncatalytic N-terminal splice variants (MAp44 and MAp19) may have similar regulatory roles.

The activation of the alternative pathway is based on the low-rate continuous spontaneous hydrolysis of C3 (C3 tick-over) (Holers 2008). Hydrolyzed C3 (C3(H_2O)) binds the serine protease factor B which is cleaved by factor D. The resulting C3(H_2O)Bb is a C3 convertase which generates C3b molecules that can attach to nearby cell surface. Deposited C3b molecules bind more factor B which after factor D-mediated cleavage form the alternative pathway C3 convertase C3bBb. Previously, it was thought that the deposition of C3b onto the cell surface is always spontaneous; recent evidence, however, suggests that properdin can act as a pattern recognition molecule initiating alternative pathway activation on apoptotic cells and certain microorganisms (Kemper and Hourcade 2008). On the surface of the host cells, the deposited C3b is rapidly degraded by the serine protease factor I (Nilsson et al. 2011). Factor I can cleave C3b only if it is in complex with factor H or other cofactors such as CR1, membrane cofactor protein (MCP), and decay-accelerating factor (DAF). Factor I can also break down the classical/lectin pathway C3 convertase C4b2a by using C4b-binding protein or one of the above-mentioned cofactors. While most of the proteases are circulating as inactive proenzymes, factor D and factor I are present as cleaved, preactivated species (Volanakis and Narayana 1996). These enzymes however recognize and cleave their substrates only in complex with other proteins (cofactors); therefore, their activity in vivo depends on the formation of transient substrate-cofactor complexes (e.g., C3bB and C3bfH). Moreover, structural studies showed that the active center of free factor D is distorted, and it adopts the active conformation only after binding to the substrate (substrate-induced catalysis). Recently, a possible new link has emerged between the lectin and the alternative pathway. It has been suggested that MASP-1 (Takahashi et al. 2010) and/or MASP-3 (Iwaki et al. 2011) are responsible for the cleavage of profactor D (see Chap. 3 in this book), a prerequisite for initiation of the alternative pathway.

Since enhanced or uncontrolled complement activation can be dangerous to the host and can contribute to the development of serious disease conditions, the complement system is tightly regulated by various natural inhibitors of different mechanisms of action. In this chapter, we focus only on the serine protease inhibitors. The initiator serine proteases of the classical and the lectin pathway are controlled by C1-inhibitor (Davis et al. 2010). The serine protease components of the C3- and C5-convertases (C2 and factor B) are not inhibited directly. Regulation of the convertases occurs through the disassembly of the complexes and proteolytic degradation of their components by factor I (Nilsson et al. 2011). In the following chapters, we discuss the basic characteristics of the natural serine protease inhibitors and also analyze the perspectives of developing novel complement protease inhibitors for therapeutic use.

2.2 Major Types of Protease Inhibitors

It is somewhat paradoxical that all the natural endogen inhibitors of the protein-degrading enzymes are themselves proteins (Bode and Huber 1992). There are two major types of natural serine protease inhibitors (Rawlings et al. 2004). The inhibitors of the first type undergo major conformational change during the inhibitory process and form irreversible complexes with the target proteases. The two classes of this type of inhibitors are the serpins (*ser*ine *p*rotease *in*hibitors) (Gettins 2002) and the macroglobulins (Sottrup-Jensen 1989) (Fig. 2.3a, b). These are large protein inhibitors and are abundant in the blood. The second major type of serine protease inhibitors consists of small proteins that form – by thermodynamic terms – reversible complexes with the target protease (Krowarsch et al. 2003) (Fig. 2.3c). Although in this case no covalent bond forms between the two proteins, the binding affinity can be so high ($K_i \leq 10^{-9}$ M) that the interaction is practically irreversible. There are two prominent classes of these tight-binding inhibitors: the canonical and noncanonical inhibitors. Both block the active site of the inhibited enzyme albeit differently. The canonical inhibitors follow the standard Laskowski mechanism (Laskowski 1986). These are typically small (usually shorter than 100 amino acids) proteins, or domains of mosaic proteins, harboring different scaffolds but possessing an exposed inhibitory loop which always shows virtually the same (canonical) main-chain conformation (Jackson and Russel 2000). This convex inhibitory loop imitates a peptide substrate and fits perfectly to the concave active site of the protease. This tight lock-and-key protein–protein interaction involves an antiparallel β-sheet. The noncanonical inhibitors do not mimic a substrate. Instead of an inhibitory loop, the N-terminal segment of the inhibitor binds to the active site of the protease forming a short parallel β-sheet. A striking characteristic of these inhibitors is that they form additional extensive secondary interactions outside the active site (exosites) resulting in a very high affinity (K_i in the fM range) (Grütter et al. 1990). These highly specialized noncanonical inhibitors are much less abundant than other types of protease inhibitors as they can be found only in blood-sucking organisms (leeches and insects) inhibiting two key enzymes of blood coagulation thrombin and factor Xa.

Proteolytic enzymes in the human body are controlled mainly by serpins and canonical inhibitors. Interestingly, signaling cascades in the blood plasma are regulated almost exclusively by serpins, and to a lesser extent, by macroglobulins. Although canonical inhibitors are abundant in the human body, they have almost no role in the regulation of the plasma cascades. Perhaps, the only exception is tissue factor pathway inhibitor (TFPI, alternative names: extrinsic pathway inhibitor, EPI, and lipoprotein-associated coagulation inhibitor, LACI) that inhibits coagulation factors VIIa and Xa (Broze et al. 1988). Studying the mechanism of action of serpins is extremely important in respect of better understanding the physiological inhibition of the proteolytic cascades and also the pathological processes associated with serpin mutations. Canonical inhibitors on the other hand are promising candidates for developing highly specific serum serine protease inhibitors for basic research as well as for pharmaceutical purposes.

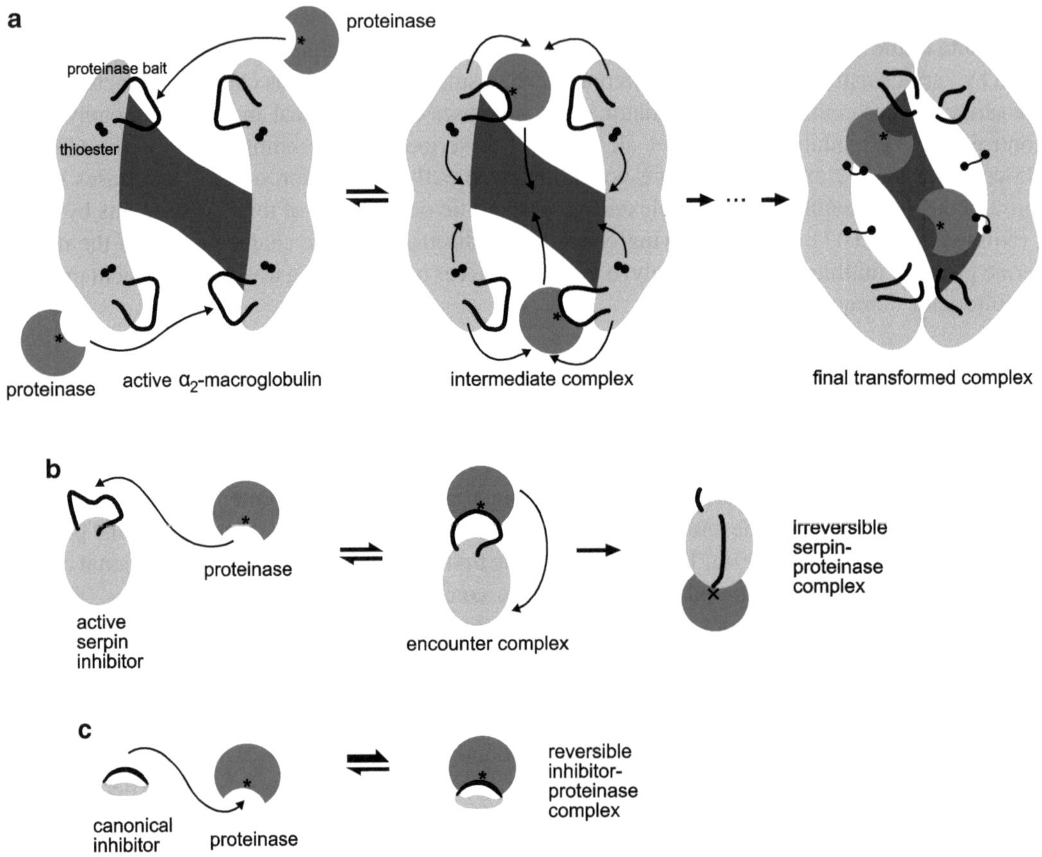

Fig. 2.3 Major types of serine protease inhibitors. Macroglobulins (**a**) and serpins (**b**) are large proteins which undergo major conformational change during the inhibition process and form irreversible complex with the target proteases. The canonical inhibitors (**c**) are small proteins which occupy the active site and form very tight but thermodynamically reversible complex with the target proteases

Since serpins are the most important protease inhibitors in the serum, it is not surprising that most serpins inhibit trypsin-like serine proteases. Intact native serpins adopt a metastable (stressed) conformation which is a prerequisite of their inhibitory activity (Silverman et al. 2001; Pike et al. 2002). The conserved serpin fold (350–500 amino acids in size) consists of three β-sheets (A, B, and C), at least 7 α-helixes (most typically 9) and loops connecting the sheets and helixes. The most important loop is the reactive center loop (RCL) of about 17 amino acids, which is tethered between β-sheets A and C and contains the scissile bond for the protease (P1–P1′). Similarly to the canonical inhibitors, serpins interact with their target proteases in a substrate-like manner. In the first step, the protease forms a noncovalent Michaelis-like complex with the serpin. The active site serine of the protease attacks the backbone carbonyl of the P1 residue forming the stable acyl-enzyme intermediate. Unlike the less flexible protease-binding loops of the canonical inhibitors, the RCLs of the serpins are able to adopt several different conformations. Immediately after cleavage, the RCL begins to insert into β-sheet A, and it carries the covalently bound protease with it. The insertion of the RCL happens so quickly that there is no time for the deacylation step (kinetic trap). After complete insertion of the RCL, the serpin adopts a very stable relaxed conformation. The protease is translocated by over 70 Å from its initial position in the Michaelis complex. Moreover, its active site is distorted as a consequence of the enzyme being crushed against the base of the serpin. The serpin-protease complex will then be cleared from the circulation long before significant complex decay could occur. The driving force of the inhibitory process is the stressed-to-relaxed transition of the serpin which provides energy for the quick distortion of the protease structure. If the insertion of the RCL is impeded, the deacylation step goes through resulting in a relaxed cleaved serpin and an active protease (noninhibitory pathway). In some cases, the uncleaved RCL can also insert into β-sheet A forming the latent conformation of the serpin. The latent conformation is inactive but can be converted back to the active (stressed) state by renaturing. The specificity of the serpins is determined primarily by the sequence of the RCL. Since the RCLs are quite flexible, it is not surprising that most serpins target several proteases. C1-inhibitor (SERPING1), besides inhibiting the classical and the lectin pathways of the complement system, also regulates the activity of proteases of the contact activation system. Antithrombin (SERPINC1) which inhibits thrombin has been shown to efficiently block the activity of complement protease MASP-1 (Dobó et al. 2009). Recently, evidences began to accumulate suggesting that besides the RCL, exosites on the body of the serpins can also play a role in the efficient recognition of the target proteases (Whisstock et al. 2010). Some serpins recognize their target proteases efficiently only in the presence of certain cofactors. A prominent example is heparin which speeds up the interaction between serpins and proteases of the blood clotting and the complement system (Gettins and Olson 2009; Pike et al. 2005). These cofactors can bind to exosites contributing to the specificity of the serpin. Mapping of these exosites is therapeutically important as one could target them with drugs and thereby modulate serpin-protease interactions.

The inhibitory action of the macroglobulins is somewhat similar to that of serpins since the proteolytic cleavage of the bait region induces major conformational change of the inhibitor. As a result of this conformational change, the protease will be buried inside the inhibitor structure and will no longer be available for substrates in spite of its active site remaining intact. The macroglobulin-inhibitor complex, like the serpin-inhibitor complex, will bind to the respective receptor and will be cleared from the circulation before any further proteolytic reaction could take place. The best characterized macroglobulin in human plasma is α_2-macroglobulin, a tetrameric glycoprotein composed of 1,451 residue subunits (Marrero et al. 2012). The reactive loop for protease attack (the bait region) is significantly larger (30–40 residues) than the RCL of the serpins. This large flexible loop can be found in the middle of the polypeptide chain and contains cleavage sites for many types of proteases. As a result, α_2-macroglobulin is a pan-protease inhibitor, inhibiting proteases of all mechanistic families. It is highly abundant in human plasma (1–2 g/L): its concentration is comparable to that of C3, the most abundant complement protein. α_2-Macroglobulin is closely related to complement components C3 and C4 arising from the same ancestral gene. These proteins contain a buried thioester bond which is exposed after proteolytic cleavage. Upon cleavage of the bait region of α_2-macroglobulin, the thioester bond between Cys949 and Gln952 becomes exposed and reacts rapidly with any available nucleophile. In this way, a covalent linkage can be formed between the inhibitor and the protease. Unfortunately, there are no detailed structural data available about α_2-macroglobulin; therefore, the mechanism of its action is only poorly characterized at the molecular level. The physiological importance of α_2-macroglobulin is also not well understood. It inhibits blood coagulation and fibrinolysis. It may also inhibit the lectin pathway of complement since it was shown to react with MASP-1 and MASP-2. Apparently, signaling cascades in the blood are regulated by the concerted action of serpins and macroglobulins.

2.3 C1-Inhibitor

C1-inhibitor is a serpin having strong anti-inflammatory activity (Davis et al. 2008, 2010), which is exerted through the regulation of both the complement and the kallikrein-kinin systems (Fig. 2.4a). It is an efficient inhibitor of the C1r and C1s components of the classical pathway and the MASP-1 and MASP-2 components of the lectin pathway of complement activation. Plasma kallikrein, the

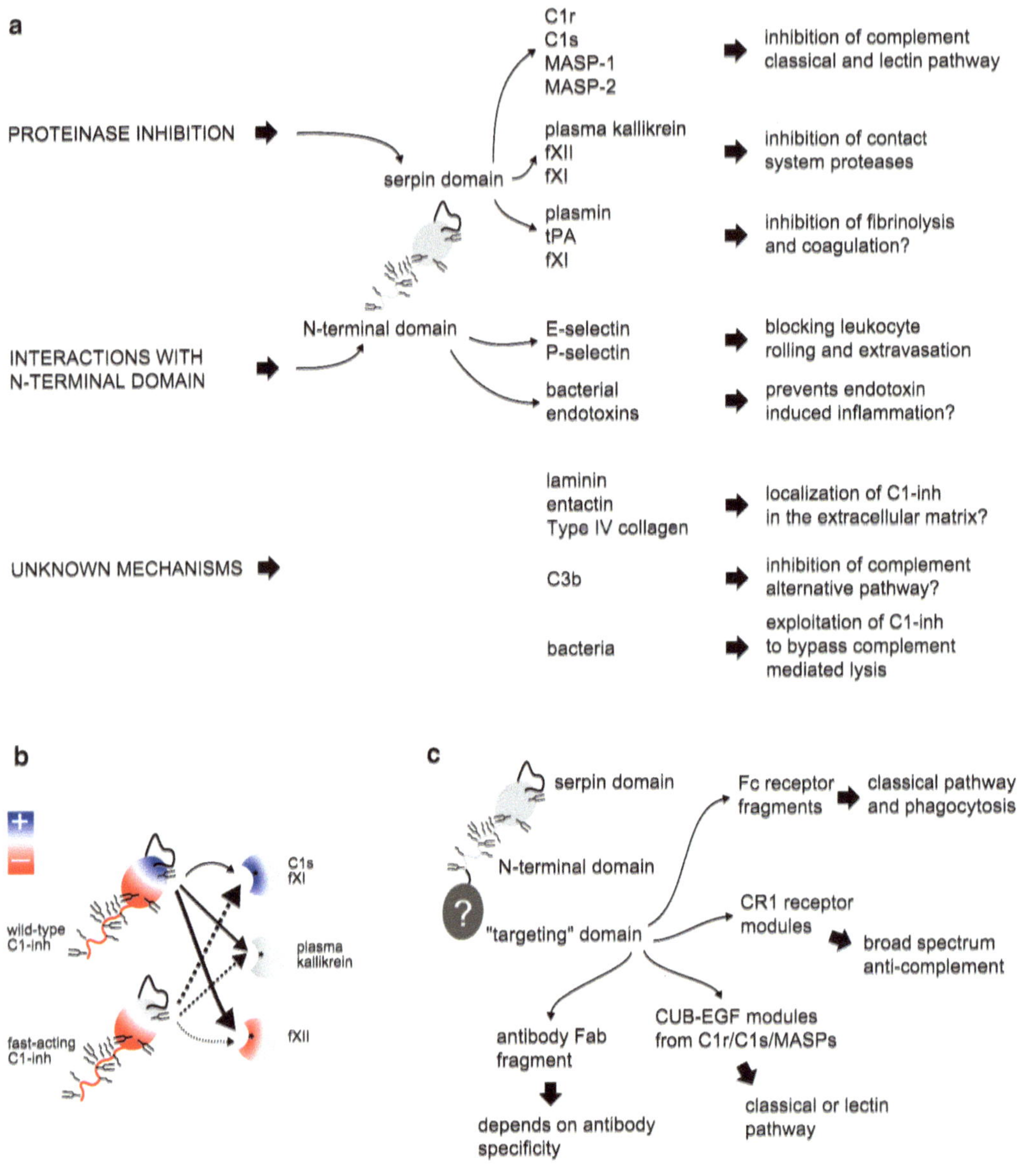

Fig. 2.4 C1-inhibitor: biological functions and pharmaceutical potentials. (**a**) C1-inhibitor mediates various biological functions. It inhibits proteases of different cascade systems, and it also exerts anti-inflammatory activities which are not related to the serpin activity. (**b**) The activity of C1-inhibitor against selected serine proteases can be tuned by altering the surface charge by mutation. Wild-type C1-inhibitor has a positive charge at the protease-contacting region. Neutralization of this positive charge will increase the reaction rate against positively charged serine proteases (e.g., fXIa, C1s). The thickness of the *arrows* indicates the relative rate of inhibition. (**c**) C1-inhibitor can be guided to selected targets by fusing it with different interacting domains ("magic bullets"). In this way, pathway-specific and/or inflammatory site-specific chimeras can be constructed

protease responsible for the generation of the highly vasoactive peptide bradykinin from high molecular weight kininogen, is also inactivated by C1-inhibitor (Kaplan and Ghebrehiwet 2010). The list of other proteases inhibited by C1-inhibitor includes factor XIIa of the contact system, factor XIa and thrombin of the coagulation system, and plasmin and tissue-type plasminogen activator of the fibrinolytic system. These proteases cleave the Arg444-Thr445 bond in the RCL which results in the formation of a covalent protease-C1-inhibitor complex. A very important role of C1-inhibitor is to prevent the spontaneous activation of the zymogen proteases in the plasma (Ziccardi 1985). C1-inhibitor may also regulate the alternative pathway of the complement through interaction with C3b (Jiang et al. 2001). This interaction does not rely on protease inhibition. C1-inhibitor is unique among serpins by containing a long, heavily glycosylated amino terminal domain. This 100-amino-acid-long nonserpin, mucin-like domain contains both N- and O-linked sugar moieties and contributes to the anti-inflammatory activity of C1-inhibitor. C1-inhibitor was shown to bind to cells, extracellular matrix component, and endogenous proteins. C1-inhibitor binds to human peripheral blood lymphocytes, monocytes, and neutrophils. It can inhibit the adhesion of human monocytic U937 cells to HUVEC (human umbilical vein endothelial cells) (Cai and Davis 2003). This effect is mediated by the sialyl Lewis^x moieties of the N-linked carbohydrate chains of the nonserpin domain. These sugars are ligands of the E- and P-selectins. In this way, C1-inhibitor, even in its cleaved, inactivated form, can inhibit leukocyte-endothelial cell adhesion and leukocyte rolling in vivo. Interaction of C1-inhibitor with these cells can significantly contribute to its anti-inflammatory effect. C1-inhibitor binds to extracellular matrix proteins type IV collagen, laminin, and entactin. These interactions may accumulate C1-inhibitor at the extravascular sites of inflammation allowing for the local regulation of the proteolytic cascades. Interestingly, some bacteria may evade complement attack by binding C1-inhibitor on their surfaces possibly through the N-terminal domain (Lathem et al. 2004).

The inhibitory activity of C1-inhibitor can be modulated by the cofactor heparin. Heparin is a naturally occurring polyanion, a sulfated glycosaminoglycan having the highest negative charge density among the biological macromolecules. It has a strong anticoagulant activity since it can facilitate the complex formation between thrombin and the serpin antithrombin. It also modulates the action of C1-inhibitor on the complement and contact system proteases. The nature and degree of this effect, however, depend on the protease. Heparin speeds up the inhibitory reaction towards factor XIa the most, while the potentiation effect is less in the case of C1s and kallikrein. Surprisingly heparin slightly impedes the reaction between C1-inhibitor and factor XIIa. The potentiation effect of heparin depends on the chain length of the polyanion in the case of most serpin-protease interactions except those involving C1-inhibitor. These and other experimental data suggest that the mode of action of heparin on the modulation of C1-inhibitor action is different from the previously described "bridging" and allostery mechanisms.

Recently, we have determined the X-ray structure of the serpin domain of human C1-inhibitor (Beinrohr et al. 2007). In order to allow crystallization, we made a construct from which most of the heavily glycosylated N-terminal domain was removed. After recombinant expression and purification, we managed to crystallize a noninhibitory form of C1-inhibitor. This noninhibitory form proved to be the latent conformation of the serpin with the uncleaved RCL being partially inserted into β-sheet A. The physiological significance of this newly discovered latent form of C1-inhibibitor is not clear, but it suggests how some disease-causing mutations may facilitate the active-to-latent transition. One of the most interesting features of the structure is the bipolar charge distribution on the surface of the serpin domain. There is a positive electric field on the "top" of the serpin, where the RCL is located, whereas the "bottom" of the serpin domain is negatively charged. This surface charge distribution can explain the unconventional effect of heparin on the C1-inhibitor-protease reactions. *In silico* docking experiments show that the negatively charged heparin binds to the positively charged region on the top of the serpin domain. This region overlaps with the area involved in the formation and stabilization of the encounter (Michaelis) complex. If heparin or other polyanion binds to this region, it neutralizes the positive charge, which diminishes the repulsion between the positively charged serpin and the positively charged protease. The more positively charged the protease, the greater the potentiation effect is. This "charge sandwich" model explains why heparin significantly facilitates the inhibition of the positively charged factor XIa while exerting an opposite effect on the inhibition of the negatively charged factor XIIa. It is also evident why the potentiation effect does not depend on the chain length of the polyanion (Rossi et al. 2010). Based on this hypothesis, it might be possible to engineer the serpin domain of C1-inhibitor in order to make fast-acting serpin by mimicking the effect of heparin (Fig. 2.4b). In this way, one could also tune the specificity of C1-inhibitor against complement proteases. Such inhibitor variants are being tested in our laboratory.

C1-inhibitor is the only approved drug on the market that inhibits complement proteases. Isolated from human plasma, it has been used for many years to treat hereditary angioedema (HAE) in Europe, and recently it has been approved by the FDA for prophylactic treatment (Epstein and Bernstein 2008). Recombinant C1-inhibitor purified from the milk of transgenic rabbits is also available (Sardana and Craig 2011). HAE is an autosomal dominant disease characterized by recurrent episodes of acute, localized increases in vascular permeability that can lead to lethal tissue swelling if it occurs in the larynx. HAE is the consequence of low C1-inhibitor activity in the patients' plasma leading to uncontrolled release of bradykinin by plasma kallikrein. Since C1-inhibitor controls the activity of the classical and lectin pathway of complement, it might be beneficial in the treatment of certain inflammatory diseases as well. Ischemia-reperfusion (IR) injury results in serious tissue damage as it occurs in myocardial infarction. The complement system recognizes the ischemic cells as altered self and initiates their elimination. The involvement of the lectin pathway in IR injury-related tissue damage has been demonstrated in different IR models (Walsh et al. 2005). Animal models and clinical studies have shown that administration of C1-inhibitor is protective against IR injury reducing infarct size and improving various cardiac parameters (Buerke et al. 1995; Fattouch et al. 2007). Other possible indications include sepsis, hyperacute transplant rejection, and other inflammatory disease models. The strong anti-inflammatory effect of C1-inhibitor is the combined result of the inhibition of the proteolytic cascades and other activities that do not require protease inhibition (e.g., blocking the activity of gram-negative bacterial endotoxin, enhancement of phagocytosis, suppression of leukocyte rolling) and are likely mediated by the N-terminal nonserpin domain. The interaction of the N-terminal domain with several proteins directs the anti-inflammatory activity of C1-inhibitor to the site of inflammation. Using this strategy, "magic bullets" can be designed, i.e., C1-inhibitor molecules fused with proteins that guide the inhibitor to the initiation complexes of the classical or lectin pathway or to the sites where these complexes bind thereby selectively eliminating an activation route (Beinrohr et al. 2008) (Fig. 2.4c). Although the applicability of this concept has not been verified experimentally, it is worth mentioning that recombinant C1-inhibitor is more effective in reducing cerebral damage in case of stroke than plasma-derived C1-inhibitor. Recombinant C1-inhibitor has a particular glycosylation pattern that enables it to bind to MBL. In the ischemic brain, recombinant C1-inhibitor colocalized with MBL in the cerebral vessels making possible the instant and selective inhibition of the lectin pathway (Gesuete et al. 2009).

2.4 Small-Molecule Inhibitors

Enzymes and receptors are the most frequent target proteins of drug development. Due to their well-defined concave substrate binding site, the serine proteases should belong to the most druggable targets in the complement system (Ricklin and Lambris 2007). In fact, early drug development attempts aimed at finding small-molecule inhibitors against the complement serine proteases. During the last decades, numerous low molecular weight compounds have been shown to efficiently inhibit complement proteases; however, none of them has been approved as an anticomplement drug (Qu et al. 2009). The main problem is that the serum contains a large number of evolutionary and therefore structurally related trypsin-like serine proteases. Low molecular weight organic compounds that bind tightly to the active site of a given serum protease usually also bind to several other serum proteases. As we have seen in the previous chapters, natural protease inhibitors are proteins which provide more extensive interactions with the protease assuring higher specificity.

A prominent example of small-molecule serine protease inhibitors, which inhibits complement activation, is FUT-175 (Nafamostat, Futhan) (Fujii and Hitomi 1981). FUT-175 inhibits C1r and C1s with high affinity, but it also blocks the activity of trypsin, plasmin, plasma kallikrein, and thrombin. In a guinea pig-to-rat ex vivo xenogeneic lung perfusion model, FUT-175 inhibited complement activation and improved lung xenograft function (Tagawa 2011). Administration of FUT-175 before reperfusion significantly reduced myocardial injury in a rabbit model of IR injury (Schwertz et al. 2008). It is not clear however whether this beneficial effect is the consequence of the complement inhibition alone or it is due to parallel inhibition of other inflammatory proteases as well.

The same rabbit model was used to test the cardioprotective effect of C1s-INH-248, a peptidomimetic C1s inhibitor ($IC_{50}=2$ nM) (Buerke et al. 2001). This compound was developed from the thrombin inhibitor D-Phe-Pro-Arg. The selectivity of this inhibitor was much better than that of FUT-175; however it may have short half-life in vivo. Another low molecular weight C1s inhibitor is BCX-1470 which also inhibits factor D ($IC_{50}=1.6$ and 96 nM, respectively) (Szalai et al. 2000). Although this compound was tested in phase I clinical trials, its development may have been discontinued. Recently a new type of C1s inhibitors has been reported based on the arylsulfonylthiophene-2-carboxamidine scaffold (Travins et al. 2008). These compounds were further optimized for dosing as an intravenous agent for the treatment of complement-related diseases. These molecules are strong ($K_i=10$ nM) and selective C1s inhibitors, but in vivo data from animal models are still lacking.

The serine protease components of the C3-/C5-convertase complexes, C2 and factor B, were also targeted with small-molecule inhibitors. The lack of selectivity is also a problem here since the inhibitors usually cross-react with the two structurally similar serine proteases. Moreover, these convertases (C4bC2a and C3bBb) are extremely unstable and short-lived complexes. This makes them ill-suited targets for high-throughput selection of protease inhibitors. Isolated C2 and factor B have negligible proteolytic activity at physiological pH. This problem has recently been overcome by showing that in the absence of cofactors, C2 and factor B have protease activity in their own under alkaline conditions. This facilitates studying their catalytic properties and developing small-molecule inhibitors against them (Halili et al. 2009; Ruiz-Gómez et al. 2009). In this way, hexapeptide aldehyde inhibitors were selected against C2 and factor B. These inhibitors blocked membrane attack complex formation and inhibited complement-mediated lysis of erythrocytes.

2.5 Engineered Protease Inhibitors

To prevent harmful activation of the complement system in disease, potent and specific complement inhibitors should be developed. A promising strategy is blocking the activity of the initiation serine proteases of the activation pathways. As the example of C1-inhibitor shows, inhibition of the early serine proteases of the classical or the lectin pathway completely and efficiently eliminates the given pathway. However, for a detailed in vivo characterization of the contribution of each individual pathway to the pathogenesis of certain diseases, even more specific inhibitors are needed. Unfortunately, neither C1-inhibitor nor the small-molecule inhibitors are specific enough. Selectivity of the C1-inhibitor may be improved by protein engineering as discussed in Chap. 3. Other options are to generate protease-specific monoclonal antibodies or to engineer canonical inhibitors.

Monoclonal antibodies can bind specifically and tightly to the target proteins. Antibodies, however, can very rarely occupy the concave active site of proteases (Farady et al. 2007) since their antigen-binding loops usually recognize only flat or convex protein surfaces. Nevertheless, one can generate antibodies that bind near enough at the active site to block the access of protein substrates and act practically as competitive inhibitors (Scott and Taggart 2010). Antibodies can also bind to exosites acting as negative allosteric effectors or can block binding sites of other allosteric effectors (Wu et al. 2007; Ganesan et al. 2009). Another possibility is to block the exosites necessary for substrate recognition and binding. In the case of C1s and MASP-2, it was shown that exosites found on the complement control protein modules are essential for efficient C4 cleavage (Rossi et al. 1998; Ambrus et al. 2003). Consequently, antibodies binding to the CCP domains can prevent the access to the exosites and inhibit C4 cleavage. A recent example of complement protease-specific monoclonal antibodies is AbyD 04211, an anti-MASP-2 antibody, which prevented lectin pathway activation in a murine model and reduced tissue damage at IR injury (Schwaeble et al. 2011).

As we mentioned in the previous chapters, canonical inhibitors are rare in the human serum. Because of their structural properties and mode of actions, canonical inhibitors can be highly specific. However, structure-based rational design of highly specific interaction sites is far from trivial. Moreover, without a high-resolution structure of the target protease engaged in a substrate-like complex, such an approach would be even more challenging. On the other hand, large libraries of canonical inhibitors can be created by combinatorial mutagenesis of inhibitory loop residues involved in the binding to the protease. Then, the most potent protease binders from this library can be selected. In this way, one could tune the specificity of a canonical inhibitor theoretically against any serine protease target assuming that its substrate binding site is different enough from that of other proteases. The most powerful directed evolution system for this particular goal is phage display (Zani and Moreau 2010). Canonical inhibitors are small proteins that can be readily displayed as functionally active molecules on bacteriophage surface. Moreover, the binding loop of these inhibitors usually contains six to seven protease-contacting surface-exposed residues. These are reasonably tolerant to point mutations, and their number is low enough to enable construction of inhibitor-phage libraries that contain all (about 10^{10}) possible variants. One prominent example of phage-display-derived recombinant canonical inhibitors is DX-88 (ecallantide) (Fig. 2.5a), which is based on the first Kunitz domain of TFPI and was selected against plasma kallikrein (Markland et al. 1996). Ecallantide has been approved by the FDA for the treatment of HAE (Lunn and Banta 2011). In the HAE patients, ecallantide successfully complements the reduced activity of C1-inhibitor preventing uncontrolled bradykinin release.

Recently we have developed inhibitors against MASP-1 and MASP-2 by means of directed evolution using phage display (Kocsis et al. 2010). As a starting scaffold, we used the sunflower trypsin inhibitor (SFTI), which is the smallest known natural serine protease inhibitor. It is a 14-amino-acid residue cyclic peptide having a disulfide bridge. SFTI is an extremely potent trypsin inhibitor having a K_i value in the subnanomolar range. From a structural point of view, SFTI represents the protease-interacting loop of the much larger canonical inhibitor scaffold of Bowman-Birk inhibitors. SFTI preserves its protease inhibitor potency in an opened-up form; therefore it can be displayed on phage in an active form. Comparative sequence analysis of different Bowman-Birk inhibitors showed us which residues should be randomized for library construction. In the SFTI scaffold, there are two cysteines (making the disulfide bond) and two prolines which have essential structural roles and are conserved in all Bowman-Birk inhibitors. In our library, we kept these structurally important residues but randomized all other potential protease-contacting residues, including the conservative P2 threonine and P1′ serine. In the P1 position, we allowed only lysine or arginine since the MASPs are trypsin-like enzymes. Our SFTI phage library contained $\sim 2 \times 10^9$ individual clones each monovalently displaying an inhibitor variant. After three rounds of panning on MASP-coated plates, two functionally distinct inhibitor sets were isolated. The first set obtained from independent selection on MASP-1 or MASP-2 inhibited both proteases, while the second set selected on MASP-2 inhibited only MASP-2. Two inhibitors, sunflower MASP inhibitor (SFMI)-1 (GICSRSLPPICIPD) and SFMI-2 (GYCSRSYPPYCIPD) were designed representing the two consensus sequences (Fig. 2.5b). We measured the inhibitory effect of SFMIs on purified recombinant MASP enzymes. SFMI-2 inhibited exclusively MASP-2 with a K_i of 180 nM whereas SFMI-1 inhibited MASP-1 in the first place ($K_i = 65$ nM), but with a lesser extent it also inhibited MASP-2 ($K_i = 1{,}030$ nM). It should be noted that SFMI-2 inhibits MASP-2 six times more efficiently than SFMI-1 does. On the other hand, SFMI-1 is 16 times more efficient against MASP-1 than MASP-2. Keeping these numbers in mind, we

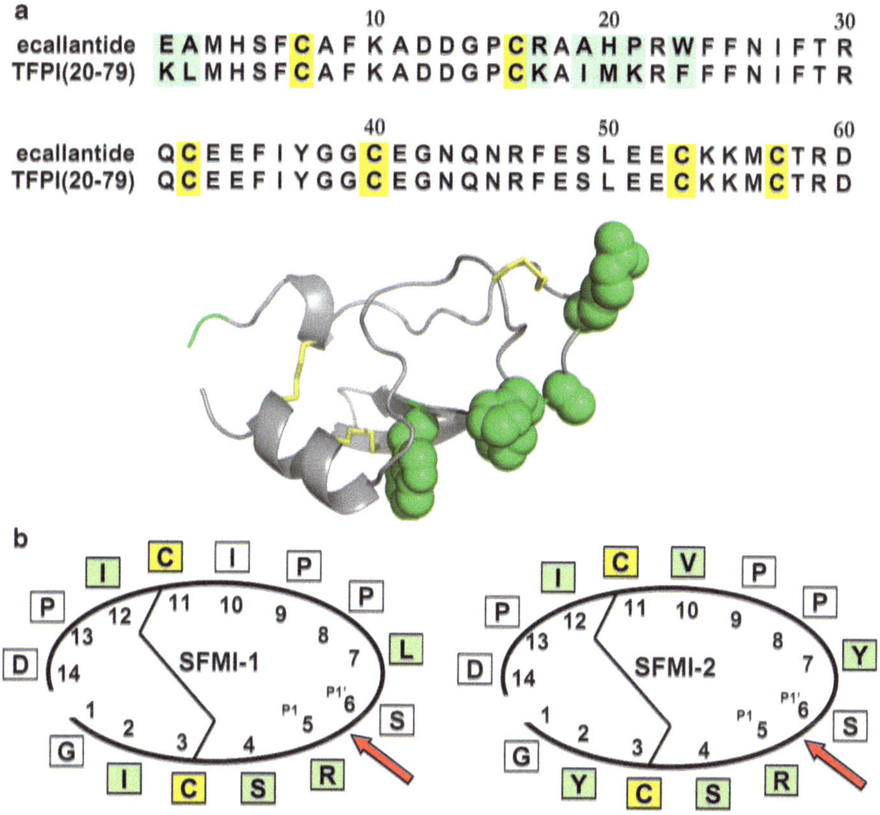

Fig. 2.5 Phage-display-selected inhibitors. The in vitro evolution method, phage display, was used to select specific and high-affinity inhibitors from large canonical inhibitor libraries against serine proteases. Ecallantide (**a**) is a powerful plasma kallikrein inhibitor which is based on the first Kunitz domain of TFPI. The structural model of ecallantide was prepared using the I-TASSER server (Roy et al. 2010). SFMI-1 and SFMI-2 (**b**) are lectin pathway-specific inhibitors which were derived from SFTI. The mutated amino acid residues are shown in *green* and the disulfide bonds in *yellow*. The *red arrows* indicate the scissile bond between the P1 and P1′ residues

performed in vitro complement activation assays. In order to test the selectivity of our inhibitors, we studied their effect on the three different activation pathways. We used the Wieslab kit which detects the formation of the terminal membrane attack complex. As we expected, SFMI-1 and SFMI-2 inhibited the lectin pathway only while leaving the classical and the alternative pathway intact and fully functional. We can draw the conclusion that SFMIs are indeed MASP-specific as they do not inhibit C1r, C1s, C2, factor B, and factor D. Surprisingly, however, SFMI-1 proved to be a more efficient lectin pathway inhibitor (IC_{50} = 3.2 μM) than SFMI-2 (IC_{50} = 9.9 μM), in spite of being a weaker MASP-2 inhibitor. The same tendency was observed at C3 deposition experiments where the IC_{50} of SFMI-1 (0.04 μM) was five times lower than that of SFMI-2 (0.22 μM). These results unambiguously demonstrated that inhibition of MASP-1 attenuates lectin pathway activity, and consequently MASP-1 significantly contributes to the activation of the lectin pathway. We know that MASP-1 alone cannot initiate lectin pathway activation since it does not cleave C4. C4 deposition experiments therefore inform us how MASP-1 influences MASP-2 activity. We performed two types of C4 deposition assays. In one experiment, we captured the MBL-MASP complexes on mannan-coated surface, washed away the other serum components, and let the MASPs become activated. Then we added purified C4 together with the inhibitors. In this case, SFMI-2 proved to be much more efficient inhibitor (IC_{50} = 2.65 μM) than SFMI-1 (IC_{50} = 25 μM) in accordance with the fact that SFMI-2 is a stronger MASP-2 inhibitor than SFMI-1. In another experimental setting, we added nonactivated serum to the mannan-coated plates and followed the effect of the inhibitors on C4 deposition. In this case, the picture changed dramatically as now SFMI-1 was one order of magnitude stronger inhibitor (IC_{50}=0.23 μM) than SFMI-2 (IC_{50}=2.7 μM). A plausible explanation of this phenomenon is that MASP-1 facilitates MASP-2 activation in normal human serum. In fact, it was shown previously that the activation of the lectin pathway is delayed and diminished in MASP-1 knockout mice and in MASP-1-depleted human serum (Takahashi et al. 2008; Moller-Kristensen et al. 2007). Our results also suggest that MASP-1 is likely to be as good target for controlling lectin pathway activation as MASP-2. As we can see, our MASP-selective inhibitors, besides being invaluable tools in complement research, can also be lead molecules for development of anticomplement drugs.

To investigate the role of MASP-1 in even more detail, we have already developed an entirely MASP-1-specific inhibitor by starting with a larger inhibitor scaffold that offered more side-chain interactions for the protease. With this unique reagent, we have corrected the hitherto inaccurate model of lectin pathway activation (Héja et al. 2012a; b). We demonstrated that MASP-2 is activated exclusively by MASP-1 in human blood.

2.6 Conclusions

Inhibition of the serine protease components is an attractive way to control complement activity in case of complement-related diseases. C1-inhibitor is the only approved drug which inhibits complement proteases. Its specificity can be tuned by protein engineering, and it can be directed to the site of inflammation by replacing its N-terminal nonserpin domain. Developing efficient and selective artificial inhibitors is a challenge because of the presence of numerous structurally and functionally related serine proteases in the blood plasma. This problem can be overcome by generating monoclonal antibodies or engineering natural serine protease inhibitors. A very promising approach to develop highly specific serine protease inhibitors is to select canonical inhibitors from large inhibitor libraries by phage display. Such libraries can be generated by mutating the protease-contacting residues of the inhibitory loop of the canonical inhibitors. The recently approved plasma kallikrein inhibitor ecallantide was developed by using this strategy. We produced lectin pathway-specific inhibitors by using the sunflower trypsin inhibitor scaffold, and we are currently working on larger scaffolds to improve the specificity and selectivity. The lectin pathway-specific inhibitors may be applied in treating various ischemia-reperfusion-related inflammatory pathologies, such as myocardial infarction or stroke.

Acknowledgments This work was supported by the Ányos Jedlik grant NKFP_07_1-MASPOK07, the Hungarian Scientific Research Fund (OTKA) grant NK77978, NK100834, K68408, NK81950 and NK100769 and the NDA grant KMOP-1.1.2-07/1-2008-0003 and the National Development Agency Grant KMOP-1.1.2-07/1-2008-0003 as well as by the European Union and the European Social Fund (TÁMOP) 4.2.1./B-09/KMR-2010-0003 grant.

References

Ambrus G, Gál P, Kojima M, Szilágyi K, Balczer J, Antal J, Gráf L, Laich A, Moffatt BE, Schwaeble W, Sim RB, Závodszky P (2003) Natural substrates and inhibitors of mannan-binding lectin-associated serine protease-1 and -2: a study on recombinant catalytic fragments. J Immunol 170:1374–1382

Beinrohr L, Harmat V, Dobó J, Lőrincz Z, Gál P, Závodszky P (2007) C1 inhibitor serpin domain structure reveals the likely mechanism of heparin potentiation and conformational disease. J Biol Chem 282:21100–21109

Beinrohr L, Dobó J, Závodszky P, Gál P (2008) C1, MBL-MASPs and C1-inhibitor: novel approaches for targeting complement-mediated inflammation. Trends Mol Med 14:511–521

Bode W, Huber R (1992) Natural protein proteinase inhibitors and their interaction with proteinases. Eur J Biochem 204:433–451

Broze GJ Jr, Warren LA, Novotny WF, Higuchi DA, Girard JJ, Miletich JP (1988) The lipoprotein-associated coagulation inhibitor that inhibits the factor VII-tissue factor complex also inhibits factor Xa: insight into its possible mechanism of action. Blood 71:335–343
Buerke M, Murohara T, Lefer AM (1995) Cardioprotective effects of a C1 esterase inhibitor in myocardial ischemia and reperfusion. Circulation 91:393–402
Buerke M, Schwertz H, Seitz W, Meyer J, Darius H (2001) Novel small molecule inhibitor of C1s exerts cardioprotective effects in ischemia-reperfusion injury in rabbits. J Immunol 167:5375–5380
Cai S, Davis AE III (2003) Complement regulatory protein C1 inhibitor binds to selectins and interferes with endothelial-leukocyte adhesion. J Immunol 171:4786–4791
Chen CB, Wallis R (2004) Two mechanisms for mannose-binding protein modulation of the activity of its associated serine proteases. J Biol Chem 279:26058–26065
Dahl MR, Thiel S, Matsushita M, Fujita T, Willis AC, Christensen T, Vorup-Jensen T, Jensenius JC (2001) MASP-3 and its association with distinct complexes of the mannan-binding lectin complement activation pathway. Immunity 15:127–135
Davis AE III, Mejia P, Lu F (2008) Biological activities of C1 inhibitor. Mol Immunol 45:4057–4063
Davis AE III, Lu F, Mejia P (2010) C1 inhibitor, a multi-functional serine protease inhibitor. Thromb Haemost 104:886–893
Degn SE, Hansen AG, Steffensen R, Jacobsen C, Jensenius JC, Thiel S (2009) MAp44, a human protein associated with pattern recognition molecules of the complement system and regulating the lectin pathway of complement activation. J Immunol 183:7371–7378
Dobó J, Harmat V, Beinrohr L, Sebestyén E, Závodszky P, Gál P (2009) MASP-1, a promiscuous complement protease: structure of its catalytic region reveals the basis of its broad specificity. J Immunol 183:1207–1214
Dommet RM, Klein N, Turner MW (2006) Mannose-binding lectin in innate immunity, past, present and future. Tissue Antigens 68:193–209
Endo Y, Matsushita M, Fujita T (2011) The role of ficolins in the lectin pathway of innate immunity. Int J Biochem Cell Biol 43:705–712
Epstein TG, Bernstein JA (2008) Current and emerging management options for hereditary angioedema. Drugs 68:2561–2573
Farady CJ, Sun J, Darragh MR, Miller SM, Craik CS (2007) The mechanism of inhibition of antibody-based inhibitors of membrane-type serine protease 1 (MT-SP1). J Mol Biol 369:1041–1051
Fattouch K, Bianco G, Speziale G, Sampognaro R, Lavalle C, Guccione F, Dioguardi P, Ruvolo G (2007) Beneficial effects of C1 esterase inhibitor in ST-elevation myocardial infarction in patients who underwent surgical reperfusion: a randomised double-blind study. Eur J Cardiothorac Surg 32:326–332
Fujii S, Hitomi Y (1981) New synthetic inhibitors of C1r, C1 esterase, thrombin, plasmin, kallikrein and trypsin. Biochim Biophys Acta 661:342–345
Gál P, Harmat V, Kocsis A, Bián T, Barna L, Ambrus G, Végh B, Balczer J, Sim RB, Náray-Szabó G, Závodszky P (2005) A true autoactivating enzyme. Structural insight into mannose-binding lectin-associated serine protease-2 activations. J Biol Chem 280:33435–33444
Gál P, Dobó J, Závodszky P, Sim RBM (2009) Early complement proteases: C1r, C1s and MASPs. A structural insight into activation and functions. Mol Immunol 46:2745–2752
Ganesan R, Eigenbrot C, Wu Y, Liang WC, Shia S, Lipari MT, Kirchhofer D (2009) Unraveling the allosteric mechanism of serine protease inhibition by an antibody. Structure 17:1614–1624
Gesuete R, Storini C, Fantin A, Stravalaci M, Zanier ER, Orsini F, Vietsch H, Mannesse ML, Ziere B, Gobbi M, De Simoni MG (2009) Recombinant C1 inhibitor in brain ischemic injury. Ann Neurol 66:332–342
Gettins PG (2002) Serpin structure, mechanism, and function. Chem Rev 102:4751–4804
Gettins PGW, Olson ST (2009) Exosite determinants of serpin specificity. J Biol Chem 284:20441–20445
Grütter MG, Priestle JP, Rahuel J, Grossenbacher H, Bode W, Hofsteenge J, Stone SR (1990) Crystal structure of the thrombin-hirudin complex: a novel mode of serine protease inhibition. EMBO J 9:2361–2365
Halili MA, Ruiz-Gómez G, Le GT, Abbenante G, Fairlie DP (2009) Complement component C2, inhibiting a latent serine protease in the classical pathway of complement activation. Biochemistry 48:8466–8472
Hansen S, Selman L, Palaniyar N, Ziegler K, Brandt J, Kliem A, Jonasson M, Skjoedt MO, Nielsen O, Hartshorn K, Jørgensen TJ, Skjødt K, Holmskov U (2010) Collectin 11 (CL-11, CL-K1) is a MASP-1/3-associated plasma collectin with microbial-binding activity. J Immunol 185:6096–6104
Hedstrom L (2002) Serine protease mechanism and specificity. Chem Rev 102:4501–4524
Héja D, Harmat V, Fodor K, Wilmanns M, Dobó J, Kékesi KA, Závodszky P, Gál P, Pál G (2012a) Monospecific inhibitors show that both mannan-binding lectin-associated serine protease (MASP)-1 and -2 are essential for lectin pathway activation and reveal structural plasticity of MASP-2. J Biol Chem 287:20290–20300
Héja D, Kocsis A, Dobó J, Szilágyi K, Szász R, Závodszky P, Pál G, Gál P (2012b) Revised mechanism of complement lectin-pathway activation revealing the role of serine protease MASP-1 as the exclusive activator of MASP-2. Proc Natl Acad Sci USA 10.1073/pnas.1202588109

Holers VM (2008) The spectrum of complement alternative pathway-mediated diseases. Immunol Rev 233:300–316
Iwaki D, Kanno K, Takahashi M, Endo Y, Matsushita M, Fujita T (2011) The role of mannose-binding lectin-associated serine protease-3 in activation of the alternative complement pathway. J Immunol 187:3751–3758
Jackson RM, Russel RB (2000) The serine protease inhibitor canonical loop conformation: examples found in extracellular hydrolases, toxins, cytokines and viral proteins. J Mol Biol 296:325–334
Jiang H, Wagner E, Zhang H, Frank MM (2001) Complement 1 inhibitor is a regulator of the alternative complement pathway. J Exp Med 194:1609–1916
Kaplan AP, Ghebrehiwet B (2010) The plasma bradykinin-forming pathways and its interrelationships with complement. Mol Immunol 47:2161–2169
Kemper C, Hourcade DE (2008) Properdin: New roles in pattern recognition and target clearance. Mol Immunol 45: 4048–4056
Kocsis A, Kékesi KA, Szász R, Végh BM, Balczer J, Dobó J, Závodszky P, Gál P, Pál G (2010) Selective inhibition of the lectin pathway of complement with phage display selected peptides against mannose-binding lectin-associated serine protease (MASP)-1 and -2: significant contribution of MASP-1 to lectin pathway activation. J Immunol 185:4169–4178
Krem MM, Di Cera E (2002) Evolution of enzyme cascades from embryonic development to blood coagulation. Trends Biochem Sci 27:67–74
Krowarsch D, Cierpicki T, Jelen F, Otlewski J (2003) Canonical protein inhibitors of serine proteases. Cell Mol Life Sci 60:2427–2444
Laskowski M Jr (1986) Protein inhibitors of serine proteinases – mechanism and classification. Adv Exp Med Biol 199:1–17
Lathem WW, Bergsbaken T, Welch RA (2004) Potentiation of C1 esterase inhibitor by StcE, a metalloprotease secreted by *Escherichia coli* O157:H7. J Exp Med 199:1077–1087
López-Otín C, Bond JS (2008) Proteases: multifunctional enzymes in life and disease. J Biol Chem 283:30433–30437
Lunn M, Banta E (2011) Ecallantide for the treatment of hereditary angioedema in adults. Clin Med Insights Cardiol 5:49–54
Markland W, Ley AC, Ladner RC (1996) Iterative optimization of high-affinity protease inhibitors using phage display. 2. Plasma kallikrein and thrombin. Biochemistry 35:8058–8067
Marrero A, Duquerroy S, Trapani S, Goulas T, Guevara T, Andersen GR, Navaza J, Sottrup-Jensen L, Gomis-Rüth FX (2012) The crystal structure of human α(2)-macroglobulin reveals a unique molecular cage. Angew Chem Int Ed Engl 10.1002/anie.201108015
Moller-Kristensen M, Thiel S, Sjoholm A, Matsushita M, Jensenius JC (2007) Cooperation between MASP-1 and MASP-2 in the generation of C3 convertase through the MBL pathway. Int Immunol 19:141–149
Nayak A, Pedenekar L, Reid KB, Kishore U (2011) Complement and non-complement activating functions of C1q: a prototypical innate immune molecule. Innate Immun. doi:10.1177/1753425910396252
Neurath H (1984) Evolution of proteolytic enzymes. Science 224:350–357
Nilsson SC, Sim RB, Lea SM, Fremeaux-Bacchi V, Blom AM (2011) Complement factor I in health and disease. Mol Immunol 48:1611–1620
Page MJ, Di Cera E (2008) Serine peptidases: classification, structure and function. Cell Mol Life Sci 65:1220–1236
Pike RN, Bottomley SP, Irving JA, Bird PI, Whisstock JC (2002) Serpins: finely balanced conformational traps. IUBMB Life 54:1–7
Pike RN, Buckle AM, le Bonniec BF, Church FC (2005) Control of the coagulation system by serpins. Getting by with a little help from glycosaminoglycans. FEBS J 272:4842–4851
Puente XS, Sanchez LM, Gutierrez-Fernandez A, Velasco G, Lopez-Otin C (2005) A genomic view of the complexity of mammalian proteolytic systems. Biochem Soc Trans 33:331–334
Qu H, Ricklin D, Lambris JD (2009) Recent developments in low molecular weight complement inhibitors. Mol Immunol 47:185–195
Rawlings ND, Tolle DP, Barret AJ (2004) Evolutionary families of peptidase inhibitors. Biochem J 378:705–716
Ricklin D, Lambris JD (2007) Complement-targeted therapeutics. Nat Biotechnol 25:1265–1275
Ricklin D, Hajishengallis G, Yang K, Lambris JD (2010) Complement: a key system for immune surveillance and homeostasis. Nat Immunol 11:785–797
Rossi V, Bally I, Thielens NM, Esser AF, Arlaud GJ (1998) Baculovirus-mediated expression of truncated modular fragments from the catalytic region of human complement serine protease C1s. Evidence for the involvement of both complement control protein modules in the recognition of the C4 protein substrate. J Biol Chem 273:1232–1239
Rossi V, Cseh S, Bally I, Thielens NM, Jensenius JC, Arlaud GJ (2001) Substrate specificities of recombinant mannan-binding lectin-associated serine proteases-1 and -2. J Biol Chem 276:40880–40887

Rossi V, Bally I, Ancelet S, Xu Y, Frémeaux-Bacchi V, Vivès RR, Sadir R, Thielens N, Arlaud GJ (2010) Functional characterization of the recombinant human C1 inhibitor serpin domain: insights into heparin binding. J Immunol 184:4982–4989

Roy A, Kucukural A, Zhang Y (2010) I-TASSER: a unified platform for automated protein structure and function prediction. Nat Protoc 5:725–738

Ruiz-Gómez G, Lim J, Halili MA, Le GT, Madala PK, Abbenante G, Fairlie DP (2009) Structure-activity relationships for substrate-based inhibitors of human complement factor B. J Med Chem 52:6042–6052

Sardana N, Craig TJ (2011) Recent advances in management and treatment of hereditary angioedema. Pediatrics 128:1173–1180

Schwaeble WJ, Lynch NJ, Clark JE, Marber M, Samani NJ, Ali YM, Dudler T, Parent B, Lhotta K, Wallis R, Farrar CA, Sacks S, Lee H, Zhang M, Iwaki D, Takahashi M, Fujita T, Tedford CE, Stover CM (2011) Targeting of mannan-binding lectin-associated serine protease-2 confers protection from myocardial and gastrointestinal ischemia/reperfusion injury. Proc Natl Acad Sci USA 108:7523–7528

Schwertz H, Carter JM, Russ M, Schubert S, Schlitt A, Buerke U, Schmidt M, Hillen H, Werdan K, Buerke M (2008) Serine protease inhibitor nafamostat given before reperfusion reduces inflammatory myocardial injury by complement and neutrophil inhibition. J Cardiovasc Pharmacol 52:151–160

Scott CJ, Taggart CC (2010) Biologic protease inhibitors as novel therapeutic agents. Biochimie 92:1681–1688

Silverman GA, Bird PI, Carrell RW, Church FC, Coughlin PB, Gettins PG, Irving JA, Lomas DA, Luke CJ, Moyer RW, Pemberton PA, Remold-O'Donnell E, Salvesen GS, Travis J, Whisstock JC (2001) The serpins are an expanding superfamily of structurally similar but functionally diverse proteins. J Biol Chem 276:33293–33296

Sim RB, Tsiftsoglou SA (2004) Proteases of the complement system. Biochem Soc Trans 32:21–27

Skjoedt MO, Hummelshoj T, Palarasah Y, Honore C, Koch C, Skjodt K, Garred P (2010a) A novel mannose-binding lectin/ficolin-associated protein is highly expressed in heart and skeletal muscle tissues and inhibits complement activation. J Biol Chem 285:8234–8243

Skjoedt MO, Palarasah Y, Munthe-Fog L, Jie Ma Y, Weiss G, Skjodt K, Koch C, Garred P (2010b) MBL-associated serine protease-3 circulates in high serum concentrations predominantly in complex with Ficolin-3 and regulates Ficolin-3 mediated complement activation. Immunobiology 215:921–931

Sottrup-Jensen L (1989) Alpha-macroglobulins: structure, shape, and mechanism of proteinase complex formation. J Biol Chem 264:11539–11542

Stover CM, Thiel S, Thelen M, Lynch NJ, Vorup-Jensen T, Jensenius JC, Schwaeble WJ (1999) Two constituents of the initiation complex of the mannan-binding lectin activation pathway of complement are encoded by a single structural gene. J Immunol 162:3481–3490

Szalai AJ, Digerness SB, Agrawal A, Kearney JF, Bucy RP, Niwas S, Kilpatrick JM, Babu YS, Volanakis JE (2000) The Arthus reaction in rodents: species-specific requirement of complement. J Immunol 164:463–468

Tagawa T (2011) Protease inhibitor nafamostat mesilate attenuates complement activation and improves function of xenografts in a discordant lung perfusion model. Xenotransplantation 18:315–319

Takahashi M, Endo Y, Fujita T, Matsushita M (1999) A truncated form of mannose-binding lectin-associated serine protease (MASP)-2 expressed by alternative polyadenylation is a component of the lectin complement pathway. Int Immunol 11:859–863

Takahashi M, Iwaki D, Kanno K, Ishida Y, Xiong J, Matsushita M, Endo Y, Miura S, Ishii N, Sugamura K, Fujita T (2008) Mannose-binding lectin (MBL)-associated serine protease (MASP)-1 contributes to activation of the lectin complement pathway. J Immunol 180:6132–6138

Takahashi M, Ishida Y, Iwaki D, Kanno K, Suzuki T, Endo Y, Homma Y, Fujita T (2010) Essential role of mannose-binding lectin-associated serine protease-1 in activation of the complement factor D. J Exp Med 207:29–37

Thiel S (2007) Complement activating soluble pattern recognition molecules with collagen-like regions, mannan-binding lectin, ficolins and associated proteins. Mol Immunol 44:3875–3888

Travins JM, Ali F, Huang H, Ballentine SK, Khalil E, Hufnagel HR, Pan W, Gushue J, Leonard K, Bone RF, Soll RM, DesJarlais RL, Crysler CS, Ninan N, Kirkpatrick J, Cummings MD, Huebert N, Molloy CJ, Gaul M, Tomczuk BE, Subasinghe NL (2008) Biphenylsulfonyl-thiophene-carboxamidine inhibitors of the complement component C1s. Bioorg Med Chem Lett 18:1603–1606

Volanakis JE, Narayana SV (1996) Complement factor D, a novel serine protease. Protein Sci 5:553–564

Vorup-Jensen T, Petersen SV, Hansen AG, Poulsen K, Schwaeble W, Sim RB, Reid KB, Davis SJ, Thiel S, Jensenius JC (2000) Distinct pathways of mannan-binding lectin (MBL)- and C1-complex autoactivation revealed by reconstitution of MBL with recombinant MBL-associated serine protease-2. J Immunol 165:2093–2100

Walport MJ (2001) Complement. First of two parts. N Engl J Med 344:1058–1066

Walsh MC, Bourcier T, Takahashi K, Shi L, Busche MN, Rother RP, Solomon SD, Ezekowitz RA, Stahl GL (2005) Mannose-binding lectin is a regulator of inflammation that accompanies myocardial ischemia and reperfusion injury. J Immunol 175:541–546

Whisstock JC, Silverman GA, Bird PI, Bottomley SP, Kaiserman D, Luke CJ, Pak SC, Reichhart JM, Huntington JA (2010) Serpins flex their muscle: II. Structural insights into target peptidase recognition, polymerization, and transport functions. J Biol Chem 285:24307–24312
Wu Y, Eigenbrot C, Liang WC, Stawicki S, Shia S, Fan B, Ganesan R, Lipari MT, Kirchhofer D (2007) Structural insight into distinct mechanisms of protease inhibition by antibodies. Proc Natl Acad Sci USA 104:19784–19789
Zani M-L, Moreau T (2010) Phage display as a powerful tool to engineer protease inhibitors. Biochemie 92:1689–1704
Ziccardi R (1985) Demonstration of the interaction of native C1 with monomeric immunoglobulins and C1 inhibitor. J Immunol 134:2559–2563
Zundel S, Cseh S, Lacroix M, Dahl MR, Matsushita M, Andrieu JP, Schwaeble WJ, Jensenius JC, Fujita T, Arlaud GJ, Thielens NM (2004) Characterization of recombinant mannan-binding lectin-associated serine protease (MASP)-3 suggests an activation mechanism different from that of MASP-1 and MASP-2. J Immunol 172:4342–4350

Chapter 3
The Role of MASP-1/3 in Complement Activation

Hideharu Sekine, Minoru Takahashi, Daisuke Iwaki, and Teizo Fujita

Abstract The complement system, which consists of more than 30 plasma and cell surface proteins, is activated by three pathways: the classical, lectin, and alternative pathways, leading to the generation of opsonins and pathogen destruction. In the lectin pathway, mannose-binding lectin (MBL) and ficolins act as pattern recognition molecules for pathogens, resulting in the activation of MBL-associated serine proteases (MASPs: MASP-1, MASP-2, and MASP-3). Among these proteases, MASP-2 is a key enzyme that cleaves C4 and C2 to assemble a C3 convertase (C4b2a). However, the physiological function of MASP-1 and MASP-3 remains unclear. To investigate the roles of MASP-1 and MASP-3, we generated a MASP-1- and MASP-3-deficient (M1/3 KO) mouse model and found that the deficient mice lacked alternative pathway activation because factor D (Df) remained as a proenzyme in the serum. MASP-1 and MASP-3 were able to convert the proenzyme of Df to an active form in vitro. In addition, MASP-1 was able to activate MASP-2 and MASP-3 as C1r activates C1s. Thus, MASP-1 and MASP-3 seem to be involved in activation of both the lectin and alternative pathways.

3.1 Introduction

Immunity to infection is mediated by two systems, the acquired (or adaptive) immune system and the innate immune system. The innate immunity system is an evolutionarily ancient form and offers the main resistance to a microbial pathogen within the first minutes, hours, or days of an infection. Innate immunity was formerly thought to be a nonspecific immune response characterized by phagocytosis. However, innate immunity has considerable specificity and is able to discriminate between pathogens and self, as well as between classes of pathogen, by pattern recognition molecules. These molecules recognize conserved pathogen-associated molecular patterns (PAMPs) shared by large group of microorganisms, thereby successfully defending animals against infection (Hoffmann et al. 1999).

H. Sekine • M. Takahashi • D. Iwaki
Department of Immunology, Fukushima Medical University School of Medicine, 1-Hikariga-oka, Fukushima 960-1295, Japan

T. Fujita (✉)
Department of Immunology, Fukushima Medical University School of Medicine, 1-Hikariga-oka, Fukushima 960-1295, Japan

Fukushima Prefectural General Hygiene Institute, Fukushima 960-1295, Japan
e-mail: tfujita@fmu.ac.jp

J.D. Lambris et al. (eds.), *Complement Therapeutics*, Advances in Experimental Medicine and Biology 735, DOI 10.1007/978-1-4614-4118-2_3, © Springer Science+Business Media New York 2013

Complement was first described in the 1890s as a heat-labile bactericidal activity in serum which was triggered after the heat-stable antibodies had recognized and bound to the invading microorganisms. Now, the complement system, consisting of more than 30 proteins in plasma and on the cell surface, plays roles in immunological responses, such as opsonization of pathogens, chemotaxis and activation of leukocytes, direct killing of pathogens, clearance of immune complexes and apoptotic cells, and enhancing the adaptive immune response as a natural adjuvant (Walport 2001a, b; Carroll 2004). Once the complement system is activated, a chain of reactions involving proteolysis and assembly occurs, resulting in cleavage of the third complement component (C3). The cascade up to C3 cleavage is called the activation pathway. There are three activation pathways: the classical, lectin, and alternative pathways. In general, the classical pathway is one of the major effector mechanisms of the acquired immune system. The other two, the lectin and alternative pathways function in innate immune defense (Fujita 2002). The classical pathway is initiated by antibodies that are bound to antigens on the pathogen surface. Upon binding of C1q to immune complexes, serine proteases C1r and C1s are activated. Activated C1s then cleaves C4 and C2 to form a C3 convertase (C4bC2a), which cleaves C3 into C3a and C3b. The alternative pathway is triggered by charge distribution patterns on a range of microorganisms and does not involve specific recognition molecules. C3 is spontaneously hydrolyzed, generating C3(H_2O) at its low level in plasma. Binding of C3(H_2O) to factor B (Bf) induces a conformation change in Bf that makes it susceptible to cleavage by factor D (Df), generating Ba and Bb. Df is thought to circulate in blood as the only activated protease in the complement system (Rosen et al. 1989). The resulting C3(H_2O)Bb complex is the initial C3 convertase. If Bf associates with C3b bound to the pathogen surface, Bf is activated by Df to form a surface-bound C3 convertase (C3bBb). This convertase generates more surface-bound C3bBb through an amplification loop, leading to the deposition of many molecules of C3b on the pathogen surface. This amplification loop also contributes complement activation triggered through the classical and lectin pathways.

The lectin pathway is initiated by binding of mannose-binding lectin (MBL) or ficolins to arrays of carbohydrates on pathogen surface and the subsequent activation of associated enzymes, MBL-associated serine proteases (MASPs) (Fujita 2002). Three distinct MASPs, MASP-1 (Matsushita and Fujita 1992), MASP-2 (Thiel et al. 1997), and MASP-3 (Dahl et al. 2001), have been identified in many species of vertebrates. Similar to C1s, MASP-2 cleaves C4 and C2 to generate the C3 convertase (C4bC2a). Although we reported that MASP-1 cleaves C3 directly (Matsushita and Fujita 1995), the activity was very weak and the physiological function of MASP-1 was unclear. Also, the function of MASP-3 is still unknown, including its activation mechanism and physiological substrates. To identify the roles of MASP-1 in vivo, we generated a MASP-1-deficient mouse in which MASP-3 is also absent (M1/3 KO). Using M1/3 KO mice, we reported that MASP-1 triggered the lectin pathway activation by promoting MASP-2 activation (Takahashi et al. 2008). Recently, we found no activation of the alternative pathway in M1/3KO mice and analyzed the underlying mechanisms (Takahashi et al. 2010). In this chapter, we focus on molecular basis of the lectin pathway and the novel function of MASP-1 and MASP-3 in relation to their roles in initiation of the alternative pathway, together with our recent progress on activation mechanism of MASP-3 (Iwaki et al. 2011).

3.2 Recognition Molecules Involved in Lectin Pathway Activation

MBL is a C-type lectin that plays a crucial role in the first line of host defense (Mizuno et al. 1981; Drickamer et al. 1986; Ezekowitz et al. 1988). The importance of this molecule is underlined by a number of clinical studies linking MBL deficiency with increased susceptibility to a variety of infectious diseases (Jack et al. 2001). MBL belongs to the collectin family of proteins that consist of collagen-like domain and carbohydrate recognition domain (CRD) (Holmskov et al. 1994). MBL is an oligomer of structural subunits each composed of three identical 32-kDa polypeptides. Human MBL has an apparent molecular mass of about 300–650 kDa, consisting of 3–6 subunits. Each subunit contains an amino-terminal region rich in cysteine, a collagen-like domain consisting of tandem repeats of Gly-*X*-*Y* triplet sequences, and a carboxy-terminal CRD. Through its CRD, MBL binds carbohydrates with 3- and 4-hydroxyl groups in the pyranose ring in the presence of calcium (Weis et al. 1992). So, prominent ligands for MBL are mannose and *N*-acetyl-glucosamine (GlcNAc), whereas carbohydrates that do not fit this steric requirement, for example, galactose and sialic acid, which usually decorate the mammalian glycoproteins, have undetectable affinity for MBL (Drickamer 1992). This steric specificity of MBL, along with differences in the spatial organization of its ligands, enables the specific recognition of carbohydrates on pathogenic microorganisms, including bacteria, fungi, parasitic protozoans, and viruses, and avoids the recognition of self (Sheriff et al. 1994).

Ficolin is a family of oligomeric proteins consisting of two major domains, an N-terminal collagen-like and a C-terminal fibrinogen-like domain. Like MBL, the collagen-like domain is involved in the oligomerization of ficolin by forming a trimeric collagenous stalk. Unlike MBL, however, the globular fibrinogen-like domain is responsible for the lectin activity specific for *N*-acetylated compounds such as GlcNAc (Garlatti et al. 2007a, b; Tanio et al. 2007; Endo et al. 2011). Ficolin was originally identified the transforming growth factor-β1-binding proteins on porcine uterus membranes (Ichijo et al. 1993). The human homologue of ficolin was isolated from the plasma, which was first designated P35 (Matsushita et al. 1996) and afterwards renamed L-ficolin or FCN-2. In addition to L-ficolin, two ficolins have been identified in humans, designated M-ficolin (FCN-1) (Endo et al. 1996; Lu et al. 1996) and H-ficolin (FCN-3) (Sugimoto et al. 1998). To date, these human ficolins have been intensively characterized and found that they recognize PAMPs on pathogens as the pattern recognition molecules. In addition to MBL, all of human ficolins have the abilities to associate with MASPs and trigger the complement activation through the lectin pathway (Matsushita et al. 2000a, 2002; Liu et al. 2005; Endo et al. 2011).

Recently, it was reported that one of the collectin family, collectin 11 (CL-11, CL-K1) associated with MASP-1/3, probably subsequently activating the complement system (Hansen et al. 2010).

3.3 Structure and Function of MASPs

In 1992, we discovered that MBL associated with a novel C1s-like proteolytic enzyme and MASP (Matsushita and Fujita 1992). Then, MASP-2 and MASP-3 were identified and first MASP was called MASP-1 (Thiel et al. 1997; Dahl et al. 2001). MASPs associate with MBL and ficolins, which are recognition molecules of the lectin pathway (Fujita 2002). The overall structure of MASPs resembles two proteolytic components of C1 in the classical pathway, C1r and C1s. Also, the gene structure of MASP-2 and C1r/C1s are similar. In contrast, MASP-1 and MASP-3 derived from the same gene; MASP-3 is produced from the *MASP1* gene by alternative splicing (Dahl et al. 2001). The *MASP1* gene has an H-chain-coding region common to MASP-1/3, followed by tandem repeats of protease domain-coding regions specific for respective MASP-3 and MASP-1 (Fig. 3.1). Thus, the *MASP1* gene is very unique in having the double protease-coding regions among the serine protease superfamily.

Five members of the MASP/C1r/C1s family are composed of clearly six domains, including two C1r/C1s/Uegf/bone morphogenetic protein 1 (CUB), an epidermal growth factor (EGF)-like domain, two complement control proteins (CCPs) or short consensus repeats (SCRs), and a serine protease domain. On binding of MBL and ficolins to carbohydrate on the surface of a pathogen, the proenzyme form of MASP is cleaved between the second CCP and protease domain, resulting in the active form comprising of two polypeptides, heavy (H) and light (L) chains, or also called A and B chains (Fig. 3.1) (Fujita 2002). In classical pathway, recognition of C1q to the Fc portion of antibodies in immune complex leads to autoactivation of C1r, followed by C1r activation of C1s. The activated C1s cleaves C4 and C2, then generating C3 convertase. In the case of the lectin pathway, however, the activation

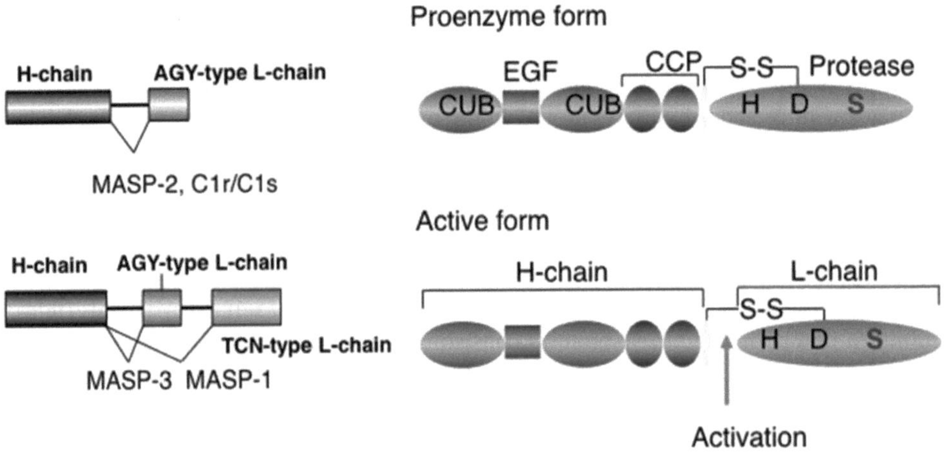

Fig. 3.1 Structure of the MASP/C1r/C1s family. Gene structure (*left*); the schematic gene structure of MASP-2 and C1r/C1s are similar (*upper*). In contrast, MASP-1 and MASP-3 derived from the same gene (*lower*): the *MASP1* gene has an H-chain-coding region common to MASP-1/3, followed by tandem repeats of protease domain-coding regions specific for respective MASP-3 and MASP-1. Domain structure (*right*); MASP-1, MASP-2, MASP-3, C1r, and C1s consist of six domains: two CUB domains, EGF-like domain, two CCP domains, and a serine protease domain. Histidine (*H*), aspartic acid (*D*), and serine (*S*) residues are essential for the formation of the active center in the serine protease domain. The proenzyme form (*upper*) is cleaved between the second CCP and the protease domain, which results in an active form that consists of two polypeptides, H and L chains (also known as A and B chains)

Table 3.1 Human MASP/C1r/C1s family

	MASP-1	MASP-2	MASP-3	C1r	C1s
Concentration (μg/mL)	6	0.5	6.4	50	50
Chromosomal location	3q27	1p36	3q27	12p3	12p3
Serine protease domain					
Number of exon	6	1	1	1	1
Number of cysteine	7	5	5	5	5
Codon of active serine	TCT	AGC	AGY	AGT	AGT
Substrate specificity	C2, C3, Df	C4, C2	Df, Bf	C1s	C4, C2

Note that underlined substrates were described in this review

mechanisms have not been fully understood. As with C1s, MASP-2 is required and essential to drive the lectin activation pathway (Thiel et al. 1997; Iwaki et al. 2006; Schwaeble et al. 2011), and activated MASP-2 cleaves C4- and C4b-bound C2, generating C4b2a, the classical and lectin pathway C3 convertase. MASP-1 appears to facilitate lectin pathway activation by either direct cleavage of complex-bound MASP-2 or cleavage of C4b-bound C2 (Matsushita et al. 2000b; Chen and Wallis 2004; Takahashi et al. 2008; Schwaeble et al. 2011), and it cleaves factor XIII and fibrinogen in vitro, suggesting the relation to the coagulation system (Hajela et al. 2002; Krarup et al. 2008). As described below, we demonstrated recently that MASP-1 and MASP-3 play a key role in the maturation and initiation of the alternative activation pathway (Takahashi et al. 2010; Iwaki et al. 2011). The properties of the MASP/C1r/C1s family are shown in Table 3.1.

In addition, two nonenzymatic splice variants of MASPs are reported. One is a truncated form of MASP-2, named small MBL-associated protein (sMAP or MAp19) (Takahashi et al. 1999; Stover et al. 1999), which is generated by alternative splicing from a single structural gene (*MASP2*), and sMAP consists of the first CUB (CUB1) domain, the EGF-like domain, and an extra four amino acids at the C-terminal end encoded by a sMAP-specific exon. The other is the product of the *MASP1* gene, named MAp44 (Degn et al. 2009) or MAP1 (Skjoedt et al. 2010), that consists of CUB1-EGF-CUB2-CCP1, and a ninth exon encodes 17 C-terminal amino acids unique to this protein. The three MASPs, sMAP, and MAp44 are considered to form homodimers, which associate with MBL and ficolins through their N-terminal domains. To clarify the role of sMAP, we have generated sMAP-deficient mice by targeted disruption of the sMAP-specific exon and found that sMAP and MASP-2 compete to bind to MBL/ficolins and sMAP has the ability to downregulate the lectin pathway (Iwaki et al. 2006). Similar results were obtained in the case of MAp44 that competes with MASP-2 for binding to MBL, resulting in inhibition of C4 deposition (Degn et al. 2009).

3.4 Character of MASP-1 and MASP-3 (M1/3) Knockout Mouse

3.4.1 Generation of MASP1/3 Knockout Mouse (M1/3KO)

To assess the roles of MASP-1 and MASP-3 in vivo, we established a gene-targeted mouse that lacks both MASP-1 and MASP-3. The *MASP1* gene has ten exons that encode the heavy chain common to both MASP-1 and MASP-3. A targeting vector was constructed to replace the second exon with a neomycin-resistant gene cassette. Finally, four neomycin-resistant ES clones were obtained, in which the homologous recombination events were confirmed by Southern blotting analysis. Two of these were independently injected into C57BL/6 blastocysts, and the founder chimeras were bred with C57BL/6J females. Southern blotting analysis of tail DNA from agouti-color pups showed germ line transmission of the targeted allele. Heterozygous mice were subsequently intercrossed to generate homozygous mice (M1/3 KO) and furthermore backcrossed with C57BL/6J for more than five generations to eliminate the 129/Sv genetic background (Takahashi et al. 2008).

M1/3 KO mice (from 4 to 10 weeks old) were significantly smaller than control mice. The average body weights of males at 6 weeks of age were 20.8 ± 1.97 g (M1/3+/+), 21.1 ± 1.34 g (M1/3+/−), and 16.8 ± 1.66 g (M1/3−/−), and those of females were 17.9 ± 1.16 g (M1/3+/+), 17.5 ± 2.38 g (M1/3+/−), and 14.8 ± 1.73 g (M1/3−/−). These data suggest that MASP-1 and/or MASP-3 might be involved, in part, in development and growth. Although they showed no abnormality in macroscopic appearance or subcutaneous adipose tissue, parametrial adipose tissue weights were significantly lower in M1/3 KO mice, compared to wild-type mice, and this tissue showed apparent atrophy by microscopic analysis (manuscript in preparation).

3.4.2 Lectin Pathway Activation in M1/3 KO Mice

To assess the activity of the lectin pathway in M1/3 KO mice, we carried out the C4-deposition assay with mannan-coated plates and found that C4 deposition from the sera of M1/3 KO mice was significantly lower than that in wild-type mice. Kinetic plots of C4 deposition also showed a very low activity in M1/3 KO mice. Next, we asked whether MASP-2 was activated in M1/3 KO serum by Western blotting using anti-MASP-2. No activation of MASP-2 was observed in M1/3 KO serum, when serum was incubated with mannan-agarose for 10 or 20 min, whereas activation of MASP-2 was observed in normal serum. MASP-2 was converted to the activated form after 30 min; however, the amount of active MASP-2 in M1/3 KO mice was lower than that in wild-type mice. We prepared recombinant MASP-1 by replacing the arginine residue at the reactive site P1 for activation of MASP-1 with lysine (rMASP-1K), and the rMASP-1K overcame the problem of self-activation and degradation of MASP-1, allowing us to produce larger amounts of the stable recombinant protein

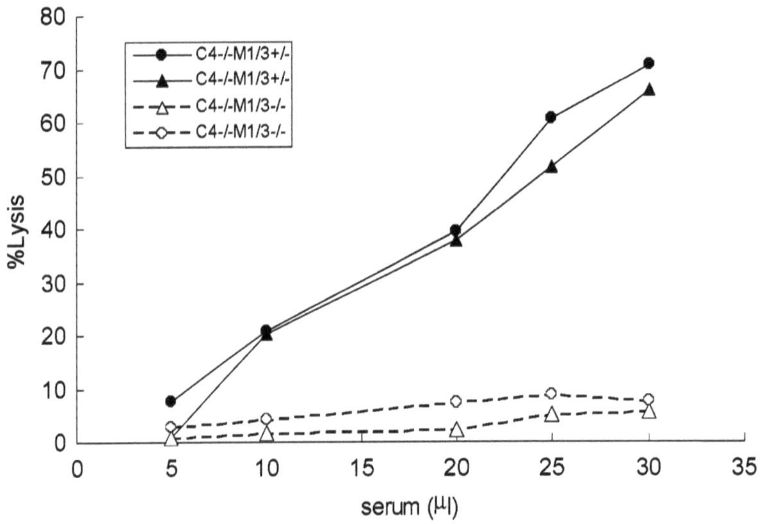

Fig. 3.2 The alternative pathway activation in M1/3 KO mice. Rabbit erythrocytes (2.5×10^6) were incubated for 1 h with mouse serum in gelatin-veronal buffer (GVB) containing Mg^{2+}-EGTA. Hemolysis was measured in a microplate reader at 405 nm. Two individual mice, C4-deficient (C4–/– M1/3+/–) and M1/3- and C4-deficient (C4–/– M1/3–/–), were analyzed. In M1/3 KO mice, no activation of the alternative pathway was observed

(Takahashi et al. 2008). The C4-deposition activity and MASP-2 activation in M1/3 KO mouse serum was restored by the reconstitution of stable rMASP-1K. These results support the possibility that MASP-1 contributes to the activation of the lectin pathway at the step of MASP-2 activation.

To confirm the above results, we assessed whether MASP-1 cleaves MASP-2 directly. Since the previous studies have shown that recombinant MASP-2 was easily autoactivated during its preparation (Vorup-Jensen et al. 2000), we prepared rMASP-2i which was not cleaved by itself because it lacks protease activity. When rMASP-2i, the mutant of MASP-2, was incubated with rMASP-1K, rMASP-1K was able to cleave rMASP-2i directly. Similarly, we performed the C3-deposition assay with mannan-coated plates and found that C3-deposition from the sera of M1/3 KO mice was significantly lower than that in wild-type mice (Takahashi et al. 2008). In addition to activation of MASP-2, MASP-1 acts to cleave C4b-bound C2, generating C3 convertase (Matsushita et al. 2000; Chen and Wallis 2004). Thus, it is likely that MASP-1 plays a pivotal role in the activation of the lectin pathway, probably through activation of MASP-2 and C2.

3.4.3 M1/3KO Mice Lack Activation of the Alternative Complement Pathway

To determine whether the alternative pathway is activated in M1/3 KO mice, the mouse sera were assayed for hemolytic activity against rabbit erythrocytes that is specific for the alternative pathway activation. In this assay, mice with a C4-deficient background (*C4*$^{-/-}$) and Mg^{2+}-EGTA-containing buffer were used to eliminate the effects of the classical and lectin pathways. Surprisingly, sera derived from M1/3 KO mice were demonstrated to have no ability to activate the alterative pathway compared to the heterozygous mice that have normal levels of MASP-1/3 (Fig. 3.2). The similar results were obtained for C3-deposition assay using zymosan-coated microwells. Therefore, these results indicate that there is some defect in the activation process of the alternative pathway in M1/3 KO mice.

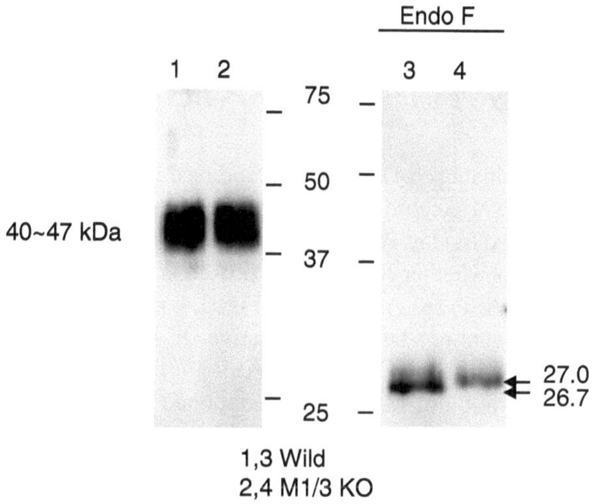

Fig. 3.3 Analysis of Df in wild-type and M1/3 KO mouse. Endogenous Df immunoprecipitated and immunoblotted with a polyclonal anti-Df antibody (*left*). Precipitated Df was incubated with N-glycosidase F (*Endo-F*) to remove N-glycosylations. Sera from wild-type and M1/3 KO mice were analyzed

To identify the cause of the above results in M1/3 KO mice, Bf cleavage was investigated by Western blotting. When serum from wild-type C57BL/6 mice was incubated at 37°C for 1 h, a fragment called Ba was cleaved from Bf by limited proteolysis. On the other hand, no Ba fragment was detected in the incubation of serum from M1/3 KO mice. Since Bf was cleaved by Df, addition of purified human Df (active form) to M1/3 KO mice sera resulted in the cleavage of Bf. A very low level of hemolytic activity against rabbit erythrocytes in the sera of M1/3 KO mice was compensated by adding active human Df. These results suggest that M1/3 KO mice carry an abnormality in the activation of Df.

3.4.4 Proenzyme of Df in M1/3KO Mice

To investigate the nature of Df in M1/3 KO serum, Df protein was isolated by immunoprecipitation with an anti-mouse Df antibody. Df was detected as 40–44-kDa broad bands in both M1/3 KO and wild-type mice. When these bands were treated with N-glycosidase F, the deglycosylated Df in M1/3 KO serum was slightly larger than that in wild-type mouse (Fig. 3.3). Df is reported to circulate in blood as an activated protease (Rosen et al. 1989). A putative activation peptide QPRGR predicted from its cDNA sequence is removed during Df maturation and secretion (Cook et al. 1985). Therefore, it is likely that Df in M1/3 KO mice still retains the activation peptide. To verify this, we performed electrospray ionization time-of-flight mass spectrometry (ESI-TOF-MS) of Df-derived fragments cleaved by V8 protease. A molecular ion peak corresponding to the mass of an N-terminal-activated Df fragment ILGGQE was detected after cleavage of Df from wild type, showing that wild-type mouse has an active Df. Among V8 protease-treated fragments of Df from M1/3 KO mice, on the other hand, a different molecular ion peak corresponded to the sequence (p)QPRGRILGGQE, including a modified N-terminal pyroglutamate. These results revealed that M1/3 KO mice failed to convert pro-Df to its active form. Thus, it seems likely that MASP-1 and/or MASP-3 is involved in the cleavage of the activation peptide from pro-Df.

Another important finding is that Df is not secreted as the active form, in contrast to several previous reports (Barnum and Volanakis 1985; White et al. 1992). Recently, we demonstrated that differentiated adipocytes secreted pro-Df with the activation peptides, clearly indicating that pro-Df is cleaved after secretion (Takahashi et al. 2010).

3.5 Novel Functions of MASP-1 and MASP-3

3.5.1 *Recombinant MASP-1 Cleaves and Activates Pro-Df*

Recombinant pro-Df has been produced in insect cells using a baculovirus expression system (Rosen et al. 1989). According to this method, we prepared recombinant mouse pro-Df containing a hexahistidine (6xHis) tag. As expected, the recombinant Df produced was pro-Df with the activation peptide. To assess whether recombinant pro-Df is a substrate for MASP-1, recombinant pro-Df was incubated with rMASP-1K. As shown in Fig. 3.4, rMASP-1K reduced the molecular weight of deglycosylated recombinant pro-Df to that of active Df, indicating that MASP-1 is able to directly cleave the activation peptide in pro-Df. Thus, MASP-1 is the enzyme for complement activation that not only enhances the activation of MASP-2 as mentioned above but also activates pro-Df in the alternative pathway (Takahashi et al. 2010).

3.5.2 *Characteristics of rMASP-3 and Its Activation*

In previous studies, the human recombinant MASP-3 was purified as a proenzyme without autoactivation (Dahl et al. 2001; Zundel et al. 2004), whereas recombinant MASP-1 and MASP-2 were autoactivated during purification steps and obtained as the activated enzymes (Thielens et al. 2001; Cseh et al. 2002; Takahashi et al. 2008). We obtained mouse rMASP-3 as a single-chain proenzyme like the recombinant human MASP-3 and tried to find substances to induce rMASP-3 activation. Mouse rMASP-3 was added to wild-type mouse serum and incubated with mannan-agarose or GlcNAc-agarose, since binding of the MBL-/ficolin-MASPs complexes to GlcNAc- or mannan-agarose-activated MASP-1 and MASP-2 (Matsushita and Fujita 1995, Matsushita et al. 2000a). However, rMASP-3 activation was hardly detected after incubation with these agarose conjugates. When rMASP-3 and wild-type mouse serum were incubated with heat-killed *Staphylococcus aureus* (*S. aureus*) at 37°C for 3 h, the L chain of activated rMASP-3 was clearly detected, showing that heat-killed *S. aureus* was an effective inducer for MASP-3 activation. We also tried to reproduce the MASP-3 activation by incubation with heat-killed *S. aureus* in combinations of the purified proteins such as rMASP-3 and rMBLs. When rMASP-3 and rMBL-A were incubated together with the bacteria, rMASP-3 was obviously activated. When rMBL-C was used instead of rMBL-A, the activation was very weak, probably, due to its low activity to form the complex with MASP-3 (Iwaki et al. 2011).

Next, we examined rMASP-3 activation in the MASP-deficient sera. We obtained MASP-2 (M2) KO mice (Iwaki et al. 2006) and all MASPs KO mice (Iwaki et al. 2011). The mixture of rMASP-3

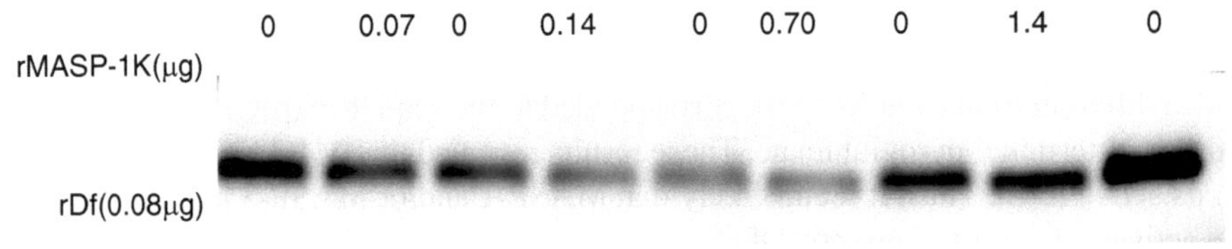

Fig. 3.4 MASP-1 activates pro-Df. Recombinant pro-Df (80 ng) was incubated with various amounts of rMASP-1K at 37°C for 1 h. As a control, pro-Df was incubated without rMASP-1K. After removing N-glycosylations by N-glycosidase F treatment, samples were separated by SDS-PAGE and immunoblotted with anti-Df antibody

and mouse serum was incubated with the bacteria and the rMASP-3 activation was observed by Western blotting. The rMASP-3 activation in all MASPs KO sera was the lowest, and the activation in M1/M3 KO serum was lower than in M2 KO serum, compared to the wild type. Therefore, MASP-1 might be more important than MASP-2 for MASP-3 activation. The activation of rMASP-3 in all MASPs KO sera was increased by addition of MASP-1, and rMASP-3 was directly cleaved by rMASP-1K. These data suggested that MASP-1 has the ability to activate MASP-3 as well as MASP-2 (Takahashi et al. 2008), as mentioned above.

3.5.3 MASP-3 Activates the Alternative Pathway

Since there is no activation of the alternative pathway in M1/3 KO sera, as mentioned above, we examined the involvement of MASP-3 in the alternative pathway activation. As previous studies have indicated (Dahl et al. 2001; Zundel et al. 2004), rMASP-3 was unable to directly cleave C3. We next examined whether MASP-3 is involved in Bf activation, a major component of the alternative pathway. After incubation of heat-killed *S. aureus* with several KO sera, Bf activation was analyzed by immunoblotting to detect fragment Ba. Bf activation in M2 KO serum was detected at almost the same level as in wild-type mouse serum. However, M1/M3 KO and all MASPs KO sera exhibited low levels of Bf activation, and this activation was restored by the addition of rMASP-3 to these sera. Then, we tried to reconstitute the Bf activation caused by MASP-3 with combinations of purified proteins and heat-killed *S. aureus*. Bf was activated in the presence of rMASP-3, rMBL-A, and $C3(H_2O)$ and the cleavage was further increased by the addition of pro-Df. By contrast, no cleavage was observed in the absence of rMASP-3.

Since the Bf cleavage was increased by the addition of pro-Df, MASP-3 may have the ability to activate Df as well as Bf. We examined Df activation by rMASP-3. When pro-Df was incubated alone or with rMASP-3, the molecular mass of Df incubated with rMASP-3 was only slightly lower than that of pro-Df incubated alone, and the same result was obtained after incubation with rMASP-1 (Fig. 3.4). These data indicate that rMASP-3, as well as rMASP-1, can directly cleave the activation peptide from pro-Df. Unlike Bf cleavage, Df cleavage by rMASP-3 was observed in the absence of MBL and bacteria, showing that MASP-3 even in a proenzyme form can activate Df. This might be a possible explanation for Df circulating in plasma in its active form (Rosen et al. 1989). In our previous study, mouse rMASP-1 was unable to activate the endogenous proenzyme Df in M1/M3 KO serum, and we considered that rMASP-1 activity was blocked by its physiological inhibitors in the serum (i.e., C1 inhibitor and alpha 2-macroglobulin) (Takahashi et al. 2010). However, we found that mouse rMASP-3 converted the endogenous proenzyme Df in all MASPs KO sera to the active form. Therefore, MASP-3, which did not react with C1 inhibitor (Zundel et al. 2004), may be more effective than MASP-1 in activating Df in plasma.

In short, MASP-3 is able to activate Bf and Df, major components of the alternative pathway, thereby playing an important role in initiating the alternative pathway activation.

3.6 Concluding Remark

In this chapter, we discussed the role of MASPs, especially MASP-1 and MASP-3 in complement activation. A putative model for MASP-mediated complement activation is shown in Fig. 3.5. We would like to emphasize the multifunction of MASP-1, that is, MASP-2 and MASP-3 activation in the lectin pathway, C2 activation in the lectin pathway, and Df activation in the alternative pathway.

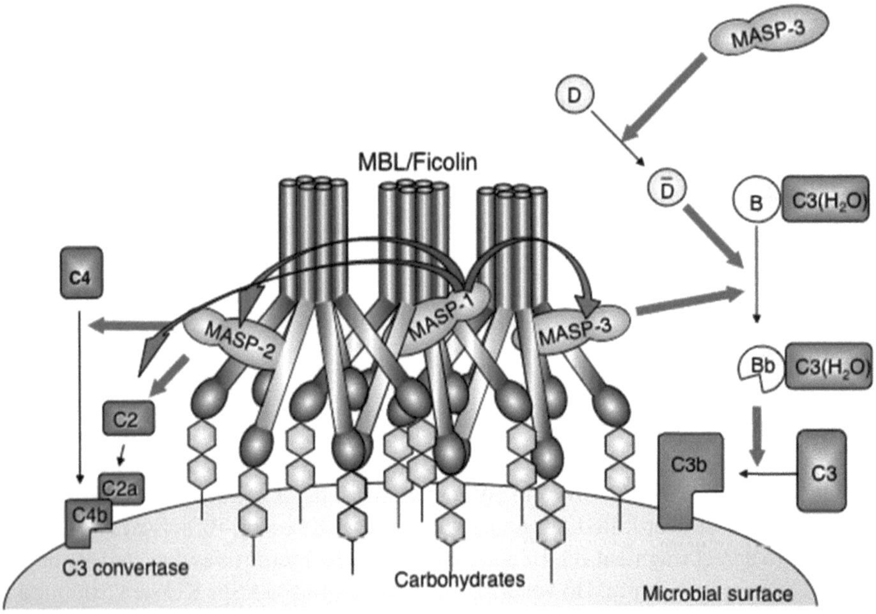

Fig. 3.5 Putative MASP-mediated pathway of complement activation. MBL is composed of trimeric subunits and each MASP forms a homodimer. Binding of MBL/ficolin to carbohydrates on the bacterial surface induces activation of MASPs in the MBL/ficolin complex. Activated MASP-1 plays a role in cleavage of MASP-2 and MASP-3, C2, and Df. MASP-2 activates C4 and C2 to generate C3 convertase. Activated MASP-3 cleaves $C3(H_2O)$-bound Bf. The resulting $C3(H_2O)Bb$ complex is the initial C3 convertase. Bacterial surface-bound C3b acts as an opsonin or associates with Bf to generate more C3 convertase (C3bBb). Proenzyme MASP-3 can activate Df

It is well known that MASP-2 activates C4 and C2 to generate C3 convertase. Proenzyme MASP-3 complexed with MBL is activated by bacteria, and activated MASP-3 induces activation of the alternative pathway by cleaving Df and Bf. Although previous studies suggested that MASP-3 inhibited MASP-2-dependent C4 deposition (Dahl et al. 2001), we propose that activation of the alternative pathway may be a more physiological function of MASP-3. Although MASP-1 and MASP-3 play a key role in the maturation and initiation of the alternative activation pathway, we do not know to what extent the MASP-1/3-mediated pathway contributes to complement activation, but it might be able to boost the alternative pathway or act as a backup pathway in the case of deficiencies of complement components, like the MBL-dependent C2 bypass pathway (Selander et al. 2006). In addition, the in vivo study revealed that phagocytic activities of M1/3 KO and all MASPs KO mice against *S. aureus* and the bacterial clearance in these mice were lower than wild-type and MASP-2-deficient mice, indicating that the MASP-1/3-mediated pathway plays an important role in host defense (Iwaki et al. 2011). On the other hand, inappropriate activation of complement affects the pathogenesis of inflammatory diseases (Walport 2001a). In this sense, there are two reports concerning developments of diseases in M1/3 KO mice such as the murine collagen Ab-induced arthritis (CAIA) model (Banda et al. 2010) and a murine model of occlusive thrombosis (La Bonte et al. 2012). These experimental models suggest the therapeutic effects of MASP-1/3 inhibition in the lectin pathway-mediated disorders, as in the case of MASP-2 where the therapeutic effects of MASP-2 inhibition were observed in the experimental model of myocardial and gastrointestinal ischemia/reperfusion injury (Schwaeble et al. 2011). Recently, mutations in the *MASP1* gene have been reported to be a cause of the Carnevale, Mingarelli, Malpuech, and Michels syndromes (so-called 3M syndrome), a disorder that includes MASP-1 and MASP-3 deficiency and craniofacial defects (Rooryck et al. 2011). It is possible that MASP-1/3-mediated pathway may also play an important role in embryonic development.

References

Banda NK, Takahashi M, Levitt B, Glogowska M, Nicholas J, Takahashi K, Stahl GL, Fujita T, Arend WP, Holers VM (2010) Essential role of complement mannose-binding lectin-associated serine proteases-1/3 in the murine collagen antibody-induced model of inflammatory arthritis. J Immunol 185:5598–5606

Barnum SR, Volanakis JE (1985) In vitro biosynthesis of complement protein D by U937 cells. J Immunol 134:1799–1803

Carroll MC (2004) The complement system in regulation of adaptive immunity. Nat Immunol 5:981–986

Chen CB, Wallis R (2004) Two mechanisms for mannose-binding protein modulation of the activity of its associated serine proteases. J Biol Chem 279:26058–26065

Cook KS, Groves DL, Min HY, Spiegelman BM (1985) A developmentally regulated mRNA from 3T3 adipocytes encodes a novel serine protease homologue. Proc Natl Acad Sci USA 82:6480–6484

Cseh S, Vera L, Matsushita M, Fujita T, Arlaud GJ, Thielens NM (2002) Characterization of the interaction between L-ficolin/p35 and mannan-binding lectin-associated serine proteases-1 and -2. J Immunol 169:5735–5743

Dahl MR, Thiel S, Matsushita M, Fujita T, Willis AC, Christensen T, Vorup-Jensen T, Jensenius JC (2001) MASP-3 and its association with distinct complexes of the mannan-binding lectin complement activation pathway. Immunity 15:127–135

Degn SE, Hansen AG, Steffensen R, Jacobsen C, Jensenius JC, Thiel S (2009) MAp44, a human protein associated with pattern recognition molecules of the complement system and regulating the lectin pathway of complement activation. J Immunol 183:7371–7378

Drickamer K (1992) Engineering galactose-binding activity into a C-type mannose-binding protein. Nature 360:183–186

Drickamer K, Dordal MS, Reynolds L (1986) Mannose-binding proteins isolated from rat liver contain carbohydrate-recognition domains linked to collagenous tails. Complete primary structures and homology with pulmonary surfactant apoprotein. J Biol Chem 261:6878–6887

Endo Y, Sato Y, Matsushita M, Fujita T (1996) Cloning and characterization of the human lectin P35 gene and its related gene. Genomics 36:515–521

Endo Y, Matsushita M, Fujita T (2011) The role of ficolins in the lectin pathway of innate immunity. Int J Biochem Cell Biol 43:705–712

Ezekowitz RA, Day LE, Herman GA (1988) A human mannose-binding protein is an acute-phase reactant that shares sequence homology with other vertebrate lectins. J Exp Med 167:1034–1046

Fujita T (2002) Evolution of the lectin-complement pathway and its role in innate immunity. Nat Rev Immunol 2:346–353

Garlatti V, Belloy N, Martin L, Lacroix M, Matsushita M, Endo Y, Fujita T, Fontecilla-Camps JC, Arlaud GJ, Thielens NM, Gaboriaud C (2007a) Structural insights into the innate immune recognition specificities of L- and H-ficolins. EMBO J 26:623–633

Garlatti V, Martin L, Gout E, Reiser JB, Fujita T, Arlaud GJ, Thielens NM, Gaboriaud C (2007b) Structural basis for innate immune sensing by M-ficolin and its control by a pH-dependent conformational switch. J Biol Chem 282:35814–35820

Hajela K, Kojima M, Ambrus G, Wong KH, Moffatt BE, Ferluga J, Hajela S, Gal P, Sim RB (2002) The biological functions of MBL-associated serine proteases (MASPs). Immunobiology 205:467–475

Hansen S, Selman L, Palaniyar N, Ziegler K, Brandt J, Kliem A, Jonasson M, Skjoedt MO, Nielsen O, Hartshorn K et al (2010) Collectin 11 (CL-11, CL-K1) is a MASP-1/3-associated plasma collectin with microbial-binding activity. J Immunol 185:6096–6104

Hoffmann JA, Kafatos FC, Janeway CA, Ezekowitz RA (1999) Phylogenetic perspectives in innate immunity. Science 284:1313–1318

Holmskov U, Malhotra R, Sim RB, Jensenius JC (1994) Collectins: collagenous C-type lectins of the innate immune defense system. Immunol Today 15:67–74

Ichijo H, Hellman U, Wernstedt C, Gonez LJ, Claesson-Welsh L, Heldin CH, Miyazono K (1993) Molecular cloning and characterization of ficolin, a multimeric protein with fibrinogen- and collagen-like domains. J Biol Chem 268:14505–14513

Iwaki D, Kanno K, Takahashi M, Endo Y, Lynch NJ, Schwaeble WJ, Matsushita M, Okabe M, Fujita T (2006) Small mannose-binding lectin-associated protein plays a regulatory role in the lectin complement pathway. J Immunol 177:8626–8632

Iwaki D, Kanno K, Takahashi M, Endo Y, Matsushita M, Fujita T (2011) The role of mannose-binding lectin-associated serine protease-3 in activation of the alternative complement pathway. J Immunol 187:3751–3758

Jack DL, Klein NJ, Turner MW (2001) Mannose-binding lectin: targeting the microbial world for complement attack and opsonophagocytosis. Immunol Rev 180:86–99

Krarup A, Gulla KC, Gal P, Hajela K, Sim RB (2008) The action of MBL-associated serine protease 1 (MASP1) on factor XIII and fibrinogen. Biochim Biophys Acta 1784:1294–1300

La Bonte LR, Pavlov VI, Tan YS, Takahashi K, Takahashi M, Banda NK, Zou C, Fujita T, Stahl GL (2012) Mannose-binding lectin-associated serine protease-1 is a significant contributor to coagulation in a murine model of occlusive thrombosis. J Immunol 188:885–891

Liu Y, Endo Y, Iwaki D, Nakata M, Matsushita M, Wada I, Inoue K, Munakata M, Fujita T (2005) Human M-ficolin is a secretory protein that activates the lectin complement pathway. J Immunol 175:3150–3156

Lu J, Tay PN, Kon OL, Reid KB (1996) Human ficolin: cDNA cloning, demonstration of peripheral blood leucocytes as the major site of synthesis and assignment of the gene to chromosome 9. Biochem J 313(Pt 2):473–478

Matsushita M, Fujita T (1992) Activation of the classical complement pathway by mannose-binding protein in association with a novel C1s-like serine protease. J Exp Med 176:1497–1502

Matsushita M, Fujita T (1995) Cleavage of the third component of complement (C3) by mannose-binding protein-associated serine protease (MASP) with subsequent complement activation. Immunobiology 194:443–448

Matsushita M, Endo Y, Taira S, Sato Y, Fujita T, Ichikawa N, Nakata M, Mizuochi T (1996) A novel human serum lectin with collagen- and fibrinogen-like domains that functions as an opsonin. J Biol Chem 271:2448–2454

Matsushita M, Endo Y, Fujita T (2000a) Cutting edge: complement-activating complex of ficolin and mannose-binding lectin-associated serine protease. J Immunol 164:2281–2284

Matsushita M, Thiel S, Jensenius JC, Terai I, Fujita T (2000b) Proteolytic activities of two types of mannose-binding lectin-associated serine protease. J Immunol 165:2637–2642

Matsushita M, Kuraya M, Hamasaki N, Tsujimura M, Shiraki H, Fujita T (2002) Activation of the lectin complement pathway by H-ficolin (Hakata antigen). J Immunol 168:3502–3506

Mizuno Y, Kozutsumi Y, Kawasaki T, Yamashina I (1981) Isolation and characterization of a mannan-binding protein from rat liver. J Biol Chem 256:4247–4252

Rooryck C, Diaz-Font A, Osborn DP, Chabchoub E, Hernandez-Hernandez V, Shamseldin H, Kenny J, Waters A, Jenkins D, Kaissi AA et al (2011) Mutations in lectin complement pathway genes COLEC11 and MASP1 cause 3MC syndrome. Nat Genet 43:197–203

Rosen BS, Cook KS, Yaglom J, Groves DL, Volanakis JE, Damm D, White T, Spiegelman BM (1989) Adipsin and complement factor D activity: an immune-related defect in obesity. Science 244:1483–1487

Schwaeble WJ, Lynch NJ, Clark JE, Marber M, Samani NJ, Ali YM, Dudler T, Parent B, Lhotta K, Wallis R et al (2011) Targeting of mannan-binding lectin-associated serine protease-2 confers protection from myocardial and gastrointestinal ischemia/reperfusion injury. Proc Natl Acad Sci USA 108:7523–7528

Selander B, Martensson U, Weintraub A, Holmstrom E, Matsushita M, Thiel S, Jensenius JC, Truedsson L, Sjoholm AG (2006) Mannan-binding lectin activates C3 and the alternative complement pathway without involvement of C2. J Clin Invest 116:1425–1434

Sheriff S, Chang CY, Ezekowitz RA (1994) Human mannose-binding protein carbohydrate recognition domain trimerizes through a triple alpha-helical coiled-coil. Nat Struct Biol 1:789–794

Skjoedt MO, Hummelshoj T, Palarasah Y, Honore C, Koch C, Skjodt K, Garred P (2010) A novel mannose-binding lectin/ficolin-associated protein is highly expressed in heart and skeletal muscle tissues and inhibits complement activation. J Biol Chem 285:8234–8243

Stover CM, Thiel S, Thelen M, Lynch NJ, Vorup-Jensen T, Jensenius JC, Schwaeble WJ (1999) Two constituents of the initiation complex of the mannan-binding lectin activation pathway of complement are encoded by a single structural gene. J Immunol 162:3481–3490

Sugimoto R, Yae Y, Akaiwa M, Kitajima S, Shibata Y, Sato H, Hirata J, Okochi K, Izuhara K, Hamasaki N (1998) Cloning and characterization of the Hakata antigen, a member of the ficolin/opsonin p35 lectin family. J Biol Chem 273:20721–20727

Takahashi M, Endo Y, Fujita T, Matsushita M (1999) A truncated form of mannose-binding lectin-associated serine protease (MASP)-2 expressed by alternative polyadenylation is a component of the lectin complement pathway. Int Immunol 11:859–863

Takahashi M, Iwaki D, Kanno K, Ishida Y, Xiong J, Matsushita M, Endo Y, Miura S, Ishii N, Sugamura K, Fujita T (2008) Mannose-binding lectin (MBL)-associated serine protease (MASP)-1 contributes to activation of the lectin complement pathway. J Immunol 180:6132–6138

Takahashi M, Ishida Y, Iwaki D, Kanno K, Suzuki T, Endo Y, Homma Y, Fujita T (2010) Essential role of mannose-binding lectin-associated serine protease-1 in activation of the complement factor D. J Exp Med 207:29–37

Tanio M, Kondo S, Sugio S, Kohno T (2007) Trivalent recognition unit of innate immunity system: crystal structure of trimeric human M-ficolin fibrinogen-like domain. J Biol Chem 282:3889–3895

Thiel S, Vorup-Jensen T, Stover CM, Schwaeble W, Laursen SB, Poulsen K, Willis AC, Eggleton P, Hansen S, Holmskov U et al (1997) A second serine protease associated with mannan-binding lectin that activates complement. Nature 386:506–510

Thielens NM, Cseh S, Thiel S, Vorup-Jensen T, Rossi V, Jensenius JC, Arlaud GJ (2001) Interaction properties of human mannan-binding lectin (MBL)-associated serine proteases-1 and -2, MBL-associated protein 19, and MBL. J Immunol 166:5068–5077

Vorup-Jensen T, Petersen SV, Hansen AG, Poulsen K, Schwaeble W, Sim RB, Reid KB, Davis SJ, Thiel S, Jensenius JC (2000) Distinct pathways of mannan-binding lectin (MBL)- and C1-complex autoactivation revealed by reconstitution of MBL with recombinant MBL-associated serine protease-2. J Immunol 165:2093–2100

Walport MJ (2001a) Complement. First of two parts. N Engl J Med 344:1058–1066

Walport MJ (2001b) Complement. Second of two parts. N Engl J Med 344:1140–1144

Weis WI, Drickamer K, Hendrickson WA (1992) Structure of a C-type mannose-binding protein complexed with an oligosaccharide. Nature 360:127–134

White RT, Damm D, Hancock N, Rosen BS, Lowell BB, Usher P, Flier JS, Spiegelman BM (1992) Human adipsin is identical to complement factor D and is expressed at high levels in adipose tissue. J Biol Chem 267:9210–9213

Zundel S, Cseh S, Lacroix M, Dahl MR, Matsushita M, Andrieu JP, Schwaeble WJ, Jensenius JC, Fujita T, Arlaud GJ, Thielens NM (2004) Characterization of recombinant mannan-binding lectin-associated serine protease (MASP)-3 suggests an activation mechanism different from that of MASP-1 and MASP-2. J Immunol 172:4342–4350

Chapter 4
Membrane-Bound Complement Regulatory Proteins as Biomarkers and Potential Therapeutic Targets for SLE

Nibhriti Das, Bintili Biswas, and Rohan Khera

Abstract For the last two decades, there had been remarkable advancement in understanding the role of complement regulatory proteins in autoimmune disorders and importance of complement inhibitors as therapeutics. Systemic lupus erythematosus is a prototype of systemic autoimmune disorders. The disease, though rare, is potentially fatal and afflicts women at their reproductive age. It is a complex disease with multiorgan involvement, and each patient presents with a different set of symptoms. The diagnosis is often difficult and is based on the diagnostic criteria set by the American Rheumatology Association. Presence of antinuclear antibodies and more specifically antidouble-stranded DNA indicates SLE. Since the disease is multifactorial and its phenotypes are highly heterogeneous, there is a need to identify multiple noninvasive biomarkers for SLE. Lack of validated biomarkers for SLE disease activity or response to treatment is a barrier to the efficient management of the disease, drug discovery, as well as development of new therapeutics. Recent studies with gene knockout mice have suggested that membrane-bound complement regulatory proteins (CRPs) may critically determine the sensitivity of host tissues to complement injury in autoimmune and inflammatory disorders. Case-controlled and follow-up studies carried out in our laboratory suggest an intimate relation between the level of DAF, MCP, CR1, and CD59 transcripts and the disease activity in SLE. Based on comparative evaluation of our data on these four membrane-bound complement regulatory proteins, we envisaged CR1 and MCP transcripts as putative noninvasive disease activity markers and the respective proteins as therapeutic targets for SLE. Following is a brief appraisal on membrane-bound complement regulatory proteins DAF, MCP, CR1, and CD59 as biomarkers and therapeutic targets for SLE.

4.1 Membrane-Bound Complement Regulatory Proteins

Exaggerated complement activation, deficiency of complement components, and modulation of complement regulatory proteins have been observed in various autoimmune disorders. While active complement peptides and membrane attack complex may cause significant cellular injury, complement regulatory proteins protect the host against complement-mediated tissue damage. Membrane-bound complement regulatory proteins (mCRPs) are ubiquitously expressed throughout the cells of the hematopoietic system and protect host cells from complement-mediated lysis. The balance between acceleration and inhibition of complement activation is critical to whether complement activation leads to host defense or tissue injury of host organs (Nangaku 2003).

N. Das (✉) • B. Biswas • R. Khera
Department of Biochemistry, All India Institute of Medical Sciences, Ansari Nagar, New Delhi 110029, India
e-mail: nibhriti@hotmail.com; nibhriti@rediffmail.com

J.D. Lambris et al. (eds.), *Complement Therapeutics*, Advances in Experimental Medicine and Biology 735,
DOI 10.1007/978-1-4614-4118-2_4, © Springer Science+Business Media New York 2013

Table 4.1 Complement regulatory proteins

Regulator	Location	Function
C1 INH	Plasma	Dissociates activated C1
C4-binding protein (C4BP)	Plasma	Dissociates classical C3 convertase Cofactor for factor I
Properdin	Plasma	Stabilizes C3bBb3b
Factor H	Plasma	Dissociates alternative C3 convertase Cofactor for factor I
Factor I	Plasma	Degrades C4b and C3b
Serum proteases	Plasma	Inactivate anaphylatoxins
S protein (vitronectin)	Plasma	Block membrane binding of soluble C567
CR1	Membrane	Dissociates C3 convertase Cofactor for factor I
Decay-accelerating factor (DAF, CD55)	Membrane	Dissociates C3 andC5 convertases Cofactor for factor I
Membrane cofactor protein (MCP, CD46)	Membrane	Cofactor for factor I
Homologous restriction factor (HRF, C8BP, MIP)	Membrane	Inhibits MAC formation
MIRL (protectin, CD59)	Membrane	Inhibits MAC formation

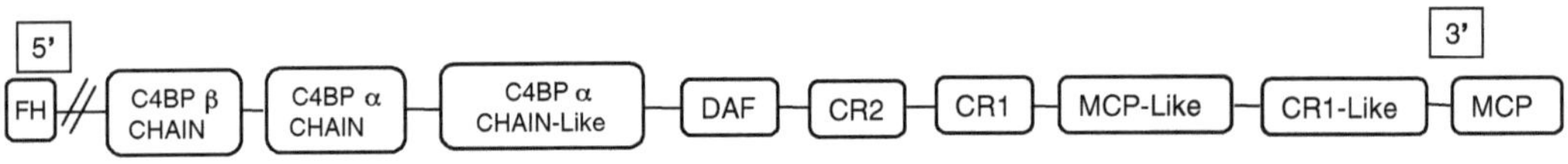

Fig. 4.1 Linkage of the regulation of complement activation (RCA) genes on chromosome 1

Therefore, the sensitivity of the host to autoimmune pathogenic mechanisms is predominantly determined by the status of the complement regulatory proteins. This forms the basis of exploring the potential of complement regulatory proteins as biomarkers and therapeutic targets for autoimmune disorders. A large number of complement regulatory proteins are known (Table 4.1), of which CR1, DAF, MCP, MIRL, and CD59 are membrane-bound. Soluble form of these membrane-bound complement regulatory proteins is also known.

In humans, a number of the regulatory proteins are encoded by a cluster of genes located on the long arm of chromosome 1 (1q32). This region is called the regulator of complement activation (RCA) gene cluster (Weis et al. 1987; Heine-Suner et al. 1997) (Fig. 4.1).

Although the proteins within the RCA family vary in size, they share significant primary amino acid structure similarities. They are organized in tandem structural units termed short consensus repeats (SCRs), which are present in multiple copies in the proteins. Each SCR consists of 60–70 highly conserved amino acids, including four cysteines. The cysteines are disulfide-bonded, holding the SCRs in a rigid triple-loop structure (Hourcade et al. 1989; Ahearn and Fearon 1989). Decay-accelerating factor (DAF), membrane cofactor protein (MCP), and complement receptor type 1 (CR1) (Rodriguez de Cordoba et al. 1985; Holers et al. 1985) are also encoded by the RCA (Fig. 4.2).

4.1.1 Decay-Accelerating Factor (DAF)

4.1.1.1 History

Decay-accelerating activity present on human erythrocytes was first described by Hoffmann (1969), and it was later described in other species like guinea pig and rabbit (Hoffmann and Etlinger 1973). DAF was first purified from guinea pig (Hoffmann and Etlinger 1973), followed by purification from humans (Nicholson-Weller et al. 1982).

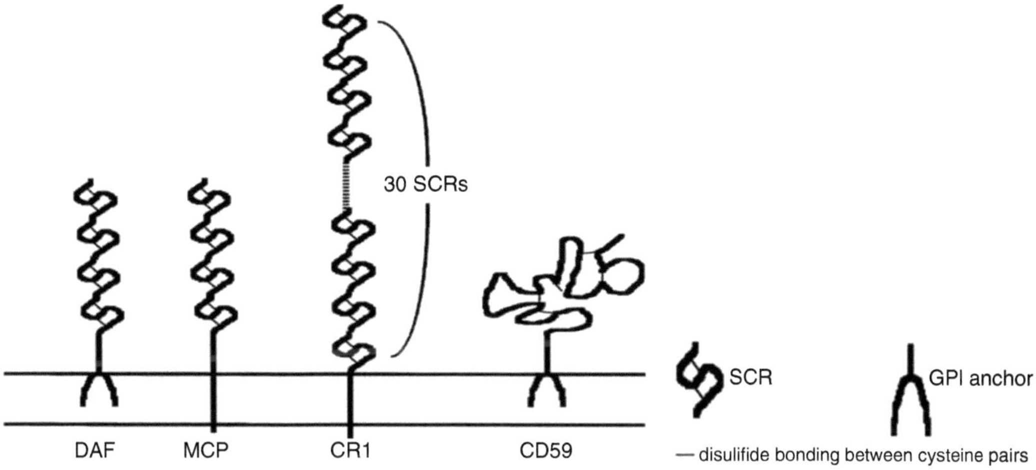

Fig. 4.2 *Membrane-bound CRPs.* While DAF, MCP, and CR1 are members of the RCA gene cluster and share SCRs, CD59 has a completely different structure

4.1.1.2 Distribution

Decay-accelerating factor is widely expressed on most circulating cells including erythrocytes, leukocytes, NK cells, vascular endothelial cells, and on many types of epithelial cells (Kinoshita et al. 1985; Medof et al. 1987b). DAF is also extensively expressed in the urinary, gastrointestinal, and exocrine systems (Medof et al. 1987b and Cosio et al. 1989); placenta; spermatozoa; and eyes and is associated with the subendothelial matrix (Hindmarsh and Marks 1998). Soluble forms of DAF have been found in many body fluids including plasma, tears, saliva, urine, synovial, and cerebrospinal fluids (Medof et al. 1987b; Cosio et al. 1989, Rooney et al. 1991). Various tumor lines overexpress CD55 including colorectal, gastric, ovarian, and osteosarcoma cells (Spendlove et al. 1999), and the expression levels can be up to 100-fold greater than normal cells (Li et al. 2001).

4.1.1.3 Structure and Molecular Biology of DAF

CD55 or decay-accelerating factor is a 70,000-MW membrane glycoprotein (Medof et al. 1987a). The amino terminal part of the protein contains four "short consensus repeats" (SCRs), each with about 60 amino acids held together in a single domain by two internal disulfide bridges. The carboxy-terminus of the protein contains 70 amino acids rich in serine, threonine, and proline residues (STP-rich region), and this region contains sites for the addition of O-linked oligosaccharides. The 17 terminal amino acids are essential for the formation of a glycophosphatidylinositol (GPI) anchor that links the functional protein to the cell membrane. The SCR domains of CD55 are commonly found within C3b/4b-binding proteins and can be cross-linked to either subunit on the cell surface, indicating that the inhibitor is in close contact with these fragments (Kinoshita et al. 1986; Reid et al. 1986). Decay-accelerating activity is located in SCR domains 2 and 3 for the classical pathway and domains 2–4 for the alternative pathways (Brodbeck et al. 1996). The CD55 gene is located within the complement regulatory locus on the long arm of chromosome 1, band q32 (Lublin et al. 1988). The gene spans approximately 40 kb and is comprised of 11 exons.

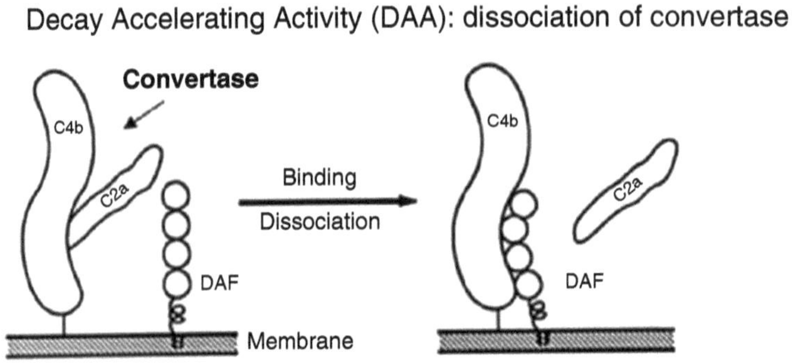

Fig. 4.3 *Inhibitory mechanisms for regulation of complement activation by decay-accelerating activity: the decay of convertases.* Illustration of the decay of the classical pathway C3 convertase, C4b2a, by DAF

4.1.1.4 DAF as Receptor for Pathogens

DAF acts as a receptor for echovirus 7 (Clarkson et al. 1995), enterovirus 70 (Karnauchow et al. 1996), and coxsackieviruses A21, B1, B3, and B5 (Bergelson et al. 1995).

4.1.1.5 Functions of DAF

Complement Regulatory Role of DAF

CD55 functions to accelerate the decay of either C4b2a or C3bBb and corresponding C5 convertases. CD55 acts intrinsically by rapidly dissociating C2a and Bb from C4b and C3b, respectively, which are bound to the cell membrane to which it is anchored, thus preventing assembly of the C3 convertase (Fujita et al. 1987). Evidence has shown that CD55 activity is predominantly upon preformed convertase complexes containing Bb or C2a and that binding either fB or C2 to C3b or C5, respectively (Fujita et al. 1987). Binding affinity analysis of CD55 and its complement-associated ligands has identified significant differences between interactions with single subunits and complete complexes. CD55 has a low affinity interaction with both C3b and C4b on the cell surface enabling their release from the regulator post convertase decay, allowing "recycling" of CD55 and enabling further interactions with newly formed convertases (Harris et al. 1999; Fig. 4.3).

DAF as a Ligand for Lymphocyte Antigen CD97

In addition to functioning as a complement inhibitor, DAF is a ligand for the activation-associated lymphocyte antigen CD97 (Hamann et al. 1996; Qian et al. 1999). CD97 is a seven-span transmembrane protein with sequence homology to type II G-protein-coupled cell surface receptors and is potentially involved in lymphocyte/macrophage activation and cell signaling (Hamann et al. 2002; Gray et al. 1996). The site of CD97 binding is SCR 1, 2, and 3 of DAF.

DAF as a Signal Transducer

A number of studies have suggested that DAF may participate in T cell function as a GPI molecule in lipid rafts (Davis et al. 1988; Shenoy-Scaria et al. 1992; Tosello et al. 1998). The GPI anchor of DAF has been shown to be vital for signaling, as when it is replaced with the transmembrane and cytoplasmic domains of membrane cofactor protein, all signaling is subsequently lost (Shenoy-Scaria et al. 1992). Although the complete signaling pathway for decay-accelerating factor is unknown, it is understood that DAF associates with *src* family tyrosine kinases in microdomains within the membrane, called lipid rafts or detergent-insoluble glycolipid-enriched domains. Davis et al. (1988) showed that a mitogenic signal could be transduced within T lymphocytes by costimulating DAF following submitogenic stimulation with PHA, resulting in T cell proliferation. Phosphatidylinositol phospholipase C treatment of T cells, cleaving the GPI anchor of DAF, abrogated this costimulation indicating that T lymphocytes are stimulated through the GPI anchor. Upregulation of glucose consumption and phagocytosis has been noted when DAF is stimulated by monoclonal antibodies on phagocytes (Shibuya et al. 1991).

DAF in T Cell Immunity

DAF has a role in the suppression of adaptive immune responses. Liu et al. (2005) reported that in both C57BL/6 and BALB/c mice, deficiency of the Daf1 gene, which encodes the murine homologue of human DAF, significantly enhanced T cell responses to active immunization. Daf1$^{-/-}$ mice exhibited hypersecretion of interferon (IFN)-γ and interleukin (IL)-2, as well as downregulation IL-10 during antigen restimulation of lymphocytes in vitro which was normalized by disabling the complement system. Compared with wild-type mice, Daf1$^{-/-}$ mice also displayed markedly exacerbated disease progression and pathology in a T cell-dependent experimental autoimmune encephalomyelitis (EAE) model which was attenuated by C disabling thus establishing a critical link between complement and T cell immunity. Spendlove et al. (2006), have shown that soluble, recombinant DAF inhibits T cell activation and proliferation, an effect that can be blocked by the addition of DAF (SCR 1-3)-specific antibodies.

4.1.2 Membrane Cofactor Protein (MCP, CD46)

MCP is a membrane-bound regulator of complement activation acting at the level C3 and C5 convertase of both classical and alternative pathways.

4.1.2.1 History

MCP was first identified in 1985 as a C3-binding membrane protein isolated from leukocytes (Cole et al. 1985). However, several groups independently discovered this protein as early as 1981 using monoclonal antibodies to cell surface antigens though its complement regulatory function was not known then (Johnson et al. 1981; Sparrow and McKenzie 1983).

4.1.2.2 Distribution

MCP is a 48–68 kDa integral membrane glycoprotein (Seya et al. 1986). In humans, it is present on all circulating cells including platelets, granulocytes, T cells, B cells, NK cells, and monocytes but is absent from erythrocytes (Cole et al. 1985; Seya et al. 1988). Leukocytes express between 4,000 and 10,000 copies of MCP per cell. MCP is expressed at a very high level on certain neoplastic hematopoietic cell lines. MCP is ubiquitously expressed on virtually every cell and tissue studied so far (Johnstone et al. 1993), and this includes epithelial cells, endothelial cells, fibroblasts, kidney, skin, respiratory tract, eye, brain, spermatozoa, and placental trophoblast cells (Cole et al. 1985; Seya et al. 1988).

4.1.2.3 Structure and Molecular Biology of MCP

MCP has an overall structure very similar to DAF. It contains four short consensus repeat (SCR) domains [also called Sushi domains or complement control protein (CCP) modules in the extracellular part (Liszewski et al. 1991)]. The binding of different ligands involves different CCP modules. For example, SCR2, SCR3, and SCR4 are necessary for C3b/C4b binding, whereas SCR1 and SCR2 are necessary for measles virus binding (Manchester et al. 1997). The SCRs are followed by a region rich in serine, threonine, and proline (STP region) with three potential glycosylation sites. However, unlike DAF, MCP has transmembrane and cytoplasmic domains (Lublin et al. 1988). The STP region of MCP is heavily glycosylated. There are two differently glycosylated forms of MCP (58–68 kDa and 48–56 kDa) that contain both O- and N-linked carbohydrate side chains with the larger form being more heavily sialylated (Ballard et al. 1987).

The MCP gene is a part of the RCA gene cluster located on the long arm of chromosome 1 (Lublin et al. 1988). The MCP gene is 43 kb in length and contains 14 exons and 13 introns. There are various isoforms of MCP, which are produced by alternative splicing, the common isoforms being STP-BC and STP-C. Alternative splicing of exon 13 gives rise to different cytoplasmic tails Cytl and Cyt2, varying in length (Liszewski et al. 1991).

4.1.2.4 MCP as Receptor for Pathogens

The first report of the association of MCP with any pathogenic element was the identification of MCP as a receptor for measles virus (Dorig et al. 1993; Naniche et al. 1993; Manchester et al. 2000). But later it was revealed that MCP acts as the receptor to as many as seven pathogenic virus and bacteria in humans and is appropriately referred to as "magnet for pathogens" (Cattaneo 2004). MCP binds to human herpesvirus 6 (Santoro et al. 1999), *Neisseria gonorrhoeae* and *Neisseria meningitidis* bacteria (Kallstrom et al. 1997), and the M protein of streptococcus (Okada et al. 1995).

4.1.2.5 Function of MCP

Complement Regulatory Role of MCP

MCP acts as a cofactor for factor I in mediating cleavage of C3b, but unlike CR1 it does not participate in the cleavage of iC3b further to C3c and C3dg. MCP is also a cofactor for the factor I-mediated cleavage of C4b resulting in the formation of C4c and C4d (Seya and Atkinson 1989). However, MCP has a comparatively weaker cofactor activity in the cleavage of C4b (Seya et al. 1986). MCP does not accelerate the decay of C3 or C5 convertase and thus complements the activity of DAF. CD46 acts intrinsically by preventing C activation on the cells on which it is expressed (Ogelsby et al. 1992), and therefore, the correct orientation of MCP to C3b on the membrane is vital for its activity. MCP preferentially inactivates the convertase of the alternative pathway (Seya et al. 1991; Kojima et al. 1993). The binding sites on MCP for C3b are located within the SCRs. More recently, it was shown that CD46 was shed from the cell surface of apoptotic and necrotic cells, allowing their recognition and efficient removal by phagocytes (Elward et al. 2005).

MCP in Signal Transduction

It has been observed that in addition to its complement regulatory function, MCP is also capable of transducing signals. MCP ligation, by cross-linking with antibodies or binding of natural ligands at the surface of several cell types, induces intracellular signaling (Astier et al. 2000) such as calcium

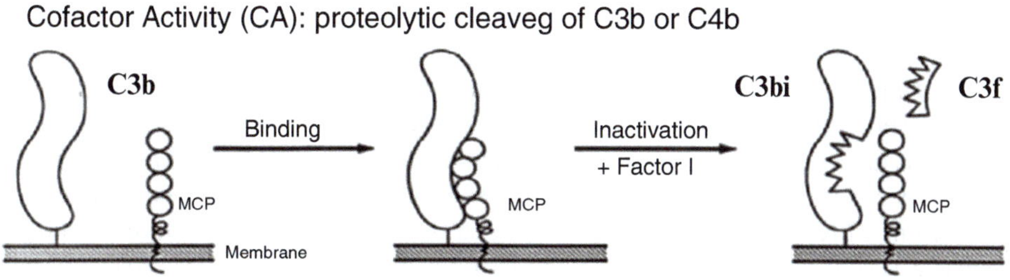

Fig. 4.4 *Inhibitory mechanisms for regulation of complement activation by cofactor activity:* C3b deposited on a target is bound by MCP which in turn serves as a cofactor for its cleavage by the serine protease FI

flux (Kallstrom et al. 1998), NO production (Hirano et al. 2002), and phosphorylation of intracellular substrates (Lee et al. 2002; Kurita-Taniguchi et al. 2000). It has been reported that measles virus-infected monocytes and dendritic cells exhibit a decreased ability to produce IL-12, necessary for activating T cells and NK cells (Karp et al. 1996; Fugier-Vivier et al. 1997). Furthermore, this suppression of IL-12 production by monocytes was also observed after C3b stimulation or MCP cross-linking. All these reports suggest active transduction pathways mediated by MCP (Fig. 4.4).

MCP in Human T Cell Immunity

It was reported that activation of MCP results in the tys-phosphorylation of two substrates, linker for activation of T cells (LAT) and p120CBL in human T cells (Astier et al. 2000). On further investigation of the role of MCP in T cell activation, it turned out that costimulation of MCP and CD3 led to potent proliferation of human T cells, with intensity comparable to that of CD28 (Astier et al. 2000). Moreover, a similar role in T cell costimulation has also been observed for Crry, the murine homologue of MCP that is expressed in mouse in testis only (Fernandez-Centeno et al. 2000; Jimenez-Perianez et al. 2005). This highlights the new role of these complement regulatory molecules in the control of T cell activation.

MCP activation also leads to morphological changes in human T cells that are suggestive of a role in migration (Zaffran et al. 2001). MCP plays a crucial role in the regulation of T cell polarity (Oliaro et al. 2006) by interacting with members of the PDZ-containing Scribble complex (Ludford-Menting et al. 2005). While MCP acts as a costimulatory molecule when coengaged with CD3, its ligation alone causes capping at the distal site in APC:T cell interaction, which prevents the immunological synapse formation, subsequent T cell activation, and natural killer cell cytotoxicity. In human primary T cells, MCP localizes at the uropod, a protrusion formed during migration, and colocalizes with DGL4, which binds specifically to Cyt1 but not Cyt2 (Ludford-Menting et al. 2002). The polarization of the MCP–DGL4 complex is also dependent on the functional expression of Scribble. Therefore, the triggering of MCP at the surface of human T cells induces a redistribution of the PDZ polarity network and regulates in fine T cell shape and migration (Ludford-Menting et al. 2005). Therefore, MCP influences T cell fate by modulating T cell proliferation and affecting T cell shape.

Apart from its role in the activation, proliferation, morphology, and polarity of T cells, MCP appears to play an important role in the differentiation of T cells. In 2003, Kemper and colleagues showed that activation of naive human $CD4^+$ T cells by MCP and CD3 in the presence of IL-2 led to differentiation towards a T regulatory cell 1 (Tr 1) phenotype (Kemper et al. 2003; Kemper and Atkinson 2007). The MCP-induced Tr1 cells secrete large amounts of IL-10 (Roncarolo et al. 2006) and are able to inhibit the proliferation of bystander CD4 effector cells. *Streptococcus pyogenes*, a natural ligand of MCP, also leads to a Tr1 phenotype, demonstrating the physiological relevance of MCP-induced Tr1 cells (Price et al. 2005). Further studies have then shown that these Tr1-induced cells secrete granzyme B as well as IL-10 and have the potential to kill autologous targets (Grossman et al. 2004).

MCP in Human B Cell Immunity

MCP has an indirect effect on human B cell immunity through its role on the differentiation of T cell into Tr1 phenotype. Fuchs and colleagues (2009) analyzed the effect of MCP-induced Treg on B cell functions in a coculture system and observed that cTreg enhanced B cell Ab production. This B cell support is dependent on cell/cell contact as well as cTreg-derived IL-10. In addition, they reported that T cells from a MCP-deficient patient were not capable of promoting B cell responses, whereas MCP-deficient B cells had no intrinsic defect in Ig production. Hence, MCP is capable of inhibiting the proliferation of $CD4^+$ T cells as well as promoting Ab production by B cells through IL-10.

MCP and Inflammation

MCP exists as different isoforms, and these include two main cytoplasmic tails, C1 and C2. Since these two isoforms are coexpressed in human cells, their individual analysis is difficult. Transgenic mice that expressed either the C1 or the C2 isoforms were produced, and this mouse model allowed the analysis of each cytoplasmic tail individually (Marie et al. 2002) When T cell-dependent inflammation was induced in a contact hypersensitivity model, striking opposite effects were observed for each cytoplasmic tail. While the C1 isoform inhibited the overall inflammation, the C2 isoform increased it. This was the result of differential effects on $CD4^+$ proliferation and $CD8^+$ cytotoxicity and of different profiles of cytokine released. In particular, IL-2 was decreased by C1 expression, whereas IL-10 secretion was decreased by C2 expression. Hence, it appears that MCP function is quite complex, and depending on which cytoplasmic tail is expressed or activated, a regulatory or proinflammatory phenotype can be observed. This implies that MCP function in humans must be thoroughly examined and understood before MCP is applied in therapies, because opposite and undesired results could be obtained (Astier 2008).

4.1.3 Complement Receptor 1 (CR1)

4.1.3.1 History

In 1953, RA Nelson initiated CR1 research, when he described binding reactions between human erythrocytes and specifically opsonized *tryponemes and pneumococci*, and coined the term "immune adherence" receptor to describe this reaction (Nelson 1953). This immune adherence receptor of the RBC was shown subsequently to be specific for C3b in the immune complex. Detailed biochemical characterization of CR1 began with Fearon's purification of a 205,000-Da glycoprotein from human RBCs (Fearon 1979). It is the best characterized receptor for the activated form of complement protein C3, C3b (Ahearn and Fearon 1989; Delibrias et al. 1992). Binding with lower affinity was also demonstrated for the degraded form of C3b, iC3b, and for complement proteins C4b, C1q, and mannan-binding lectins (Tas et al. 1999; Ghiran et al. 2000).

4.1.3.2 Structure and Molecular Biology of CR1

Human CR1 is a single chain, type 1 transmembrane glycoprotein composed of an extracellular domain of 1,930 residues organized into 30 short repeats of a consensus element and anchored in plasma membrane through a hydrophobic transmembrane domain of 25 residues with a 43 residue cytoplasmic domain (Klickstein et al. 1987).

Four polymorphic forms have been identified with relative molecular weights of 160, 000 (C), 190, 000 (A), 220, 000 (B), and 250, 000 (D) kDa, respectively, controlled by 4 autosomal codominant alleles A, B, C, and D (Dykman et al. 1985). Most common alleles of CR1 (A and B) have gene frequencies of 0.8 and 0.2. Rare alleles (C and D) occur at a frequency of less than 5% (Moulds et al. 1996). It belongs to the family of proteins encoded by RCA gene cluster located on chromosome 1q32.

4.1.3.3 Functions of CR1

Regulation of Complement Cascade

CR1 is involved in the regulation of both the classical and the alternate pathways (CP and AP, respectively) and encompasses the functions of both DAF and MCP by acting as a decay-accelerating factor for C3 and C5 convertases and a cofactor for factor I-mediated cleavage of C3b and C4b.

CR1 binds C3b and C4b and also iC3b and C3dg with a lower affinity (Fearon 1980). Ligand binding by CR1 leads to complement regulation by accelerating convertase decay. Through binding to C4b, CR1 accelerates the decay of the classical C3 convertase (C4b2a) by displacing C2a (Iida and Nussenzweig 1981). In a similar manner, through the binding of C3b, CR1 accelerates the decay of the alternative pathway C3 convertase (C3bBb) by displacing Bb (Fearon 1980). The binding also decays the classical pathway C5 convertase (C4b2a3b) and the alternative C5 convertase (C3bBb3b) (Iida and Nussenzweig 1981). CR1 acts mainly on extrinsic convertases on nearby cells and on immune complexes (ICs), whereas MCP and DAF act mainly on C3/C5 convertases on the same cell.

CR1 also contributes to complement regulation by providing cofactor activity for factor I, a serine esterase that cleaves C3b and C4b to hemolytically inactive forms. Specifically, CR1 promotes the cleavage of C3b, by factor I, to iC3b and further promotes the cleavage of iC3b to C3dg and C3c (Medof et al. 1982; Ross et al. 1982). The CR1 cofactor activity for the generation of C3dg is based on the ability of CR1 to bind iC3b and represents a function not exhibited by any other cofactor under physiological conditions. CR1 also provides cofactor activity for the factor I-mediated cleavage of C4b to C4c and C4d (Iida and Nussenzweig 1981).

Immune Complex Clearance

Erythrocyte CR1 (E-CR1) serves as an immune adherence receptor for C3b/C4b-opsonized immune complexes, which, following adherence, are transported to liver and spleen where they are transferred to and processed by fixed macrophages. CR1 on erythrocytes acts as a vehicle for clearance of ICs. The ICs altered this way become less pathogenic (Hebert 1991). The liver is the main site for removal of C3b-bearing ICs (Cosio et al. 1990). Kupffer cells trap immune complexes after cleavage of C3b into iC3b or C3dg.

Phagocytosis

CR1 expressed on the surface of phagocytic cells may bind soluble polymeric C3b, which is covalently fixed to immune complexes or particles and enhances their phagocytosis (Fearon et al. 1981). The CR1 and Fc gamma receptors cooperate for phagocytosis of targets that have been coated with suboptimal amounts of IgG. The cross-linking of these receptors elicits a number of secondary responses in phagocytic cells. These include neosynthesis and release of arachidonic acid metabolites, stimulation of oxidative burst, and release of toxic oxygen derivatives and lysosomal enzymes (Weiss et al. 1989).

Role in Mediating Inhibitory Signals to B Cells

Fingeroth et al. (1989) reported that cocross-linking of surface-Ig and CR1 on resting splenic cells results in inhibition of the anti-IgM-induced proliferation. Jozsi et al. (2002) also showed that clustering of CR1 via its natural ligand on anti-μ-activated human B cells results in generation of inhibitory signals. The mechanism proposed for this inhibition was that inhibitory molecules (such as FcγRIIb) cocluster with CR1 upon engagement of the latter by its activation threshold of B cells. This mechanism may ensure additional level of regulation, which depending on the composition of ICs and the degradation stage of C3 might reduce nonspecific B cell activation.

Regulation of the Immune Responses

The stimulation of human monocytes with C3b in vitro induces the intracellular production and extracellular release of IL-1 in serum free conditions. It enhances the differentiation of B cells but does not have any effect on memory responses.

4.1.4 CD59

The human CD59-antigen (protectin) is the most important membrane-bound complement regulatory protein at the level of MAC assembly.

4.1.4.1 History

CD59 was isolated independently in a number of laboratories (Sugita et al. 1988; Davies et al. 1989; Holguin et al. 1989; Okada et al. 1989). This protein has been termed as the membrane-attack-complex-inhibitory factor (MACIF) (Sugita et al. 1988), homologous restriction factor 20 (HRF20) (Okada et al. 1989), and membrane inhibitor of reactive lysis (MIRL) (Holguin et al. 1989), protectin (Meri et al. 1990).

4.1.4.2 Distribution

CD59 is widely expressed in human cells and tissues. It is present on all circulating cells (Davies et al. 1989), endothelial cells (Nose et al. 1990), in most epithelial cells (Meri et al. 1991), and spermatozoa (Rooney et al. 1992). The average number of CD59 molecules on human erythrocytes is 25,000–50,000 per cell (Davies et al. 1989; Meri et al. 1990). The protein is expressed also on Schwann cell sheath of peripheral nerve fibers, neurons, microglia, oligodendrocytes, astrocytes, ependymal cells, and certain epithelial cells such as acinar cells of the salivary gland, bronchial epithelium, renal tubules, and squamous epithelium (Nose et al. 1990; Vedeler et al. 1994; Hideshima et al. 1990).

4.1.4.3 Structure of CD59

CD59 is a membrane-bound small glycoprotein made of 77 amino acids and has a molecular weight of 18–25 kDa (McKusick 2001).

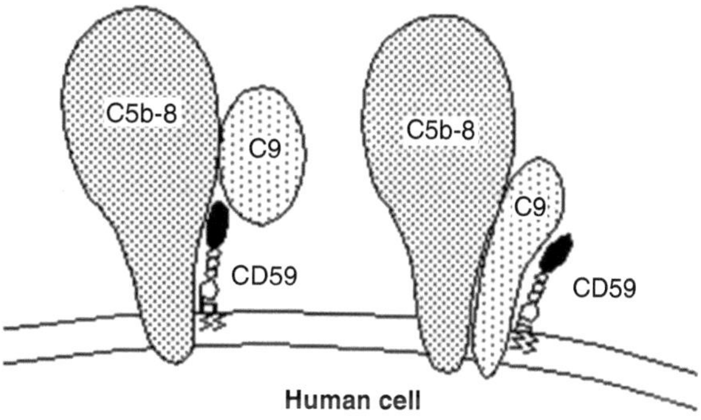

Fig. 4.5 Inhibition of MAC assembly by CD59 (Adapted from Morgan and Meri 1994)

Human CD59 is closely related to mouse Ly6 antigen (Petranka et al. 1992). The human gene gives rise to more than four different mRNA molecules, which are generated by alternative polyadenylation (Tone et al. 1992).

CD59 is located on short arm of chromosome 11 in the human genome, 11p13 and 11p14 (McKusick 2001). The CD59 gene consists of 4 exons. The first exon is untranslated, the second encodes the hydrophobic leader sequence of the protein, the third exon encodes the N-terminal portion of the mature protein, and the fourth exon encodes the remainder of the mature protein including the glycosylphosphatidylinositol (GPI) anchor attachment in the plasma membrane. Since CD59 does not belong to RCA gene family, its structure is very different from MCP and DAF. The structure of the CD59 protein is very similar to the Ly6 family. This structure is characterized as a single cysteine-rich domain, having a hydrophobic core with three loops and a small fourth loop, which is helical (Yu et al. 1997). The helical loop is made of two antiparallel beta sheets and the other three loops are three antiparallel beta sheets, which form the core. Connecting each of these loops are five disulfide-bonded cysteine pairs.

4.1.4.4 Functions of CD59

Complement Regulatory Function of CD59

The function of CD59 is to inhibit final steps of MAC assembly on cell membranes by activated terminal complement proteins C5b to C9 and to protect the cell from complement-mediated cell lysis. By binding to the C5b-8 complex, CD59 limits C9 input and prevents formation of the polymeric C9 complex (Rollins et al. 1990) and unfolding and polymerization of the final C9 pore (Farkas et al. 2002; Longhi et al. 2006). The details of this process are not well-understood. The activity of CD59 has been suggested to be species-restricted because of its apparent ability to inhibit primarily human MAC and to a lesser degree MAC from other species (Rollins et al. 1991; Fig. 4.5).

CD59 in T Cell Immunity

A less common function of CD59 can influence the proliferation capacity of T cells and their ability to produce cytokines, which can influence how T cells respond to a given antigen entering the bloodstream. CD59 works in the innate immune system but has also been implicated in the adaptive immune system (Longhi et al. 2006). In addition, CD59 influences the outcome of T cell response to a given antigen. Longhi et al. reported that direct interaction between CD59 on a T cell and a specific receptor on an antigen-presenting cell (APC) results in an inhibitory signal being transmitted to both the T cell and the APC. Due to this inhibitory signal, downmodulation of APC activity and consequently T cell activity results. Longhi et al. also reported that CD59 on T cells may reduce the strength of the positive signal transduction pathway delivered through the T cell receptor (TCR). The CD59 surface protein can undergo cross-linking on lymphocytes to stimulate the T cell, cause organization of the cytoskeleton, and enhance proliferation. The GPI anchors on the CD59 protein are responsible for this clustering in a lipid raft type manner (Longhi et al. 2006).

4.1.5 *Relationship with SLE*

SLE remains the prototypic example of human autoimmune disease with diverse clinical features encompassing every organ system. It attacks the bones, joints, skin, serous membranes (pleura, pericardium, and peritoneum), kidneys, gastrointestinal tract, cardiovascular system, brain, and clotting system. The organ systems may be affected alone or in combination exhibiting different symptoms and signs, depending on the system affected. In SLE, some organs are affected at greater frequencies than others. There is also marked variability in the clinical features depending upon ethnicity. The disease, though rare, is potentially fatal and afflicts women in the reproductive age group severely affecting their quality of life.

Like any other autoimmune disorder, SLE is an outcome of immune dysregulation. Immune dysregulation in SLE is multifaceted – it involves loss of immune tolerance, increased antigenic load, excess T cell help, defective B cell suppression, and the shifting of T helper 1 (Th1) to Th2 immune responses. All these lead to B cell hyperactivity and the production of pathogenic autoantibodies. The production of abnormal antibodies such as the antidouble-stranded DNA antibody (anti-dsDNA Ab) by B cells remains the hallmark of lupus erythematosus. Defective immune regulatory mechanisms including exaggerated complement activation, impaired clearance of apoptotic cells, and immune complex overload are important contributors to the manifestations of SLE. Genetic, environmental, and hormonal factors contribute to such immune dysregulation. Pathophysiology of SLE, however, is not completely understood (Liu and Ahearn 2009).

4.1.5.1 Decay-Accelerating Factor (DAF, CD55)

Not much is known about the role of DAF in the disease process of SLE; however, evidence strongly indicates its possible role in autoimmune disease modulation. DAF is inducible on human endothelial cells during inflammation and provides increased protection against complement attack (Miwa et al. 2002). A study of the autoimmune disease in DAF knockout MRL (DAFKO MRL/lprIgh) mice has shown more aggressive disease development in such mice characterized by accelerated and aggravated dermatitis and lymphadenopathy as well as increased antichromatin autoantibody production (Takashi et al. 2007). Another study on gene knockout mice showed that injection of a subnephritogenic dose of rabbit antimouse glomerular basement membrane (GBM) serum induced glomerular disease in DAF knockout mice but not in wild-type controls, with DAF knockout mice displaying an increased glomerular volume (Sogabe et al. 2001) and severe dermatitis when crossed with SLE-prone mice. Enhanced expression of DAF has been documented in glomerular cells in the presence of glomerular injury (Cosio et al. 1989), indicating an enhanced protective role of this complement regulatory protein to complement attack. On neutralizing DAF with antibody, the glomerular epithelial cells have been demonstrated to be more susceptible to complement-mediated cytotoxicity (Quigg et al. 1989; Shibata et al. 1991), which elucidates the functional importance of DAF on glomerular cells. However, there is a contradictory report stating a slight but statistically significant decrease of DAF level on the erythrocytes of SLE patients with diffuse proliferative glomerulonephritis, which was shown to be dependent upon the stage of the disease activity (Jones et al. 1994). Another study reported a deficiency of red cell-bound DAF in SLE patients with autoimmune hemolytic anemia (Richaud-Patin et al. 2003), indicating the modulatory role played by DAF in the different clinical manifestations of SLE.

An increase in erythrocyte DAF has been reported in SLE patients. Enhanced expression of DAF had been observed in the glomeruli of SLE patients (Arora et al. 2000). DAF may also have a role in the induction phase of SLE since cross-linking CD55, as for many GPI-anchored molecules, induces signal transduction in T cells. DAF provides a costimulatory signal for the activation of T lymphocytes and favors a Th2 lymphokine secretion pattern.

4.1.5.2 Membrane Cofactor Protein (MCP, CD46)

There are very few studies elucidating the role of MCP in the modulation of SLE. Elevated serum levels of soluble MCP (sMCP) in patients with SLE have been reported by Kawano et al. (1999). Also, the level was significantly higher in case of patients with active SLE when compared to those with inactive SLE, and the level declined with cortisol treatment. Longitudinal analysis of sMCP showed that its levels decreased in parallel with the levels of anti-dsDNA and correlated with reduced CH50 levels. However, the function of serum sMCP is not clear, nor is its source. An immunohistochemical study of the expression of MCP in the kidney tissues of patients with renal diseases, including lupus nephritis, has shown significantly increased intensity of MCP in the affected glomeruli when compared to normal (Endoh et al. 1993). Since there is a massive deposition of C3 and immunoglobulin in the glomeruli of lupus nephritis patients (Kashgarian et al. 2002), the C3 regulatory protein MCP might have a protective role in the glomerular tissue injury, and the increased MCP could be a result of upregulated synthesis of MCP in the glomerular cells.

A possible role of MCP in the etiology of SLE cannot be ruled out especially in the light of the finding that MCP plays an important role in inducing the activation of T regulatory cells, which is important in the control of autoreactive T cells (Astier 2008).

4.1.5.3 Complement Receptor Type 1 (CR1, CD35)

The observation of low CR1 and deposition of C4 and C3 fragments on erythrocytes in active SLE was made by Walport and colleagues (Walport et al. 1987; Davies et al. 1990). In another study, Manzi et al. (1996) demonstrated that the combined detection of high levels of erythrocyte-bound C4d and low levels of CR1 on erythrocytes had high sensitivity (72%) and specificity (79%) for SLE. E-CR1 has received much attention due to the fact that it plays an important role in the clearance of immune complexes, overload of which is a major cause of the pathological manifestations in SLE (Wilson et al. 1986; Katyal et al. 2003). Later, in addition to erythrocyte CR1, lower than normal levels of CR1 on leukocytes and glomerular podocytes were also reported in patients with SLE (Ross et al. 1985; Wilson et al. 1986a, b; Mitchell et al. 1989; Marquart et al. 1995; Arora et al. 2000; Birmingham et al. 2006). Studies by Wilson et al. (1986) and Marquart et al. (1995) documented a decline in B cell and neutrophil CR1, which correlated positively with the decline in E-CR1 in SLE. Oudin et al. (2000) showed that recombinant sCR1 could restore the immune complex binding ability of CR1-deficient erythrocytes. These findings strongly suggest a key role of deficient CR1 expression in the pathogenesis of SLE. Administration of a recombinant soluble CR1 molecule was highly effective in alleviating the inflammatory reactions and disease manifestations (Weisman et al. 1990).

Studies from our lab have shown a drastic decline in the levels of u-CR1 in SLE (Sivasankar et al. 2004) which correlate with the expression of glomerular CR1. Therefore, u-CR1 has been suggested as a potential marker to diagnose glomerular involvement in SLE. Earlier, our studies have suggested that E-CR1 can serve as a prognostic marker for glomerulonephritis and rheumatoid arthritis (Arora et al. 2000). More recently, we have found a marked decline in the levels of CR1 transcript and protein and negative correlations between the levels of leukocyte CR1 transcript and protein with the disease activity and severity of SLE. This suggested L-CR1 transcript can potentially serve as a disease activity marker for SLE. Further studies are needed in this direction to have more conclusive evidence about the usefulness of leukocyte CR1 as a biomarker for SLE (Arora et al. 2004).

4.1.5.4 CD59 (Protectin)

Several studies have demonstrated increased serum levels of MAC C5b-9 in active SLE patients (Falk et al. 1985; Garwryl et al. 1988) indicating its potential role in the complement-mediated tissue injury in the disease process. Since CD59 is a very important surface molecule protecting the autologous cells from MAC-mediated lysis, it is likely that it plays an important role in the modulation of the complement-mediated injury in SLE. Tamai et al. (1991) first reported an upregulation of CD59 in the glomerular cells of lupus nephritis patients, which was followed by a few other reports of increased CD59 in the glomeruli of SLE patients. Suppression of CD59 by monoclonal antibody has shown a dose-dependent increase MAC deposition and exacerbated tubulointerstitial injury in rats (Matsuo et al. 1994; Nangaku et al. 1998; Watanabe et al. 2000) further elucidating the role of CD59 in the maintenance of normal structural and functional integrity of the kidney. Cultured glomerular epithelial, endothelial, and mesangial cells have been shown to exhibit increased susceptibility to complement-mediated lysis in the presence of neutralizing antibodies in vitro (Quigg et al. 1995; Rooney et al. 1991). In a study on CD59a gene knockout mice, Turnberg et al. (2004) demonstrated that mice lacking the mCd59a gene were more susceptible to accelerated nephrotoxic nephritis than matched controls. These mice developed greater glomerular cellularity early in the disease process and more severe glomerular thrombosis and proteinuria at later time points. The excess C9 deposition reflects the presence of a greater quantity of MAC and most likely represents the mechanism whereby the absence of CD59a caused greater tissue injury. Most of the studies on the role of CD59 in autoimmunity deal with the expression of CD59 in the kidney of patients with renal diseases. Expression of CD59 is significantly decreased on the synovial lining, stromal cells, and endothelial cells (Tarkowski et al. 1992). Moreover, injection of antirat CD59 mAb into a rat knee joint induces a spontaneous complement-dependent arthritis (Mizuno et al. 2001), and CD59-deficient mice are prone to enhanced antigen-induced arthritis (Williams et al. 2004).

Arora et al. (2000) have reported an increased expression of CD59 on the erythrocytes of SLE patients with diffuse proliferative glomerulonephritis. There is a contradictory report stating the decreased level of CD59 on the red cells of SLE patients with autoimmune hemolytic anemia. Thus, the expression of CD59 on erythrocytes may vary with the different manifestations of SLE.

4.1.5.5 Scope of DAF, MCP, CR1, and CD59 as Biomarkers for SLE

The above account on relationship between these membrane-bound complement regulatory proteins and SLE indicates an intimate association of each of these proteins with the pathophysiology of the disease and the disease progression. A role of these proteins in the etiology of SLE is also suggested. Hence, these proteins hold promise as novel biomarkers for SLE. However, over the years, most studies have focused mainly on E-CR1 as a disease biomarker of SLE. Interest in other membrane-bound CRPs is comparatively recent. Most of the studies have been carried out in animal models. For validating any of these proteins as biomarkers, extensive studies on human subjects are required.

4.1.6 Therapeutic Implications

Several approaches have been adopted to develop complement-based therapeutics, and these have been reviewed extensively by Ricklin and Lambris (2007). Identification and structural analyses of complement inhibitors, targeting of the functional domains, direct blocking of activation pathways and active complement peptides by inhibitors or antibodies, and development of chimeric molecules inhibiting the cascade at two or more points are some of the approaches to develop therapeutics against inflammatory diseases. Replenishment of defective or deficient complement regulators is another approach where soluble forms of membrane-bound complement regulatory proteins hold promise. Deficiency of complement components C1q, C2, and C4 complement activation is associated with the etiology and pathogenesis of SLE in some patients. Paradoxically, disease-acquired impairment of CR1 expression and exaggerated complement activation also are evident in a large number of patients. Therefore, in SLE, the replenishment of defective and deficient complement regulators and prevention of cell lysis by membrane attack complex may be a more desirable therapeutic approach. Recombinant soluble forms of membrane-bound CRPs hold promise. The strategy is to increase the amount of soluble protein and also to target it to the cells where their function is important in controlling complement-mediated autoinjury. Several strategies have been employed including preparation of antibody-complement inhibitor constructs with antibodies specific to disease-relevant antigens.

4.1.6.1 DAF

Three forms of recombinant DAF have been purified from transfected Chinese hamster ovary cells: glycophosphatidylinositol (GPI)-linked membrane DAF (mDAF) extracted from cell membranes, spontaneously shed soluble DAF (sDAF) derived from mDAF, and a novel-secreted protein (seDAF), generated by deletion of the signal for GPI attachment. All three molecules inhibit both the classical and the alternative pathways of complement activation. mDAF extracted from Chinese hamster ovary cells reincorporates into RBC membranes via its GPI anchor. sDAF and seDAF, which lack a GPI anchor, do not associate with cell membranes. mDAF is a more potent inhibitor of complement-mediated hemolysis than either sDAF or seDAF, suggesting that incorporation into cell membranes greatly enhances the efficiency with which DAF inhibits complement activation on the cell surface. In contrast, complement activation in the fluid phase is inhibited by sDAF and seDAF but not by mDAF, possibly due to interference by serum lipoproteins. A reversed passive Arthus reaction in guinea pigs was used to evaluate the ability of recombinant seDAF to inhibit C activation in vivo. When administered at dermal sites, seDAF reduced the severity of immune complex-mediated inflammatory reactions induced by a reversed passive Arthus reaction, as judged by both gross and histologic examination. These data indicate that seDAF may be useful as an anti-inflammatory therapeutic (Moran et al. 1992). Targeting DAF to a cell surface by means of an antibody combining site is feasible and that targeted DAF provides cells with enhanced protection from complement.

4.1.6.2 MCP

A recombinant sMCP has been shown to inhibit immune complex-mediated inflammation in the reverse passive Arthus reaction model in rats (Christiansen et al. 1996). As in the case of sDAF, the single activity of sMCP limits its potential as an effective therapeutic reagent. However, sMCP may prove to be a valuable reagent in combination with other complement inhibitors.

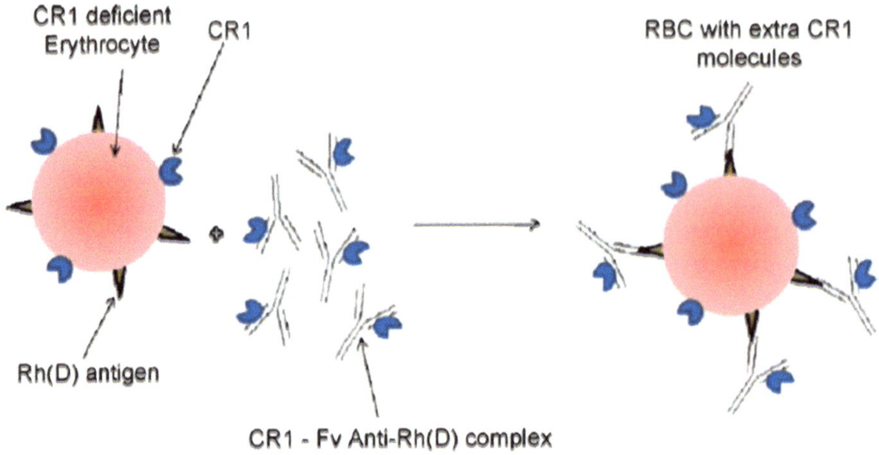

Fig. 4.6 CR1-Fv anti-Rh(D) complex-mediated increase in E-CR1 levels in individuals with E-CR1 deficiency (Adapted from Khera and Das (2009))

4.1.6.3 Complement Receptor 1

A recombinant form of sCR1 has been designed to assess the therapeutic potential of CR1. This recombinant CR1, the rCR1, has the C3b and C4b binding sites and the activity as a cofactor for factor I. Its therapeutic efficacy has been assessed for various autoimmune and inflammatory disorders (Weisman et al. 1990a). The animal model experiments have validated the resolution of inflammatory changes accompanying inflammatory lung injury (Rabinovici et al. 1992), myocardial infarction (Weisman et al. 1990b), autoimmune thyroiditis (Metcalfe et al. 1996), and glomerulonephritis (Couser et al. 1995) with the use of recombinant soluble CR1. Use of sCR1 for managing incompatible blood transfusion by inhibition of complement activation and subsequent immune hemolysis had been suggested (Yazdanbakhsh and Scaradavou 2004). Since sCR1 can potentially prevent E-rosetting and clear immune complexes in patients, it is envisaged as a therapeutic agent for severe falciparum malaria (Cockburn et al. 2004). Soluble CR1 has long been expected to ameliorate the suffering in various disease conditions, but most of the studies, which promise such a revolution in health care strategies, have been based on animal models of these diseases (Weisman et al. 1990b; Rabinovici et al. 1992; Couser et al. 1995), which may have limited application in human subjects. Based on these encouraging results, sCR1 was developed as a therapeutic (TP10; Avant Immunotherapeutics, Needham, MA, USA) for use after coronary artery bypass graft surgery. TP10 is expressed as a 240-kDa glycoprotein in Chinese hamster ovary cells, has a plasma half-life of ~55 h, and is safe and well-tolerated in both adult and infant patients. However, its use was more fruitful for male patients and not the females, and hence, the extrapolation of its therapeutic potential for SLE, a female-biased disease, is highly questionable (Li et al. 2006). This formulation has already been withdrawn by the manufacturer. However, newer strategies are currently being explored (Ricklin and Lambris 2007).

CR1-Fv anti-Rh(D) complex is designed to target and add extra CR1 molecules on CR1 deficient erythrocytes in various disorders (Oudin et al. 2000) (Fig. 4.6) to augment immune complex clearance and inhibit complement pathways.

As a method to fight growing antibiotic resistance in bacteria, CR1-based immunotherapeutic strategies have been evaluated. For the treatment of resistant *Staphylococcus aureus* infections, a bispecific monoclonal antibody complex (heteropolymer, HP) specific to E-CR1 and type 5 capsular polysaccharide of the T5 isolate of *S. aureus* has been designed. This promotes binding of *S. aureus* to erythrocytes

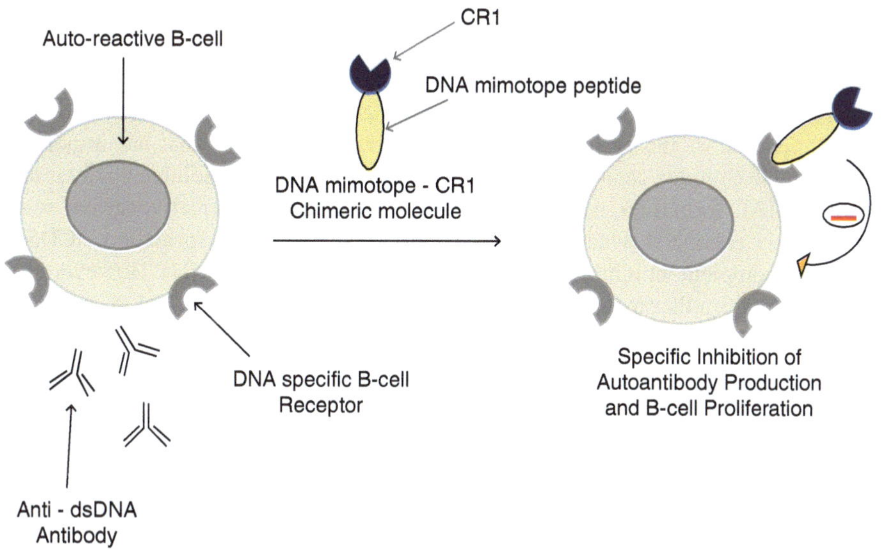

Fig. 4.7 Specific inhibition of autoreactive B cells (here dsDNA) by chimeric CR1-DNA mimotope (Adapted from Khera and Das (2009))

(via E-CR1) with subsequent transfer to monocytes or macrophages for internalization and bacterial killing (Gyimesi et al. 2004). CR1 has also been studied as an important molecule to circumvent hyperacute tissue rejection, as occurring in allogenic transplants. Cells transfected with recombinant CR1 (Manzi et al. 2006) or phosphatidylinositol-bound mini CR1 (Mikata et al. 1998) showed enhanced survival when subjected to conditions simulating hyperacute rejection, and such results have aroused great interest in scientists all over the world. Another recent molecule, Sialyl Lewis(x) hybridized soluble complement receptor 1, *sCR1-sLe*(*x*) (Picard et al. 2000), has been developed. In this, the sCR1 fragment acts as the complement inhibitor, and the sLe(x) acts to inhibit the selectin-mediated interaction of lymphocytes and neutrophils with endothelium. *sCR1-sLe*(*x*) has been proposed to be useful in ischemia-reperfusion injury, thermal injury, and immune complex-mediated injury (Asghar and Pasch 2000). Its efficacy in preventing lung ischemia-reperfusion injury as well as stroke reperfusion injury has been studied extensively (Stammberger et al. 2000; Ducruet et al. 2007). Preclinical evaluation for *sCR1-sLe*(*x*) (TP20 Avant Immune) conducted in nonhuman primate models of reperfused stroke had to be terminated prematurely, as not only was it unable to perform the proposed function of reducing the infarct volume but also caused a few unfavorable reactions (Ducruet et al. 2007), thus emphasizing the need for further research for this molecule. Neither sCR1 with sialyl Lewis^{X}-glycosylation for targeting sites of inflammation (Rittershaus et al. 1999) nor Mirococept/APT070 (a truncated sCR1 with a membrane-tethering motif (Smith 2002) Inflazyme Pharmaceuticals, Richmond, BC, Canada) has been tested in clinical trials (Ricklin and Lambris 2007).

A strategy specifically developed for SLE involves the use of an antigen-based heteropolymer (AHP), which is a dsDNA molecule bound to a monoclonal antibody against CR1. This molecule would specifically bind to the anti-dsDNA and target it to the E-CR1 through anti-E-CR1 to facilitate the removal of anti-ds DNA from circulation at a faster rate (Ferguson et al. 1995). For the purpose of specifically inhibiting the autoreactive cells characteristic of autoimmune disorders, a DNA mimotope–CR1complex has been designed which specifically inhibits the autoreactive B cells. The DNA mimotope has been included for specific binding ability to the anti-dsDNA antibody secreting B cell, while CR1 acts as a B cell inhibitor (Fig. 4.7; Voynova et al. 2008).

4.1.6.4 CD59

In diseases in which the MAC is involved in pathogenesis, CD59-based inhibitors offer a potential advantage over inhibitors of complement activation because CD59 will only block MAC formation and not the generation of the earlier complement activation products that are important for host defense and immune complex catabolism (Zhang et al. 1999). Although soluble CD59 is not an effective inhibitor of MAC formation, we have shown previously that the effectiveness of soluble CD59 can be significantly enhanced when it is targeted to the site of MAC formation. sCD59 has been shown to possess complement inhibitory activity in vitro (Sugita et al. 1994). However, the potential usefulness of sCD59 as a therapeutic complement blocker is limited by its lack of certain functional properties as discussed previously (Kalli et al. 1994). Although the inhibition of MAC assembly would be of benefit in inflammation, the late stage in the complement cascade at which CD59 leaves the generation of anaphylatoxins and their pathological sequelae unaffected. Therapeutic modulation of CD59 is being explored more aggressively to facilitate complement-mediated lysis of cancer cells (Rossen et al. 2008; Huang et al. 2005). Gene therapy targeting CD59 is under investigation for advanced macular degeneration (Cashman et al. 2011).

4.1.7 Molecular Fusion of Different Complement Regulatory Proteins

The molecular fusion of different complement regulatory proteins has been used to create chimeric molecules endowed with novel functions. Fodor and colleagues (1995) constructed two such chimeric complement inhibitors for cell surface expression using a GPI anchor: CD (NH2-CD59-DAF-GPI) and DC (NH2-DAF-CD59-GPI). The rationale behind this work was to create a single protein that blocks C3 and C5 convertase activity as well as the assembly of the MAC. Of the two molecules, CD retained DAF function but did not inhibit C5b-9 assembly. The DC chimera, however, exhibited both DAF and CD59 activity. The DC chimera may have utility in the production of transgenic organs for the inhibition of hyperacute rejection in xenotransplantation (Kennedy et al. 1994; Fodor et al. 1994; McCurry et al. 1995a, b; Miyagawa et al. 1995; Heckl-Östreicher et al. 1996; Kroshus et al. 1996; Diamond et al. 1996; Byrne et al. 1997).

The molecular fusion of membrane cofactor protein (MCP) and decay-accelerating factor (DAF) brings together the complementary activities of these two regulatory molecules to create a single protein that has both factor I cofactor activity and decay-accelerating activity. A membrane-bound chimeric MCP-DAF has been expressed in CHO cells, and its activity has been compared with that of transfectants expressing MCP or DAF or MCP plus DAF (Iwata et al. 1994). In vitro studies in stably transfected swine endothelial cells exposed to human complement (Miyagawa et al. 1994) have shown that the surface-expressed MCP-DAF hybrid inhibits cell lysis more effectively than MCP alone and apparently as effectively as DAF. A soluble version of chimeric MCP-DAF, referred to as complement activation blocker-2 (CAB-2), possesses factor I cofactor activity and decay-accelerating activity and has been shown to inactivate both classical and alternative C3 and C5 convertases in vitro as measured by assays of inhibition of cytotoxicity and anaphylatoxin generation (Higgins et al. 1997). This hybrid was shown to inhibit complement activation in vivo, in the reversed passive Arthus reaction and in the direct passive Arthus reaction, as well as in the Forssman shock model in guinea pigs. The $t_{1/2}\beta$ of CAB-2 in rats was 8 h (Higgins et al. 1997), which is suitable for human therapy. It is possible that the half-life of CAB-2 may be longer in humans than in rats, as has been the case for sCR1. One potential limitation of CAB-2 as a therapeutic is its potential immunogenicity. The molecular fusion of two otherwise natural proteins is likely to create novel epitopes, which might trigger an immune response. In this case, CAB-2 might be useful in acute indications, depending on the severity of the anti-CAB-2 response.

We recently conducted a case-control study and a follow-up investigation on leukocyte DAF, MCP, CR1, and CD59 transcript and protein levels to evaluate their significance as biomarkers and, consequently, as therapeutic targets of SLE. Levels of leukocyte MCP, DAF, and CD59 transcripts were increased and that of CR1 were decreased significantly in patients with active disease. The follow-up study revealed that the levels of leukocyte MCP, DAF, and CD59 transcripts decreased and that of CR1 increased significantly in patients during remission when compared to active disease. These observations suggest an intimate relation between the level of these four CRP transcripts and the disease activity in SLE. Based on comparative evaluation of our data on these four membrane-bound complement regulatory proteins, we envisaged CR1 and MCP transcripts as putative noninvasive disease activity markers and the respective proteins as therapeutic targets for SLE (Das, Biswas, Kumar et al. – Meeting abstract, Fifth International Complement Therapeutics Conference). However, the increases in the levels of DAF, MCP, and CD59 transcripts did not correlate with levels of respective proteins on the leukocyte cell surface. Relationship of soluble forms and cell surface expression of these proteins and those with the levels of respective transcript levels need to be investigated. Upregulation of CR1 and MCP together at the protein level may be an effective strategy to control complement- and immune complex-mediated self tissue destruction in SLE. Studies pertaining to the pleiotropic functions and regulation of the expression of these membrane-bound proteins in human subjects are under progress.

To conclude, the membrane-bound complement regulatory proteins hold promise as biomarkers of SLE. Our case-control and follow-up studies have shown intimate relations of leukocyte or neutrophils CR1 and MCP transcripts with disease activity in patients. These findings however need further confirmation and validation to recommend the application of CR1 and MCP transcripts as clinically applied biomarkers for SLE. Despite decades of extensive work in the understanding of the etiopathogenesis of systemic lupus erythematosus, few biomarkers have been validated and widely accepted for this disease. The lack of reliable, specific biomarkers not only hampers clinical management of systemic lupus erythematosus but also impedes development of new therapeutic agents. Therefore, an extensive search for candidate biomarkers of SLE needs to be continued.

Studies need to be extended to evaluate the significance of soluble forms of these proteins as biomarkers for SLE. Information on therapeutic implications of DAF, MCP, and CD59 specifically for SLE is limited. Further, with growing knowledge of pleiotropic role of complement regulatory proteins and complement-cytokine nexus, insight into the overall implication of blocking or stimulating complement proteins needs to be taken into consideration while targeting any of the complement regulatory proteins for developing a therapeutic. Moreover, there is a need to identify the endogenous and exogenous factors, which influence the normal expression of these proteins and modulate the same in SLE. Although mice models are useful, the results may not be same with human subjects. It is high time now that more studies be focused to understand the significance of membrane-bound complement proteins in human subjects to identify valid biomarkers and effective therapeutic targets of SLE.

References

Ahearn JM, Fearon DT (1989) Structure and function of the C receptors, CR1 (CD35) and CR2 (CD21). Adv Immunol 46:183–219

Astier AL (2008) T-cell regulation by CD46 and its relevance in multiple sclerosis. Immunology 124(2):149–154

Astier A, Trescol-Biemont MC, Azocar O, Lamouille B, Rabourdin-Combe C (2000) Cutting edge: CD46, a new costimulatory molecule for T cells, that induces p120CBL and LAT phosphorylation. J Immunol 164:6091–6095

Arora M, Arora R, Tiwari SC, Das N, Srivastava LM (2000) Expression of complement regulatory proteins in diffuse proliferative glomerulonephritis. Lupus 9:127–131

Asghar SS, Pasch MC (2000) Therapeutic inhibition of the complement system. Y2K update. Front Biosci 5:E63–E81

Ballard L, Seya T, Teckman J, Lublin DM, Atkinson JP (1987) A polymorphism of the complement regulatory protein MCP (membrane cofactor protein or gp45-70). J Immunol 138(11):3850–3855

Bergelson JM, Mohanty JG, Crowell RL, St. John NF, Lublin DM, Finberg RW (1995) Coxsackievirus B3 adapted to growth in RD cells binds to decay-accelerating factor (CD55). J Virol 69(3):1903–1906
Birmingham DJ, Gavit KF, McCarty SM, Yu CY, Rovin BH, Nagaraja HN, Hebert LA (2006) Consumption of erythrocyte CR1 (CD35) is associated with protection against systemic lupus erythematosus renal flare. Clin Exp Immunol 143:274–280
Brodbeck WG, Liu D, Sperry J et al (1996) Localization of classical and alternative pathway regulatory activity within the decay accelerating factor. J Immunol 156:2528–2533
Byrne JL, Carter GI, Ellis I, Haynes AP, Russell NH (1997) Autologous GVHD following PBSCT, with evidence for a graft-versus-myeloma effect. Bone Marrow Transplant 20(6):517–520
Cashman SM, Ramo K, Kumar-Singh R (2011) A non membrane-targeted human soluble CD59 attenuates choroidal neovascularization in a model of age related macular degeneration. PLoS One 6(4):e19078
Cattaneo R (2004) Four viruses, two bacteria, and one receptor: membrane cofactor protein (CD46) as pathogens' magnet. J Virol 78:4385–4388
Christiansen D, Milland J, Thorley BR, McKenzie IF, Loveland BE (1996) A functional analysis of recombinant soluble CD46 in vivo and a comparison with recombinant soluble forms of CD55 and CD35 in vitro. Eur J Immunol 26(3):578–585
Clarkson NA, Kaufman R, Lublin DM, Ward T, Pipkin PA, Minor PD, Evans DJ, Almond JW (1995) Characterization of the echovirus 7 receptor: domains of CD55 critical for virus binding. J Virol 69(9):5497–5501
Cockburn IA, Mackinnon MJ, O'Donnell A, Allen SJ, Moulds JM, Baisor M, Bockarie M, Reeder JC, Rowe JA (2004) A human complement receptor 1 polymorphism that reduces *Plasmodium falciparum* rosetting confers protection against severe malaria. Proc Natl Acad Sci USA 101(1):272–277
Cole J, Housley GA, Dykman TR, MacDermott RP, Atkinson JP (1985) Identification of an additional class of C3-binding membrane proteins of human peripheral blood leukocytes and cell lines. Proc Natl Acad Sci USA 83:859
Cosio FG, Sedmak DD, Mahan JD, Nahman NS (1989) Localization of decay accelerating factor in normal and diseased kidneys. Kidney Int 36:100
Cosio FG, Shen XP, Birmingham DJ, Van Aman M, Hebert LA (1990) Evaluation of the mechanisms responsible for the reduction in erythrocyte complement receptors when immune complexes form in vivo in primates. J Immunol 145:4198–4206
Couser WG, Johnson RJ, Young BA, Yeh CG, Toth CA, Rudolph AR (1995) The effects of soluble recombinant complement receptor 1 on complement-mediated experimental glomerulonephritis. J Am Soc Nephrol 5:1888–1894
Davies A, Simmons DL, Hale G, Harrison RA, Tighe H, Lachmann PJ, Waldmann H (1989) CD59, an Ly-6 like protein expressed in human lymphoid cells, regulates the action of the complement membrane attack complex on homologous cells. J Exp Med 170:637
Davies KA, Hird V, Stewart S et al (1990) A study of in vivo immune complex formation and clearance in man. J Immunol 144:4613–4620
Davis LS, Patel SS, Atkinson JP, Lipsky PE (1988) Decay-accelerating factor functions as a signal transducing molecule for human T cells. J Immunol 141:2246–2252
Delibrias CC, Fischer E, Bismuth G, Kazatchkine MD (1992) Expression, molecular association, and functions of C3 complement receptors CR1 (CD35) and CR2 (CD21) on the human T cell line HPB-ALL. J Immunol 149:768–774
Diamond LE, McCurry KR, Martin MJ, McClellan SB, Oldham ER, Platt JL, Logan JS (1996) Characterization of transgenic pigs expressing functionally active human CD59 on cardiac endothelium. Transplantation 61(8):1241–1249
Dorig RE, Marcil A, Chopra A, Richardson CD (1993) The human CD46 molecule is a receptor for measles virus (Edmonston strain). Cell 75:295
Ducruet AF, Mocco J, Mack WJ, Coon AL, Marsh HC, Pinsky DJ, Hickman ZL, Kim GH, Connolly ES Jr (2007) Preclinical evaluation of an sLe x-glycosylated complement inhibitory protein in a non-human primate model of reperfused stroke. J Med Primatol 36:375–380
Dykman TR, Hatch JA, Aqua MS, Atkinson JP (1985) Polymorphism of the C3b/C4b receptor (CR1): characterization of a fourth allele. J Immunol 134:1787–1789
Elward K, Griffiths M, Mizuno M, Harris CL, Neal JW, Morgan BP, Gasque P (2005) CD46 plays a key role in tailoring innate immune recognition of apoptotic and necrotic cells. J Biol Chem 280:36342–36354
Endoh M, Yamashina M, Ohi H, Funahashi K, Ikuno T, Yasugi T, Atkinson JP, Okada H (1993) Immunohistochemical demonstration of membrane cofactor protein (MCP) of complement in normal and diseased kidney tissues. Clin Exp Immunol 94(1):182–188
Falk RJ, Dalmasso AP, Kim Y, Lam S, Michael A (1985) Radioimmunoassay of the attack complex of complement in serum from patients with systemic lupus erythematosus. N Engl J Med 312:1594–1599
Farkas I, Baranyi L, Ishikawa Y, Okada N, Bohata C, Budai D, Fukuda A, Imai M, Okada H (2002) CD59 blocks not only the insertion of C9 into MAC but inhibits ion channel formation by homologous C5b-8 as well as C5b-9. J Physiol 539(Pt 2):537–545

Fearon DT (1979) Regulation of the amplification C3 convertase of human complement by an inhibitory protein isolated from human erythrocyte membrane. Proc Natl Acad Sci USA 76:5867–5871
Fearon DT (1980) Identification of the membrane glycoprotein that is the C3b receptor of the human erythrocyte, polymorphonuclear leukocyte, B lymphocyte, and monocyte. J Exp Med 152:20–30
Fearon DT, Kaneko I, Thomson GG (1981) Membrane distribution and adsorptive endocytosis by C3b receptors on human polymorphonuclear leukocytes. J Exp Med 153:1615–1628
Ferguson PJ, Reist CJ, Martin EN, Johnson C, Greene KL, Kuhn S, Niebur J, Emlen W, Taylor RP (1995) Antigen-based heteropolymers. A potential therapy for binding and clearing autoantibodies via erythrocyte CR1. Arthritis Rheum 38:190–200
Fernandez-Centeno E, de Ojeda G, Rojo JM, Portoles P (2000) Crry/p65, a membrane complement regulatory protein, has costimulatory properties on mouse T cells. J Immunol 164:4533–4542
Fingeroth JD, Heath ME, Ambrosino DM (1989) Proliferation of resting B cells is modulated by CR2 and CR1. Immunol Lett 21:291–301
Fodor WL, Williams BL, Matis LA, Madri JA, Rollins SA, Knight JW, Velander W, Squinto SP (1994) Expression of a functional human complement inhibitor in a transgenic pig as a model for the prevention of xenogeneic hyperacute organ rejection. Proc Natl Acad Sci USA 91(23):11153–11157
Fodor WL, Rollins SA, Guilmette ER, Setter E, Squinto SP (1995) A novel bifunctional chimeric complement inhibitor that regulates C3 convertase and formation of the membrane attack complex. J Immunol 155(9):4135–4138
Fuchs A, Atkinson JP, Fremeaux-Bacchi V, Kemper C (2009) CD46-induced human Treg enhance B-cell responses. Eur J Immunol 39(11):3097–3109
Fugier-Vivier I, Servet-Delprat C, Rivailler P, Rissoan MC, Liu YJ, Rabourdin-Combe C (1997) Measles virus suppresses cell-mediated immunity by interfering with the survival and functions of dendritic and T cells. J Exp Med 186:813
Fujita T, Inoue T, Ogawa K, Lida K, Tamura N (1987) The mechanism of DAF. DAF inhibits the assembly of the C3 convertases by dissociating C2a and Bb. J Exp Med 166:1221–1228
Gawryl MS, Chudwin DS, Langlois PF, Lint TF (1988) The terminal complement complex, C5b-9, a marker of disease activity in patients with systemic lupus erythematosus. Arthritis Rheum 31(2):188–195
Ghiran I, Barbashov SF, Klickstein LB, Tas SW, Jensenius JC, Nicholson-Weller A (2000) Complement receptor 1/CD35 is a receptor for mannan-binding lectin. J Exp Med 192(12):1797–1808
Gray JX, Haino M, Roth MJ, Maguire JE, Jensen PN, Yarme A, Stetler-Stevenson MA, Siebenlist U, Kelly K (1996) CD97 is a processed, seven-transmembrane, heterodimeric receptor associated with inflammation. J Immunol 157:5438–5447
Grossman WJ, Verbsky JW, Barchet W, Colonna M, Atkinson JP, Ley TJ (2004) Human T regulatory cells can use the perforin pathway to cause autologous target cell death. Immunity 21:589–601
Gyimesi E, Bankovich AJ, Schuman TA, Goldberg JB, Lindorfer MA, Taylor RP (2004) Staphylococcus aureus bound to complement receptor 1 on human erythrocytes by bispecific monoclonal antibodies is phagocytosed by acceptor macrophages. Immunol Lett 95:185–192
Hamann J, Vogel B, van Schijndel GM, van Lier RA (1996) The seven-span transmembrane receptor CD97 has a cellular ligand (CD55, DAF). J Exp Med 184:1185–1189
Hamann L, Stamme C, Ulmer AJ, Schumann RR (2002) Inhibition of LPS-induced activation of alveolar macrophages by high concentrations of LPS-binding protein. Biochem Biophys Res Commun 295(2):553–560
Harris CL, Rushmere NK, Morgan BP (1999) Molecular and functional analysis of mouse decay accelerating factor (CD55). Biochem J 341(Pt 3):821–829
Hebert LA (1991) The clearance of immune complexes from the circulation of man and other primates. Am J Kidney Dis 17:352–361
Heckl-Ostreicher B, Wosnik A, Kirschfink M (1996) Protection of porcine endothelial cells from complement-mediated cytotoxicity by the human complement regulators CD59, C1 inhibitor, and soluble complement receptor type 1. Analysis in a pig-to-human in vitro model relevant to hyperacute xenograft rejection. Transplantation 62(11):1693–1696
Heine-Suner D, Diaz-Guillen MA, de Villena FP, Robledo M, Benitez J, Rodriguez de Cordoba S (1997) A high-resolution map of the regulator of the complement activation gene cluster on 1q32 that integrates new genes and markers. Immunogenetics 45:422–427
Hideshima T, Okada N, Okada H (1990) Expression of HRF20, a regulatory molecule of complement activation, on peripheral blood mononuclear cells. Immunology 69(3):396–401
Higgins PJ, Ko JL, Lobell R, Sardonini C, Alessi MK, Yeh CG (1997) A soluble chimeric complement inhibitory protein that possesses both decay-accelerating and factor I cofactor activities. J Immunol 158:2872–2881
Hindmarsh EJ, Marks RM (1998) Decay accelerating factor is a component of subendothelial extracellular matrix in vitro, and is augmented by activation of endothelial protein kinase C. Eur J Immunol 28:1052
Hirano A, Kurita-Taniguchi M, Katayama Y, Matsumoto M, Wong TC, Seya T (2002) Ligation of human CD46 with purified complement C3b or F(ab)2 of monoclonal antibodies enhances isoform-specific interferon gamma-dependent nitric oxide production in macrophages. J Biochem 132:83–91

Hoffmann EM (1969) Inhibition of complement by a substance isolated from human erythrocytes. I. Extraction from human erythrocyte stroma. Immunochemistry 6:391
Hoffmann EM, Etlinger HM (1973) Extraction of complement inhibitory factors from erythrocytes of non-human species. J Immunol 111:946
Holers VM, Cole JL, Lublin DM, Seya T, Atkinson JP (1985) Human C3b- and C4b-regulatory proteins: a new multigene family. Immunol Today 6:188
Holguin MH, Fredrick LR, Bernshaw NJ, Wilcox LA, Parker CJ (1989) Isolation and characterization of a membrane protein from normal human erythrocytes that inhibits reactive lysis of the erythrocytes of paroxysmal nocturnal hemoglobinuria. J Clin Invest 84(1):7–17
Hourcade D, Holers VM, Atkinson JP (1989) The regulators of complement activation (RCA) gene cluster. Adv Immunol 45:381–416
Huang Y, Smith CA, Song H, Morgan BP, Abagyan R, Tomlinson S (2005) Insights into the human CD59 complement binding interface toward engineering new therapeutics. J Biol Chem 280(40):34073–34079
Iida K, Nussenzweig V (1981) Complement receptor is an inhibitor of the complement cascade. J Exp Med 153:1138–1150
Iwata K, Seya T, Ariga H, Nagasawa S (1994) Expression of a hybrid complement regulatory protein, membrane cofactor protein decay accelerating factor on Chinese hamster ovary: comparison of its regulatory effect with those of decay accelerating factor and membrane cofactor protein. J Immunol 152:3436–3444
Jimenez-Perianez A, Ojeda G, Criado G, Sanchez A, Pini E, Madrenas J, Rojo JM, Portoles P (2005) Complement regulatory protein Crry/p65-mediated signaling in T lymphocytes: role of its cytoplasmic domain and partitioning into lipid rafts. J Leukoc Biol 78:1386–1396
Johnson NM, McNicol MW, Burton-Kee EJ, Mowbray JF (1981) Circulating immune complexes in tuberculosis. Thorax 36(8):610–617
Johnstone RW, Loveland BE, McKenzie IF (1993) Identification and quantification of complement regulator CD46 on normal human tissues. Immunology 79(3):341–347
Jones DR, Thompson RJ, Rao SA, Imrie H (1994) An enzyme-linked immunosorbent assay for complement regulatory proteins and membrane-bound immunoglobulins on intact red blood cells. J Immunol Methods 177(1–2):235–242
Jozsi M, Prechl J, Bajtay Z, Erdei A (2002) Complement receptor type 1 (CD35) mediates inhibitory signals in human B lymphocytes. J Immunol 168:2782–2788
Kalli KR, Hsu P, Fearon DT (1994) Therapeutic uses of recombinant complement protein inhibitors. Springer Semin Immunopathol 15:417–431
Kallstrom H, Liszewski MK, Atkinson JP, Jonsson AB (1997) Membrane cofactor protein (MCP or CD46) is a cellular pilus receptor for pathogenic *Neisseria*. Mol Microbiol 25:639
Kallstrom H, Islam MS, Berggren PO, Jonsson AB (1998) Cell signaling by the type IV pili of pathogenic *Neisseria*. J Biol Chem 273:21777
Karnauchow TM, Tolson DL, Harrison BA, Altman E, Lublin DM, Dimock K (1996) The HeLa cell receptor for enterovirus 70 is decay-accelerating factor (CD55). J Virol 70(8):5143–5152
Karp CL, Wysocka M, Wahl LM, Ahearn JM, Cuomo PJ, Sherry B, Trinchieri G, Griffin DE (1996) Mechanism of suppression of cell-mediated immunity by measles virus. Science 273:228
Katyal M, Sivasankar B, Ayub S, Das N (2003) Genetic and structural polymorphism of complement receptor 1 in normal Indian subjects. Immunol Lett 89:93–98
Kawano M, Seya T, Koni I, Mabuchi H (1999) Elevated serum levels of soluble membrane cofactor protein (CD46, MCP) in patients with systemic lupus erythematosus (SLE). Clin Exp Immunol 116:542–546
Kemper C, Atkinson JP (2007) T-cell regulation: with complements from innate immunity. Nat Rev Immunol 7:9–18
Kemper C, Chan AC, Green JM, Brett KA, Murphy KM, Atkinson JP (2003) Activation of human CD41cells with CD3 and CD46 induces a T-regulatory cell 1 phenotype. Nature 421:388–392
Kennedy SP, Rollins SA, Burton WV, Sims PJ, Bothwell AL, Squinto SP, Zavoico GB (1994) Protection of porcine aortic endothelial cells from complement-mediated cell lysis and activation by recombinant human CD59. Transplantation 57:1494–1501
Khera R, Das N (2009) Complement receptor 1: disease associations and therapeutic implications. Mol Immunol 46(5):761–772
Kinoshita T, Medof ME, Silber R, Nussenzweig V (1985) Distribution of decay accelerating factor in the peripheral blood of normal individuals and patients with paroxysmal nocturnal hemoglobinuria. J Exp Med 162:75
Kinoshita T, Medof ME, Nussenzweig V (1986) Endogenous association of decay-accelerating factor (DAF) with C4b and C3b on cell membranes. J Immunol 136(9):3390–3395
Klickstein LB, Wong WW, Smith JA, Weis JH, Wilson JG, Fearon DT (1987) Human C3b/C4b receptor (CR1): demonstration of long homologous repeating domains that are composed of the short consensus repeats characteristics of C3/C4 binding proteins. J Exp Med 165:1095–1112
Kojima A, Iwata K, Seya T, Matsumoto M, Ariga H, Atkinson JP, Nagasawa S (1993) Membrane cofactor protein (CD46) protects cells predominantly from alternative complement pathway-mediated C3-fragment deposition and cytolysis. J Immunol 151:1519

Kroshus TJ, Bolman RM III, Dalmasso AP, Rollins SA, Guilmette ER, Williams BL, Squinto SP, Fodor WL (1996) Expression of human CD59 in transgenic pig organs enhances organ survival in an ex vivo xenogeneic perfusion model. Transplantation 61:1513–1521

Kurita-Taniguchi M, Fukui A, Hazeki K et al (2000) Functional modulation of human macrophages through CD46 (measles virus receptor): production of IL-12 p40 and nitric oxide in association with recruitment of protein-tyrosine phosphatase SHP-1 to CD46. J Immunol 165:5143–5152

Lee SW, Bonnah RA, Higashi DL, Atkinson JP, Milgram SL, So M (2002) CD46 is phosphorylated at tyrosine 354 upon infection of epithelial cells by *Neisseria gonorrhoeae.* J Cell Biol 156:951–957

Li J, McMurray RW (2006) Effects of estrogen receptor subtype-selective agonists on immune functions in ovariectomized mice. Int Immunopharmacol 6(9):1413–1423

Li L, Spendlove I, Morgan J, Durrant LG (2001) CD55 is over-expressed in the tumour environment. Br J Cancer 84:80–86

Liszewski MK, Post TW, Atkinson JP (1991) Membrane cofactor protein (MCP or CD46): newest member of the regulators of complement activation gene cluster. Annu Rev Immunol 9:431–455

Liu CC, Ahearn JM (2009) The search for lupus biomarkers. Best Pract Res Clin Rheumatol 23(4):507–523

Liu J, Miwa T, Hilliard B, Chen Y, Lambris JD, Wells AD et al (2005) The complement inhibitory protein DAF (CD55) suppresses T cell immunity in vivo. J Exp Med 201(4):567–577

Longhi MP, Harris CL, Morgan BP, Gallimore A (2006) Holding T cells in check–a new role for complement regulators? Trends Immunol 27(2):102–108

Lublin DM, Liszewski MK, Post TW, Arce MA, Le Beau MM, Rebentisch MB, Lemons RS, Seya T, Atkinson JP (1988) Molecular cloning and chromosomal localization of human membrane cofactor protein (MCP): evidence for inclusion in the multi-gene family of complement regulatory proteins. J Exp Med 168:181

Ludford-Menting MJ, Thomas SJ, Crimeen B, Harris LJ, Loveland BE, Bills M, Ellis S, Russell SM (2002) A functional interaction between CD46 and DLG4: a role for DLG4 in epithelial polarization. J Biol Chem 277:4477–4484

Ludford-Menting MJ, Oliaro J, Sacirbegovic F et al (2005) A network of PDZ-containing proteins regulates T cell polarity and morphology during migration and immunological synapse formation. Immunity 22:737–748

Manchester M, Gairin JE, Patterson JB, Alvarez J, Liszewski MK, Eto DS, Atkinson JP, Oldstone MB (1997) Measles virus recognizes its receptor, CD46, via two distinct binding domains within SCR1-2. Virology 233(1):174–184

Manchester M, Eto D, Valsamakis A, Liton P, Fernandez-Munoz R, Rota PA, Bellini WJ, Forthal DN, Oldstone MB (2000) Clinical isolates of measles virus use CD46 as a cellular receptor. J Virol 74:3967

Manzi S, Rairie JE, Carpenter AB, Kelly RH, Jagarlapudi SP, Sereika SM et al (1996a) Sensitivity and specificity of plasma and urine complement split products as indicators of lupus disease activity. Arthritis Rheum 39:1178–1188

Manzi S, Rairie JE, Carpenter AB, Kelly RH, Jagarlapudi SP, Sereika SM, Medsger TA Jr, Ramsey-Goldman R (1996b) Sensitivity and specificity of plasma and urine complement split products as indicators of lupus disease activity. Arthritis Rheum 39(7):1178–1188

Manzi L, Montaño R, Abad MJ, Arsenak M, Romano E, Taylor P (2006) Expression of human soluble complement receptor 1 by a pig endothelial cell line inhibits lysis by human serum. Xenotransplantation 13(1):75–79

Marie JC, Astier AL, Rivailler P, Rabourdin-Combe C, Wild TF, Horvat B (2002) Linking innate and acquired immunity: divergent role of CD46 cytoplasmic domains in T cell induced inflammation. Nat Immunol 3:659–666

Marquart HV, Svendsen A, Rasmussen JM, Nielsen CH, Junker P, Svehag SE, Leslie RG (1995) Complement receptor expression and activation of the complement cascade on B lymphocytes from patients with systemic lupus erythematosus (SLE). Clin Exp Immunol 101:60–65

Matsuo S, Nishikage H, Yoshida F, Nomura A, Piddlesden SJ, Morgan BP (1994) Role of CD59 in experimental glomerulonephritis in rats. Kidney Int 46:191–200

McCurry KR, Kooyman DL, Diamond LE, Byrne GW, Logan JS, Platt JL (1995a) Transgenic expression of human complement regulatory proteins in mice results in diminished complement deposition during organ xenoperfusion. Transplantation 59(8):1177–1182

McCurry KR, Kooyman DL, Diamond LE, Byrne GW, Martin MJ, Logan JS, Platt JL (1995b) Human complement regulatory proteins in transgenic animals regulate complement activation in xenoperfused organs. Transplant Proc 27(1):317–318

McKusick VA (2001) The anatomy of the human genome: a neo-Vesalian basis for medicine in the 21st century. JAMA 286(18):2289–2295

Medof ME, Iida K, Mold C, Nussenzweig V (1982) Unique role of the complement receptor CR1 in the degradation of C3b associated with immune complexes. J Exp Med 156:1739–1754

Medof ME, Lublin DM, Holers VM, Ayers DJ, Getty RR, Leykam JF, Atkinson JP, Tykocinski ML (1987a) Cloning and characterization of cDNAs encoding the complete sequence of decay accelerating factor of human complement. Proc Natl Acad Sci USA 84:2007

Medof ME, Walter EI, Rutgers JL, Knowles DM, Nussenzweig V (1987b) Identification of the complement decay-accelerating factor (DAF) on epithelium and glandular cells and in body fluids. J Exp Med 165:848

Meri S, Morgan BP, Davies A, Daniels RH, Olavesen MG, Waldmann H, Lachmann PJ (1990) Human protectin (CD59), an 18,000–20,000 MW complement lysis restricting factor, inhibits C5b–8 catalysed insertion of C9 into lipid bilayers. Immunology 71:1

Meri S, Waldmann H, Lachmann PJ (1991) Distribution of protectin (CD59), a complement membrane attack inhibitor, in normal human tissues. Lab Invest 65(5):532–537

Metcalfe RA, McIntosh RS, Morgan BP, Levin JL, Weetman AP (1996) The effect of soluble complement receptor 1 (sCR1) and human thyroid antibodies on the course of experimental autoimmune thyroiditis in rats. Autoimmunity 23:1–8

Mikata S, Miyagawa S, Yoshitatsu M, Ikawa M, Okabe M, Matsuda H, Shirakura R (1998) Prevention of hyperacute rejection by phosphatidylinositol-anchored mini-complement receptor type 1. Transplant Immunol 6:107–110

Mitchell JA, Sim RB, Sim E (1989) CR1 polymorphism in hydralazine-induced systemic lupus erythematosus: DNA restriction fragment length polymorphism. Clin Exp Immunol 78:354–358

Miwa T, Maldonado MA, Zhou L, Sun X, Luo HY, Cai D, Werth VP, Madaio MP, Eisenberg RA, Song WC (2002) Deletion of decay-accelerating factor (CD55) exacerbates autoimmune disease development in MRL/lpr mice. Am J Pathol 161:1077–1086

Miyagawa S, Shirakura R, Matsumiya G, Nakata S, Matsuda H, Hatanaka M, Matsumoto M, Seya T (1994) Possibility of prevention of hyperacute rejection by DAF and CD59 in xenotransplantation. Transplant Proc 26(3):1235–1238

Miyagawa S, Shirakura R, Nakata S, Izutani H, Matsuda H, Iwata K, Nagasawa S, Terado A, Hatanaka M, Matsumoto M et al (1995) Effect of transfected MACIF (CD59) on complement-mediated swine endothelial cell lysis, compared with those of membrane cofactor protein (CD46) and decay-accelerating factor (CD55). Transplant Proc 27(1):328–329

Mizuno M, Nishikawa K, Spiller OB, Morgan BP, Okada N, Okada H, Matsuo S (2001) Membrane complement regulators protect against the development of type II collagen-induced arthritis in rats. Arthritis Rheum 44(10):2425–2434

Moran P, Beasley H, Gorrell A, Martin E, Gribling P, Fuchs H, Gillett N, Burton LE, Caras IW (1992) Human recombinant soluble decay accelerating factor inhibits complement activation in vitro and in vivo. J Immunol 149(5):1736–1743

Morgan BP, Meri S (1994) Membrane proteins that protect against complement lysis. Springer Semin Immunopathol 15:369–396

Moulds JM, Reveille JD, Arnett FC (1996) Structural polymorphisms of complement receptor 1 (CR1) in systemic lupus erythematosus (SLE) patients and normal controls of three ethnic groups. Clin Exp Immunol 105:302–305

Nangaku M (2003) Complement regulatory proteins: are they important in disease? J Am Soc Nephrol 14(9):2411–2413

Nangaku M, Alpers CE, Pippin J, Shankland SJ, Kurokawa K, Adler S, Morgan BP, Johnson RJ, Couser WG (1998) CD59 protects glomerular endothelial cells from immune-mediated thrombotic microangiopathy in rats. J Am Soc Nephrol 9:590–597

Naniche D, Varior-Krishnan G, Cervoni F, Wild TF, Rossi B, Rabourdin-Combe C, Gerlier D (1993) Human membrane cofactor protein (CD46) acts as a cellular receptor for measles virus. J Virol 67:6025

Nelson RA Jr (1953) The immune-adherence phenomenon; an immunologically specific reaction between microorganisms and erythrocytes leading to enhanced phagocytosis. Science 118(3077):733–737

Nicholson-Weller A, Burge J, Fearon DT, Weller PF, Austen KF (1982) Isolation of a human erythrocyte membrane glycoprotein with decay-accelerating activity for C3 convertases of the complement system. J Immunol 129:184

Nose M, Katoh M, Okada N, Kyogoku M, Okada H (1990) Tissue distribution of HRF20, a novel factor preventing the membrane attack of homologous complement, and its predominant expression on endothelial cells in vivo. Immunology 70:145

Ogelsby TJ, Allen CJ, Liszewski MK, Whiti DJG, Atkinson JP (1992) Membrane cofactor protein (CD46) protects cells from complement-mediated attack by an intrinsic mechanism. J Exp Med 175:1547

Okada N, Harada R, Fujita T, Okada H (1989) A novel membrane glycoprotein capable of inhibiting membrane attack by homologous complement. Int Immunol 1:205

Okada N, Liszewski MK, Atkinson JP, Caparon M (1995) Membrane cofactor protein (CD46) is a keratinocyte receptor for the M protein of the group A streptococcus. Proc Natl Acad Sci USA 92:2489

Oliaro J, Pasam A, Waterhouse NJ, Browne KA, Ludford-Menting MJ, Trapani JA, Russell SM (2006) Ligation of the cell surface receptor, CD46, alters T cell polarity and response to antigen presentation. Proc Natl Acad Sci USA 103:18685–18690

Oudin S, Libyh MT, Goossens D, Dervillez X, Philbert F, Reveil B, Bougy F, Tabary T, Rouger P, Klatzmann D, Cohen JH (2000) A soluble recombinant multimeric anti-Rh(D) single-chain Fv/CR1 molecule restores the immune complex binding ability of CR1-deficient erythrocytes. J Immunol 164:1505–1513

Petranka JG, Fleenor DE, Sykes K, Kaufman RE, Rosse WF (1992) Structure of the CD59-encoding gene: further evidence of a relationship to murine lymphocyte antigen Ly-6 protein. Proc Natl Acad Sci USA 89(17):7876–7879, Erratum in: Proc Natl Acad Sci USA 1993 Jun 15; 90(12):5878

Picard MD, Pettey CL, Marsh HC, Thomas LJ (2000) Characterization of N-linked oligosaccharides bearing sialyl lewis xmoieties on an alternatively glycosylated form of soluble complement receptor type 1 (sCR1). Biotechnol Appl Biochem 31(Pt 1):5–13
Price JD, Schaumburg J, Sandin C, Atkinson JP, Lindahl G, Kemper C (2005) Induction of a regulatory phenotype in human CD4+ T cells by streptococcal M protein. J Immunol 175:677–684
Qian YM, Haino M, Kelly K, Song WC (1999) Structural characterization of mouse CD97 and study of its specific interaction with the murine decay-accelerating factor (DAF, CD55). Immunology 98:303–311
Quigg RJ, Nicholson-Weller A, Cybulsky AV, Badalamenti J, Salant DJ (1989) Decay accelerating factor regulates complement activation on glomerular epithelial cells. J Immunol 142(3):877–882
Quigg RJ, Morgan BP, Holers VM, Adler S, Sneed AE 3rd, Lo CF (1995) Complement regulation in the rat glomerulus: crry and CD59 regulate complement in glomerular mesangial and endothelial cells. Kidney Int 48(2):412–421
Rabinovici R, Yeh CG, Hillegass LM, Griswold DE, DiMartino MJ, Vernick J, Fong KL, Feuerstein G (1992) Role of complement in endotoxin/plateletactivating factor-induced lung injury. J Immunol 149:1744–1750
Reid KBM, Bentley DR, Campbell RD, Chung LP, Sim RB, Kristensen T, Tack BF (1986) Complement system proteins which interact with Cab or C4b. A superfamily of structurally related proteins. Immunol Today 7:230
Richaud-Patin Y, Pérez-Romano B, Carrillo-Maravilla E, Rodriguez AB, Simon AJ, Cabiedes J, Jakez-Ocampo J, Llorente L, Ruiz-Argüelles A (2003) Deficiency of red cell bound CD55 and CD59 in patients with systemic lupus erythematosus. Immunol Lett 88(2):95–99
Ricklin D, Lambris JD (2007) Complement-targeted therapeutics. Nat Biotechnol 25(11):1265–1275
Rittershaus CW, Thomas LJ, Miller DP, Picard MD, Geoghegan-Barek KM, Scesney SM, Henry LD, Sen AC, Bertino AM, Hannig G, Adari H, Mealey RA, Gosselin ML, Couto M, Hayman EG, Levin JL, Reinhold VN, Marsh HC Jr (1999) Recombinant glycoproteins that inhibit complement activation and also bind the selectin adhesion molecules. J Biol Chem 274(16):11237–11244
Rodriguez de Cordoba S, Lublin DM, Rubenstein P, Atkinson JP (1985) Human genes for three complement components that regulate the activation of C3 are tightly linked. J Exp Med 161:1189
Rollins SA, Sims PJ (1990) The complement-inhibitory activity of CD59 resides in its capacity to block incorporation of C9 into membrane C5b-9. J Immunol 144(9):3478–3483
Rollins SA, Zhao J, Ninomiya H, Sims PJ (1991) Inhibition of homologous complement by CD59 is mediated by a species-selective recognition conferred through binding to C8 within C5b-8 or C9 within C5b-9. J Immunol 146(7):2345–2351
Roncarolo MG, Gregori S, Battaglia M, Bacchetta R, Fleischhauer K, Levings MK (2006) Interleukin-10-secreting type 1 regulatory T cells in rodents and humans. Immunol Rev 212:28–50
Rooney A, Davies A, Griffiths D, Williamst JD, Daviest M, Merit S, Lachmannt PJ, Morgan BP (1991) The complement-inhibiting protein, protectin (CD59 antigen), is present and functionally active on glomerular epithelial cells. Clin Exp Immunol 83:251–256
Rooney IA, Davies A, Morgan BP (1992) Membrane attack complex (MAC)-mediated damage to spermatozoa: protection of the cells by the presence on their membranes of MAC inhibitory proteins. Immunology 75(3):499–506
Ross GD, Lambris JD, Cain JA, Newman SL (1982) Generation of three different fragments of bound C3 with purified factor I or serum. I. Requirements for factor H vs CR1 cofactor activity. J Immunol 129:2051–2060
Ross GD, Yount WJ, Walport MJ, Winfield JB, Parker CJ, Fuller CR, Taylor RP, Myones BL, Lachmann PJ (1985) Disease-associated loss of erythrocyte complement receptors (CR1, C3b receptors) in patients with systemic lupus erythematosus and other diseases involving autoantibodies and/or complement activation. J Immunol 135:2005–2014
Santoro F, Kennedy PE, Locatelli G, Malnati MS, Berger EA, Lusso P (1999) CD46 is a cellular receptor for human herpesvirus 6. Cell 99:817
Seya T, Atkinson JP (1989) Functional properties of membrane cofactor protein of complement. Biochem J 264:581
Seya T, Turner JR, Atkinson JP (1986) Purification and characterization of a membrane protein (gp45-70) that is a cofactor for cleavage of C3b and C4b. J Exp Med 163:837
Seya T, Ballard L, Bora N, MacNearney T, Atkinson JP (1988) Distribution of membrane cofactor protein (MCP) of complement on human peripheral blood cells. Eur J Immunol 18:1289
Seya T, Okada M, Matsumoto M, Hong KS, Kinoshita T, Atkinson JP (1991) Preferential activation of the C5 convertase of the alternative complement pathway by factor I and membrane cofactor protein (MCP). Mol Immunol 28:1137
Shenoy-Scaria AM, Kwong J, Fujita T, Olszowy MW, Shaw AS, Lublin DM (1992) Signal transduction through decay-accelerating factor. Interaction of glycosyl-phosphatidylinositol anchor and protein tyrosine kinases p56lck and p59fyn 1. J Immunol 149:3535–3541
Shibata T, Cosio FG, Brimingham DJ (1991) Complement activation induces expression of decay accelerating factor on human mesangial cells. J Immunol 147:3901–3908
Shibuya T, Izuchi K, Kuroiwa A, Harada H, Kumamoto A, Shirakawa K (1991) Study on nonspecific immunity in pregnant women: II. Effect of hormones on chemiluminescence response of peripheral blood phagocytes. Am J Reprod Immunol 26(2):76–81

Sivasankar B, Dinda AK, Tiwari SC, Srivastava LM, Das N (2004) Modulation of urinary CR1 in systemic lupus erythematosus. Lupus 13:228–233
Smith RA (2002) Targeting anticomplement agents. Biochem Soc Trans 30(Pt6):1037–1041
Sogabe H, Nangaku M, Ishibashi Y, Wada T, Fujita T, Sun X, Miwa T, Madaio MP, Song WC (2001) Increased susceptibility of decay-accelerating factor deficient mice to anti-glomerular basement membrane glomerulonephritis. J Immunol 167:2791–2797
Sparrow RL, McKenzie IF (1983) Human 'Ia' antigen populations defined by monoclonal antibodies. J Immunogenet 10(3):179–189
Spendlove I, Li L, Carmichael J, Durrant L (1999) Decay accelerating factor(CD55): a target for cancer vaccines? Cancer Res 59:2282–2286
Spendlove I, Ramage JM, Bradley R, Harris C, Durrant LG (2006) Complement decay accelerating factor (DAF)/CD55 in cancer. Cancer Immunol Immunother 55(8):987–995
Stammberger U, Hamacher J, Hillinger S, Schmid RA (2000) sCR1sLe ameliorates ischemia/reperfusion injury in experimental lung transplantation. J Thorac Cardiovasc Surg 120:1078–1084
Sugita Y, Nakano Y, Tomita M (1988) Isolation from human erythrocytes a new membrane protein which inhibits the formation of complement transmembrane channels. J Biochem 147:633, Tokyo
Sugita Y, Ito K, Shiozuka K, Suzuki H, Gushima H, Tomita M, Masuho Y (1994) Recombinant soluble CD59 inhibits reactive haemolysis with complement. Immunology 82:34–41
Tamai H, Matsuo S, Fukatsu A, Nishikawa K, Sakamoto N, Yoshioka K, Okada N, Okada H (1991) Localization of 20-kDa homologous restriction factor (HRF20) in diseased human glomeruli. An immunofluorescence study. Clin Exp Immunol 84:256–262
Tarkowski A, Trollmo C, Seifert PS, Hansson GK (1992) Expression of decay-accelerating factor on synovial lining cells in inflammatory and degenerative arthritides. Rheumatol Int 12:201–205
Tas SW, Klickstein LB, Barbashov SF, Nicholson-Weller A (1999) C1q and C4b bind simultaneously to CR1 and additively support erythrocyte adhesion. J Immunol 163:5056–5063
Tone M, Walsh LA, Waldmann H (1992) Gene structure of human CD59 and demonstration that discrete mRNAs are generated by alternative polyadenylation. J Mol Biol 227(3):971–976
Tosello AC, Mary F, Amiot M, Bernard A, Mary D (1998) Activation of T cells via CD55: recruitment of early components of the CD3-TCR pathway is required for IL-2 secretion. J Inflamm 48:13–27
Turnberg D, Botto M, Lewis M, Zhou W, Sacks SH, Morgan BP, Walport MJ, Cook HT (2004) 59a deficiency exacerbates ischemia-reperfusion injury in mice. Am J Pathol 165(3):825–832
Vedeler C, Ulvestad E, Bjørge L, Conti G, Williams K, Mørk S, Matre R (1994) The expression of CD59 in normal human nervous tissue. Immunology 82(4):542–547
Voynova E, Tchorbanov A, Prechl J, Nikolova M, Baleva M, Erdei A, Vassilev T (2008) An antibody-based construct carrying DNA-mimotope and targeting CR1(CD35) selectively suppresses human autoreactive B-lymphocytes. Immunol Lett 116:168–173
Walport M, Ng YC, Lachmann PJ (1987) Erythrocytes transfused into patients with SLE and hemolytic anemia lose complement receptor type 1 from their cell surface. Clin Exp Immunol 69:501–505
Watanabe M, Morita Y, Mizuno M, Nishikawa K, Yuzawa Y, Hotta N, Morgan BP, Okada N, Okada H, Matsuo S (2000) CD59 protects rat kidney from complement mediated injury in collaboration with crry. Kidney Int 58(4):1569–1579
Weis JH, Morton CC, Bruns GAP, Weis JJ, Klickstein LB, Wong WW, Fearon DT (1987) A complement receptor locus: genes encoding C3b/C4b receptor and C3d/Epstein Barr virus receptor map to 1q32. J Immunol 138:312–315
Weisman HF, Bartow T, Leppo MK, Marsh HC Jr, Carson GR, Concino MF, Boyle MP, Roux KH, Weisfeldt ML, Fearon DT (1990a) Soluble human complement receptor type 1: in vivo inhibitor of complement suppressing post-ischemic myocardial inflammation and necrosis. Science 249(4965):146–151
Weisman HF, Bartow T, Leppo MK, Marsh HC Jr, Carson GR, Concino MF, Boyle MP, Roux KH, Weisfeldt ML, Fearon DT (1990b) Soluble human complement receptor type 1: in vivo inhibitor of complement suppressing post-ischemic myocardial inflammation and necrosis. Science 249:146–151
Weiss L, Fischer E, Haeffner-Cavaillon N, Jouvin MH, Appay MD, Bariety J, Kazatchkine M (1989) The human C3b receptor (CR1). Adv Nephrol Necker Hosp 18:249–269
Williams AS, Mizuno M, Richards PJ, Holt DS, Morgan BP (2004) Deletion of the gene encoding CD59a in mice increases disease severity in a murine model of rheumatoid arthritis. Arthritis Rheum 50:3035–3044
Wilson JG, Murphy EE, Wong WW, Klickstein LB, Weis JH, Fearon DT (1986a) Identification of a restriction fragment length polymorphism by a CR1 cDNA that correlates with the number of CR1 on erythrocytes. J Exp Med 164:50–59
Wilson JG, Ratnoff WD, Schur PH, Fearon DT (1986b) Decreased expression of the C3b/C4b receptor (CR1) and the C3d receptor (CR2) on B lymphocytes and of CR1 on neutrophils of patients with systemic lupus erythematosus. Arthritis Rheum 29:739–747

Yazdanbakhsh K, Kang S, Tamasauskas D, Sung D, Scaradavou A (2003) Complement receptor 1 inhibitors for prevention of immune-mediated red cell destruction: potential use in transfusion therapy. Blood 101(12):5046–5052
Yu J, Abagyan R, Dong S, Gilbert A, Nussenzweig V, Tomlinson S (1997) Mapping the active site of CD59. J Exp Med 185(4):745–753
Zaffran Y, Destaing O, Roux A, Ory S, Nheu T, Jurdic P, Rabourdin-Combe C, Astier AL (2001) CD46/CD3 costimulation induces morphological changes of human T cells and activation of Vav, Rac, and extracellular signal-regulated kinase mitogenactivated protein kinase. J Immunol 167:6780–6785
Zhang H-F, Yu J, Bajwa E, Morrison SL, Tomlinson S (1999) Targeting of functional antibody-CD59 fusion proteins to a cell surface. J Clin Invest 103(1):55–61

Chapter 5
DAF as a Therapeutic Target for Steroid Hormones: Implications for Host–Pathogen Interactions

Bogdan Nowicki and Stella Nowicki

Abstract In this chapter, we present a concise historic prospective and a summary of accumulated knowledge on steroid hormones, DAF expression, and therapeutic implication of steroid hormone treatment on multiple pathologies, including infection and the host–pathogen interactions. DAF/CD55 plays multiple physiologic functions including tissue protection from the cytotoxic complement injury, an anti-inflammatory function due to its anti-adherence properties which enhance transmigration of monocytes and macrophages and reduce tissue injury. DAF physiologic functions are essential in many organ systems including pregnancy for protection of the semiallogeneic fetus or for preventing uncontrolled infiltration by white cells in their pro- and/or anti-inflammatory functions. DAF expression appears to have multiple regulatory tissue-specific and/or menstrual cycle-specific mechanisms, which involve complex signaling mechanisms. Regulation of DAF expression may involve a direct or an indirect effect of at least the estrogen, progesterone, and corticosteroid regulatory pathways. DAF is exploited in multiple pathologic conditions by pathogens and viruses in chronic tissue infection processes. The binding of *Escherichia coli* bearing Dr adhesins to the DAF/CD55 receptor is DAF density dependent and triggers internalization of *E. coli* via an endocytic pathway involving CD55, lipid rafts, and microtubules. Dr^+ *E. coli* or Dr antigen may persist in vivo in the interstitium for several months. Further understanding of such processes should be instrumental in designing therapeutic strategies for multiple conditions involving DAF's protective or pathologic functions and tailoring host expression of DAF.

5.1 Decay Accelerating Factor (DAF): Excess or Inadequate Levels and Pathophysiology

DAF, originally discovered by Hoffman in 1969, was identified as an inhibitor molecule that protected red blood cells from lysis by the complement cascade (Hoffmann 1969). The clinical significance of DAF's protective function was originally implicated in the pathogenesis of paroxysmal nocturnal

B. Nowicki (✉)
Department of Obstetrics and Gynecology, Meharry Medical College,
1005 Dr DB Todd Jr. Bulvd, Nashville, TN 37208, USA

Department of Microbiology and Immunology, Meharry Medical College,
1005 Dr DB Todd Jr. Bulvd, Nashville, TN 37208, USA
e-mail: bnowicki@mmc.edu

S. Nowicki
Department of Microbiology and Immunology, Meharry Medical College,
1005 Dr DB Todd Jr. Bulvd, Nashville, TN 37208, USA

J.D. Lambris et al. (eds.), *Complement Therapeutics*, Advances in Experimental Medicine and Biology 735,
DOI 10.1007/978-1-4614-4118-2_5, © Springer Science+Business Media New York 2013

hemoglobinuria (PNH), a disease of chronic anemia. PNH patients are low or deficient in the glycosyl-phosphoinositol (GPI) anchoring system which is involved in the attachment of DAF to the red cell membrane (Nicholson-Weller et al. 1983; Pangburn et al. 1983). Thus, the low or absent levels of DAF in abnormal PNH III cells have been associated with complement-mediated lysis of red blood cells (RBCs), chronic anemia, and multiple health complications in these patients including renal dysfunction and death at a young age. On the other hand, DAF overexpression also has been implicated in several normal and disease conditions. For example, upregulation of DAF in the feto-placental trophoblast protects the fetus from maternal complement injury and prevents rejection of the fetal graft (Holmes et al. 1992). In contrast, overexpression of DAF in endometrial cancer has been implicated in a cancer's capacity to escape a complement attack (Nowicki et al. 2001).

The biology of the reproductive tract, including the endometrium or placenta, is controlled by estrogen (E2) and progesterone (P4). Therefore, we previously postulated that DAF, which plays an essential function in reproductive tract biology and infection processes, was most likely affected by E2 and P4 levels (Holmes et al. 1992; Nowicki et al. 1988c). Thus, we and others have explored, to some extent, the contribution of steroid hormones to DAF expression.

In this chapter, we present a concise historic prospective and a summary of accumulated knowledge on steroid hormones, DAF expression, and therapeutic implication of E2 and P4 steroid treatment on multiple pathologies, including infection and the host–pathogen interactions.

5.2 DAF, Cromer Complex, Inab–Freiberg Blood Group Antigen and Blood Serology

DAF was rediscovered in the field of blood group serology but under different names (Cromer complex, Inab–Freiberg, and/or Dr(a-) blood group antigen) due to its various antigenic functions. Blood serology focuses on the identification of major blood group antigens, such as ABO, and is crucial for blood typing in patients requiring transfusions. There are more than 260 recognized blood group antigens, of which 60 or more remain to be classified; 215 of these antigens are classified in 23 distinct systems, and in 12 of these systems, there exists a null cell that is totally devoid of any antigen (Moulds et al. 1996). Of these rare, null phenotypic individuals were Inab and Freiberg, two patients with severe enteropathy and red cell membranes missing a 75-Kd protein called the Inab–Freiberg or the Cromer blood group complex. In the 1980s, the Cromer complex had not been well characterized as yet and remained to be classified as a blood group system. Nowicki and Moulds discovered that red cells lacking the Inab–Freiberg blood group antigen (or Cromer complex) and PNH III red cells were all missing the same DAF molecule (Nowicki et al. 1988a). Thus, this DAF research contributed to progress on the classification of blood group antigens and facilitated the cross talk between several clinical and basic science disciplines.

5.3 DAF, Dr(a) Blood Group Antigen, and Dr Receptors in Human Tissues

Another name used to describe DAF in human tissues in the field of bacteriology is the Dr receptor. Nowicki and Moulds discovered that *E. coli* bearing 075X adhesin did not hemagglutinate RBCs that were missing the Inab–Freiberg/Cromer complex and the Dr(a-) (null phenotype) or that were PNH III RBCs. In controls, *E. coli* 075X hemagglutinin (an adherence factor) allowed *E. coli* to attach to RBCs and hemagglutinate RBC, while purified DAF blocked *E. coli* 075X hemagglutination of RBCs that expressed the above antigens. We concluded that *E. coli* 075X recognize DAF on all tested cells (Nowicki et al. 1988a, b). The discovery that *E. coli* 075X bound to DAF was the first description of a pathogen using DAF, a complement inhibitor, as the tissue receptor for colonization, and as later described, for invasion of human tissues. Due to the receptor specificity of Dr(a), the *E. coli* 075X adhesin was renamed Dr adhesin, and the tissue receptor recognized by the *E. coli* Dr adhesin (originally detected in the kidney and uterine endometrium) was frequently called a Dr receptor. Later, these receptors were referred to as either DAF or more recently, CD55 (Nowicki et al. 1988c; Kaul et al. 1994). Subsequently, researchers accumulated evidence that DAF is recognized by several viruses such as echo, coxsackie, and others (Bergelson et al. 1994, 1995). Most recently, we found that Helicobacter pylori also recognizes DAF (O'Brien et al. 2006). Although the role of the DAF receptors in the pathogenesis of chronic diseases is not altogether clear as yet, the majority of DAF-specific viruses and pathogens can cause chronic and/or recurrent diseases (Śledzińska et al. 2011). The broad distribution of DAF expression in epithelial and endothelial cell types and multiple organs including kidney, stomach, placenta, gastrointestinal tract, or blood cells may account for the broad tissue tropism of DAF-specific viruses and bacteria. *E. coli* or viral invasion of tissues promoted by DAF has been implicated in the intracellular persistence of such pathogens and therefore in chronic disease processes (O'Brien et al. 2006; Śledzińska et al. 2011; Nowicki 2002; Nowicki et al. 2011a).

5.4 Host–Pathogen Interactions and the Mechanism of Persistent Infection

The binding of Dr^+ *E. coli* to the DAF/CD55 receptor is DAF density dependent and triggers internalization of *E. coli* via an endocytic pathway involving CD55, lipid rafts, and microtubules. Dr^+ *E. coli* or Dr antigen may persist in vivo in the interstitium for several months. A simplified hypothetical model of attachment and invasion process of human tissues and possible involvement of hormones is presented in Fig. 5.1 (Nowicki 2002; Nowicki et al. 2011a; Goluszko et al. 1997). Unlike *E. coli* bearing mannose-specific type 1 adhesin which multiplies in bladder cells, Dr^+*E. coli* can reside protected within the cell or interstitium without multiplication (Nowicki 2002; Anderson et al. 2003). Persistent gastritis induced by *H. pylori* is the strongest known risk factor for peptic ulcer disease and distal gastric adenocarcinoma, a process in which adherence of *H. pylori* to gastric epithelial cells is critical. Our studies showed that *H. pylori* adhered to Chinese hamster ovary cells expressing human DAF but not to vector controls. In cultured gastric epithelial cells, *H. pylori*-induced transcriptional upregulation of DAF and a genetic deficiency of DAF attenuated the development of inflammation among *H. pylori*-infected mice. These results indicated that DAF–*H. pylori*–epithelial cell interactions may contribute to a chronic inflammatory process (O'Brien et al. 2006). Although P4 has been implicated in upregulation of DAF, the exact impact of P4 on the invasion and the persistence process remains to be investigated.

5.5 Steroid Hormones Regulate DAF Expression

A recognition of the role of steroid hormones in regulation of DAF expression resulted from our studies on the expression of DAF in human reproductive tissues (Kaul et al. 1994, 1996; Young et al. 2002). In addition to finding that DAF is expressed in epithelial cells covering the urinary tract and therefore supporting the theory of receptor–adhesin-mediated ascending urinary tract infection (UTI), DAF was discovered in the lumen of endometrial glands. Intriguingly, immunohistochemical staining of uterine cryostat preparations showed that only some of the uterine glands were positively stained with Dr adhesin or anti-DAF IH4 monoclonal antibody (Kaul et al. 1994). This observation implied a variable expression of Dr receptors/DAF in the endometrium. Because the endometrium undergoes cyclic changes controlled by steroid hormones, variable DAF staining in endometrial glands suggested indirectly that DAF expression in these glands might be under steroid control. This concept was supported by our results indicating a variable expression level (low, high) of Dr receptors/DAF in

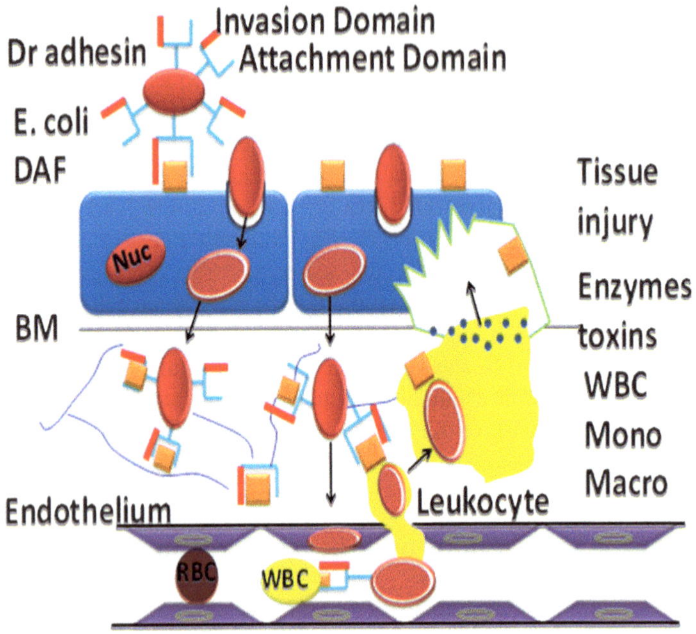

Fig. 5.1 *Depicts our proposed DAF receptor – Dr ligand-mediated invasion by Dr⁺ E. coli.* The invasion domain of Dr adhesin (in *blue*) bound to *E. coli* (in *red*) recognizes cell surface DAF receptors (in *orange*). The attachment, inflammatory process, and steroid hormones enhance DAF densities and increase invasion rates. The invasion domain of Dr adhesin facilitates safe entry into cells, thereby avoiding destruction by phagolysosomal fusion. The *E. coli* Dr adhesin collagen IV domain (not shown) allows the bacteria to cross the basement membrane (*BM*) and establish an interstitial colonization, followed by invasion of the endothelium, and dissemination into the blood stream. The resulting IL8 inflammation and transmigration of leukocytes/monocytes is unable to eliminate the Dr⁺ *E. coli* resulting in tissue damage. The invasion process is effected by steroid hormones including progesterone, which increase DAF density and enhance colonization, and corticosteroids, which decrease DAF density and/or intracellular *E. coli* survival

endometrial biopsies of several patients (Kaul et al. 1994). The clinical implications of variable receptor density were that as a result, these individuals may have either a high or low sensitivity to *E. coli* and/or viral colonization and infection.

5.6 DAF Expression and Regulation by Progesterone

In order to explore the possible steroid hormones' regulatory roles in DAF expression, Nowicki, Nagamani, and colleagues performed a study on endometrial biopsies of healthy control women during the menstrual cycle. The menstrual cycle of the female endometrium is divided into two major phases: menstrual cycle day 1–14 is the proliferative phase, and days 15–28, the secretory phase. Several biopsies that were blindly evaluated by a pathologist for DAF expression using purified Dr adhesin or anti-DAF antibodies showed that the proliferative phase was associated with very low Dr receptor/DAF levels; only a minor proportion of the glands were stained. During the secretory phase, there is an upregulation of Dr receptors/DAF that reached a maximum between days 20 and 26, a time when 100% of glands stained the glandular luminal epithelial compartment (Fig. 5.2). Since the secretory phase is under the control of P4, we proposed that Dr receptor/DAF expression is most likely regulated by progesterone (Fig. 5.3) (Kaul et al. 1996; Young et al. 2002). These findings had a much

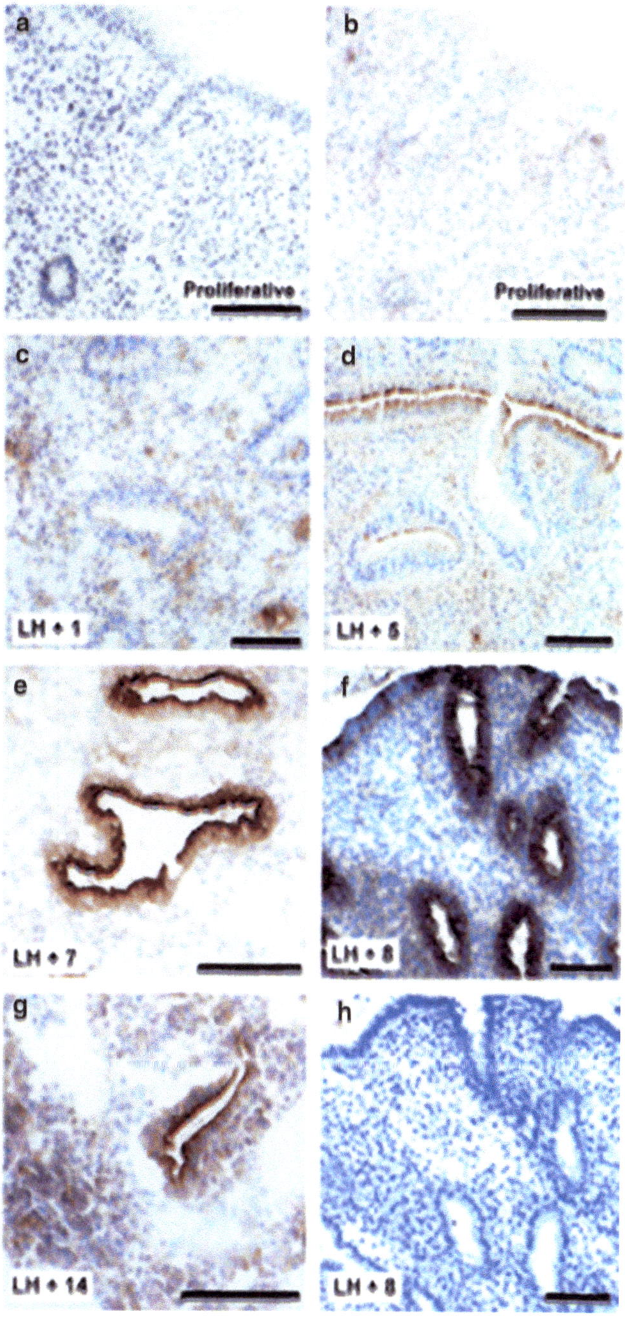

Fig. 5.2 *Immunohistochemical detection of DAF in human endometrium showing an increase DAF density with consecutive days of the secretory phase of the menstrual cycle.* (**a**) Very weak endometrial staining for DAF in the late proliferative epithelium as well as gradual increase of DAF staining on the indicated days of secretory phase dated after urinary detection of the LH surge (**c–g**). Staining controls using no primary antibody (a) or an isotype control primary antibody (**h**) were performed on sections semi-adjacent to (b) and (f), respectively. Note that staining first appears on the luminal epithelium (**d**), and then intensity of staining gradually increase in consecutive days of secretory phase identified as LH + day number (**e–g**). *Scale bar*, 100 μm (From Young et al. 2002)

broader clinical implication than that for menstrual cycle expression of DAF alone. They showed the first example of a relatively rapid cyclic change of a tissue receptor potentially involved in the pathogenesis of urogenital infections and suggested that this cyclic change may impact the temporary resistance of human tissues to cytotoxic complement injury. Therefore, the same individual may change their susceptibility to infection and/or complement injury, secondary to extant steroid hormone(s) levels and the resulting DAF receptor expression levels.

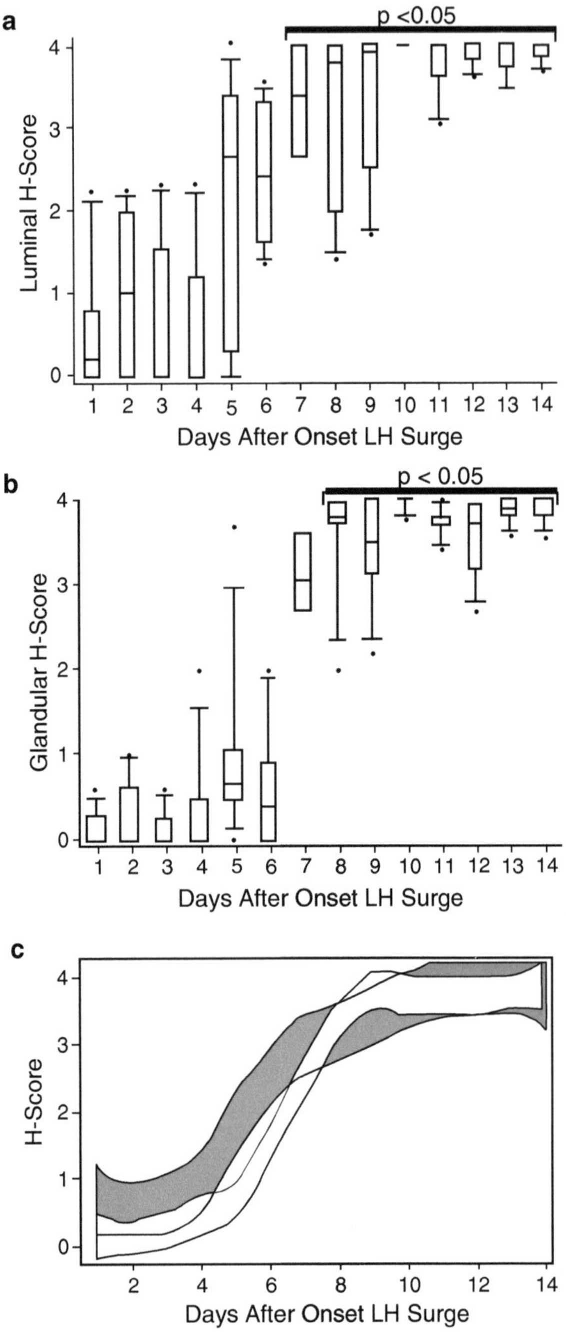

Fig. 5.3 *Graphic representation of endometrial epithelial staining for DAF.* Note the gradual increase in DAF densities with progression of the secretory phase. (H-score on "y" axis represents semiquantitative histology score.)The secretory phase day is expressed by assigning the onset of the LH surge to day = 0. Box plot representations of (**a**) luminal staining and (**b**) glandular staining are shown. On each box, the median (*middle line*), 25th and 75th percentiles (*box edges*), and 10th and 90th percentiles (*whiskers*) are represented. (**c**) A smoothed curve with 95th percentile ranges on which the comparison of the curves is based 0 (From Young et al. 2002)

5.7 DAF, Luteal Phase Defect (LPD), and Progesterone-Regulated DAF Expression

LPD is a clinical syndrome that affects certain infertile patients and involves dysregulation of endometrial function. Often low P4 levels accompany the syndrome, resulting in a delay in the endometrial secretory phase. Treatment with P4 frequently improves and/or results in a resolution of LPD and/or infertility. We interrogated the hypothesis that pregnancy losses in LPD patients may be associated with low DAF expression and complement-mediated injury to the endometrial pregnancy implant (rejection). In an attempt to better understand the suspected role of DAF in LPD, we investigated several patients with LPD before and after P4 treatment (Kaul et al. 1995). Using computer image quantitative analysis of DAF-stained biopsies, we found that before treatment, LPD patients had a low DAF staining pattern, while after treatment with P4, DAF expression was restored to the high level typical for the secretory phase. Several patients also resolved their medical issues posttreatment. Mean DAF receptor density in LPD patients was 15% compared to 60% in the control group. However, the mean DAF receptor density in LPD patients was 88% after treatment with P4. Results support an interpretation that P4 upregulates DAF, while a decreased P4 may be associated with a reduced expression of the DAF in LPD patients. The conclusion of this study was that DAF expression in the endometrium is under direct or an indirect control of P4 (Kaul et al. 1995). The clinical implication of this study was that in the presence of low DAF, complement mediates injury to the implant, resulting in pregnancy loss, and that P4 treatment restores protective DAF levels and contributes to successful pregnancy outcome.

5.8 Paracrine Regulation of DAF Expression with HB-EGF (Heparin-Binding EGF-like Growth Factor)

The physiology of the menstrual cycle and the P4 responsiveness of the endometrium depend upon the presence of progesterone receptors (PRs). Intriguingly, in the secretory phase of the menstrual cycle, PRs are not present in the glandular epithelium of endometrial glands. Instead, they are expressed in the subepithelial compartment of the endometrium, namely, the stroma. This suggests that there may be an indirect regulatory mechanism of P4 on epithelial expression of DAF. We therefore interrogated whether or not there was a paracrine mechanism in which P4 would act first on PRs in the endometrial stroma; in turn, the P4-responsive stroma could secrete epidermal growth factor (EGF), which would act in a paracrine fashion on the endometrial epithelium and upregulate DAF expression. An endometrium-specific subclass of EGF is HB-EGF. Therefore, the expression of DAF was tested in the endometrial cell line RL95-2. Results showed a HB-EGF dose-dependent expression of DAF (Fig. 5.4) (Young et al. 2002). This experimental strategy suggested a P4–EGF interaction and, as a result, a paracrine regulation of DAF. This requires further study.

5.9 DAF, Endometrial Carcinoma, and Estrogen

Although the biology of the reproductive tract is controlled by E2 and P4, the exact mechanisms involved in DAF regulation are not well studied (Kaul et al. 1994). Some mechanistic explanations of E2-mediated regulation of DAF stream from experimental research using either an in vivo mouse cancer model or examining DAF expression in vitro using human-derived cancer cells. Investigators found that neonatal mice exposed to E2 leads to an increased occurrence of uterine tumors (Song et al. 1996). Experimental studies were performed to screen for estrogen-regulated genes in the uterus of the CD-1 neonatal mouse, and the DAF gene in this neonatal mouse model appeared to be estrogen inducible. The induction of mouse DAF by E2 was tissue specific and was mimicked by the antiestrogen tamoxifen. Furthermore, the regulation of uterine DAF expression by E2 was limited to the GPI-anchored DAF gene. This suggests that GPI-anchored DAF may play an important role in E2 responses and other physiologic or pathophysiologic processes in the female reproductive system.

A human example of upregulation of DAF in pathologic conditions is overexpression in endometrial adenocarcinoma (Nowicki et al. 2001). DAF expression on epithelial cells varies between 1% and

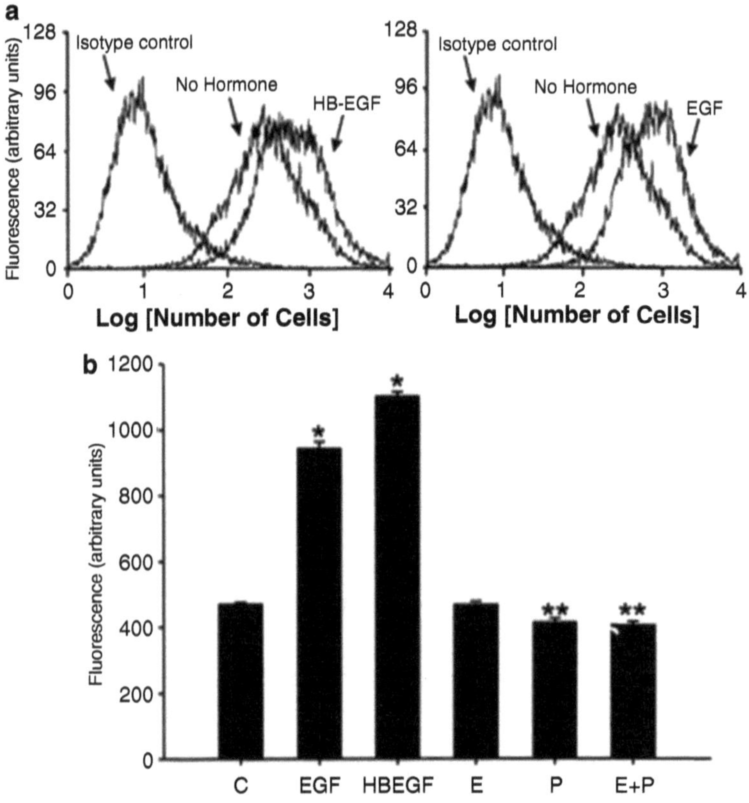

Fig. 5.4 *HB-EGF and hormonal regulation of DAF expression assessed by flow cytometry, measured using fluorescent staining with anti-DAF IH4 antibody.* RL95-2 cells were serum-starved for 48 h and then treated for 24 h with carrier (*C*), 10 ng/ml EGF, 10 ng/ml HB-EGF, 10 nM E2 (*E*), 100 nM progesterone (*P4*), or 10 nM E2 and 100 nM P (*E + P*). (**a**) Representative histograms showing specificity of antibody (isotype control *vs.* untreated) and effects of HB-EGF or EGF treatment. (**b**) Quantitation of triplicate experiments performed in parallel. The median of each peak was used as a measure of DAF expression, and the *error bars* represent ±1 SD. ANOVA with Tukey *post hoc* comparison was used (differences from control: $*P<0.0001$; $**P<0.005$) (From Young et al. 2002)

30% in normal glandular endometrial epithelium. In contrast, DAF density in patients with stage I cancer was in the range 56–98% (mean 78%), and stage III values varied from 16% to 28% (mean 21%), $P<0.05$. DAF density in the well-differentiated Ishikawa cell line was also twofold higher than in metastatic cell line AN3CA. These findings were consistent with the hypothesis that early stage endometrial adenocarcinoma exposed to complement attack may upregulate DAF to protect malignant cells from complement lysis. Again, the endometrial cancer tissue appears to upregulate DAF to escape complement destruction, but the mechanisms of DAF regulation and the roles of E2 and/or P4 in this malignant process are not well understood.

5.10 DAF, RBC, and Estrogen

In addition to Dr^+*E. coli*, DAF is also an RBC receptor recognized by multiple viruses, including group B coxsackieviruses (CVB). A direct contact of Dr^+*E. coli* or a CBV virus results in hemagglutination of RBCs. Therefore, studies were performed, using blood samples from young females, to investigate the role of hormones in DAF RBC expression and the capacity of CVB to hemagglutinate

RBC (Sartini et al. 2004). RBCs were collected on days 11 and 22 of the menstrual cycle. Samples were categorized into luteal and follicular phases based on serum P4 (either <2.0 ng/mL for follicular or ≥2.0 ng/mL for luteal) and analyzed by flow cytometry for DAF expression on RBC and CD21+ B lymphocytes. Tested subjects showed significant variations in CVB-induced hemagglutination and % RBC or CD19+ cells that were DAF+. The authors reported a correlation between the levels of E2 and the level of DAF expressed on RBCs ($P<0.01$) in the luteal phase but not in the follicular phase, suggesting that females may show E2-dependent expression of DAF in the circulatory system. These findings may fit with an apparent P4-dependent DAF expression in the human endometrium in the secretory (luteal) phase. Unfortunately, the role of P4 on DAF expression during the luteal phase was not studied by these authors. The interplay of E2-/P4-mediated responses underscores the complexity of the hormonal regulation of DAF which appears to be both E2 and/or P4 specific in human RBCs versus P4 and/or E2 in the endometrium and menstrual cycle phases.

5.11 DAF, Hormone Replacement Therapy, and Risk of Infection

Hormone replacement therapy (HRT) is broadly used in postmenopausal patients and is thought to reduce the risk for urinary tract infection (UTI) and to improve vaginal atrophy in these subjects. In particular, E2 has been implicated in this protective role. On the other hand, some clinical epidemiologic studies have suggested, to the contrary, that there is an increased risk for UTI following E2 replacement (Orlander et al. 1992). Therefore, using ovariectomized mice, we developed an experimental protocol to study the role of E2 replacement in protection from UTI (Curran et al. 2007). To our surprise, mice with experimental ascending UTI established by an introduction of a single dose of *E. coli* via a bladder catheterization developed more severe levels of infection as defined by higher *E. coli* colony-forming unit (CFU) counts if given E2 than without E2. The E2 increased CFU counts of virulent Dr^{+} *E. coli* which recognized DAF but not if we used *E. coli* Dr adhesin mutants, suggesting an E2 virulence-specific process. Although the status of DAF expression in E2-treated ovariectomized animals was not tested, the studies alluded above using the neonatal cancer mouse model suggested that E2 supplementation upregulates DAF expression (Song et al. 1996). While the full implications of E2 supplementation are not clear as yet, studies to date suggest a need for careful clinical evaluation of patients' risk versus benefit for UTI in the context of E2 replacement.

5.12 DAF, P4, Nitric Oxide (NO), and Sensitivity to Experimental Infection in Pregnancy

P4 and subsequently NO are upregulated in pregnancy, leading to uterine muscle relaxation and fetal protection from infection, preterm contractions, and/or preterm labor. Interestingly, the NO system may contribute to the maintenance of uterine quiescence during pregnancy, when P4 levels are elevated, but not during delivery (Yallampalli et al. 1994). We discovered in both epithelial cell in vitro models and in a pregnant rat model of uterine infection that NO regulates DAF expression (Nowicki et al. 1997; Fang et al. 2004, 2001). The inhibition of NO in vitro and in vivo resulted in twofold increase of DAF expression, a twofold increase of E. coli attachment and invasion of epithelial cells, and increased CFU counts in infected pregnant rats, in contrast to providing a source of NO (L-arginine)-downregulated DAF expression in both epithelial HeLa cells and the rat endometrium and reduced *E. coli* invasion rates. We observed that *E. coli*-infected nonpregnant animals had an increase in endometrial NO production and NO-reduced DAF expression, and the infection was resolved. In contrast, in pregnant animals *E. coli* infection did not result in upregulation of NO; we observed a NO nonresponsiveness and no change in DAF levels which led to an increase in

E. coli-induced mortality (Nowicki et al. 1997; Fang et al. 2004, 2001). These results suggest that host–pathogen interactions and the outcome of infection involve DAF expression and multiple signaling molecules including hormones that play a significant role, either directly or indirectly, by regulation of NO, HB-EGF, or other signaling molecules.

5.13 DAF, Progesterone, and Human Sensitivity to Renal Infection

Progesterone and DAF expression have been indirectly implicated in human sensitivity to pyelonephritis. We performed clinical studies on pregnant patients with pyelonephritis and noted that *E. coli* bearing Dr adhesin which recognized DAF caused up to 40% of the pyelonephritis cases diagnosed during the third trimester of pregnancy (Fig. 5.5) (Hart et al. 2001). In contrast, the first and the second trimester pyelonephritis cases were less likely to be associated with DAF-specific Dr$^+$*E. coli.*

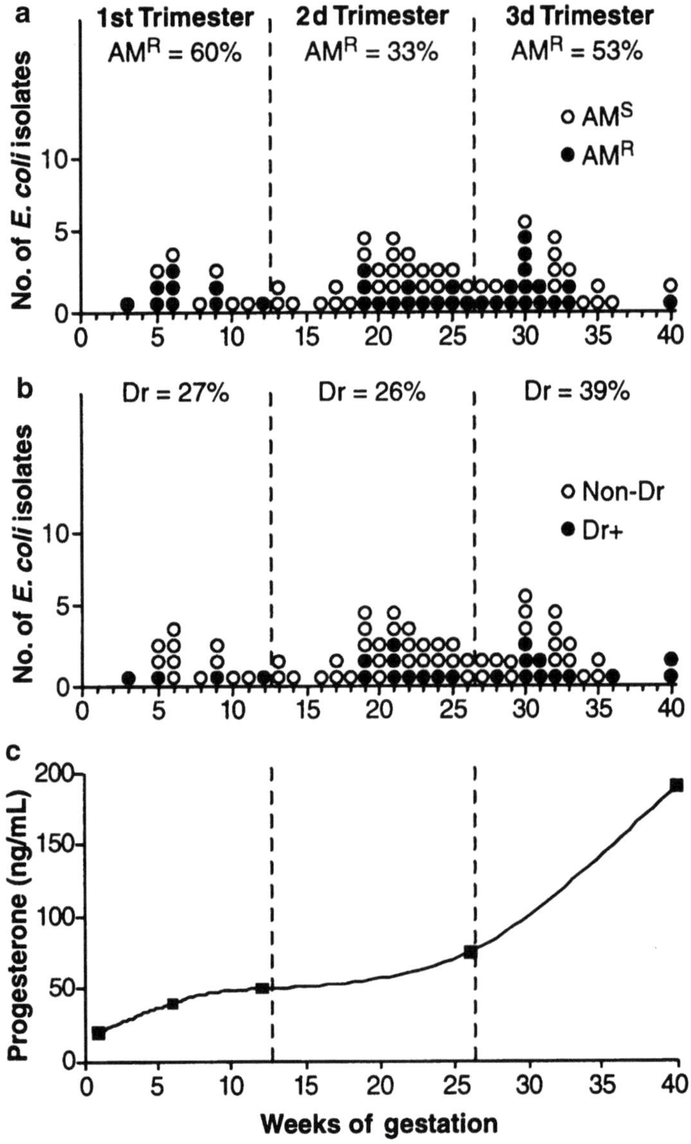

Fig. 5.5 *Pregnant patients with pyelonephritis: increase rate of infection with Dr$^+$ E. coli in third trimester coincides with an increase in P4 levels.* (**a**) The gestational distribution of ampicillin-resistant (*AM^R*) and ampicillin-sensitive (*AM^S*) *E. coli*. (**b**) Dra+and Dra−*Escherichia coli* isolated from pregnant patients with pyelonephritis. (**c**) The expected increase of P4 serum levels during pregnancy (From Hart et al. 2001)

Interestingly, the level of P4 sharply increased in the third trimester. Because both in vitro and in vivo pregnant animal studies showed that the increase in DAF density is associated with an increase of Dr^+*E. coli* attachment and invasion of uroepithelial cells, we postulated that the increased frequency of pyelonephritis late in pregnancy could at least be in part associated with the sharp increase in P4 levels and DAF densities (Holmes et al. 1992; Nowicki et al. 1988c; Selvarangan et al. 2000). In fact, DAF density sharply increases in the placental trophoblast during pregnancy to protect the fetus from maternal complement attack, a process that appears to be P4 dependent (Holmes et al. 1992). We have experimentally proven that both the in vitro attachment and invasion of epithelial cells are directly related to DAF density (Selvarangan et al. 2000). The implications of these findings are that while the maternal hormonal milieu upregulates DAF for fetal protection, pathogens increase DAF-mediated colonization and the risk of infection. This paradoxical phenomenon of increased protection paralleled by a simultaneous increased risk of infection should heighten the level of clinical attention to pregnant patients at risk for UTI and associated gestational complications.

5.14 DAF and Preterm Labor (PTL)

Intrauterine inflammation and subclinical infection is a frequent and significant factor associated with the pathogenesis of preterm labor/preterm birth (PTL/PTB). However, it remains unclear whether the intrauterine inflammatory responses activate the maternal peripheral circulation. We explored the association between PTL/PTB and the "activation" of the peripheral circulatory system by determining whether DAF/CD55 mRNA expression within peripheral WBCs differed between PTL and control patients not in labor (Nowicki et al. 2009; Pacheco et al. 2011). The mean DAF mRNA level within the PTL group (0.77±0.03) was 1.48-fold higher than that observed (0.52±0.02) within the control group ($P<0.0001$); 71% of PTL patients and only 6.7% of control subjects expressed elevated DAF mRNA. In the patient population that delivered prematurely (before 37 weeks), 81% expressed elevated DAF mRNA levels with a mean of 0.78±0.03 and 95% CI of 0.71–0.84. Therefore, we reported for the first time that DAF mRNA expression was elevated in the peripheral WBCs of subjects with PTL compared with control gestationally matched pregnant women and that elevated leukocyte DAF may be a useful predictor of subsequent PTB (Fig. 5.6). The mechanism of DAF

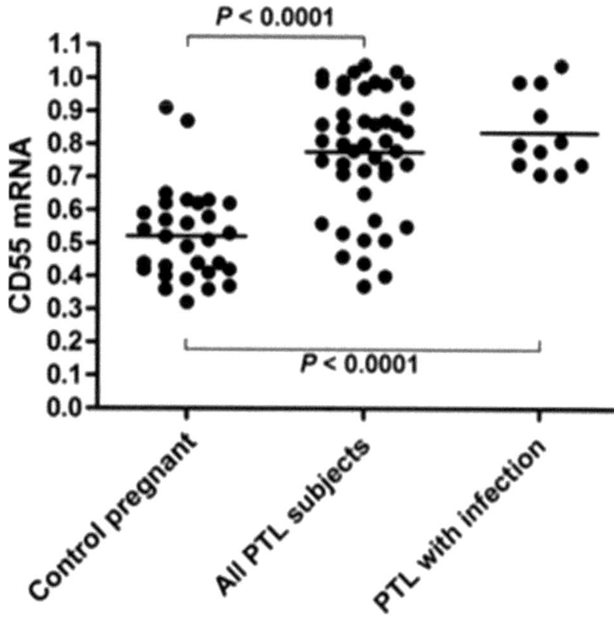

Fig. 5.6 *DAF/CD55 mRNA levels are elevated in peripheral white blood cells of infection-associated and idiopathic preterm labor (PTL) subjects compared to controls not in labor.* The normalized integrated density values for CD55 mRNA levels are plotted for PTL (weeks 24–37 of gestation) and gestationally matched control subjects as indicated; the *horizontal bar* indicates the mean value. The *P* value was determined using a two-tailed *t*-test (From Nowicki et al. 2009)

upregulation on WBCs in PTL may be multifactorial. Of interest is that patients at risk for PTL treated with 17-OH-progesterone showed a reduction of PTL. The reduction of PTL with 17-OH-progesterone, in the context of other studies with P4, may suggest a possible link with DAF's responsiveness to steroids and/or protective role in PTL (Pacheco et al. 2011).

5.15 Corticosteroids, DAF, and Host–Pathogen Interactions

Corticosteroids such as dexamethasone are often used to stimulate fetal lung development in pregnant women with PTL. Infection, including urinary tract infections, increases the risk for PTL, but the roles of hormones, pregnancy, and the risk for infection remain underinvestigated. *E. coli* bearing Dr adhesins (Dr^+*E. coli*) often cause pyelonephritis in pregnant women. The strains bearing adhesins of the Dr family bind to DAF, thereby invading epithelial cells and causing chronic or recurrent gestational infections (Hoffmann 1969). Considering that steroid hormones often enhance the infection process, we hypothesized that treatment with the anti-inflammatory dexamethasone will increase DAF, leading to increase in the invasion of host cells by *E. coli*. Thus, we treated HeLa cells with dexamethasone (1 μM) for 24 h, and DAF levels were checked by real-time PCR, western blotting, and immunofluorescence. Treated and untreated HeLa cells were then exposed to *E. coli*, and results of invasion and survival were compared. In contrast to our prediction, our preliminary results suggest that DAF mRNA and protein levels significantly decreased in HeLa cells treated with dexamethasone. *E. coli* internalization assays in treated cells also showed a significant reduction in invasion capacity and intracellular survival as compared to control cells. Unexpectedly, these preliminary results indicated that dexamethasone reduced DAF expression and E. coli intracellular persistence and therefore, exhibited a promising therapeutic property that potentially could be beneficial for management of some gestational infections (Nowicki et al. 2011b).

5.16 Regulation of DAF

Although DAF is ubiquitously expressed on epithelial cells and within the peripheral blood system, a mechanistic understanding of E2-/P4-regulated expression in leukocytic cell types remains relatively unexplored. Complement activation is associated with the pathogenesis of PTL and inflammatory cardiovascular and irritable bowel diseases. Studies of DAF/CD55 regulation in vascular endothelial and quiescent colonic epithelial cell lines have indicated that DAF/CD55 expression is modulated by the activation of PLCγ/PKC, p38-MAPK, PKA, and PI3K/Akt cell signaling pathways and by a hypoxia-induced stress response (Mason et al. 2004; Louis et al. 2005a; Holla et al. 2005; Spiller et al. 2000; Andoh et al. 1997). A lipopolysaccharide (LPS) signaling cascade in monocytic cell types utilizes the bacterial molecule pattern receptor-associated MyD88/interleukin-1 receptor-associated kinases which results in an early phase activation of NFκB (Akira 2003). LPS-induced expression of CD55 has been observed using a heterogeneous population of primary mononuclear cells, and CD55 has been proposed as an NFκB target gene in monocytic lineages in global LPS-induced NFκB genome-binding studies (Schreiber et al. 2006). Our laboratory currently is exploring the role of steroid hormones including E2, P4, and vitamin D3 in transcriptional regulation of DAF expression. Thus far, we have observed that P4 plus LPS may potentiate temporary CD55 response, a change that occurs within 2 h of treatment but does not reach statistical significance with the concentrations used. Further studies are in progress to characterize DAF/CD55 monocytic responses to steroid hormones in pregnant patients with the long-term goal for designing supplemental therapeutic strategies that would improve chronic inflammatory/infectious processes including PTL.

5.17 Summary: Implications of Hormone Therapy for DAF Expression and Infection

DAF/CD55 plays multiple physiologic functions including tissue protection from the cytotoxic complement injury, an anti-inflammatory function due to its anti-adherence properties which enhance transmigration of monocytes and macrophages and reduce tissue injury (Louis et al. 2005b). DAF physiologic functions are essential in pregnancy for protection of the semiallogeneic fetus or for prevention of uncontrolled infiltration by white cells in their pro- and/or anti-inflammatory functions. DAF expression appears to have multiple regulatory tissue-specific and/or menstrual cycle-specific mechanisms, which involve complex signaling mechanisms. Regulation of DAF expression may involve a direct or an indirect effect of at least the E2, P4, and corticosteroid regulatory pathways. DAF is exploited in multiple pathologic conditions by pathogens and viruses in chronic tissue infection processes. Further understanding of such processes should be instrumental in designing therapeutic strategies for multiple conditions involving DAF protective or pathologic functions and tailoring host expression of DAF.

References

Akira S (2003) Toll-like receptor signaling. J Biol Chem 278:38105–38108

Anderson GG, Palermo JJ, Schillin JD, Roth R, Heuser J, Hultgren SJ (2003) Intracellular bacterial biofilm-like pods in urinary tract infections. Science 301:105–107

Andoh A, Fujiyama Y, Sumiyoshi K, Sakumoto H, Okabe H, Bamba T (1997) Tumour necrosis factor-alpha up-regulates decay-accelerating factor gene expression in human intestinal epithelial cells. Immunology 90:358–363

Bergelson JM, Chan M, Solomon KR, St John NF, Lin H, Finberg RW (1994) Decay-accelerating factor (CD55), a glycosylphosphatidylinositol-anchored complement regulatory protein, is a receptor for several echoviruses. Proc Natl Acad Sci USA 91:6245–6249

Bergelson JM, Mohanty JG, Crowell RL, St John NF, Lublin DM, Finberg RW (1995) Coxsackievirus B3 adapted to growth in RD cells binds to decay-accelerating factor (CD55). J Virol 69:1903–1906

Curran E, Tassell AH, Judy BM, Nowicki B, Montgomery-Rice V, Estes DM, Nowicki S (2007) Estrogen increases menopausal host susceptibility to experimental ascending urinary tract infection. J Infect Dis 195:680–683

Fang L, Nowicki B, Yallampalli C (2001) Differential expression of uterine nitric oxide system in pregnant and non-pregnant rats with intrauterine bacterial infection. Am J Physiol Regul Integr Comp Physiol 280:R1356–R1363

Fang L, Nowicki BJ, Urvil P, Goluszko P, Nowicki S, Young SL, Yallampalli C (2004) Epithelial invasion by *Escherichia coli* bearing Dr fimbriae is controlled by nitric oxide-regulated expression of CD55. Infect Immun 72:2907–2914

Goluszko P, Moseley S, Truong LD, Kaul A, Nowicki S, Nowicki B (1997) Development of experimental model of chronic pyelonephritis with *Escherichia coli* O75:K5:H- bearing Dr fimbriae: mutation in the dra region prevented tubulointerstitial nephritis. J Clin Invest 99:1–11

Hart A, Nowicki BJ, Reisner B, Pawelczyk E, Goluszko P, Urvil P, Anderson G, Nowicki S (2001) Ampicillin-resistant *Escherichia coli* in gestational pyelonephritis: increased occurrence and association with the colonization factor Dr adhesin. J Infect Dis 183:1526–1529

Hoffmann EM (1969) Inhibition of complement by a substance isolated from human erythrocytes. I. Extraction from human erythrocyte stroma. Immunochemistry 6:391–403

Holla VR, Wang D, Brown JR, Mann JR, Katkuri S, DuBois RN (2005) Prostaglandin E2 regulates the complement inhibitor CD55/decay accelerating factor in colorectal cancer. J Biol Chem 280:476–483

Holmes CH, Simpson KL, Okada H, Okada N, Wainwright SD, Purcell DF, Houlihan JM (1992) Complement regulatory proteins at the feto-maternal interface during human placental development: distribution of CD59 by comparison with membrane cofactor protein (CD46) and decay accelerating factor (CD55). Eur J Immunol 22: 1579–1585

Kaul AK, Nowicki BJ, Martens MG, Goluszko P, Hart AR, Nagamani M, Kumar D, Pham T, Nowicki S (1994) Decay accelerating factor (DAF) is expressed in the human endometrium and may serve as the natural ligand for Dr pili of *Escherichia coli*. Am J Reprod Immunol 32:194–199

Kaul A, Nagamani M, Nowicki B (1995) Decreased expression of endometrial decay accelerating factor (DAF), a complement regulatory protein, in patients with luteal phase defect. Am J Reprod Immunol 34:236–240

Kaul A, Goluszko PG, Hart A, Martens M, Pham T, Nowicki S, Nowicki B (1996) Rapid cycle changes in density and accessibility of endometrial ligands for *Escherichia coli* Dr fimbriae. Infect Immun 64:611–615

Louis NA, Kong T, Hamilton KE, Colgan SP (2005a) HIF-dependent induction of apical CD55 coordinates epithelial clearance of neutrophils. FASEB J 19:950–959

Louis NA, Hamilton KE, Kong T, Colgan SP (2005b) HIF-dependent induction of apical CD55 coordinates epithelial clearance of neutrophils. FASEB J 19:950–959

Mason JC, Steinberg R, Lidington EA, Kinderlerer AR, Ohba M, Haskard DO (2004) Decay-accelerating factor induction on vascular endothelium by vascular endothelial growth factor (VEGF) is mediated via a VEGF receptor-2 (VEGF-R2)- and protein kinase C-alpha/epsilon (PKCalpha/epsilon)-dependent cytoprotective signaling pathway and is inhibited by cyclosporin A. J Biol Chem 279:41611–41618

Moulds JM, Nowicki S, Nowicki BJ, Moulds JJ (1996) Human blood groups: incidental receptors for viruses and bacteria. Transfusion 36:362–374

Nicholson-Weller A, March JP, Rosenfeld SI, Austen KF (1983) Affected erythrocytes of patients with paroxysmal nocturnal hemoglobinuria are deficient in the complement regulatory protein, decay accelerating factor. Proc Natl Acad Sci USA 80:5066–5070

Nowicki B (2002) Urinary tract infection in pregnant women: old dogmas and current concepts. Curr Infect Dis Rep 4:529–535

Nowicki B, Hull R, Moulds J (1988a) Use of the Dr hemagglutinin of uropathogenic *Escherichia coli* to differentiate normal from abnormal red cells in paroxysmal nocturnal hemoglobinuria. N Eng J Med 319:1289–1290

Nowicki B, Moulds J, Hull R, Hull S (1988b) A hemagglutinin of uropathogenic *E. coli* recognized the Dr blood group antigen. Infect Immun 56:1057–1060

Nowicki B, Truong L, Moulds J, Hull R (1988c) Presence of the Dr receptor in normal human tissues and its possible role in pathogenesis of ascending urinary tract infection. Am J Pathol 133:1–4

Nowicki B, Fang L, Singhal S, Nowicki S, Yallampalli C (1997) Lethal outcome of uterine infection in pregnant but not in nonpregnant rats and increased death rate with inhibition of nitric oxide. Am J Reprod Immunol 38:309–312

Nowicki S, Nowicki B, Pham T, Hasan R, Nagamani M (2001) Expression of decay accelerating factor (DAF) in endometrial adenocarcinoma is inversely related to the stage of tumor. Am J Reprod Immunol 46:144–148

Nowicki S, Izban M, Pawelczyk E, Agboto V, Pratap S, Olson G, Nowicki B (2009) Preterm labor: CD55 in maternal blood leukocytes. Am J Rep Immunol 61:360–367

Nowicki B, Sledzinska A, Samet A, Nowicki S (2011a) Pathogenesis of gestational UTI: urinary obstruction vs. immune adaptation and microbial virulence. BJOG 218:109–112

Nowicki S, Rana T, Singh R, Nowicki B (2011) Dexamethasone treatment downregulates DAF and decrease intracellular survival of Dr + *E. coli* in HeLa cells. In: Proceedings of the 5th international conference on complement therapeutics, abstract 18, Rhodes, 22–27 June 2011

O'Brien DP, Israel DA, Krishna U, Romero-Gallo J, Nedrud J, Medof ME, Lin F, Redline R, Lublin DM, Nowicki BJ, Franco AT, Ogden S, Williams AD, Polk DB, Peek RM Jr (2006) The role of decay-accelerating factor as a receptor for *Helicobacter pylori* and a mediator of gastric inflammation. J Biol Chem 281:13317–13323

Orlander JD, Jick SS, Dean AD, Jick H (1992) Urinary tract infections and estrogen use in older women. J Am Geriatr Soc 40:817–820

Pacheco LD, Hankins GD, Costantine MM, Anderson GD, Pawelczyk E, Nowicki S, Nowicki BJ (2011) The role of human decay-accelerating factor in the pathogenesis of preterm labor. Am J Perinatol 28:565–570

Pangburn MK, Schreiber RD, Muller-Eberhard HJ (1983) Deficiency of an erythrocyte membrane protein with complement regulatory activity in paroxysmal nocturnal hemoglobinuria. Proc Natl Acad Sci USA 80:5430–5434

Sartini D, Moussawi M, Sallam R, Bernstein I, Huber S (2004) Correlation between serum estradiol in the follicular phase of the ovarian cycle and decay acceleration factor (DAF) expression on red blood cells and coxsackievirus B-3 induced hemagglutination in young cycling women. Am J Reprod Immunol 51:180–187

Schreiber J, Jenner RG, Murray HL, Gerber GK, Gifford DK, Young RA (2006) Coordinated binding of NF-κB family members in the response of human cells to lipopolysaccharide. Proc Natl Acad Sci USA 103:5899–5904

Selvarangan R, Goluszko P, Popov V, Singhal J, Pham T, Lublin DM, Nowicki S, Nowicki B (2000) Role of decay-accelerating factor domains and anchorage in internalization of Dr. fimbriated *Escherichia coli*. Infect Immun 68:1391–1399

Śledzińska A, Mielech A, Krawczyk B, Samet A, Nowicki B, Nowicki S, Jankowski Z, Kur J (2011) Fatal sepsis in a pregnant woman with chronic pyelonephritis caused by *Escherichia coli* bearing Dr and P adhesins: diagnosis based on postmortem strain genotyping. BJOG 18:266–269

Song WC, Deng C, Raszmann K, Moore R, Newbold R, McLachlan JA, Negishi M (1996) Mouse decay-accelerating factor: selective and tissue-specific induction by estrogen of the gene encoding the glycosylphosphatidylinositol-anchored form. J Immunol 157:4166–4172

Spiller OB, Criado-Garcia O, Rodriguez DC, Morgan BP (2000) Cytokine-mediated up-regulation of CD55 and CD59 protects human hepatoma cells from complement attack. Clin Exp Immunol 121:234–241

Yallampalli C, Izumi H, Byam-Smith M, Garfield RE (1994) An L-arginine-nitric oxide-cyclic guanosine monophosphate system exists in the uterus and inhibits contractility during pregnancy. Am J Obstet Gynecol 170:175–185

Young SL, Lessey BA, Fritz MA, Meyer WR, Murray MJ, Speckman PL, Nowicki BJ (2002) In vivo and in vitro evidence suggests that HB-EGF regulates endometrial expression of human decay-accelerating factor. J Clin Endocrinol Metab 87:1368–1375

Chapter 6
Targeting gC1qR Domains for Therapy Against Infection and Inflammation

Berhane Ghebrehiwet, Jolyon Jesty, Rama Vinayagasundaram, Uma Vinayagasundaram, Yan Ji, Alisa Valentino, Nithin Tumma, Kinga H. Hosszu, and Ellinor I.B. Peerschke

Abstract The receptor for the globular heads of C1q, gC1qR/p33, is a widely expressed cellular protein, which binds to diverse ligands including plasma proteins, cellular proteins, and microbial ligands. In addition to C1q, gC1qR also binds high molecular weight kininogen (HK), which also has two other cell surface sites, namely, cytokeratin 1 and urokinase plasminogen activator receptor (uPAR). On endothelial cells (ECs), the three molecules form two closely associated bimolecular complexes of gC1qR/cytokeratin 1 and uPAR/cytokeratin 1. However, by virtue of its high affinity for HK, gC1qR plays a central role in the assembly of the kallikrein–kinin system, leading to the generation of bradykinin (BK). BK in turn is largely responsible for the vascular leakage and associated inflammation seen in angioedema patients. Therefore, blockade of gC1qR by inhibitory peptides or antibodies may not only prevent the generation of BK but also reduce C1q-induced or microbial-ligand-induced inflammatory responses. Employing synthetic peptides and gC1qR deletion mutants, we confirmed previously predicted sites for C1q (residues 75–96) and HK (residues 204–218) and identified additional sites for both C1q and HK (residues190–202), for C1q (residues 144–162), and for HIV-1 gp41 (residues 174–180). With the exception of residues 75–96, which is located in the αA coiled-coil N-terminal segment, most of the identified residues form part of the highly charged loops connecting the various β-strands in the crystal structure. Taken together, the data support the notion that gC1qR could serve as a *novel* molecular target for the design of antibody-based and/or peptide-based therapy to attenuate acute and/or chronic inflammation associated with vascular leakage and infection.

Abbreviations

gC1qR	Receptor for the globular heads of C1q
cC1qR	Receptor for the collagen tail of C1q
HK	High molecular weight kininogen
KKS	Kallikrein–kinin system
BK	Bradykinin

B. Ghebrehiwet (✉) • J. Jesty • R. Vinayagasundaram • U. Vinayagasundaram • Y. Ji • A. Valentino
N. Tumma • K.H. Hosszu
The Department of Medicine, Stony Brook University School of Medicine, Health Sciences Center, T-16; Room 040, Stony Brook, New York, NY 11794-8161, USA
Berhane.Ghebrehiwet@stonybrook.edu

E.I. Peerschke
The Department of Laboratory Medicine, Memorial Sloan-Kettering Cancer Center, New York, NY 10065, USA

J.D. Lambris et al. (eds.), *Complement Therapeutics*, Advances in Experimental Medicine and Biology 735, DOI 10.1007/978-1-4614-4118-2_6, © Springer Science+Business Media New York 2013

ATIII	Antithrombin III
VN	Vitronectin
TAT	Thrombin–antithrombin complex
VNTAT	Vitronectin–thrombin–antithrombin complex

6.1 Introduction

The receptor for the globular heads of C1q, gC1qR, is a ubiquitously distributed, multifunctional cellular protein, which modulates a plethora of immunological processes including inflammation, autoimmunity, infection, and cancer (Ghebrehiwet and Peerschke 2004; Peerschke and Ghebrehiwet 2001). Although first characterized as a receptor for C1q (Ghebrehiwet et al. 1994), experimental evidence accumulated to date convincingly shows that in addition to C1q, gC1qR binds various plasma ligands including high molecular weight kininogen (Joseph et al. 1996; Herwald et al. 1996), factor XII (Joseph et al. 1996), multimeric vitronectin (Lim et al. 1996), fibrinogen (Lu et al. 1999), and thrombin (Ghebrehiwet and Peerschke 2004). More importantly, gC1qR is able to bind diverse pathogen-associated molecular ligands and as such is considered to be a significant pathogen recognition receptor. Pathogenic microorganisms such as *S. aureus* (Nguyen et al. 2000), *B. cereus* (Ghebrehiwet et al. 2007), or *L. monocytogenes* (Braun et al. 2000) in turn bind gC1qR either to gain access into the cell or to induce cellular responses that enhance their survival.

Because gC1qR is involved in either triggering or enhancing inflammatory processes that accompany various disease conditions, including infection, cancer, and autoimmune diseases (Fig. 6.1), identification of the structural domains that enable it to interact with such a diversity of molecules could be useful for the design of therapeutic interventions. First at sites of atherosclerotic and inflammatory and vascular lesions, where gC1qR is heavily expressed and both C1q and HK are present (Peerschke et al. 2004), both ligands participate in the inflammatory process: C1q via activation of complement and inducing the release of proinflammatory cytokines from monocytes and macrophages and HK by FXII-dependent generation of bradykinin. This is particularly relevant in HAE patients, where deficiency in C1-INH—a protease inhibitor, which controls the critical enzymes of both the classical pathway and the kinin-generating system—can exacerbate the vascular leakage and inflammation due to uncontrolled activation of both pathways (Agostoni and Cicardi 1992;

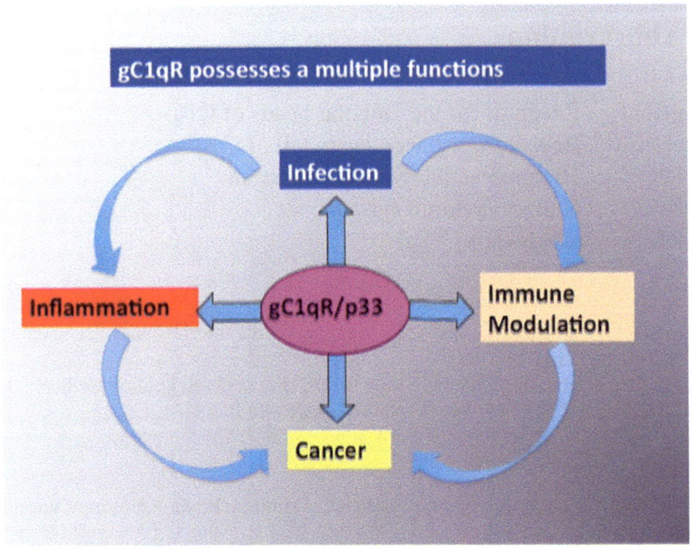

Fig. 6.1 Schematic representation of the multiple functions of gC1qR. While inflammation is at the center of most pathogenic conditions, the crosstalk triggered between infection, cancer, and autoimmune diseases in which gC1qR plays a role is underscored in this model

Kaplan 2004b). In addition, enzymes generated during contact activation, such as FXIIa and FXIIf, can directly activate C1 (Gigli et al. 1971; Ghebrehiwet et al. 1981) and thus enhance the crosstalk between both proinflammatory pathways. Therefore, inhibition of gC1qR interaction with either of its primary ligands can prevent C1q- or HK-induced inflammatory processes. Second, blockade of the interaction of gC1qR with a wide range of pathogenic microorganisms could prevent bacterial entry and dissemination. Therefore, using knowledge derived from analysis of its structure, we generated several gC1qR deletion mutants in order to confirm the previously identified binding sites and/or identify new ones. Identification and refinement of the precise interaction sites between gC1qR and its various ligands, in turn, will allow us to translate this knowledge into novel small molecule-based, peptide-based, or antibody-based diagnostic and/or therapeutic strategies to block the generation of bradykinin and other vasoactive molecules that have been shown to contribute significantly to inflammatory diseases, vascular permeability, and edema (Bossi et al. 2009, 2011).

6.2 Structure of gC1qR

Mature gC1qR (residues 74–282) is extremely acidic with a calculated pI of 4.15 (Ghebrehiwet et al. 1994; Jiang et al. 1999). It has one Cys at residue 186 and thus does not have any intrachain disulfide bonding. It does not dimerize by interchain disulfide bonding either: on SDS-PAGE, it migrates as a 33-kDa band under both reducing and nonreducing conditions. However, it behaves as a trimer on gel filtration in non-dissociating conditions, and evidence from our laboratory suggests that multimer formation is an essential process that increases the affinity of gC1qR for multivalent ligands such as C1q and HK (Ghebrehiwet et al. 1994). The noncovalent trimeric structure has since been confirmed by X-ray crystallography (Jiang et al. 1999), which reveals a doughnut-shaped quaternary structure with asymmetric charge distribution—with one side containing a high distribution of negatively charged residues and which we refer to as the solution face (S-face) and the other side containing a more or less neutral net charge and referred to as the membrane face (M-face). More importantly, the 3D structure of the molecule reveals the presence of several highly charged domains with potential to play a role in ligand binding and cell attachment (Ghebrehiwet et al. 2002).

Although gC1qR is synthesized as a pre-protein of 282 residues by virtue that the 73-aa pre-sequence contains a mitochondrial-targeting sequence, recent evidence suggests that the full-length protein (residues 1–282) may actually be expressed on the membrane rafts, together with CD44, monosialoganglioside, and actin (Kim et al. 2011). This fact was confirmed by experiments in which an antibody generated to a peptide in the pre–pro sequence stained unpermeabilized cells as assessed by flow cytometry (Hosszu et al. 2010a). The cleavage at Ser 73 that forms the putative mature protein in the sequence CGSLHT is probably processed by the same pathway as the intracellular protein, although the enzyme responsible for this cleavage is still unknown (van Leuwen and O'Hare 2001). The primary structure of gC1qR also reveals several important details that may give insight into its function. It has one Cys at residue 186, and although is deeply embedded in the structure and does not make intrachain disulfide bonding, it can nonetheless engage in the formation of supramolecular complexes between trimers under certain conditions. Evidence from our laboratory suggests that multimer formation is an essential process that increases the affinity of gC1qR for multivalent ligands such as C1q and HK (Ghebrehiwet and Peerschke 1998, 2004).

The gC1qR amino acid sequence contains three consensus N-glycosylation sites at residues 114 (NGT), 136 (NNS), and 223 (NYT), and we have shown that cell-surface gC1qR contains carbohydrate (3). It also contains a PKC phosphorylation site at residue 207, a tyrosine kinase recognition site at position 268, and a myristoylation site at position 252. What is more intriguing is the fact that there is a homologous consensus for an ITAM and ITIM motif around the tyrosines at positions 224 and 236, respectively. An important feature of the mature gC1qR molecule is that its translated amino acid sequence does not predict for the presence of either a conventional transmembrane segment or a consensus site for a GPI anchor, and the lack of any phospholipid-binding domains is confirmed by the crystal structure. Nonetheless, the lack of a direct conduit into the intracellular compartment is mitigated by the fact that can associate with transmembrane molecules including CD44, $\alpha_4\beta_1$ integrin, and DC-SIGN (Feng et al. 2002; Agostinis et al. 2010; Hosszu et al. 2010a). Furthermore, there are six Cys residues within the 73 residues of the full-length gC1qR, with a "motif-like" sequence—CACGCGCG—which may be involved in membrane attachment in the lipid raft.

Comparison of the cDNA-derived amino acid sequences reveal that rat and mouse gC1qR are 97.6% identical, whereas both rodent sequences are 89.9% identical to the human. The human and mouse gC1qR genes have been characterized (Guo et al. 1997; Lim et al. 1998, 2001) and are essentially similar in their exon/intron organization comprising six exons and five introns each within a total length of approximately 6 kb, which is spliced into a 1.5 kb mRNA. The first exon encodes the 73-aa pre-sequence, including the putative signal peptide, plus four amino acid residues that comprise the N-terminus of the mature protein. Crystal structure of gC1qR (Jiang et al. 1999) shows that the termini of exons II and VI code for a pair of antiparallel terminal helices that form the majority of the circumference of the native trimeric doughnut, while the remainder of exon II plus exons III–V code mainly for a highly twisted 7-strand β sheet that makes up the core of the trimeric doughnut. The molecule features several conspicuous surface loops that join β strands, and it is largely these loops that are the potential targets for drug design.

6.3 Plasma Ligands of gC1qR

Although the gC1qR molecule was first identified and characterized as a receptor that recognizes the globular heads of C1q, data accumulated to date suggest that it can bind to a number of other plasma ligands. Because interaction of gC1qR with each of these molecules is of biological significance, some are discussed below in more detail.

6.3.1 C1q

Binding of C1q to cells is known to induce and modulate a number of C1q-mediated cellular responses including inositol trisphosphate (IP3) production in, expression of P-selectin on, and generation of procoagulant activity on platelets (Peerschke et al. 1993); suppression of B and T cell proliferation (Ghebrehiwet et al. 1990), trophoblast cell migration (Agostinis et al. 2010), regulation of dendritic cells (Castellano et al. 2004a; Hosszu et al. 2010a), and angiogenesis (Bossi et al. 2011); chemotaxis of eosinophils, mast cells, neutrophils, and dendritic cells (Ghebrehiwet et al. 1995; Kuna et al. 1996; Leigh et al. 1998; Vegh et al. 2005); adhesion of fibroblasts and endothelial cells (Bordin et al. 1990; Feng et al. 2002); activation and expression of the adhesion molecules E-selectin, ICAM-1, and VCAM-1 (Lozada et al. 1995); and production of IL-6, IL-8, and monocyte chemoattractant protein-1 (MCP-1) on endothelial cells (ECs) (van den Berg et al. 1998). While the C1q site on gC1qR has been identified to be at the N-terminus and comprises of residues 76–93, the gC1qR site on C1q has been identified to be in the A-chain comprising of residues 155–164, with the two arginine residues at positions 162 and 163 playing a critical role (Marques et al. 1993; Ghebrehiwet and Peerschke 2004). Recently, we have also identified two potential C1q sites on gC1qR: these are residues 190–202 and 144–162 (Xu et al 2009). Although several cell surface molecules—known to bind C1q including cC1qR/CR—could potentially contribute to these biologic functions, the binding of free C1q or C1q within macromolecular C1 occurs predominantly via gC1qR, as the majority of the functions described

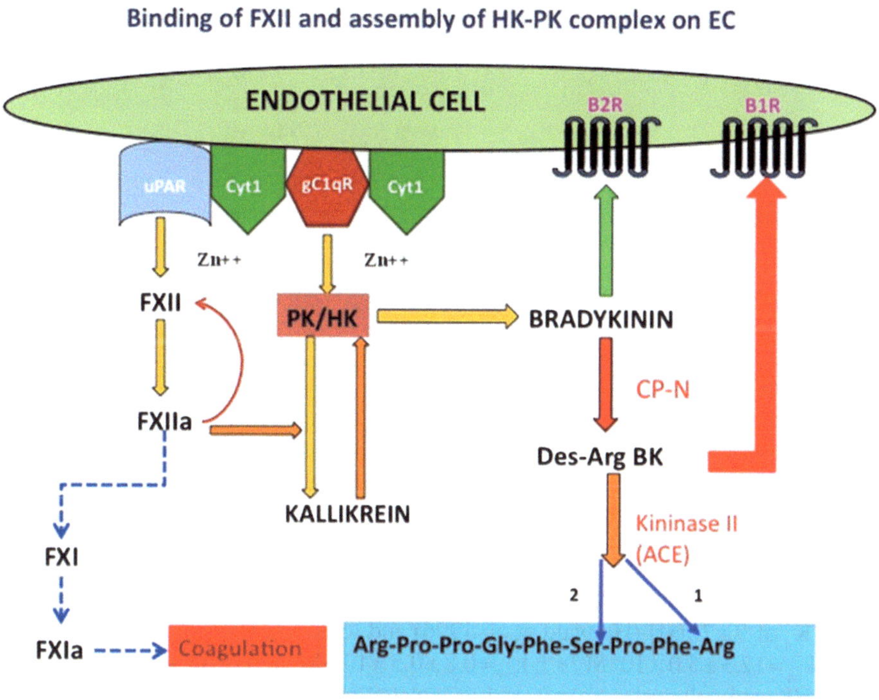

Fig. 6.2 Macromolecular assembly of the kinin–kallikrein system (KKS) on the endothelial cell surface. The receptor complex comprises of two bimolecular complexes of uPAR-cytokeratin 1 and gC1qR-cytokeratin 1. Because of the affinity of FXII for the uPAR-cytokeratin 1 complex, binding of FXII to this complex results in FXII autoactivation to generate FXIIa. This enzyme then cleaves prekallikrein (*PK*), which is bound—in complex with and via high molecular weight kininogen (*HK*)—to the gC1qR-cytokeratin complex. Kallikrein then cleaves HK to generate bradykinin (*BK*). BK binds to bradykinin receptor 2 (*B2R*)—a constitutively expressed receptor—to trigger the initial vascular leakage. BK is regulated by carboxypeptidase N (*CP-N*), which converts it to des-Arg BK. Des-Arg BK then binds to bradykinin receptor 1 (*B1R*), an inducible receptor, to enhance and exacerbate the vascular leakage. Des-Arg BK is further degraded by kininase II or angiotensin-converting enzyme (*ACE*)

could be substantially inhibited by anti-gC1qR. Because soluble gC1qR can activate the classical pathway, gC1qR secreted and/or released from infected cells or cells undergoing active proliferation can exacerbate or even trigger the local inflammatory process.

6.3.2 Factor XII (FXII) and High Molecular Weight Kininogen (HK)

The plasma kinin-forming pathway consists of three essential proteins (Fig. 6.2) that interact in vivo in a complex fashion once bound to a macromolecular complex formed during inflammatory response or bound to proteins along cell surfaces (Kaplan et al. 2004a). These are coagulation factor XII [FXII (or Hageman Factor, HF)], prekallikrein (PK), and high molecular weight kininogen (HK). Once factor XII is activated to factor XIIa, it converts prekallikrein, which circulates as complex with HK, to kallikrein and the latter in turn digests HK to generate the nonapeptide BK (H-Arg-Pro-Pro-Gly-Phe-Ser-Pro-Phe-Arg-OH) (Bhoola et al. 1992; Kaplan 2004a). Bradykinin belongs to the kinin family of proinflammatory peptides and is among the most potent vasodilator agonists known (Stewart 1980; Regoli and Barabe 1980).

Table 6.1 Plasmon resonance BIAcore binding studies

Ligand	Rate Constants	K_D
C1q	K_{on} 5 × 10^4 $M^{-1}s^{-1}$	13.5 nM
	K_{off} 6.8 × 10^{-4} s^{-1}	
gC1q	K_{on} 2.3 × 10^{-3} $M^{-1}s^{-1}$	710 nM
	K_{off} 6.8 × 10^{-4} s^{-1}	
HK	K_{on} 1.1 × 10^5 $M^{-1}s^{-1}$	0.539 nM
	K_{off} 5.9 × 10^3 s^{-1}	
HKa	K_{on} 3.3 × 10^5 $M^{-1}s^{-1}$	0.0012 nM
	K_{off} 40 s^{-1}	

We have shown previously that gC1qR binds to HK and that this binding is strictly zinc dependent (Joseph et al. 1996; Herwald et al 1996). HK binds to endothelial-cell-surface-expressed gC1qR with high affinity [K_d of 9±2 nM] (Herwald et al. 1996) and to isolated gC1qR (Table 6.1) with a K_d of 0.8±0.7 nM (k_{on}=12.3±5.0×10^4 M^{-1} s^{-1}; k_{off}=0.8±0.5×10^{-4} M^{-1} s^{-1}) (Pixley et al. 2011). The interaction between HK and cell-surface-expressed or isolated gC1qR is inhibited by mAb 74.5.2, which recognizes the C terminal half of gC1qR comprising residues 204–218 (Joseph et al. 1996; Ghebrehiwet et al. 2001). However, in addition to gC1qR, HK has two other cell surface binding sites, namely, cytokeratin 1 and urokinase plasminogen activator receptor (uPAR) (Colman et al. 1997; Hasan et al. 1998; Mahdi et al. 2001). Therefore, on endothelial cells, the three surface proteins form two bimolecular receptor complexes—gC1qR-cytokeratin 1 and uPAR-cytokeratin 1— in which gC1qR, by virtue of its high affinity for HK (gC1qR>cytokeratin1>uPAR), plays a central role in the assembly of the kallikrein–kinin system (Fig. 6.2), leading to the generation of the nonapeptide, bradykinin (BK). BK generated at the endothelial cell surface in turn is largely responsible for the vascular permeability and inflammation seen in the majority, not only of angioedema due to C1-INH deficiency but also angioedema induced as a consequence of treatment with ACE inhibitors (Fig. 6.3). Domain 5 of HK located at the N-terminus of the light chain, which is rich in histidine and arginine residues, contains the site for interaction with gC1qR, and a 20-aa peptide within this termed HKH20 has been shown to be a site of interaction of HK with ECs (Hasan et al. 1995; Herwald et al. 1995). Although there is another binding site in the HK heavy chain that involves cytokeratin 1, its affinity (8.48 nM) is much less. Thus, it is the light chain binding of gC1qR that appears to be important for the assembly and/or triggering of the kinin-generating system (Kaplan 2004a). Furthermore, gC1qR has also been shown to bind FXII, and since the binding of HK is inhibited by FXII (Reddigari et al. 1993a, b), it is likely that they bind to the same or overlapping sites made up of residues that form a cluster in space (Ghebrehiwet et al. 2011).

Employing synthetic peptides and gC1qR deletion mutants, we have identified and confirmed two domains of gC1qR that contain HK binding sites: one at position 190–202 and the other at position 204–218. Whereas HK binding to domain 190–202 was inhibited by mAbs 48 and 83, the binding of HK was inhibited by mAb 74.5.2. More importantly, using in vivo and in vitro models of vascular permeability, we confirmed that blockade of gC1qR with mAb 74.5.2 not only inhibits vascular permeability (Bossi et al. 2009) but also reduces *S. aureus* colonization of distal target tissues (Peerschke et al. 2006; Sethi et al. 2011). Therefore, blockade of gC1qR by inhibitory peptides or antibodies may prevent the generation of BK and thus attenuate the vascular leakage, which is known to be the trigger of edema formation and acute and/or chronic inflammation.

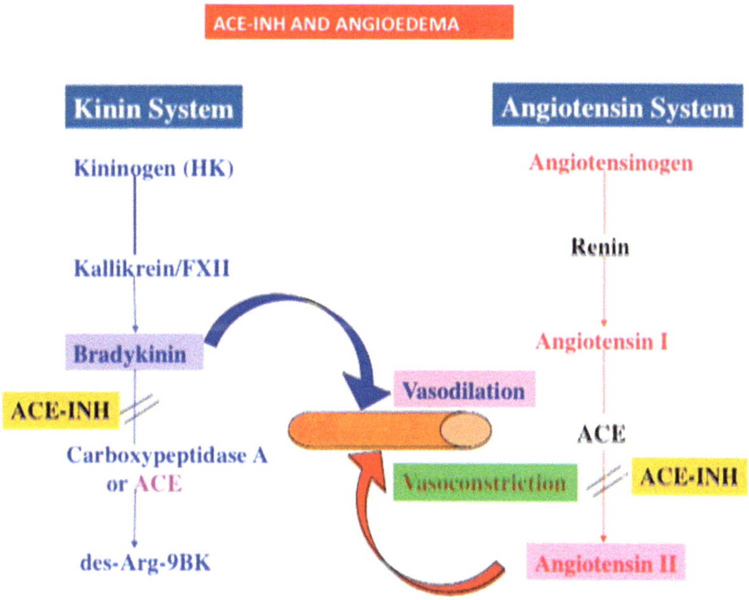

Fig. 6.3 Schematic model of ACE-inhibitor-mediated angioedema. In the angiotensin system, rennin converts angiotensinogen to generate angiotensin I, and the latter is converted by angiotensin-converting enzyme or ACE to angiotensin II. Angiotensin II is responsible for vasoconstriction and elevated blood pressure within the vessels. While treatment with ACE inhibitor (*ACE-INH*) can slow down the effect of angiotensin II-induced high blood pressure, the same treatment can have deleterious side effect; sine ACE inhibitor treatment can cause angioedema by negating the very enzyme, ACE, which regulates bradykinin (kinin system)

6.3.3 Vitronectin and Thrombin

Using recombinant-gC1qR-coated microtiter plates, we also searched for additional plasma ligands unrelated to C1q. Surprisingly, vitronectin (VN) or complexes containing vitronectin were identified (Lim et al. 1996). Subsequent analysis revealed the specific binding of the ternary complex of vitronectin–thrombin–antithrombin III (VNTAT) to gC1qR. Interestingly, although gC1qR was shown to bind thrombin (factor II), no binding was observed to either antithrombin III (ATIII) alone or a preformed complex of thrombin–antithrombin III (TAT) indicating that gC1qR and ATIII may compete for the same or overlapping site on thrombin (Ghebrehiwet et al. 2001). Because gC1qR binds to ternary complex of VNTAT but does not interact with the thrombin–antithrombin III complex, direct binding of gC1qR to vitronectin was performed. These experiments showed that the heparin-binding multimeric VN but not the plasma form of VN was found to bind to either recombinant gC1qR or the native form of gC1qR isolated from Raji cell membranes (Lim et al. 1996) in a manner that was specific and saturable ($K_D = 20$ nM). Furthermore, the interaction was inhibited by glycosaminoglycans such as heparin but not by chondroitin sulfate. The binding sites for VN and C1q are in separate location of the gC1qR molecule since VN bound equally well to a gC1qR, a deletion mutant lacking the N-terminal 22 residues, which contains the C1q-binding domain. Although the biological significance of this interaction has yet to be elucidated, one could speculate that gC1qR may participate in the clearance of VN-containing complexes or, in conjunction with VN, participate in the inhibition of complement-mediated cytolysis (Lim et al. 1996).

6.3.4 *Fibrinogen*

Another plasma ligand for gC1qR is fibrinogen. Human fibrinogen is a 340-kDa soluble plasma glycoprotein synthesized in the liver as a product of three genes located on chromosome 4 (Lu et al. 1999). It is composed of two Aα and two Bβ and two γ chains arranged in a symmetrical fashion. The molecule is made up of three domains: two outer D domains containing the carboxy termini of the Aα, Bβ, and γ chains and a central E domain comprising of the amino termini of all six polypeptide chains (Hettasch and Greenberg 1998). During blood coagulation, fibrinogen is converted to fibrin by thrombin, and fibrin is then cross-linked by factor XII to form a clot. Using microplate-bound fibrinogen, we have shown that gC1qR interacts with fibrinogen in a manner that was specific and inhibitable by excess fibrinogen. More interestingly, gC1qR was able to inhibit fibrin polymerization in a dose-dependent manner. Reptilase-induced fibrin clot formation was completely inhibited by gC1qR at a 2:1 (gC1qR: fibrinogen) molar ratio. Furthermore, at equivalent molar concentrations, gC1qR appeared to be a more potent inhibitor of fibrin polymerization than fibrinogen, a well-known inhibitor. Using plasmin-derived fibrinogen degradation fragments, we also showed that gC1qR binds to the fibrinogen fragment D-100 but not to the E domain indicating that gC1qR binds to the D domain of fibrinogen/fibrin (Lu et al. 1999). Therefore, at local sites of injury or inflammation, gC1qR may play a role in modulating fibrin formation.

6.4 Pathogen-Associated Ligands

6.4.1 *Bacterial Ligands*

Data accumulated over the past 20 years support the concept that gC1qR is an emerging pathogen-associated pattern-recognition receptor. This conclusion is based on the fact that a wide variety of pathogenic microorganisms (Table 6.2) are known to use gC1qR either to gain access into cells or to suppress immune response so that their survival is guaranteed. For example, *S. aureus*—which causes a range of illnesses, from minor skin infections, such as pimples and abscesses, to life-threatening diseases such as pneumonia, meningitis, sepsis, and endocarditis—interacts with gC1qR through the virulence factor, protein A. While protein A expressing *S. aureus* Cowan I strain was able to bind both purified gC1qR and gC1qR on platelets, the protein A-deficient Wood 46 strain was not (Nguyen et al. 2000). The adhesion of *S. aureus* to platelets in turn is a major determinant of virulence in the pathogenesis of endovascular infections. Therefore, the adherence of bacteria to platelets at the surface of abnormal valvular endothelium may contribute to *S. aureus* localization to sites of vascular injury and thrombosis. Indeed, this concept was validated by experiments in which blockade of gC1qR with mAb 74.5.2, which recognizes and blocks the HK site on gC1qR—and to a lesser extent by mAb

Table 6.2 Interaction of gC1qR with bacterial antigens

Bacterial Ag	gC1qR residues (aa)	Function	Reference
L. monocytogenes InlB	76-93 (?)	Serves as receptor for the invasion protein InlB and promotes entry of L. monocytogenes	Braun L, et al, 2000
S. aureus Protein A	ND	Interacts with platelet gC1qR and is transported to distal sites of infection	Nguyen T, et al 2000
B. cereus, exosporium	ND	Serves as spore attachment site	Ghebrehiwet B et al. 2007

60.11, which recognizes the C1q site on gC1qR—reduced *S. aureus* colonization of target tissues in experimental rat animal model of *S. aureus* endocarditis (Peerschke et al. 2006) and inhibition of adherence of *S. aureus* in a Syrian golden hamster dorsal skin model (Sethi et al. 2011). This in turn supports the hypothesis that platelet gC1qR, which serves as a binding site for *S. aureus* protein A, may play a role in the pathogenesis of endocarditis by providing a surface for the initial adhesion and shuttling of the pathogen to distal targets of infection (Peerschke et al. 2000).

Another bacterium that employs gC1qR to enter host cells is *L. monocytogenes* (Braun et al. 2000). Entry of *L. monocytogenes* into target host cells is mediated by interaction between the bacterial internalin A (InlA) and the host cell E-cadherin, which is required for intestinal and placental barriers in a species-specific manner. In human cells, however, *L. monocytogenes* uses another bacterial virulence factor called InlB and two host cell receptors, Met and gC1qR, which have an additive effect (Khelef et al. 2006).

The exosporium of *B. cereus* also contains a putative binding site for gC1qR (Ghebrehiwet et al. 2007). *B. cereus* is a member of a genus of aerobic, gram-positive, spore-forming rod-like bacilli, which includes the deadly *B. anthracis* the causative agent of anthrax. It is involved in several types of food poisoning and tissue destruction, including bacterial endophthalmitis. Employing microtiter-plate-fixed *B. cereus* spores, we showed that gC1qR could bind to the spores in a specific manner. Subsequent experiments designed to examine the adherence of *B. cereus* spores to monolayers of human colon carcinoma (Caco-2) cells showed that spore attachment is mediated by gC1qR as assessed by electron microscopy using mAb to gC1qR that had been conjugated to carbon nanotubes. What is more interesting is that the exosporia of the activated and germinating spores were often decorated with gC1qR, indicating that attachment of the spores via gC1qR induces overexpression of gC1qR. The exosporium-expressed molecule, which serves as a ligand for gC1qR, is not known. However, a major structural protein called BclA (**B**acillus **c**ollagen-**l**ike protein of **a**nthracis), which shares a remarkable similarity to the C1q/TNF family of proteins, was recently identified in *B. anthracis* (Rety et al. 2005). Because of the similarity between *B. cereus* and *B. anthracis*, it is possible to speculate that the latter may also have a BclA-like structure that allows it to interact with gC1qR.

6.4.2 *Viral Ligands*

As summarized in Table 6.3, the gC1qR molecule is targeted by a wide range of viruses. What is intriguing however is that, some of the viral envelope proteins such as the outer envelope protein of HIV, gp120, share some structural and functional similarities with C1q (Szabo et al. 2001). Therefore, proteins that are able to interact with C1q may also interact with isolated gp120 as well as the whole HIV-1 virus. Based on this hypothesis, we studied the potential ability of gC1qR, to inhibit the growth of different HIV-1 strains in cell cultures. Data obtained from these studies showed gC1qR to effectively and dose-dependently inhibit the production of one T-lymphotropic (X4) and one macrophage-tropic (R5) strain in human T cell lines (MT-4 and H9) and human monocyte-derived macrophage cultures, respectively. At a concentration range of 5–25 μg/mL, gC1qR caused a marked and prolonged suppression of virus production. The extent of inhibition was enhanced when gC1qR was first incubated with and then removed from the target cell cultures before virus infection, compared to that when the cells were infected with gC1qR-HIV mixtures (Szabo et al. 2001). The extent of inhibition was comparable to that of the Leu3a anti-CD4 antibody. In ELISA experiments, gC1qR did not bind to a solid-phase recombinant gp120, while strong and dose-dependent binding of gC1qR to solid-phase CD4 was observed. However, although gp120 does not appear to bind gC1qR directly, there is a strong interaction between gp41, another HIV protein, and gC1qR. During the initial weeks post HIV-1 infection, the great majority of dying cells are the uninfected $CD4^+$ T cells. An insight into the mechanism of this cell death was recently gained when it was shown that the interaction of gp41 and

Table 6.3 Interaction of gC1qR/p32/p33 with viral antigens

Viral Ag	gC1qR (aa residues)	Encoding Exon	Function	Reference
EBNA-1	244-282	VI (234-282)	Intracellular: Regulates splicing	Wang Y et al.1997
HIV-1 Tat	244-260**	VI (234-282)	Intracellular: Regulates HIV1 splicing through interaction with p32 (gC1qR)	Yu L, et al, 95
HIV-1 Rev	196-208	V (193-233)	Intracellular: regulates viral transcription	Luo Y, et al, 1994
HIV-1 gp41	174-180	IV (159-192)	gp41 3S motif binds to CD4+ T cells-to induce NKp44L, a cellular ligand for an activating NK receptor	Fausther-Bovendo H et al.2010
Herpes simplex ORF P	ND	ND	Intracellular: regulates RNA splicing	Bruni R, et al, 1996
Core Protein V-Adenovirus	ND	ND	Intracellular-imports proteins to the nucleus and CPVA hijacks this process to deliver its genome to the nucleus	Mathews DA, et al. 1998
Core Protein HCV	188-259	IV-VI (159-282)	Binds to gC1qR on T cells and suppress cell proliferation and immune response	Kittlesen DJ, et al, 2000
Capsid protein-Rubella virus :N-terminal 28aa	212-282	V-VI (193-282)	Interacts with gC1qR and enhances viral infectivity	Mohan, KVK, et al. 2002
Hanta virus	ND	ND	Requires gC1qR for binding and infection	Choi Y et al 2008

gC1qR plays a central role. The 3S motif of HIV-1 gp41 binds to gC1qR on uninfected bystander CD4+ T cells and induces surface expression of NKp44L, a cellular ligand for an activating NK receptor, rendering them susceptible to autologous NK killing (Fausther-Bonvendo et al. 2010).

The hepatitis C virus (HCV) core protein is another viral ligand with affinity for gC1qR. By binding to gC1qR on T cells, HCV core protein is able to inhibit T cell proliferation (Kittlesen et al. 2000), in a manner similar to that of C1q-induced antiproliferative response reported earlier (Chen et al. 1994). Suppression of the host immune response in turn contributes to an extremely high rate of viral persistence and severe disease progression. Although the binding of HCV core protein has been identified to be within a long stretch of gC1qR comprising of residues 188–259, the precise binding site is yet to be identified.

6.5 Role of gC1qR in Cancer

There is a single copy of the gC1qR gene in both the human and mouse genomes, and the human gene is located on chromosome 17p.13.3 in a region conserved with mouse chromosome 11 (Guo et al. 1997). The localization of gC1qR to the short arm of chromosome 17 band 13.3 is intriguing since several tumor suppressor genes deleted in ovarian cancer, distinct from Tp53, are located in the same region. Although by virtue of its colocalization with the tumor suppressor genes, it was predicted that gC1qR may also have tumor suppressor functions and thus would be downregulated in cancer cells, the results of immunohistochemical staining of various adenocarcinomas however showed the opposite. First, histochemical staining with anti-gC1qR mAb revealed marked differential expression of gC1qR in thyroid, colon, pancreatic, gastric, esophageal, and lung adenocarcinomas compared to the nonmalignant histologic counterparts. In contrast, differential expression was not observed in endometrial, renal, and prostate carcinoma (Rubinstein et al. 2004) results, which were corroborated more recently (Peerschke et al. 2011). In addition, recent experiments performed by others showed that cell-surface-expressed gC1qR is a receptor for the tumor-homing peptide, LyP-1 (Fogal et al. 2008). Furthermore, the interaction of LyP-1 and gC1qR is inhibited by mAb 60.11, which recognizes the N-terminal gC1qR residues 76–93 encoded by exon 2 (Peerschke and Ghebrehiwet 2001). The involvement of gC1qR in carcinogenesis is further substantiated by more recent studies, which used human lung carcinoma A549 cells (Kim et al. 2011). These experiments showed that the cell-surface gC1qR is a key regulator for lamellipodia formation and cancer metastasis via receptor tyrosine kinase activation. The growth factor-induced lamellipodia formation and cell migration were significantly decreased in gC1qR-depleted cells, with a concomitant blunt activation of the focal adhesion kinase and the respective receptor tyrosine kinases (Kim et al. 2011). These studies collectively suggest that gC1qR, which is highly upregulated on the surface of tumor cells and inflammatory cells, can serve not only as a tumor cell marker but also can be a target for therapeutic intervention (Peerschke et al. 2011).

6.6 Conclusion

By interacting with a diverse array of microbial antigens as well as key proteins of both the complement and kinin-generating systems, gC1qR contributes significantly to the inflammatory processes especially those associated with microbial infection, cancer, and autoimmunity. Therefore, identification of the gC1qR domains that enable it to interact with the various ligands is likely to generate useful information for the design of peptide-based or antibody-based therapy to prevent gC1qR-induced inflammation. Using in vivo models of vascular permeability, we have confirmed that blockade of gC1qR with mAb 74.5.2 not only inhibits vascular permeability but also reduces *S. aureus* colonization of distal target tissues. Together, the data therefore identify gC1qR as a *novel* molecular target for the design of antibody-based and/or peptide-based therapy in which vascular leakage is known to be the trigger of edema formation and acute and/or chronic inflammation.

Acknowledgement This work was supported in part by grants from the National Institutes of Health (R01 AI 060866 and R01 AI-084178).

References

Agostinis CR, Bulla C, Tripodo A, Gismondi H, Stabile F, Bossi C, Guarnotta C, De Garlanda F, Seta P, Spessotto A, Santoni B, Ghebrehiwet GG, Tedesco F (2010) An alternative role of C1q in cell migration and tissue remodelling: contribution to trophoblast invasion and placental development. J Immunol 185:4420–4429

Agostoni A, Cicardi M (1992) Hereditary and acquired C1-inhibitor deficiency. Biological and clinical characteristics in 235 patients. Medicine 71:206–215

Bhoola KD, Figueroa CD, Worthy K (1992) Bioregulation of kinins: kallikrein, kininogen and kinases. Pharmacol Rev 44:1–80

Bordin S, Ghebrehiwet B, Page RC (1990) Participation of C1q and its receptor in adherence of human diploid fibroblasts. J Immunol 145:2520–2526

Bossi F, Fischetti F, Regoli D, Durigutto P, Frossi B, Jr F, Gobeil B, Ghebrehiwet EIB, Peerschke MC, Tedesco F (2009) Novel pathogenic mechanism and therapeutic approaches to angioedema associated with C1 inhibitor deficiency. J Allergy Clin Immunol 124:1303–1310

Bossi F, Peerschke EI, Ghebrehiwet B, Tedesco F (2011) Cross-talk between the complement and the kinin system in vascular permeability. Immunol Lett 105:1053–1059

Braun L, Ghebrehiwet B, Cossart P (2000) gC1q-R/p32, a C1q-binding protein is a novel receptor for *Listeria monocytogenes*. EMBO J 19:1458–1466

Castellano G, Woltman AM, Nauta AJ, Roos A, Trouw LA, Seelen M, Schena FB, Daha MR, van Kooten C (2004a) Maturation of dendritic cells abrogates C1q production in vivo and in vitro. Blood 103:3813–3820
Castellano G, Woltman AM, Schena FP, Roos A, Daha MR, van Kooten C (2004b) Dendritic cells and complement: at the crossroad of innate and adaptive immunity. Mol Immunol 41:133–140
Chen A, Gaddipati S, Volkman DJ, Peerschke EIB, Ghebrehiwet B (1994) Human T cells possess specific receptors for C1q: role in activation and proliferation. J Immunol 153:1430–1440
Colman RW, Pixley RA, Najamunnisa S, Yan W, Wang J, Mazar A, McRae KR (1997) Binding of high molecular weight kininogen to human umbilical vein endothelial cells is mediated via a site within domains 2 and 3 of the urokinase receptor. J Clin Invest 100:1481–1487
Fausther-Bonvendo H, Vieillard V, Sagan S, Bismuth G, Debre P (2010) HIV gp41 engages gC1qR on CD4+ T cells to induce the expression of an NK ligand through the PIP3/H2O2 pathway. PLoS Pathog 6(7):e1000975. doi:10.1371/journal.ppat.1000975
Feng X, Tonnesen MG, Peerschke EIB, Ghebrehiwet B (2002) Cooperation of C1q receptors and integrins in C1q-mediated endothelial cell adhesion and spreading. J Immunol 168:2441–2448
Fogal V, Zhang L, Krajewski S, Ruoslahti E (2008) Mitochondrial/cell-surface protein p32/gC1qR as a molecular target in tumor cells and tumor stroma. Cancer Res 68:7210–7218
Ghebrehiwet B, Peerschke EIB (1998) Structure and function of gC1q-R a multiligand binding membrane protein. Immunobiology 199:225–238
Ghebrehiwet B, Peerschke EIB (2004) cC1qR (calreticulin) and gC1q-R/p33: ubiquitously expressed multi-ligand binding cellular proteins involved in inflammation and infection. Mol Immunol 41:173–183
Ghebrehiwet B, Silverberg M, Kaplan AP (1981) Activation of the classical pathway of complement by Hageman factor fragment. J Exp Med 153:665–676
Ghebrehiwet B, Beck G, Habicht GS (1990) Interaction of C1q with its receptor (C1qR) on cultured cell lines induces an antiproliferative response. Clin Immunol Immunopathol 54:148–160
Ghebrehiwet B, Lim B-L, Peerschke EIB, Willis AC, Reid KBM (1994) Isolation cDNA cloning, and overexpression of a 33-kDa cell surface glycoprotein that binds to the globular 'heads' of C1q. J Exp Med 179:1809–1821
Ghebrehiwet B, Lim B-L, Kumar R, Feng X, Peerschke EIB (2001) gC1q-R/p33, a member of a new class of multifunctional and multicompartmental cellular proteins is involved in inflammation and infection. Immunol Rev 180:65–77
Ghebrehiwet B, Jesty J, Peerschke EIB (2002) gC1q-R/p33: structure-function predictions from the crystal structure. Immunobiology 205:421–432
Ghebrehiwet B, Tantral L, Titmus MA, Panessa-Warren B, Tortora GT, Wong SS, Warren J (2007) Current topics in innate immunity, vol 598. Springer, Berlin, pp 180–197
Ghebrehiwet B, Jesty J, Xu S, Vinayagasundaram R, Vinayagasundaram U, Ji Y, Valentino A, Hosszu KK, Mathew S, Joseph K, Kaplan AP, Peerschke EIB (2011) Structure-function studies using deletion mutants identify domains of gC1qR/p33 as potential therapeutic targets for vascular permeability and inflammation. Front Immunol 2:1–9
Gigli I, Kaplan AP, Austen KF (1971) Modulation of function of the activated first component of complement by a fragment derived from serum. J Exp Med 134:1466–1484
Guo N, Weremowicz S, Lynch N, Schwaeble W, Lim B-L, Morton CC, Peerschke EIB, Reid KBM, Ghebrehiwet B, Sastry KN (1997) Assignment of gC1q-R, the C1q globular domain binding protein (C1qBP) to human chromosome band 17p13.3 by in situ hybridization. Cytogenet Cell Genet 77:283–284
Hasan AAK, Cines DB, Herwald H, Schmaier AH, Müller-Esterl W (1995) Mapping the cell binding site on high molecular weight kininogen domain 5. J Biol Chem 270:19256–119261
Hasan AAK, Zisman T, Schmaier AH (1998) Identification of cytokeratin 1 as a binding protein and presentation receptor for kininogens on endothelial cells. Proc Natl Acad Sci USA 95:3615–3620
Herwald H, Hasan AAK, Godovac-Zimmermann J, Schmaier AH, Müller-Esterl W (1995) Identification of an endothelial cell binding site on kininogen domain D3. J Biol Chem 270:14634–14642
Herwald H, Dedio J, Kellner R, Loos M, Müller-Esterl W (1996) Isolation and characterization of the kininogen-binding protein p33 from endothelial cells. Identity with the gC1q receptor. J Biol Chem 271:13040–13047
Hettasch JM, Greenberg CS (1998) Fibrin formation and stabilization. In: Loscalzo J, Schafer AI (eds) Thrombosis and hemorrhage. William & Willkins, Philadelphia, pp 129–154
Hosszu KK, Vinayagasundaram U, Habiel D, Ji Y, Vinayagasundaram R, Peerschke EIB, Ghebrehiwet B (2010a) Evidence that the full-length gC1qR (residues 1–282) is expressed on the cell surface. Mol Immunol 47:2223, XXIIIrd, (abstract)
Hosszu K, Vinayagasundaram U, Vinayagasundaram R, Santiago-Schwarz F, Peerschke E (2010b) Evidence that gC1qR and DC-SIGN associate on the surface of immature dendritic cells. J Immunol 184:136.24 (abstract)
Jiang J, Zhang Y, Krainer A, Xu R-M (1999) Crystal structure of p32, a doughnut-shaped acidic mitochondrial matrix protein. Proc Natl Acad Sci USA 96:3572–3577
Joseph K, Ghebrehiwet B, Peerschke EIB, Reid KBM, Kaplan AP (1996) Identification of the zinc-dependent endothelial cell binding protein for high molecular weight kininogen and factor XII: identity with the receptor, which binds to the globular "heads" of C1q (gC1q-R). Proc Natl Acad Sci USA 93:8552–8557

Kaplan AP (2004a) Mechanisms of bradykinin formation. In: Greaves MW, Kaplan AP (eds) Urticaria and angioedema. Marcel Dekker Inc., New York, pp 51–72

Kaplan AP (2004b) C1 inhibitor deficiency. In: Greaves MW, Kaplan AP (eds) Urticaria and angioedema. Marcel Dekker Inc., New York, pp 303–320

Khelef N, Lecuit M, Bierne H, Cossart P (2006) Species specificity of the Listeria monocytogenes InlB protein. Cell Microbiol 8:457–470

Kim K-B, Yi J-S, Nguyen N, Lee J-H, Kwon Y-C, Ahn B-Y, Cho H, Kim YK, Yoo H-J, Lee J-S, Ko Y-G (2011) Cell-surface receptor for complement component C1q (gC1qR) is a key regulator for lamellipodia formation and cancer metastasis. J Biol Chem 286:23093–23101

Kittlesen DJ, Chianese-Bullock KA, Yao ZQ, Braciale TJ, Hahn YS (2000) Interaction between complement receptor gC1q-R and hepatitis C virus core protein inhibits T-lymphocyte proliferation. J Clin Invest 106:1239–1249

Kuna P, Iyer M, Peerschke EIB, Kaplan AP, Reid KBM, Ghebrehiwet B (1996) Human C1q induces eosinophil migration. Clin Immunol Immunopathol 81:48–54

Leigh LE, Ghebrehiwet B, Perera TP, Bird IN, Strong P, Kishore U, Reid KBM, Eggleton P (1998) C1q-mediated chemotaxis by human neutrophils: involvement of gClqR and G-protein signalling mechanisms. Biochem J 1998(330):247–254

Lim B-L, Preissner K, Ghebrehiwet B, Leigh LEA, Reid KBM (1996) The binding protein for globular "heads" of complement C1q, gC1q-R: functional expression and characterization as a novel vitronectin binding factor. J Biol Chem 271:26739

Lim B-L, White RA, Hummel GS, Mak S-C, Schwaeble WJ, Reid KBM, Peerschke EIB, Ghebrehiwet B (1998) Characterization of the murine gene for gC1q-BP (gC1q-R), a novel cell protein that binds the globular heads of C1q, vitronectin, high molecular weight kininogen and factor XII. Gene 209:229–237

Lim B-L, Tye AJ, Guo N, Sastry KN, Peerschke EIB, Ghebrehiwet B (2001) The Human gC1q-R/p33 gene (C1qBP): genomic organization and promoter analysis. J Biol Chem 276:17069–17075

Lozada C, Levin RI, Huie MR, Hirschhorn M, Naime D, Whittlow M, Recht PA, Golden B, Cronstein B (1995) Identification of C1q as the heat-labile serum cofactor required for immune complexes to stimulate endothelial expression of the adhesion molecules E-selectin and intracellular and vascular cell adhesion molecules 1. Proc Natl Acad Sci USA 92:8378–8382

Lu PD, Galanakis D, Ghebrehiwet B, Peerschke EIB (1999) The receptor for the globular 'heads' of C1q, gC1q-R binds to fibrinogen and impairs its polymerization. Clin Immunol Immunopathol 90:360–367

Mahdi F, Shariat-Madar Z, Todd RF III, Figueroa CD, Schmaier AH (2001) Expression and Localization of cytokeratin 1 and urokinase plasminogen activator receptor on endothelial cells. Blood 97:2342–2350

Marques G, Anton LC, Barrio E, Sanchez A, Ruiz S, Gavilanes F, Vivanco F (1993) Arginine residues of the globular regions of human C1q involved in the interaction with immunoglobulin G. J Biol Chem 268:10393–10402

Nguyen T, Ghebrehiwet B, Peerschke EIB (2000) *Staphylococcus aureus* protein A recognizes platelet gC1q-R/p33: a novel mechanism for staphylococcal interactions with platelets. Infect Immun 68:2061–2068

Peerschke EIB, Ghebrehiwet B (2001) Human blood platelet gC1qR/p33. Immunol Rev 180:56–64

Peerschke EIB, Reid KBM, Ghebrehiwet B (1993) Platelet activation by C1q results in the induction of αIIb/β3 integrins (GP IIb/IIIa) and the expression of P-selectin and procoagulant activity. J Exp Med 178:579–587

Peerschke EIB, Minta JO, Zhou SZ, Bini A, Gotlieb A, Colman RW, Ghebrehiwet B (2004) Expression of gC1qR/p33 and its major ligands in human atherosclerotic lesions. Mol Immunol 41:759–766

Peerschke EI, Bayer AS, Ghebrehiwet B, Xiong YQ (2006) gC1qR/p33 blockade reduces *Staphylococcus aureus* colonization of target tissues in an animal model of infective endocarditis. Infect Immun 74:4418–4423

Peerschke EIB, Dembitzer FR, Kinoshita Y, Burstein D, Phelps R, Ghebrehiwet B (2011) Differential expression of gC1qR in normal and pathologic human tissue. Mol Immunol 48:1718–2011 (abstract)

Pixley RA, Espinola RG, Ghebrehiwet B, Joseph K, Kao A, Cines DB, Colman RW (2011) Interaction of high molecular-weight kininogen with endothelial cell binding proteins suPAR, gC1qR and cytokeratin 1 determined by surface plasmon resonance (BiaCore). Thromb Hemost 105:1053–1059

Reddigari SR, Shibayama Y, Brunnee T, Kaplan AP (1993a) Human Hageman factor (factor XII) and high molecular weight kininogen compete for the same binding site on human umbilical vein endothelial cells. J Biol Chem 268:11982–11987

Reddigari SR, Kuna P, Miragliotta GF, Shibayama Y, Nishikawa K, Kaplan AP (1993b) Human high molecular weight kininogen binds to human umbilical vein endothelial cells via its heavy and light chains. Blood 81:1306–1311

Regoli D, Barabe J (1980) Pharmacology of bradykinin and related kinins. Pharmacol Rev 32:1–46

Rety S, Salamitou S, Garcia-Verdugo I, Hulmes DJS, Le Hagarat F, Chaby R, Lewit-Bentley A (2005) The crystal structure of the Bacillus anthracis spore surface protein BclA shows remarkable similarity to mammalian proteins. J Biol Chem 280:43073–43078

Rubinstein DB, Stortchevoi A, Boosalis M, Ashfaq R, Ghebrehiwet B, Peerschke EIB, Calvo F, Gillaume T (2004) Receptor for the globular heads of C1q (gC1q-R; p33); Hyaluronan-binding protein is preferentially expressed by adenocarcinoma cells. Int Cancer Res 110:741–750

Sethi S, Herrmann M, Roller J, von Müller L, Peerschke EI, Ghebrehiwet B, Bajric I, Menger MD, Laschke MW (2011) Inhibition of gC1qR/p33, a receptor for C1q, decreases adherence of *Staphylococcus aureus* to the microvascular endothelium in vivo. Microvasc Res 82:66–72
Stewart JM (1980) Chemistry and biologic activity of peptides related to bradykinin. In: Erdos EG (ed) Bradykinin, kallidin and kallikrein. Springer, New York, pp 227–272
Szabo J, Cervenak L, Toth FD, Prohaszka Z, Horvath L, Kerekes K, Beck Z, Bacsi A, Erdei A, Peerschke EI, Füst G, Ghebrehiwet B (2001) Soluble gC1q-R/p33, a cell protein that binds to the globular "heads" of C1q, effectively inhibits the growth of HIV-1 strains in cell cultures. Clin Immunol 99(2):222–231
van den Berg RH, Faber-Krol MC, Sim RB, Daha MR (1998) The first subcomponent of complement C1q triggers the production of IL-6, IL-8 and chemoattractant peptide 1 by human umbilical vein endothelial cells. J Immunol 161:6924–6934
van Leuwen HC, O'Hare P (2001) Retargeting of the mitochondrial protein p32/gC1qR to a cytoplasmic compartment and the cell surface. J Cell Sci 114:2115–2123
Vegh-Goyarts Z, Kew R, Gruber B, Ghebrehiwet B (2005) Chemotaxis of human monocyte derived dendritic cells toward C1q is mediated by gC1qR and cC1qR. Mol Immunol 43:1402–1407
Xu S, Menzies CA, Vinayagasundaram R, Mathew S, Vinayagasundaram U, Peerschke EIB, Jesty J (2009) Structure-function analysis of gC1qR/p33 using deletion mutants show additional binding sites for C1q. J Immunol 182:134.69 (abstract)

Chapter 7
The Alternative C5a Receptor Function

Hiroshi Nishiura

Abstract When acute inflammatory states are induced by treatment with chemical mediators in C5-deficient mice, neutrophil influxes are commonly decreased. Therefore, the neutrophil C5a receptor (C5aR) is believed to be a member of the pro-inflammatory receptors. However, C5aR deficiency endows mouse neutrophils with increased sensitivity to *Pseudomonas aeruginosa*. We have demonstrated that C5aR accepts not only C5a but also ribosomal protein S19 (RP S19) oligomers. RP S19 oligomers released from apoptotic cells promote apoptosis or induce dual agonistic and antagonistic effects on the chemotaxis of macrophages and neutrophils in an autocrine or paracrine manner, respectively. We assumed that the function of C5aR in apoptotic cells is almost the same as that in neutrophils infiltrating acute inflammatory lesions. Therefore, we believe that RP S19 oligomers can explain the opposite response of neutrophils in C5aR-deficient mice. In the present study, we found that anti-human RP S19 rabbit IgG cross-reacted with mouse RP S19 monomers and oligomers in plasma and serum, respectively, whereas anti-human C5a rabbit IgG only cross-reacted with mouse RP S19 oligomers in serum. To examine a role of RP S19 oligomers in vivo, we injected carrageenan (50 μg/100 μL) into the thoracic cavities of mice in the simultaneous presence of rabbit IgG and anti-human C5a rabbit IgG (100 μg/100 μL). Before 4 h and after 24 h, we did not observe any inflammatory cues in pleural exudates and lung substances from control mice. However, infiltrating neutrophils were detected in pleural exudates and lung tissues at 4 h after the addition of anti-human RP S19 rabbit IgG. Moreover, anti-human C5a rabbit IgG retards the initiation phase of carrageenan-induced mouse plurality. Many of the neutrophils infiltrating the thoracic cavities of the mice remained annexin V-negative. Neutrophil infiltration into pneumonic lesions became more severe, as alveolar septal destruction and haemorrhage concomitant with increased numbers of neutrophils in the pleural exudates were observed. These in vivo data demonstrate that the neutrophil C5aR acts as a dual pro-inflammatory and pro-apoptosis receptor during the initiation and the resolution phases of acute inflammation, respectively.

7.1 The C5aR-Mediated Classical Function

The structure of *the C5a receptor (C5aR) gene* includes two exons, and the translated portion is encoded by exon 2 (Gerard et al. 1993). This indicates that C5aR coupled with the Gαi protein and/or Gα16 protein is expressed on leukocytes (Lo et al. 2003; Zhang et al. 2007). There is a one ligand/

H. Nishiura (✉)
Department of Molecular Pathology, Kumamoto University Graduate School,
Honjyo 1-1-1, Kumamoto 860-8556, Japan
e-mail: seino@kumamoto-u.ac.jp

J.D. Lambris et al. (eds.), *Complement Therapeutics*, Advances in Experimental Medicine and Biology 735,
DOI 10.1007/978-1-4614-4118-2_7,

one receptor dogma regarding ligand-receptor relationships. Therefore, C5aR-mediated downstream signals have been investigated by stimulation with C5a, and the interaction of Gαi protein with phosphatidylinositol-3 kinase (PI3K) and phospholipase C (PLC) has been detected (Monk et al. 2007). Phosphatidylinositol 4,5-bisphosphate (PIP2) is cleaved into diacylglycerol (DAG) and inositol 1,4,5-trisphosphate (IP3) by PLC. Intracellular Ca^{2+} release from the endoplasmic reticulum (ER) through IP3R induces extracellular Ca^{2+} entry through Ca^{2+} release–activated Ca^{2+} (CRAC) channels. Ras and Raf are sequentially phosphorylated by intracellular Ca^{2+}-dependent activation of protein kinase C (PKC) and are connected to PI3K and mitogen-activated kinase kinase1/2 (MEK1/2) upstream of extracellular signal-regulated kinase1/2 ERK1/2, respectively. Activation of C5aR is limited by the binding of arrestin to its C-terminal intracellular regions, which are sequentially phosphorylated by PKC and G protein-coupled receptor kinase2 (GRK2) downstream of ERK1/2 (Elorza et al. 2000). It is generally thought that the C5aR-mediated classical PI3K signal is controlled by the ERK1/2 pathway. Moreover, neutrophil influx is decreased following the injection of bacterial endotoxin into the peritoneal cavity in C5-deficient mice (Mullick et al. 2011; Snyderman et al. 1971). It is believed that the neutrophil C5aR works as a pro-inflammatory receptor during the initiation phase of acute inflammation. However, neutrophil influx is increased by intrapulmonary infection of *Pseudomonas aeruginosa* in C5aR-deficient mice (Hopken et al. 1996, 1997), suggesting that the neutrophil C5aR alternatively works as an anti-inflammatory receptor. This discrepancy represents a difficulty in the complement field.

7.2 The C5aR-Mediated Alternative Function

7.2.1 *RP S19 Oligomers in Serum*

We have demonstrated that the C5aR binds to not only C5a but also ribosomal protein S19 (RP S19) oligomers and the *E. coli* chaperone protein Skp (Jia et al. 2010; Nishiura et al. 1998). G protein-coupled receptor (GPCR) ligands are classified into classical and pure chemoattractants with or without Ca^{2+} mobilisation, respectively. We recently found that RP S19 monomers (apparent molecular weight of 17 kDa) form a complex with prothrombin (apparent molecular weight of 70 kDa), preventing them from being subjected to the filtration system of the kidney (Nishiura et al. 2011b). RP S19 monomer associates with phosphatidylserine on activated platelets during plasma clotting (Fig. 7.1) (Semba et al. 2010). Platelets are commonly activated by thrombin. RP S19 monomers are subsequently cross-linked by activated coagulation factor XIII (FXIIIa), resulting in the formation of oligomers. FXIII is also activated by thrombin. In contrast, C3BbC3b acts as a C5 convertase to hydrolyse the α chain, liberating C5a and C5b (Woodruff et al. 2011). When clots prepared in vitro are injected into the abdominal cavity of guinea pigs, they are cleared by infiltrating macrophages. Phagocytic clearance of guinea pig clots is blocked when anti-RP S19 rabbit IgG is present (Ota et al. 2010). Therefore, we have demonstrated that C5a and RP S19 oligomers exhibit different actions as classical chemoattractants associated with neutrophil and monocyte C5aRs during the initiation and the resolution phases of acute inflammation.

7.2.2 *RP S19 Oligomers in Apoptotic Cells*

We have also demonstrated that RP S19 monomers in ribosomes associate with phosphatidylserine on the plasma membrane during apoptosis, resulting in oligomer formation via tissue transglutaminases

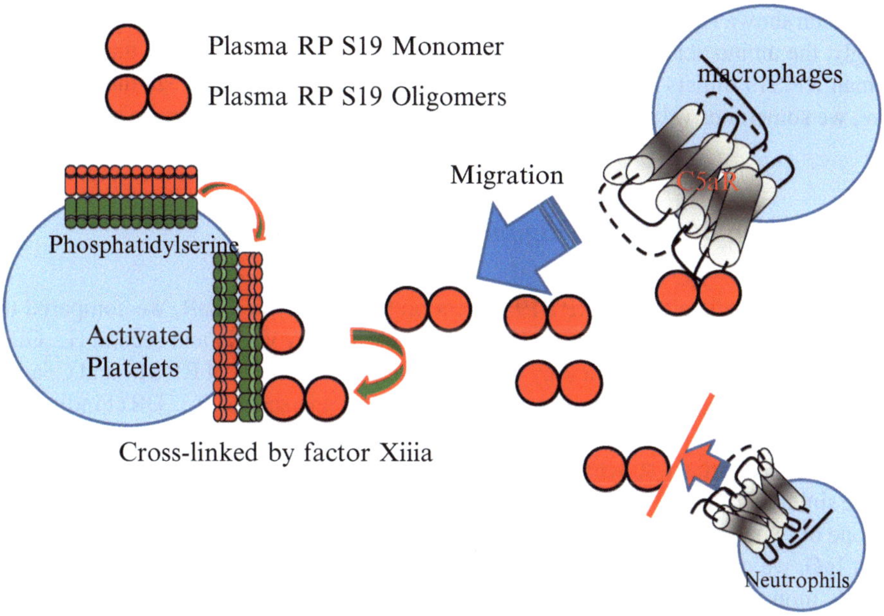

Fig. 7.1 *Mechanism of the generation of plasma RP S19 oligomers*. Plasma RP S19 monomers associate with phosphatidylserine on activated platelets during blood clotting. Plasma RP S19 monomers are cross-linked by FXIIIa

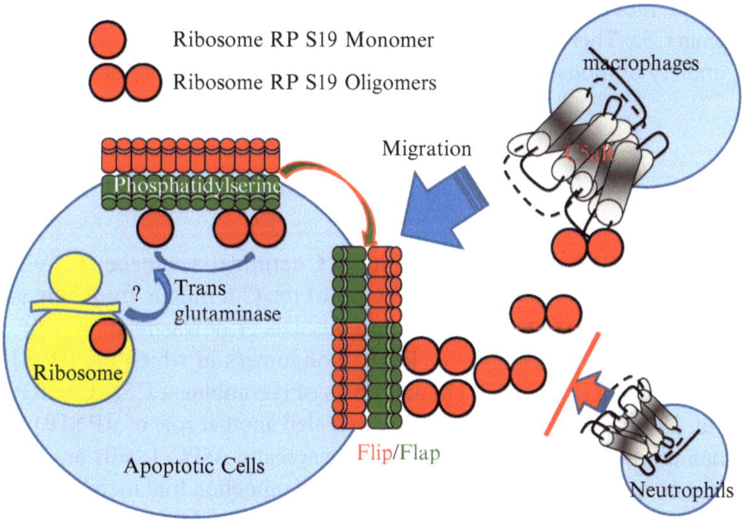

Fig. 7.2 *Mechanism of the generation of RP S19 oligomers*. RP S19 monomers in ribosomes associate with phosphatidylserine on the plasma membrane during apoptosis. Ribosome RP S19 monomers are cross-linked by tissue transglutaminases

(Fig. 7.2) (Horino et al. 1998; Nishimura et al. 2001). In addition to the generation of RP S19 oligomers, apoptotic cells exhibit de novo synthesis of their receptor, C5aR (Nishiura et al. 2005). To examine the roles of RP S19 oligomers, we generally prepare recombinant RP S19 monomers and stimulate the formation of dimers with FXIIIa (Nishiura et al. 1999). Using RP S19 dimers, we found that RP S19 oligomers bind to the monocyte C5aR and function as a classical chemoattractant, as has commonly been shown for C5a. However, the neutrophil C5aR does not transmit downstream signals. Interestingly, the amino acid sequence of human RP S19 100% overlaps with that of guinea pig RP S19. Human RP S19 dimers exhibit the same effect as RP S19 oligomers via the guinea pig C5aR. Therefore, we sometimes validate the function of RP S19 oligomers in guinea pigs.

7.2.3 Interaction of RP S19 Oligomers with the C5aR

To study the antagonistic effect of RP S19 dimers on the neutrophil C5aR, we compared the amino acid sequences of RP S19 and C5a. For C5a, the first (Y_{13}KHSVVKK) and second (L_{72}GR) binding moieties bind to the C5aR. The homology of the amino acid sequences of RP S19 and C5a is less than 4%. However, we have identified similar first (K_{48}LAKHK) and second (L_{131}DR) binding moieties in RP S19 (Shibuya et al. 2001). Moreover, we developed a simulated tertiary structure model of human RP S19 based on the tertiary structure of archaebacterial RP S19 (Nishiura et al. 2010). The components of the simulated RP S19 are similar to those of hC5a, indicating that the C5a-like structure likely provides one of two receptor-binding moieties via each RP S19 subunit. In addition to anti-human RP S19 rabbit IgG, we prepare anti-human C5a rabbit IgG. Anti-human RP S19 rabbit IgG cross-reacts with not only monomers but also with oligomers of mouse RP S19 in serum, whereas anti-human C5a rabbit IgG cross-reacts only with mouse RP S19 oligomers. The amino acid sequence homologies between human and mouse RP S19, C5a and C5aR are 99%, 64% and 68%, respectively (*BLASTP program search of protein subjects using a protein query*: Table 7.1). Mouse macrophages infiltrate mouse skin when human RP S19 dimers are injected into the skin, but this effect is not observed for human C5a. However, mouse C5a shares a conserved Cys residue position and second binding moiety (L_{72}GR) with human C5a. Therefore, we assumed that the A_{63}NISHKDM_{70} moiety of human C5a and K_{63}ESPHKPV_{70} moiety of mouse C5a are crucial for discrimination from other C5aRs.

7.2.4 The Pro-apoptosis Potency of RP S19 Oligomers

Compared with C5a, RP S19 exhibits an additional C-terminal sequence (I_{134}AGQVAAANKKH) (Shrestha et al. 2003). Interestingly, a truncated form of the C-terminus from human RP S19 dimers binds to the neutrophil C5aR and rescues the transmission of downstream signals. Based on these data, we prepared a functional analogue for RP S19 oligomers in which the RP S19 C-terminus is connected to the C-terminus of the G73D mutant form of recombinant C5a (C5a/RP S19) (Oda et al. 2008; Revollo et al. 2005). Using C5a/RP S19, we revealed another role of RP S19 oligomers in apoptosis. When human promyelocytic HL-60 cells and pancreatic AsPC-1 cells are subjected to hyperthermia at 43 °C for 60 min together with treatment with a mitochondrial membrane voltage exchange agent (0.5 mM $MnCl_2$) and extracellular Ca^{2+} chelator (1 μM thapsigargin), the cells undergo apoptosis. Under these apoptotic conditions, the process of programmed cell death is sometimes divided into initiation, effector and execution phases (Nishiura et al. 2009). The period from the effector phase to the execution phase is shortened or lengthened based on the simultaneous presence of C5a/RP S19 or anti-human RP S19 rabbit IgG and C5a, respectively. The C5aR selectively activates the ERK1/2 or p38 MAPK signalling pathway in response to stimulation with C5a or C5a/RP S19, respectively. Therefore, we sometimes refer to the C5aR-mediated ERK1/2 and p38 MAPK signals, anti-apoptotic and pro-apoptotic signals, respectively.

Table 7.1 Homologies at amino acid sequences between human and mouse RP S19, C5a and C5aR

```
RP S19

Human    1  MPGVTVKDVNQQEFVRALAAFLKKSGKLKVPEWVDTVKLAKHKELAPYDENWFYTRAAST  60
            MPGVTVKDVNQQEFVRALAAFLKKSGKLKVPEWVDTVKLAKHKELAPYDENWFYTRAAST
Mouse    1  MPGVTVKDVNQQEFVRALAAFLKKSGKLKVPEWVDTVKLAKHKELAPYDENWFYTRAAST  60

Human   61  ARHLYLRGGAGVGSMTKIYGGRQRNGVMPSHFSRGSKSVARRVLQALEGLKMVEKDQDGG  120
            ARHLYLRGGAGVGSMTKIYGGRQRNGV PSHFSRGSKSVARRVLQALEGLKMVEKDQDGG
Mouse   61  ARHLYLRGGAGVGSMTKIYGGRQRNGVRPSHFSRGSKSVARRVLQALEGLKMVEKDQDGG  120

Human  121  RKLTPQGQRDLDRIAGQVAAANK  143
            RKLTPQGQRDLDRIAGQVAAANK
Mouse  121  RKLTPQGQRDLDRIAGQVAAANK  143


C5a

Human    1  MLQKKIEEIAAKYKHSVVKKCCYDGACVNNDETCEQRAARISLGPRCIKAFTECCVVASQ  60
            +L++KIEE AAKYKHSV KKCCYDGA VN  ETCE+R AR+++GP CI+AF ECC +A++
Mouse    1  LLRQKIEEQAAKYKHSVPKKCCYDGARVNFYETCEERVARVTIGPLCIRAFNECCTIANK  60

Human   61  LRANISHKDMQLGR  74
            +R    HK +QLGR
Mouse   61  IRKESPHKPVQLGR  74


C5aR

Human   13  HYDDKDTLDLNTPVDKTSNTLRVP-DILALVIFAVVFLVGVLGNALVVWVTAFEAKRTIN  71
            +YD   T+D N P D      R P D+ AL+I++VVFLVGV GNALVVWVTAFEA+R +N
Mouse   12  NYDHYGTMDPNIPADGIHLPKRQPGDVAALIIYSVVFLVGVPGNALVVWVTAFEARRAVN  71

Human   72  AIWFLNLAVADFLSCLALPILFTSIVQHHHWPFGGAACSILPSLILLNMYASILLLATIS  131
            AIWFLNLAVAD LSCLALP+LFT+++ H++W F   AC +LPSLILLNMYASILLLATIS
Mouse   72  AIWFLNLAVADLLSCLALPVLFTTVLNHNYWYFDATACIVLPSLILLNMYASILLLATIS  131

Human  132  ADRFLLVFKPIWCQNFRGAGLAWIACAVAWGLALLLTIPSFLYRVVREEYFPPKVLCGVD  191
            ADRFLLVFKPIWCQ  RG GLAW+AC VAW LALLLTIPSF+YR   ++++    +CG++
Mouse  132  ADRFLLVFKPIWCQKVRGTGLAWMACGVAWVLALLLTIPSFVYREAYKDFYSEHTVCGIN  191

Human  192  YSHDK-RRERAVAIVRLVLGFLWPLLTLTICYTFILLRTWSRRATRSTKTLKVVVAVVAS  250
            Y      +E+AVAI+RL++GF+ PLLTL ICYTF+LLRTWSR+ATRSTKTLKVV+AVV
Mouse  192  YGGGSFPKEKAVAILRLMVGFVLPLLTLNICYTFLLLRTWSRKATRSTKTLKVVMAVVIC  251

Human  251  FFIFWLPYQVTGIMMSFLEPSSPTFLLLNKLDSLCVSFAYINCCINPIIYVVAGQGFQGR  310
            FFIFWLPYQVTG+M+++L PSSPT   + KL+SLCVS AYINCC+NPIIYV+AGQGF GR
Mouse  252  FFIFWLPYQVTGVMIAWLPPSSPTLKRVEKLNSLCVSLAYINCCVNPIIYVMAGQGFHGR  311

Human  311  LRKSLPSLLRNVLTEESVVRESKSFTRSTVDTMAQKTQAV  350
            L +SLPS++RN L+E+SV R+SK+FT ST DT  +K+QAV
Mouse  312  LLRSLPSIIRNALSEDSVGRDSKTFTPSTTDTSTRKSQAV  351
```

Moreover, the interaction of C5a/RP S19 with the neutrophil C5aR induces de novo synthesis of regulator of G protein signalling 3 (RGS3) (Nishiura et al. 2009). RGS3 is a member of the R4 family (Bansal et al. 2007), which exhibits structural domains that activate the Gα subunit of GTPases. Acceleration of the GTPase activity of the Gα subunit promotes GTP hydrolysis, resulting in inactivation the G protein and rapidly switching off GPCR-mediated downstream signals. We observed that the C5aR-mediated ERK1/2 signal in apoptotic HL-60 cells is decreased by overexpression of RGS3 (Nishiura et al. 2009). Additionally, we have demonstrated that RP S19 oligomers released from apoptotic cells promote apoptosis.

7.3 Mechanisms for Synchronising the Effects of RP S19 Oligomers Between Promoting the Apoptosis of Infiltrating Neutrophils and the Migration of Macrophages During the Resolution Phase of Acute Inflammation

7.3.1 Role of RP S19 Oligomers in Acute Inflammation

Based on all of the data presented above, we hypothesise that RP S19 oligomers released from infiltrating neutrophils into acute inflammatory lesions bind to the neutrophil C5aR and promote the apoptosis of neutrophils in an autocrine manner. In contrast, RP S19 oligomers bind to the monocyte C5aR and attract macrophages to infiltrating neutrophils in a paracrine manner. We have proposed a method for the synchronisation effects of RP S19 oligomers between promoting the apoptosis of infiltrating neutrophils and the migration of macrophages during the resolution phase of acute inflammation.

7.3.2 Reprogramming of the Lifespan of Infiltrating Neutrophils in Mouse Pleurisy

Monomeric human RP S19 does not bind to the human C5aR. However, RP S19 oligomers bind to both the human C5aR and the mouse C5aR. Conversely, human C5a binds to the human C5aR, although not to the mouse C5aR. Therefore, anti-human RP S19 rabbit IgG cross-reacts with mouse RP S19 monomers and oligomers, but not with the C5a of any other species. In contrast, anti-human C5a rabbit IgG specifically cross-reacts with mouse RP S19 oligomers, whereas it does not react with the C5a of any other species (Nishiura et al. 2010).

To confirm the role of RP S19 oligomers in vivo, we used a carrageenan-induced acute pulmonary inflammation mouse model. Samples of lung, bone marrow, peripheral blood and pleural fluid were collected 4 and 24 h post-carrageenan injection. The injections were performed concurrently with the application of control rabbit IgG, anti-human RP S19 rabbit IgG or anti-human C5a rabbit IgG. Following the injections, pathological and FACS analyses were performed. In the left lung, we observed weak oedema and neutrophil infiltration of the surrounding bronchial tissue and large pulmonary arteries 4 h post-injection. These observations indicated short-term acute pleurisy and localised bronchopneumonia, which were resolved after 24 h (Figs. 7.3a and 7.4a). The right lung and pleural cavity remained intact. Administration of anti-human RP S19 rabbit IgG resulted in functional loss of mouse RP S19 monomers and oligomers, whereas anti-human C5a rabbit IgG specifically induced functional loss of mouse RP S19 oligomers.

At 4 h post-injection, compared with the administration of control IgG, the administration of anti-human RP S19 rabbit IgG induced leukocyte accumulation and haemorrhaging in the surrounding large blood vessels and peripheral airway, without an increase in oedema in the surrounding bronchi. Consistent with these pathological findings, following the administration of anti-human RP S19 rabbit IgG, the number of leukocytes that infiltrated from the circulation into the thoracic cavity through the lung parenchyma was 2.6 times higher than that observed in the control IgG-treated mice (Fig. 7.3b).

In contrast to the control injection of distilled water into the right thoracic cavity, carrageenan injection into the left thoracic cavity induced weak neutrophil infiltration as well as oedema surrounding the large blood vessels and bronchi after 24 h (Fig. 7.4a). No alveolar haemorrhaging was observed.

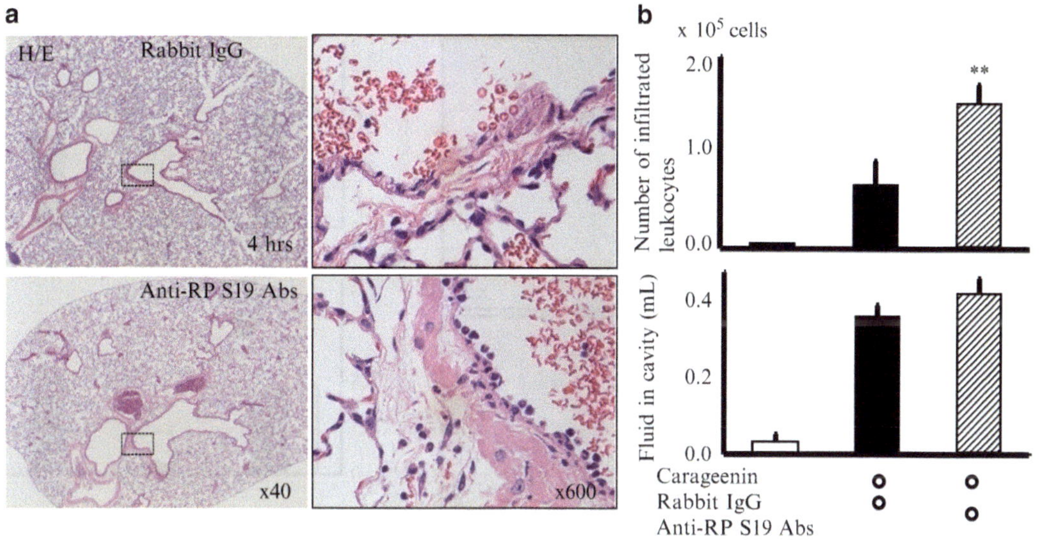

Fig. 7.3 *Enhancement of neutrophil influx in carrageenan-induced mouse pleurisy by anti-human RP S19 rabbit IgG.* Pathogen-free female Crij:CD1(ICR) mice (15–20 g body weight range) were purchased from Charies River (Yokohama, Japan). Mice were maintained in the Center for Animal Resources and Development, Kumamoto University. Animal experiments were performed two separate times in triplicate, under the control of the Ethical Committee for Animal Experiment, Kumamoto University School of Medicine. Mice were killed by euthanasia while under ether anaesthesia. (**a**) *Haematoxylin and eosin staining* A sample consisting of 0.5% carrageenin in 100 μL of distilled water was injected with 27-gauge needles into the left side of the mouse thoracic cavity along with control rabbit IgG, anti-human RP S19 rabbit IgG or anti-human C5a rabbit IgG (100 μg IgG/100 μL PBS). The animals were exsanguinated 4 and 24 h later under ether anaesthesia, and then, the lungs were immediately collected and fixed in 10% formalin for paraffin sectioning. Paraffin sections of 4 μm thickness were stained with haematoxylin and eosin (*H/E*) using typical methods. Lung sections are stained with hematoxylin and eosin (*H/E*). Cells were observed using an automatic microscope, Provis AX (Olympus, Tokyo, Japan) and a digital camera, Penguin 600CL (Pixera, Los Gatos, CA, USA). (**b**) *Leukocyte subset analysis* Mouse cells in the bone marrow, peripheral blood and thoracic cavity were harvested at 4 and 24 h after treatment with control rabbit IgG, anti-human RP S19 rabbit IgG or anti-human C5a rabbit IgG. Red cells were depleted in a 0.83% NH_4Cl solution, and the cell numbers were counted using trypan blue dye exclusion. Percentages of granulocytes, monocytes and T cells were measured on the basis of the forward scatter (FSC) and side scatter (SSC) values. The numbers of leukocytes infiltrating pleural effusions in carrageenan-induced mouse pleurisy after 4 h were measured using FACS analysis following administration of control rabbit IgG or anti-human RP S19 rabbit IgG. Data are expressed as the mean ± SD (n = 4). *Statistical analysis* Statistical significance was calculated by non-parametric or parametric tests offered in a two-way analysis of variance window ($^{*}p<0.05$; $^{**}p<0.01$)

Administration of anti-human C5a rabbit IgG into the left thoracic cavity resulted in excessive leukocyte accumulation, severe oedema and considerable haemorrhaging surrounding the large blood vessels, bronchi and peripheral airway. Infiltrating neutrophils were a predominant component of the observed leukocyte accumulation, and alveolar septum degradation was observed in areas of excessive haemorrhaging. In accordance with this pathological finding, the number of leukocytes infiltrating the thoracic cavity significantly increased at 24 h post-injection (Fig. 7.4b). Specifically, the numbers of granulocytes and monocytes increased 31 and 11 times, respectively, based on the forward and side scatter values determined by FACS analysis. Compared with the total number of granulocytes present in the thoracic cavity, the total number of granulocytes present in the peripheral blood was decreased by 50%.

Interestingly, at 24 h post-injection, the infiltrating neutrophils in the lung parenchyma still did not react with Apo-Taq solution (Fig. 7.4a). These data indicate that the observed tissue destruction could be due to the long lifespan of the neutrophils.

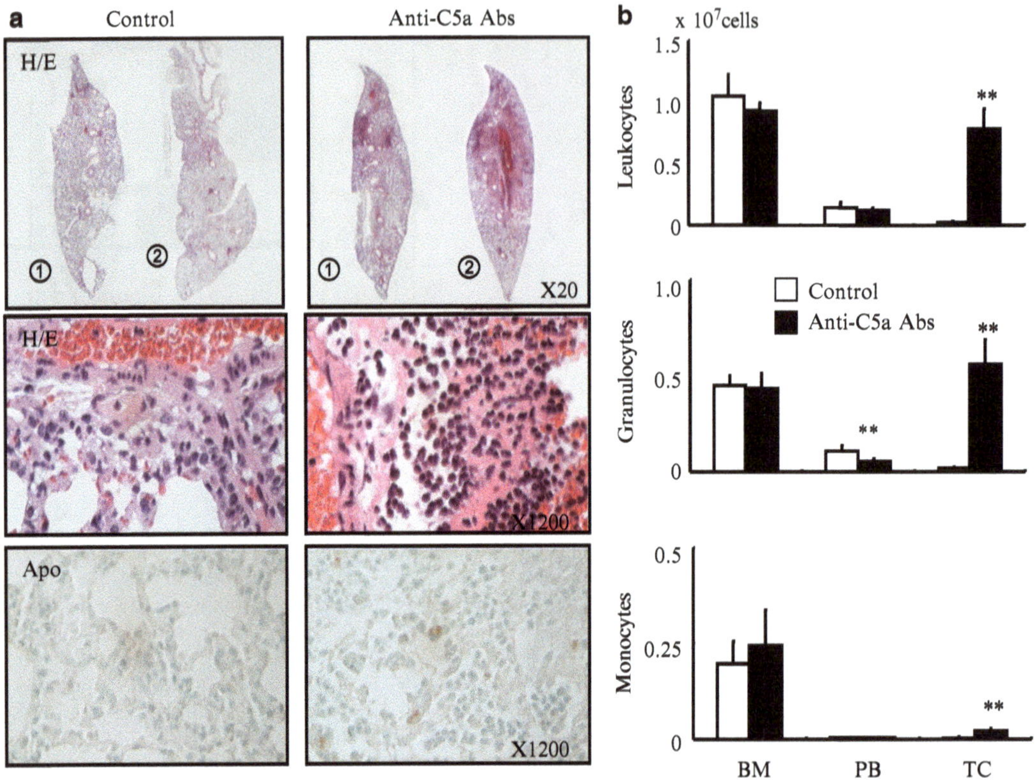

Fig. 7.4 *Reprogramming mouse neutrophils to increase their lifespan via the neutralisation of mouse RP S19 oligomers with anti-human C5a rabbit IgG.* (**a**) *DNA staining* Fragmentation of the 3′-termini of double-stranded DNA was visualised by the ApoMark DNA Fragmentation Detection Kit (Funakoshi, Tokyo, Japan). Lung sections are stained with H/E and using an Apo-Taq kit. (**b**) In addition to the leukocyte subsets in bone marrow (*BM*) and peripheral blood (*PB*), the subsets among the leukocytes infiltrating pleural effusions in the mouse thoracic cavity (*TC*) in carrageenan-induced mouse pleurisy at 24 h were determined using FACS analysis following the administration of control rabbit IgG or anti-human C5a rabbit IgG. Data are expressed as the mean ± SD (n = 6)

7.3.3 Dual Functions of the Neutrophil C5aR in Acute Inflammation

Complement activation takes place during the initiation phase of acute inflammation. The neutrophil C5aR, which is the receptor for C5a, is involved in the chemotactic migration of neutrophils during this phase. During the resolution phase, C5aR serves as a receptor for RP S19 oligomers in neutrophils undergoing apoptosis (Fig. 7.5). Therefore, the neutrophil C5aR plays roles in both the initiation and resolution phases of acute inflammation. Conflicting experimental data have been reported in acute inflammation research using mouse and guinea pig models. Treatment with anti-human C5a rabbit IgG and C5a/RP S19 results in increased neutrophil influx in carrageenan-induced mouse pleurisy versus decreased influx in an anti-ovalbumin Ab-induced reverse passive Arthus reaction guinea pig model (Oda et al. 2008). The latter result has also been previously reported in C5aR-deficient mice (Hopken et al. 1996, 1997). In light of the data obtained in the present study, both of these results can now be understood. RP S19 oligomers are predominantly present during the early resolution phase, and thus, the neutrophil C5aR functions as a pro-apoptosis receptor during the early resolution phase. However, C5a is predominantly present during the early initiation phase, and the neutrophil C5aR therefore acts as a chemotaxis/secretion receptor during the early initiation phase.

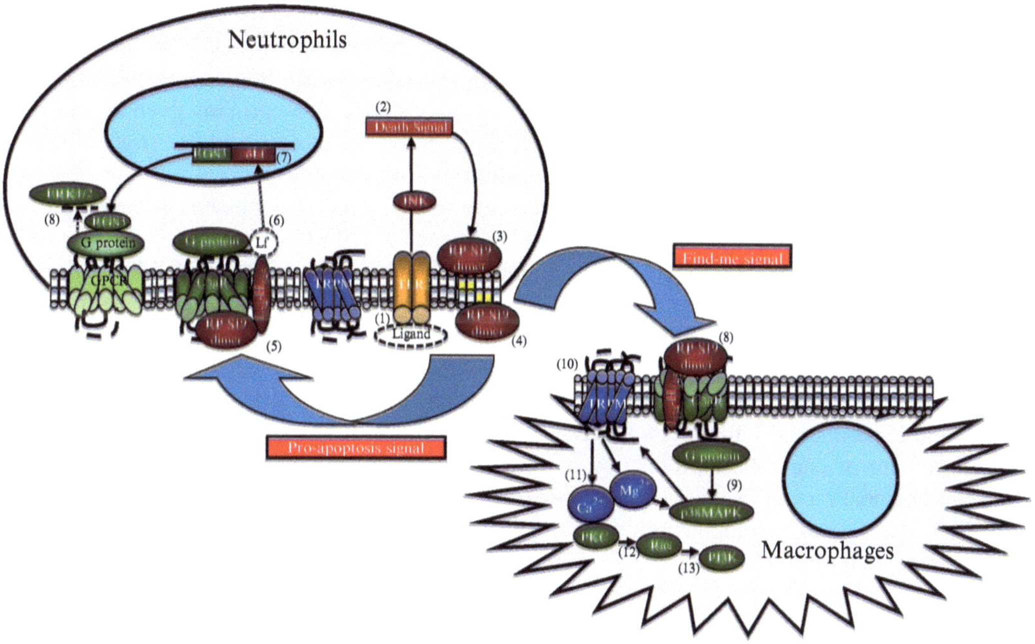

Fig. 7.5 *Mechanisms for synchronising the effects of RP S19 oligomers in an autocrine/paracrine manner.* In infiltrating neutrophils, which are committed to apoptosis, RP S19 is cross-linked, and its oligomers are released during the early resolution phase of acute inflammation. The interaction of RP S19 oligomers with C5aRs on the infiltrating neutrophils promotes apoptosis via RGS3 generation, resulting in progression from the early resolution phase to the late resolution phase of acute inflammation in an autocrine manner. Then, RP S19 oligomers attract macrophages to the infiltrating apoptotic cells in a paracrine manner

7.3.4 Other Functions of RP S19 in Acute Inflammation

We detected mouse RP S19 monomers not only in a complexed form in plasma but also in free form in serum. In contrast to the concentration of plasma prothrombin (50 μg/mL), the concentration of plasma RP S19 in ICR mice is generally on the order of several hundred ng/mL, as determined by a sandwich enzyme-linked immunosorbent assay (data not shown). Therefore, the formation of complexes between plasma RP S19 monomers and prothrombin did not affect the rate of the complement cascade. In contrast, free serum RP S19 monomers are thought to be immediately filtered by the renal glomeruli. Recently, free serum RP S19 monomers complexed with macrophage migration inhibitory factor were suggested to function as promoters of macrophage migration during the early resolution phase of acute inflammation (Filip et al. 2009). Thus, the results of our experiments involving the neutralisation of mouse serum RP S19 monomers with anti-human RP S19 rabbit IgG suggest an inhibitory effect of RP S19 on monocyte migration.

7.3.5 C5aR-Mediated Leukocyte Infiltration in Acute Inflammation

Carrageenan-induced pleurisy and air-pouch inflammation models have been well characterised in rats (Garcia-Ramallo et al. 2002; Murai et al. 2003). The predominant infiltration of monocytes into acute and chronic inflammatory sites is thought to be initiated by chemokines and/or classical chemoattractants that are released from apoptotic infiltrating neutrophils. We usually refer to chemoattractants as "find me" signals (Chekeni et al. 2010). Recently, the mechanism underlying the predominant attraction of monocytes by chemoattractants was described. It was suggested that lactoferrin released from apoptotic cells inhibits the expression of a special integrin involved in neutrophil migration (Bournazou et al. 2009). We suggest that the inhibition of lactoferrin release from infiltrating neutrophils might result in sustained neutrophil infiltration (Fig. 7.4a).

Inhibition of "find me" signals resulted in a decrease in the influx of monocytes in the mouse thoracic cavity after 24 h. We previously examined the influence of RP S19 oligomers on the absorption rate of intraperitoneally inserted coagula in a guinea pig model. Treatment with RP S19 oligomer-neutralising antibodies slowed the absorption of coagula via suppression of monocyte migration (Nishiura et al. 2010). However, compared with the administration of control rabbit IgG, the number of infiltrating mouse monocytes was observed to be increased 11-fold by the administration of anti-human C5a rabbit IgG in the current study (Fig. 7.4b). The mouse C5a protein and C5aR remained active in our experimental setting. It was recently suggested that the C5a–C5aR interaction results in an enhancement of chemokine production and activation of macrophage inflammatory protein-2, monocyte chemoattractant protein-1 and stromal-derived growth factor-1 (Laudes et al. 2002; Lee et al. 2009). Thus, such chemokine production-enhancing mechanisms might participate in the initiation of monocyte infiltration.

Acknowledgments This work was supported by Grant-in-Aid for Scientific Research (C) (Nishiura; KAKENHI 22590362) from the Ministry of Education, Culture, Sports, Science and Technology, Japan.

References

Bansal G, Druey KM, Xie Z (2007) R4 RGS proteins: regulation of G-protein signaling and beyond. Pharmacol Ther 116:473–495

Bournazou I, Pound JD, Duffin R, Bournazos S, Melville LA, Brown SB, Rossi AG, Gregory CD (2009) Apoptotic human cells inhibit migration of granulocytes via release of lactoferrin. J Clin Invest 119:20–32

Chekeni FB, Elliott MR, Sandilos JK, Walk SF, Kinchen JM, Lazarowski ER, Armstrong AJ, Penuela S, Laird DW, Salvesen GS et al (2010) Pannexin 1 channels mediate 'find-me' signal release and membrane permeability during apoptosis. Nature 467:863–867

Elorza A, Sarnago S, Mayor F Jr (2000) Agonist-dependent modulation of G protein-coupled receptor kinase 2 by mitogen-activated protein kinases. Mol Pharmacol 57:778–783

Filip AM, Klug J, Cayli S, Frohlich S, Henke T, Lacher P, Eickhoff R, Bulau P, Linder M, Carlsson-Skwirut C et al (2009) Ribosomal protein S19 interacts with macrophage migration inhibitory factor and attenuates its pro-inflammatory function. J Biol Chem 284:7977–7985

Garcia-Ramallo E, Marques T, Prats N, Beleta J, Kunkel SL, Godessart N (2002) Resident cell chemokine expression serves as the major mechanism for leukocyte recruitment during local inflammation. J Immunol 169:6467–6473

Gerard NP, Bao L, Xiao-Ping H, Eddy RL Jr, Shows TB, Gerard C (1993) Human chemotaxis receptor genes cluster at 19q13.3–13.4. Characterization of the human C5a receptor gene. Biochemistry 32:1243–1250

Hopken UE, Lu B, Gerard NP, Gerard C (1996) The C5a chemoattractant receptor mediates mucosal defence to infection. Nature 383:86–89

Hopken UE, Lu B, Gerard NP, Gerard C (1997) Impaired inflammatory responses in the reverse arthus reaction through genetic deletion of the C5a receptor. J Exp Med 186:749–756

Horino K, Nishiura H, Ohsako T, Shibuya Y, Hiraoka T, Kitamura N, Yamamoto T (1998) A monocyte chemotactic factor, S19 ribosomal protein dimer, in phagocytic clearance of apoptotic cells. Lab Invest 78:603–617

Jia N, Semba U, Nishiura H, Kuniyasu A, Nsiama TK, Nishino N, Yamamoto T (2010) Pivotal advance: interconversion between pure chemotactic ligands and chemoattractant/secretagogue ligands of neutrophil C5a receptor by a single amino acid substitution. J Leukoc Biol 87:965–975

Laudes IJ, Chu JC, Huber-Lang M, Guo RF, Riedemann NC, Sarma JV, Mahdi F, Murphy HS, Speyer C, Lu KT et al (2002) Expression and function of C5a receptor in mouse microvascular endothelial cells. J Immunol 169:5962–5970

Lee JS, Yang EJ, Kim IS (2009) The roles of MCP-1 and protein kinase C delta activation in human eosinophilic leukemia EoL-1 cells. Cytokine 48:186–195

Lo RK, Cheung H, Wong YH (2003) Constitutively active Galpha16 stimulates STAT3 via a c-Src/JAK- and ERK-dependent mechanism. J Biol Chem 278:52154–52165

Monk PN, Scola AM, Madala P, Fairlie DP (2007) Function, structure and therapeutic potential of complement C5a receptors. Br J Pharmacol 152:429–448

Mullick A, Tremblay J, Leon Z, Gros P (2011) A novel role for the fifth component of complement (C5) in cardiac physiology. PLoS One 6:e22919

Murai N, Nagai K, Fujisawa H, Hatanaka K, Kawamura M, Harada Y (2003) Concurrent evolution and resolution in an acute inflammatory model of rat carrageenan-induced pleurisy. J Leukoc Biol 73:456–463

Nishimura T, Horino K, Nishiura H, Shibuya Y, Hiraoka T, Tanase S, Yamamoto T (2001) Apoptotic cells of an epithelial cell line, AsPC-1, release monocyte chemotactic S19 ribosomal protein dimer. J Biochem 129:445–454

Nishiura H, Shibuya Y, Yamamoto T (1998) S19 ribosomal protein cross-linked dimer causes monocyte-predominant infiltration by means of molecular mimicry to complement C5a. Lab Invest 78:1615–1623

Nishiura H, Tanase S, Sibuya Y, Nishimura T, Yamamoto T (1999) Determination of the cross-linked residues in homodimerization of S19 ribosomal protein concomitant with exhibition of monocyte chemotactic activity. Lab Invest 79:915–923

Nishiura H, Tanase S, Shibuya Y, Futa N, Sakamoto T, Higginbottom A, Monk P, Zwirner J, Yamamoto T (2005) S19 ribosomal protein dimer augments metal-induced apoptosis in a mouse fibroblastic cell line by ligation of the C5a receptor. J Cell Biochem 94:540–553

Nishiura H, Nonaka H, Revollo IS, Semba U, Li Y, Ota Y, Irie A, Harada K, Kehrl JH, Yamamoto T (2009) Pro- and anti-apoptotic dual functions of the C5a receptor: involvement of regulator of G protein signaling 3 and extracellular signal-regulated kinase. Lab Invest 89:676–694

Nishiura H, Chen J, Ota Y, Semba U, Higuchi H, Nakashima T, Yamamoto T (2010) Base of molecular mimicry between human ribosomal protein S19 dimer and human C5a anaphylatoxin. Int Immunopharmacol 10:1541–1547

Nishiura H, Tanase S, Tsujita K, Sugiyama S, Ogawa H, Nakagaki T, Semba U, Yamamoto T (2011) Maintenance of ribosomal protein S19 in plasma by complex formation with prothrombin. Eur J Haematol 86:436–441

Oda Y, Tokita K, Ota Y, Li Y, Taniguchi K, Nishino N, Takagi K, Yamamoto T, Nishiura H (2008) Agonistic and antagonistic effects of C5a-chimera bearing S19 ribosomal protein tail portion on the C5a receptor of monocytes and neutrophils, respectively. J Biochem 144:371–381

Ota Y, Chen J, Shin M, Nishiura H, Tokita K, Shinohara M, Yamamoto T (2010) Role of ribosomal protein S19-like plasma protein in blood coagulum resorption. Exp Mol Pathol 90:19–28

Revollo I, Nishiura H, Shibuya Y, Oda Y, Nishino N, Yamamoto T (2005) Agonist and antagonist dual effect of the cross-linked S19 ribosomal protein dimer in the C5a receptor-mediated respiratory burst reaction of phagocytic leukocytes. Inflamm Res 54:82–90

Semba U, Chen J, Ota Y, Jia N, Arima H, Nishiura H, Yamamoto T (2010) A plasma protein indistinguishable from ribosomal protein S19: conversion to a monocyte chemotactic factor by a factor XIIIa-catalyzed reaction on activated platelet membrane phosphatidylserine in association with blood coagulation. Am J Pathol 176:1542–1551

Shibuya Y, Shiokawa M, Nishiura H, Nishimura T, Nishino N, Okabe H, Takagi K, Yamamoto T (2001) Identification of receptor-binding sites of monocyte chemotactic S19 ribosomal protein dimer. Am J Pathol 159:2293–2301

Shrestha A, Shiokawa M, Nishimura T, Nishiura H, Tanaka Y, Nishino N, Shibuya Y, Yamamoto T (2003) Switch moiety in agonist/antagonist dual effect of S19 ribosomal protein dimer on leukocyte chemotactic C5a receptor. Am J Pathol 162:1381–1388

Snyderman R, Phillips JK, Mergenhagen SE (1971) Biological activity of complement in vivo. Role of C5 in the accumulation of polymorphonuclear leukocytes in inflammatory exudates. J Exp Med 134:1131–1143

Woodruff TM, Nandakumar KS, Tedesco F (2011) Inhibiting the C5–C5a receptor axis. Mol Immunol 48:1631–1642

Zhang C, Xu R, Wang J, Han G, Chen G, Wang R, Wei H, Shen B, Ma Y, Li Y (2007) Functional identification of the stable transfection C5aR cell line Molt-4. Cell Mol Immunol 4:461–465

Chapter 8
The Effects of Selective Complement and CD14 Inhibition on the *E. coli*-Induced Tissue Factor mRNA Upregulation, Monocyte Tissue Factor Expression, and Tissue Factor Functional Activity in Human Whole Blood

O.-L. Brekke, C. Waage, D. Christiansen, H. Fure, H. Qu, John D. Lambris, B. Østerud, E.W. Nielsen, and T.E. Mollnes

Abstract *Background*: The complement pathway and CD14 play essential roles in inflammation, but little is known about the relative roles of complement and CD14 in *E. coli*-induced tissue factor (TF) mRNA upregulation, expression by monocytes, and functional activity in human whole blood. *Methods*: Whole *E. coli* bacteria were incubated for up to 4 h in human whole blood containing the anticoagulant lepirudin, which does not affect complement activation. TF mRNA levels were analyzed using reverse transcription, quantitative real-time PCR (RT-qPCR), and the expression of TF on the cell surface was analyzed using flow cytometry. Complement was selectively inhibited using the C3 convertase inhibitor compstatin or a C5a receptor antagonist (C5aRa), while CD14 was blocked by an anti-CD14 $F(ab')_2$ monoclonal antibody. *Results*: The *E. coli*-induced TF mRNA upregulation was reduced to virtually background levels by compstatin, whereas anti-CD14 had no effect. Monocyte TF expression and TF activity in plasma microparticles were significantly reduced by C5aRa. Anti-CD14 alone only slightly reduced *E. coli*-induced monocyte TF expression but showed a modest additive effect when combined with the complement inhibitors. Inhibiting complement and CD14 efficiently reduced the expression of the *E. coli*-induced cytokines IL-1β, IL-6, IL-8, and platelet-derived growth factor bb. *Conclusion*: Our results indicate that *E. coli*-induced TF mRNA upregulation is mainly dependent on complement activation, while CD14 plays a modest role in monocyte TF expression and the plasma TF activity in human whole blood.

8.1 Introduction

Sepsis almost invariably leads to hemostatic abnormalities ranging from insignificant blood test changes to severe disseminated intravascular coagulation (DIC), which may cause bleeding or microvascular dysfunction and contributes to organ failure (Levi et al. 2006). Gram-negative bacteria such as *Escherichia coli* are important pathogens that cause urinary tract infections and sepsis in humans.

Tissue factor (TF) is the main initiator of blood coagulation in vivo. TF is a 47 kDa, type I transmembrane glycoprotein that initiates coagulation by binding to factor VII/VIIa and the active serine protease FVIIa (Siegbahn 2000). The TF/factor VIIa complex has both procoagulant capabilities and

O.-L. Brekke • C. Waage • D. Christiansen • H. Fure • H. Qu • J.D. Lambris
B. Østerud • E.W. Nielsen • T.E. Mollnes (✉)
Department of Laboratory Medicine, Nordland Hospital, Bodø, N-8092, Norway

Institute of Medical Biology, University of Tromsø, Tromsø, Norway
e-mail: t.e.mollnes@medisin.uio.no

J.D. Lambris et al. (eds.), *Complement Therapeutics*, Advances in Experimental Medicine and Biology 735,
DOI 10.1007/978-1-4614-4118-2_8, © Springer Science+Business Media New York 2013

signaling activities (Mackman 2009). TF is constitutively expressed in the cells surrounding the blood vessels and large organs, forming a hemostatic barrier (Mackman 2009). The total lethality of homozygous TF knockout mouse embryos indicates that TF is indispensable for life (Siegbahn et al. 2005). During infection, bacteria and lipopolysaccharides (LPS) also induce TF expression in endothelial cells and monocytes (Siegbahn 2000). TF activity may also be present in plasma microparticles derived from activated platelets or apoptotic monocytes in the blood (Henriksson et al. 2005). These intravascular microparticles can fuse with activated platelets (Del Conde et al. 2005) and bind to endothelial cells, resulting in their concentration at sites of injury (Coughlin 2000). In healthy individuals, monocytes express traces of TF encrypted in the cell membrane (Osterud et al. 2008).

The most studied inducer of TF expression in vivo and in vitro is LPS, the main component of the outer membrane of Gram-negative bacteria (Hiller et al. 1977). LPS-induced TF expression by monocytes is dependent on the recognition molecule CD14 (Meszaros et al. 1994). LPS binds to LPS-binding protein (LBP) in the plasma, and the LBP-LPS complex binds to the cell surface receptor CD14 (Schumann et al. 1990). Membrane-bound CD14 then transports LPS to the co-receptor MD-2, which induces Toll-like receptor-4 (TLR-4) dimerization and activates a signaling cascade via myeloid differentiation protein 88 (MyD88) (Malarstig and Siegbahn 2007). Thereafter, the NF*k*B subunits p50 and p65 translocate into the nucleus for the initiation of transcription, leading to the upregulation of TF and many other inflammatory mediators, including a number of cytokines (Oeth et al. 1994). Other proinflammatory mediators, such as immune complexes (Lyberg and Prydz 1982), complement (Prydz et al. 1977), oxidized low density lipoprotein (Eligini et al. 2002), tumor necrosis factor-α (TNF)-α, IL-1α and IL-1β (Osnes et al. 1996), monocyte chemoattractant protein-1 (Ernofsson and Siegbahn 1996), platelet-derived growth factor bb (Ernofsson and Siegbahn 1996), C-reactive protein (Ernofsson et al. 1996), and tethering adhesion molecules (Kappelmayer et al. 1993), also induce TF expression in monocytes in vitro. In a whole blood environment, however, only LPS, immune complexes, and adhesion induce TF expression in monocytes (Osterud 1995).

The activation of the complement system is a hallmark of severe Gram-negative sepsis and is related to disease severity (Brandtzaeg et al. 1989). *E. coli* bacteria mainly activate the alternative pathway (Mollnes et al. 2002). Blocking the C5a-receptor during experimental, sepsis greatly improves survival in rodents, supporting a key role of complement activation in sepsis (Riedemann et al. 2003). LPS-induced septic shock in rats can be mimicked by the injection of C5a, and blocking C5a with an antibody attenuates LPS-induced responses (Ward 2004). However, whole *E. coli* bacteria activate the complement system more potently than *E. coli*-derived LPS (Brekke et al. 2007). *E. coli*-induced complement activation results in the release of C5a, rapidly upregulates CD11b and enhances the oxidative burst in human whole blood granulocytes and monocytes (Mollnes et al. 2002). This *E. coli*-induced CD11b upregulation and oxidative burst were completely blocked by inhibiting C5a and CD14 (Brekke et al. 2007).

Therefore, the purpose of the present study was to examine the separate and combined effects of complement and CD14 inhibition on *E. coli*-induced TF mRNA upregulation, monocyte TF expression, and plasma TF activity in human whole blood. The anticoagulant lepirudin was used because, in contrast to heparin and calcium-binding anticoagulants, it does not affect complement activation (Mollnes et al. 2002).

8.2 Materials and Methods

8.2.1 Antibodies and Reagents

All equipment, including tubes, tips, and solutions, were endotoxin-free. Polypropylene tubes from Nalgene Nunc (Roskilde, DK) were used to obtain low background activation of complement. Sterile phosphate-buffered saline (PBS) with or without Ca^{2+} and Mg^{2+} was from Life Technologies (Paisley, UK).

Lepirudin (Refludan®) was obtained from Hoechst (Frankfurt am Main, Germany). Ultrapure (up) *E. coli* LPS (strain O111:B4) was purchased from InvivoGen (San Diego, CA). Recombinant human apo-tissue factor was purchased from Calbiochem® Merck Biosciences Gmbh (Swalbach, Germany). A potent analog (Ac-I[CV(1MeW)QDWGAHRC]T-NH_2) of the C3 convertase inhibitor compstatin (Katragadda et al. 2006), its corresponding control peptide (Ac-IAVVQDWGAHRAT-NH_2), and the specific C5a-receptor-antagonist (AcF[OPdChaWR]) (C5aRa) were synthesized, as previously described (Strey et al. 2003). Murine antihuman CD14 $F(ab')_2$ (clone 18D11) was obtained from Diatec AS (Oslo, Norway). The optimal inhibitor concentration was determined in separate dose-response experiments (data not shown). All inhibitors were checked for LPS contamination using an endotoxin kit from Cape Cod Inc. (East Falmouth, MA) or QCL-1000 from BioWhittaker (Walkersville, MD), with the final LPS contamination measuring in the low pg/mL range.

8.2.2 Bacterial Preparation

Nonopsonized *E. coli* strain LE392 (ATCC 33572) was obtained from the American Type Culture Collection (Manassas, VA). The bacteria were grown, heat-inactivated, and washed nine times in PBS without Ca^{2+} and Mg^{2+} to remove extracellular LPS, as previously described (Mollnes et al. 2002). A stock solution of 0.5×10^9 bacteria per mL PBS was stored at +4°C. The LPS concentration in the bacterial supernatant was 5 ng/mL and was diluted to 580 pg LPS/mL when the final *E. coli* concentration was 14.2×10^6 cells/mL.

8.2.3 Whole Blood Model of Sepsis

Blood from healthy adult donors was obtained according to guidelines from the Regional Ethics Committee after informed consent. Experiments with each blood sample were performed as single experiments at different time points. In brief, fresh venous human blood was drawn from an antecubital vein into sterile polypropylene tubes containing lepirudin (50 μg/mL). Aliquots (500 μL) of whole blood were then preincubated in PBS with Ca^{2+} and Mg^{2+} or antibodies or inhibitors in PBS (100 μL or 14.3% (v/v)) for 4 min in sterile polypropylene tubes as indicated. *E. coli, E. coli* LPS or PBS (100 μL or 14.3% (v/v)), was then added, and the samples were incubated at 37°C for the time indicated. After incubation, the samples were immediately processed for flow cytometry. The plasma was harvested by centrifugation (3,220 × *g*, 15 min), and samples were stored at −80°C.

8.2.4 RT-qPCR of TF mRNA Expression

Total RNA was extracted from lepirudin-treated whole blood using the total RNA chemistry protocol from Applied Biosystems (ABI, Foster City, CA). Whole blood samples were centrifuged (3,220 × *g*, 15 min), and two volumes of 1× nucleic acid purification lysis solution (ABI) were added to the blood cells. The lysate was stored at −70°C. Total RNA was extracted using an ABI PRISM™ 6100 Nucleic Acid PrepStation in accordance with the manufacturer's protocol. The RNA was quantified by UV spectroscopy using the Gene Quant II (Pharmacia Biotech, Cambridge, UK). Thereafter, cDNA was synthesized from approximately 0.2 μg RNA using the High-Capacity cDNA Archive Kit (ABI) and the 2720 Thermal Cycler (ABI). RT-qPCR was performed with TaqMan Universal PCR Master Mix (ABI) and predeveloped TaqMan® gene expression assays (ABI). The gene expression assays used were TF (Hs00175225_m1) and beta-2-microglobulin (assay ID 4333766F). Beta-2-microglobulin was used as the endogenous control. The PCR was performed in triplicate in a single-plex reaction containing a 25-μL final volume using the 7500 Real-Time PCR System (ABI) in accordance with the manual. The threshold cycles (Ct) were determined by comparing all samples with the zero time calibrator (T0). The T0 value was set to one using the delta-Ct method for relative quantification.

8.2.5 *Flow Cytometry*

TF and CD14 were detected simultaneously using a FITC-conjugated murine antihuman TF antibody (product no. 4507CJ, clone VD8, American Diagnostica Inc., Stamford, CT) and a PE-conjugated anti-CD14 antibody (Becton Dickinson, San Jose, CA). A mouse IgG1-FITC antibody (Becton Dickinson) served as an isotype Fc gamma control. The antibodies were added to 25-μL blood and incubated for 15 min in the dark. Thereafter, 500-μL EasyLyse (Dako, Glostrup, DK) was added, and samples were incubated for another 15 min in the dark. The samples were centrifuged (290 × *g*, 5 min), 300-μL PBS was added, and samples were analyzed using a FACScalibur cytometer from BD. The monocytes were gated using a CD14 PE/SSC dot plot, and the median fluorescence intensity (MFI) for the gamma control was subtracted from the MFI indicating TF expression.

8.2.6 *Quantification of TF Activity in Plasma Microparticles*

Whole blood samples were centrifuged as described above, and the TF functional activity in plasma microparticles from platelet-poor plasma was analyzed as described by Engstad et al. (1995). Briefly, microparticles were isolated from previously isolated platelet-poor plasma by ultracentrifugation at 40,000 × *g* for 1 h 30 min at 4°C, followed by resuspension in 200-μL 0.15 M NaCl and freezing at −70°C. Samples were then thawed and tested for TF activity. TF was measured in a two-stage amidolytic assay based on the ability of TF to accelerate the activation of FX by FVIIa, followed by the FXa conversion of prothrombin to thrombin in the presence of activated FV. Thrombin was quantified using the Th-1 substrate, and the amount of color generated was determined spectrophotometrically at 405 nm using a microplate reader. Crude rabbit brain extract was used as a standard for TF activity, with an undiluted activity assigned at 1,000 mU/mL.

8.2.7 *ELISA of the Terminal Complement Complex (TCC)*

The activation of the complement cascade was analyzed by measuring the soluble C5b-9 terminal complement complex (TCC) using an ELISA as previously described (Mollnes et al. 1985). The monoclonal antibody aE11, which reacts with the C9 neoepitope exposed after incorporation in the C5b-9 complex, was used as the capture antibody, and a biotinylated anti-C6 monoclonal antibody (Quidel Corporation, San Diego, CA) was used as the detection antibody. The results were expressed in arbitrary units/mL (AU/mL), using human serum activated with zymosan as a positive control set to 1,000 AU/mL.

8.2.8 *Cytokine Analysis*

Lepirudin plasma was prepared by centrifugation, as described above. The cytokine levels in the plasma were analyzed using the microsphere-based Bio-Plex Human Cytokine 27-plex (27 different cytokines, interleukins, and growth factors) Assay (Bio-Rad, Hercules, CA), as previously described (Brekke et al. 2008).

8.2.9 *Statistics*

The data were analyzed using the SigmaStat version 3.5 statistical program (Systat Software Inc., Chicago, IL). The data were analyzed by one-way, repeated measures ANOVA using Dunnett's post hoc multiple comparisons. Data were compared to *E. coli* alone after 2 h. The data were transformed logarithmically if the normality test failed. $p<0.05$ was considered statistically significant.

8.3 Results

8.3.1 *The Effect of Selective Complement and CD14 Inhibition on E. coli-Induced TF mRNA Upregulation in Human Whole Blood*

Incubation with *E. coli* (14.2×10^6/mL) for 2 h in fresh human whole blood increased TF mRNA levels approximately sixfold compared to the spontaneous control (Fig. 8.1). The C3 convertase inhibitor compstatin abolished *E. coli*-induced TF mRNA upregulation ($p<0.05$). C5aRa also markedly reduced *E. coli*-induced TF mRNA upregulation, although the reduction did not reach statistical significance. Furthermore, inhibiting both complement and CD14 had no additional effect. In comparison, selective CD14 inhibition using anti-CD14 had no effect on *E. coli*-induced TF mRNA upregulation. Similar results were found using a higher *E. coli* concentration (data not shown). The incubation of whole blood with PBS only for 2 h led to the spontaneous upregulation of TF mRNA (Fig. 8.1, first left column) as compared to baseline levels at time zero, which were set to RQ = 1 in the assay.

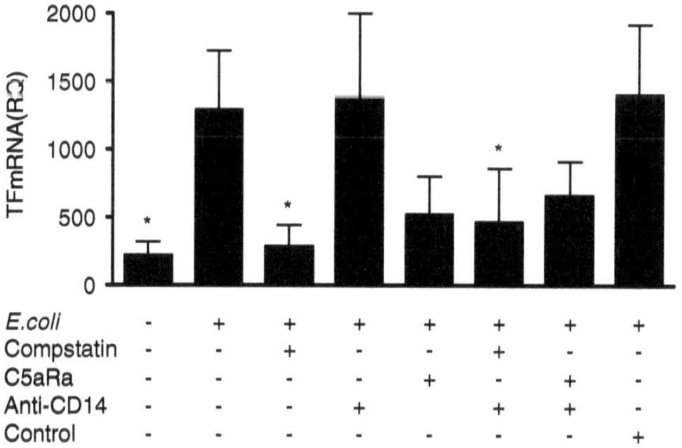

Fig. 8.1 *The effect of selective complement and CD14 inhibition on E. coli-induced TF mRNA upregulation in human whole blood. E. coli* (14.2×10^6/mL) was incubated with human whole blood in the presence of compstatin (25 μM), a C5a receptor antagonist (C5aRa, 10 μg/mL), anti-CD14 F(ab')$_2$ (anti-CD14, 50 μg/mL), or combination thereof. A control peptide (10 μg/mL) and a control F(ab')$_2$ (50 μg/mL) were added in combination as a control (Control). Total RNA was extracted from whole blood and TF expression was analyzed using RT-qPCR. The TF mRNA level in the spontaneous control after incubation with PBS only is indicated (first bar from the *left*). The fold change in TF mRNA relative to the baseline sample at time zero is expressed as relative quantitation (*RQ*). Values are means ±SD (n = 4) using separate blood donors in each experiment. *$p<0.05$ compared to *E. coli* in the absence of inhibitors

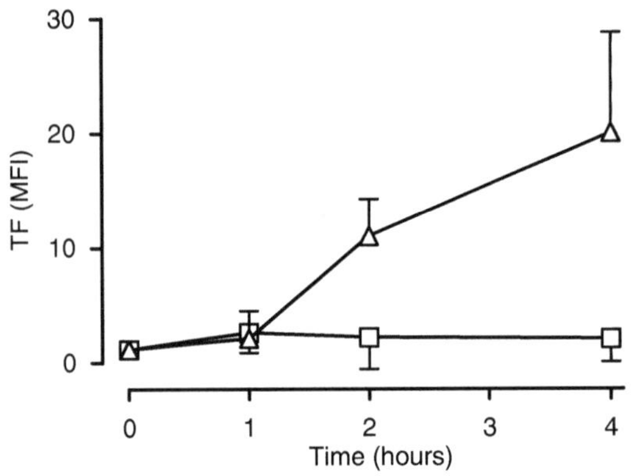

Fig. 8.2 *Time course of E. coli-induced TF expression in whole blood monocytes.* Human whole blood was incubated in the absence (*open square*) or presence (*open triangle*) of *E. coli* (14.2×10^6/mL). The expression of TF on the surface of monocytes was analyzed using flow cytometry at the time points indicated and is expressed as MFI. Values are means ±SD from two to three separate experiments using different blood donors each time

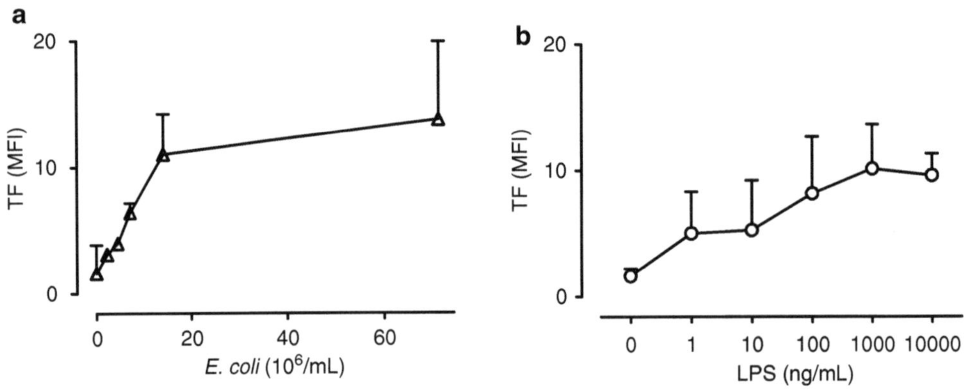

Fig. 8.3 *The dose-response effect of E. coli bacteria and ultra-purified E. coli LPS on monocyte TF expression in human whole blood.* (**a**) The dose-response effect of *E. coli* bacteria (*open triangles*) and (**b**) ultra-purified *E. coli* LPS (*open circles*) on monocyte TF expression. TF surface expression was analyzed using flow cytometry and is expressed as MFI. Values are means ±SD from two to three separate experiments using different blood donors each time

8.3.2 Time Course of E. coli-Induced TF Cell Surface Expression on Monocytes and Dose-Response Effect of LPS

The incubation of whole blood with *E. coli* (14.2×10^6/mL) was continued for up to 4 h (Fig. 8.2). A time-dependent increase in the monocyte TF cell surface expression was observed after 1 h (Fig. 8.2). TF expression by monocytes did not increase in whole blood incubated in the absence of bacteria (Fig. 8.2). The expression of TF on the surface of whole blood granulocytes was not detected using flow cytometry (data not shown); however, cytoplasmic TF staining was not performed in this study.

E. coli dose dependently increased monocyte expression of TF (Fig. 8.3a). Ultra-purified LPS was then added to whole blood at doses ranging from 1 to 10,000 ng/mL, and the incubation continued for

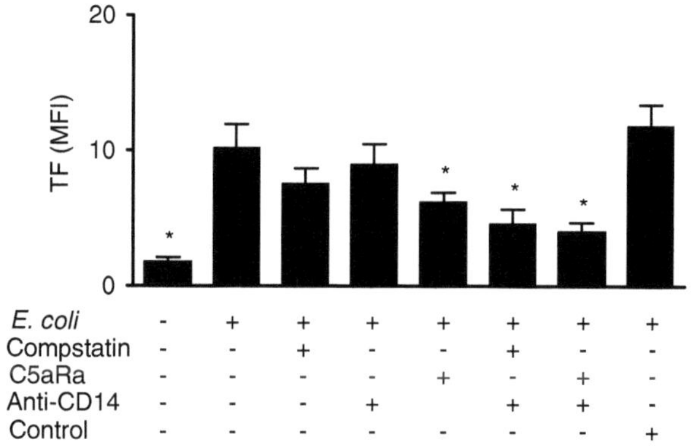

Fig. 8.4 *The effect of selective complement and CD14 inhibition on E. coli-induced TF expression by human monocytes.* *E. coli* (14.2 × 10^6/mL) was added to human whole blood in the presence of compstatin, C5aRa, anti-CD14, or a combination thereof as indicated. The concentrations and abbreviations of inhibitors and controls are the same as indicated in the legend for Fig. 8.1. The expression of TF on the surface of monocytes after 2 h of incubation was analyzed using flow cytometry and is expressed as MFI. The TF expression in the baseline sample after 2 h of incubation with PBS only is shown in the first bar from the left. Values are given as means ±SD (n = 6) using separate blood donors each time. $^*p < 0.05$ compared to *E. coli* in the absence of inhibitors

4 h (Fig. 8.3b). LPS dose dependently increased TF expression by human whole blood monocytes (Fig. 8.3b). Notably, LPS at a dose below 1,000 ng/mL did not activate complement measured as TCC in plasma (data not shown). In comparison, *E. coli* bacteria significantly increased the TCC levels threefold compared to the PBS control after 2 h of incubation (data not shown), which is in accordance with previous findings (Brekke et al. 2007).

To confirm the specificity of the flow cytometry used to detect the expression of TF in monocytes, recombinant human TF (rhTF) was added to human whole blood. The addition of increasing concentrations of rhTF dose dependently blocked the *E. coli*-induced TF MFI signal detected by flow cytometry (data not shown), confirming the specificity of this antibody in this particular method.

8.3.3 The Effect of Selective Complement and CD14 Inhibition on E. coli-Induced TF Surface Expression in Monocytes

Incubating monocytes with *E. coli* for 2 h significantly ($p < 0.05$) enhanced monocyte TF (MFI) expression (Fig. 8.4). The selective inhibition of complement using compstatin or CD14 using anti-CD14 nonsignificantly reduced *E. coli*-induced TF expression by 31% and 14%, respectively. In comparison, treating monocytes with C5aRa alone significantly reduced *E. coli*-induced TF expression by 47% ($p < 0.05$). The combination of compstatin and anti-CD14 significantly reduced the *E. coli*-induced expression of TF in monocytes by 67% ($p < 0.05$), and the combination of C5aRa and anti-CD14 significantly reduced the *E. coli*-induced TF expression by 73% ($p < 0.05$) (Fig. 8.4). Similar effects were also obtained when TF expression was calculated as the percent of all CD14 positive cells that were also TF positive (data not shown). Thus, treatment with anti-CD14 and the complement inhibitors exhibited an additive effect when TF expression was calculated either as MFI or the percentage of TF + cells. The combined effect of increasing doses of compstatin and anti-CD14 on *E. coli*-induced TF expression was then examined. Compstatin alone and compstatin plus anti-CD14 both dose dependently reduced *E. coli*-induced TF expression by monocytes in human whole blood (data not shown).

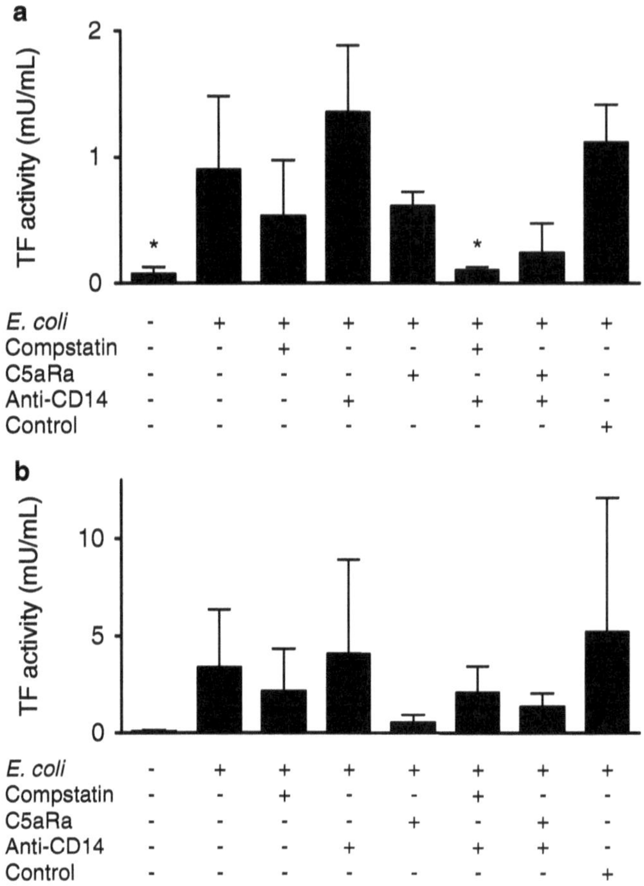

Fig. 8.5 *The effect of selective complement and CD14 inhibition on E. coli-induced TF functional activity in plasma microparticles.* (**a**) A low dose (14.2×10^6/mL) or (**b**) high dose (71×10^6/mL) of *E. coli* was added to human whole blood in the presence of compstatin, C5aRa, anti-CD14, or a combination thereof. The concentrations of inhibitors and controls are indicated in the legend for Fig. 1. The TF activity in plasma microparticles from platelet-poor plasma was analyzed in a two-stage amidolytic assay. The TF activity after 2 h of incubation with PBS only is shown in the first bar from the left. TF activity is expressed as mUnits/mL (mU/mL). Values are means ±SD (n = 2–4). $^*p<0.05$ compared to *E. coli* in the absence of inhibitors

8.3.4 The Effect of Selective Complement and CD14 Inhibition on TF Functional Activity in Plasma Microparticles

Very low baseline levels of TF activity in plasma microparticles were detected at time zero (data not shown). There was no increase in TF activity in samples incubated with PBS between baseline and two hours (Fig. 8.5a, b, first left column). After 2 h of incubation, a low dose of *E. coli* (14.2×10^6/mL) enhanced TF activity 13-fold compared to the spontaneous control (Fig. 8.5a). Selective complement inhibition using compstatin or C5aRa reduced *E. coli*-induced TF activity in plasma microparticles by 44% and 34%, respectively. However, compstatin plus anti-CD14 reduced low-dose *E. coli*-induced TF activity in plasma microparticles by 96%. In comparison, C5aRa and anti-CD14 in combination reduced the *E. coli*-induced TF activity by 79%. In contrast, anti-CD14 alone had no effect on *E. coli*-induced TF activity in plasma microparticles. A high dose of *E. coli* (71×10^6/mL) enhanced TF functional activity approximately 41-fold (Fig. 8.5b). C5aRa was the most efficient treatment, reducing *E. coli*-induced TF activity by approximately 86% ($p<0.05$). Compstatin and anti-CD14 in combination reduced high-dose *E. coli*-induced TF activity by 38%. In comparison, the combination of C5aRa and anti-CD14 reduced *E. coli*-induced TF activity by 60%. Again, anti-CD14 alone had no effect, indicating that CD14 does not play a major role in *E. coli*-induced TF activity in plasma.

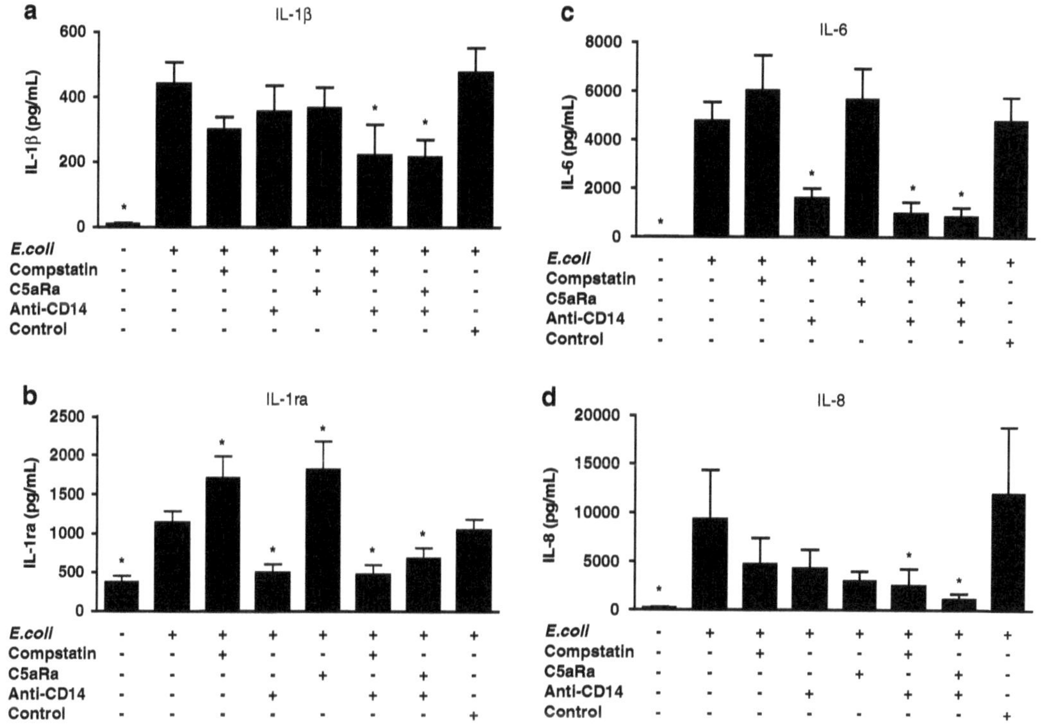

Fig. 8.6 *The effect of selective complement and CD14 inhibition on E. coli-induced IL-1β, IL-1ra, IL-6, and IL-8 synthesis in whole blood.* A low dose of *E. coli* (14.2×10^6/mL) was added to human whole blood in the presence of compstatin, C5aRa, anti-CD14, or a combination thereof and incubated for 2 h. The concentrations of inhibitors and controls are indicated in the legend for Fig. 8.1. (**a**) IL-1β, (**b**) IL-1ra, (**c**) IL-6, and (**d**) IL-8 levels in plasma were analyzed using a multiplex ELISA. Results are expressed as pg/mL. Values are means ±SD (n=6). $^*p<0.05$ compared to *E. coli* in the absence of inhibitors

8.3.5 The Effect of Selective Complement and CD14 Inhibition on E. coli-Induced Cytokine Synthesis

We next examined the effect of complement and CD14 inhibition on *E. coli* (14.2×10^6/mL)-induced cytokine levels in the plasma from six healthy blood donors after 2 h of incubation (Fig. 8.6). The *E. coli*-induced synthesis of IL-1β was only reduced by compstatin or C5aRa and anti-CD14 in combination (Fig. 8.6a). The C3 convertase inhibitor compstatin and C5aRa both significantly reduced *E. coli*-induced IL-8 synthesis but had no effect on the other cytokines examined (Fig. 8.6d). In comparison, anti-CD14 significantly reduced *E. coli*-induced IL-1ra, IL-6, IL-8 (Fig. 8.6b–d), and IP-10 synthesis (Fig. 8.7c). The combination of complement and CD14 inhibition almost completely blocked *E. coli*-induced cytokine synthesis (Figs. 8.6 and 8.7).

8.4 Discussion

This chapter indicates an essential role of complement in *E. coli*-induced TF mRNA upregulation, a combined role of complement and CD14 in TF surface expression by human whole blood monocytes and a major role of complement in TF functional activity in plasma.

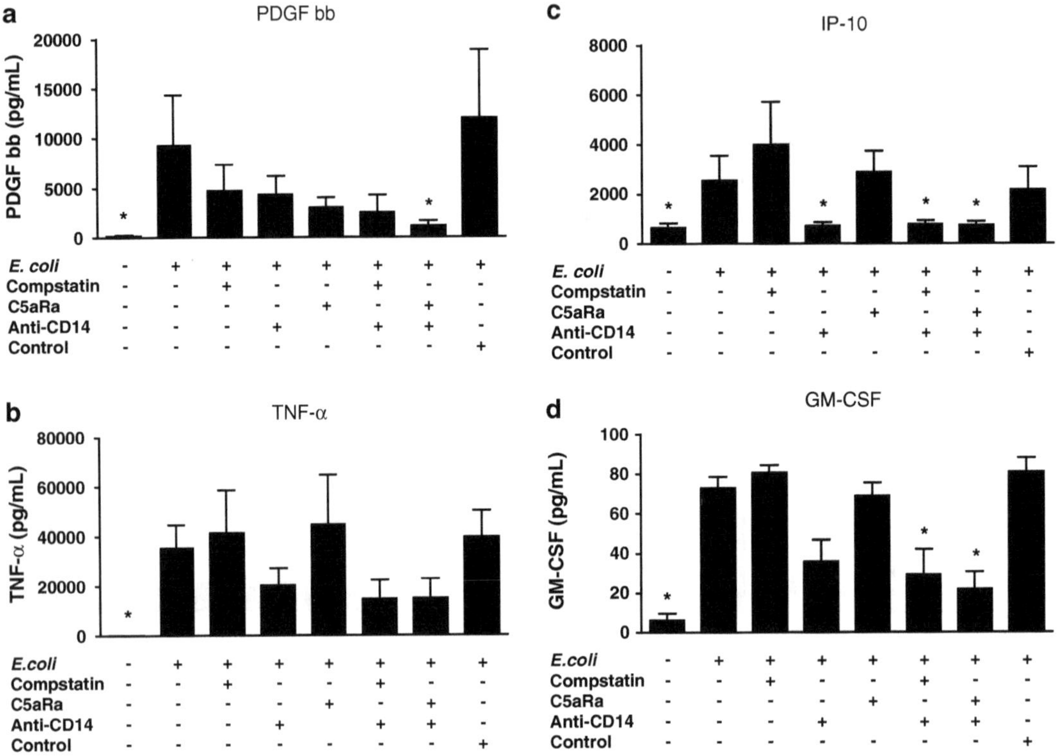

Fig. 8.7 *The effect of selective complement and CD14 inhibition on E. coli-induced PDGFbb, TNF-α, IP-10, and GM-CSF*. A low dose of *E. coli* (14.2×10^6/mL) was added to human whole blood in the presence of compstatin, C5aRa, anti-CD14, or a combination thereof and incubated for 2 h. The concentrations of inhibitors and controls are indicated in the legend for Fig. 8.1. (**a**) PDGFbb, (**b**) TNF-α, (**c**) IP-10, and (**d**) GM-CSF levels in plasma were analyzed using a multiplex ELISA. Results expressed as pg/mL. Values are means ±SD (n=6). $^*p<0.05$ compared to *E. coli* in the absence of inhibitors

We analyzed TF mRNA upregulation, TF expression on the surface of monocytes, and TF functional activity in plasma microparticles and found a distinct TF signal after 2 h of incubation with *E. coli*. The time course for *E. coli*-induced TF expression suggests that new TF is synthesized by increased transcription and potentially the translation of nascent TF molecules as well. This result is supported by the *E. coli*-induced TF mRNA upregulation we observed in whole blood and is in agreement with previous data generated using LPS (Gregory et al. 1989). The upregulation of TF mRNA was analyzed in total mRNA from whole blood in this study, and it is therefore not possible to determine whether this upregulation occurred in monocytes, granulocytes (Kambas et al. 2008), or other cell types.

The significant effect that compstatin had on *E. coli*-induced TF mRNA upregulation indicates that complement activation plays a major role in TF mRNA upregulation. Furthermore, the effect of C5aRa on *E. coli*-induced TF mRNA upregulation suggests that the anaphylatoxin C5a may be involved. The spontaneous increase in TF mRNA during the incubation of otherwise untreated samples indicates that whole blood exposed to polypropylene plastic tubes upregulates TF mRNA in a manner similar to other bio-incompatibility reactions of immunocompetent cells (Nilsson et al. 2007). However, this spontaneous TF mRNA upregulation was not followed by a spontaneous increase in TF expression by monocytes. We also found that anti-CD14 did not inhibit *E. coli*-induced TF mRNA upregulation. Compared to the significant effect compstatin had on this process, our data suggest that the *E. coli*-induced TF mRNA upregulation in human whole blood is almost completely dependent on complement and is CD14 independent. This conclusion is supported by the finding that compstatin alone inhibited the upregulation of TF mRNA better than the combination of compstatin and anti-CD14. Studies using LPS as a stimulant make it difficult to identify the important role played by complement in *E. coli*-induced TF upregulation because LPS only activates complement at high concentrations (Brekke et al. 2007). Compstatin binds to and neutralizes native C3, thereby inhibiting the C3 convertase-mediated cleavage of C3 to C3a and C3b (Sahu et al. 1996). The finding that the combination of compstatin and anti-CD14 significantly reduced TF expression by monocytes indicates that anti-CD14, which did not significantly reduce TF expression on its own, inhibits a different pathway leading to TF expression. Recent studies have shown that TF receives several posttranslational modifications, and we speculate that these processes may be affected by LPS (Mohan Rao and Mackman 2010). Anti-CD14 probably blocks the binding of LPS to membrane-bound CD14 on monocytes, although its effects on soluble CD14 in the plasma may also be involved. The finding that we more than doubled the inhibitory effect on monocyte TF expression by combining compstatin and anti-CD14 suggests that complement and CD14 synergistically affect TF surface expression. The mechanism behind this combined complement and CD14 inhibition is not known, although previous data indicates that it blocks both phagocytosis of *E. coli* and cytokine synthesis (Brekke et al. 2008). In addition, CD14 is an important recognition molecule associated with the functions of TLR2 and TLR3. The significant effect that inhibiting complement and CD14 has on *E. coli*-induced cytokine expression was confirmed by this report after we switched to using Nunc tubes (Brekke et al. 2008). The finding that the C5aRa efficiently reduced TF expression supports the idea that complement plays an important role in mediating TF expression by monocytes. However, treating cells with C5aRa and anti-CD14 in combination further reduced the *E. coli*-induced TF expression by monocytes, suggesting a synergistic cross talk between the C5aR and the TLR-4/CD14/MD2 receptor complex. The main mechanism functioning in this process is probably the complement-mediated inhibition of TF mRNA upregulation. The complement-dependent expression of TF in monocytes that we identified is in accordance with a previous study showing that *N. meningitidis*-induced TF expression was more than 50% dependent on complement (Lappegard et al. 2009). Because the C3 convertase inhibitor compstatin completely blocked *E. coli*-induced mRNA upregulation but only slightly reduced *E. coli*-induced TF expression by monocytes, it is possible that presynthesized TF is exposed on the monocyte surface in response to *E. coli* in a process that is partly dependent on C3 activation. However, the C5aRa significantly reduced *E. coli*-induced TF expression, suggesting that C5a might be involved. Because compstatin inhibits complement and is a small nonimmunogenic peptide, it is a potential therapeutic agent (Janssen et al. 2007). The role of excessive complement activation in experimental sepsis models is further supported by the pivotal role of the C5a–C5a receptor interaction in the development of the systemic inflammatory response and multiorgan failure (Ward 2004; Huber-Lang et al. 2002). In human sepsis, harmful outcomes have been correlated with increased plasma levels of the complement activation products C3a, C4a, and C5a (Nakae et al. 1996). The potential beneficial effects of complement inhibitors in the prevention of tissue factor-induced thrombosis in vivo remain to be elucidated.

Huber-Lang et al. have described a new pathway of complement activation that is C3-independent (Huber-Lang et al. 2006). Their study showed that in the genetic absence of C3, thrombin substitutes for the C3-dependent C5 convertase. The potential role of this pathway in our in vitro sepsis model is difficult to assess due to the fact that we used lepirudin, a thrombin inhibitor, as the anticoagulant. It is possible that lepirudin inhibits this new pathway by using thrombin as an alternative C5 convertase. Whether TF or related factors are activated concomitantly with complement to convert prothrombin to thrombin remains to be determined.

This study indicates that anti-CD14 alone has no effect on *E. coli*-induced TF mRNA upregulation, TF expression by monocytes, or TF activity in whole blood. This may be due to the low concentration of free LPS in our *E. coli* bacteria preparation, which was at the pg/mL level. However, ultra-purified LPS at high concentrations significantly increased TF expression on monocytes, as expected. A study by Steinemann et al. showed that anti-CD14 inhibited LPS-induced TF expression on monocytes, supporting that CD14 is important in LPS-induced TF expression (Steinemann et al. 1994). Although crude LPS preparations were used in their study, Osterud et al. previously showed that complement may play a role in LPS-induced TF (Osterud et al. 1984). However, the addition of low doses of *E. coli* LPS did not stimulate complement activation (Brekke et al. 2007). To the best of our knowledge, there is no previous report on the relative roles of complement and CD14 on *E. coli*-induced TF expression in human whole blood using lepirudin as an anticoagulant.

The combined inhibition of complement and anti-CD14 efficiently reduced *E. coli*-induced cytokine release. The release of IL-1β in the plasma indicates that caspase-1 is activated through the inflammasome, resulting in the subsequent synthesis of pro-IL-1β and the release of IL-1β into the plasma (Van de Veerdonk et al. 2011). The inhibitory effect that compstatin or C5aRa and anti-CD14 in combination have on *E. coli*-induced IL-1β and TF expression indicates that complement activation acts upstream of inflammasome activation. Interestingly, the C3 convertase inhibitor compstatin and C5aRa both increased the synthesis of IL-1ra, which binds to and thereby inhibits the effect of IL-1β (Gabay et al. 2010). We speculate that the C3 convertase inhibitor compstatin partly inhibits inflammation through increased IL-1ra synthesis in addition to inhibiting complement activation. Furthermore, the level of platelet-derived growth factor bb (PDGF bb) was also reduced by the combination of C5aRa and anti-CD14, indicating that adding *E. coli* to whole blood activates platelets and that this activation is reduced by blocking C5aR and CD14. However, the reduction of PDGF bb levels by compstatin or C5aRa alone was not statistically significant. In summary, the inhibition of complement and CD14 significantly reduced the levels of *E. coli*-induced IL-1β and other cytokines in addition to reducing TF expression, suggesting that complement and CD14 block signaling pathways that are upstream of inflammasome activation.

In conclusion, the present data indicate that complement is essential for *E. coli*-induced TF mRNA upregulation and that complement participates in *E. coli*-induced TF expression by monocytes. Adding CD14-inhibition to complement inhibition adds moderately to the expression of TF on human monocytes and, to some extent, plasma TF functional activity. This process seems to be upstream of the actual coagulation cascade itself. The potentially beneficial effects that inhibiting complement has on TF expression during *E. coli* sepsis requires clarification in future studies.

Acknowledgments The technical assistance of Grethe Bergseth is greatly acknowledged. This work was supported by NIH grants AI-068730 and GM 62134 and grants from Helse Nord RHF.

Conflict of Interest Statement The author Professor J.D. Lambris has submitted several patent applications on complement inhibitors. None of the other authors have conflicts of interest.

References

Brandtzaeg P, Mollnes TE, Kierulf P (1989) Complement activation and endotoxin levels in systemic meningococcal disease. J Infect Dis 160(1):58–65

Brekke OL, Christiansen D, Fure H, Fung M, Mollnes TE (2007) The role of complement C3 opsonization, C5a receptor, and CD14 in *E. coli*-induced up-regulation of granulocyte and monocyte CD11b/CD18 (CR3), phagocytosis, and oxidative burst in human whole blood. J Leukoc Biol 81(6):1404–1413

Brekke OL, Christiansen D, Fure H, Pharo A, Fung M, Riesenfeld J et al (2008) Combined inhibition of complement and CD14 abolish *E. coli*-induced cytokine-, chemokine- and growth factor-synthesis in human whole blood. Mol Immunol 45(14):3804–3813

Coughlin SR (2000) Thrombin signalling and protease-activated receptors. Nature 407(6801):258–264

Del Conde I, Shrimpton CN, Thiagarajan P, Lopez JA (2005) Tissue-factor-bearing microvesicles arise from lipid rafts and fuse with activated platelets to initiate coagulation. Blood 106(5):1604–1611

Eligini S, Brambilla M, Banfi C, Camera M, Sironi L, Barbieri SS et al (2002) Oxidized phospholipids inhibit cyclooxygenase-2 in human macrophages via nuclear factor-kappaB/IkappaB- and ERK2-dependent mechanisms. Cardiovasc Res 55(2):406–415

Engstad CS, Lia K, Rekdal O, Olsen JO, Osterud B (1995) A novel biological effect of platelet factor 4 (PF4): enhancement of LPS-induced tissue factor activity in monocytes. J Leukoc Biol 58(5):575–581
Ernofsson M, Siegbahn A (1996) Platelet-derived growth factor-BB and monocyte chemotactic protein-1 induce human peripheral blood monocytes to express tissue factor. Thromb Res 83(4):307–320
Ernofsson M, Tenno T, Siegbahn A (1996) Inhibition of tissue factor surface expression in human peripheral blood monocytes exposed to cytokines. Br J Haematol 95(2):249–257
Gabay C, Lamacchia C, Palmer G (2010) IL-1 pathways in inflammation and human diseases. Nat Rev Rheumatol 6(4):232–241
Gregory SA, Morrissey JH, Edgington TS (1989) Regulation of tissue factor gene expression in the monocyte procoagulant response to endotoxin. Mol Cell Biol 9(6):2752–2755
Henriksson CE, Klingenberg O, Ovstebo R, Joo GB, Westvik AB, Kierulf P (2005) Discrepancy between tissue factor activity and tissue factor expression in endotoxin-induced monocytes is associated with apoptosis and necrosis. Thromb Haemost 94(6):1236–1244
Hiller E, Saal JG, Riethmuller G (1977) Procoagulant activity of activated monocytes. Haemostasis 6(6):347–350
Huber-Lang MS, Riedeman NC, Sarma JV, Younkin EM, McGuire SR, Laudes IJ et al (2002) Protection of innate immunity by C5aR antagonist in septic mice. FASEB J 16(12):1567–1574
Huber-Lang M, Sarma JV, Zetoune FS, Rittirsch D, Neff TA, McGuire SR et al (2006) Generation of C5a in the absence of C3: a new complement activation pathway. Nat Med 12(6):682–687
Janssen BJ, Halff EF, Lambris JD, Gros P (2007) Structure of compstatin in complex with complement component C3c reveals a new mechanism of complement inhibition. J Biol Chem 282(40):29241–29247
Kambas K, Markiewski M, Pneumatikos I, Rafail S, Theodorou V, Konstantonis D et al (2008) C5a and TNF-alpha up-regulate the expression of tissue factor in intra-alveolar neutrophils of patients with the acute respiratory distress syndrome. J Immunol 180:7368–7375
Kappelmayer J, Bernabei A, Edmunds LH Jr, Edgington TS, Colman RW (1993) Tissue factor is expressed on monocytes during simulated extracorporeal circulation. Circ Res 72(5):1075–1081
Katragadda M, Magotti P, Sfyroera G, Lambris JD (2006) Hydrophobic effect and hydrogen bonds account for the improved activity of a complement inhibitor, compstatin. J Med Chem 49(15):4616–4622
Lappegard KT, Christiansen D, Pharo A, Thorgersen EB, Hellerud BC, Lindstad J et al (2009) Human genetic deficiencies reveal the roles of complement in the inflammatory network: lessons from nature. Proc Natl Acad Sci USA 106:15861–15866
Levi M, de Jonge E, van der Poll T (2006) Plasma and plasma components in the management of disseminated intravascular coagulation. Best Pract Res Clin Haematol 19(1):127–142
Lyberg T, Prydz H (1982) Thromboplastin (factor III) activity in human monocytes induced by immune complexes. Eur J Clin Invest 12(3):229–234
Mackman N (2009) The many faces of tissue factor. J Thromb Haemost 7(Suppl 1):136–139
Malarstig A, Siegbahn A (2007) The intersubject variability of tissue factor mRNA production in human monocytes–relation with the toll-like receptor 4. Thromb Res 120(3):407–413
Meszaros K, Aberle S, Dedrick R, Machovich R, Horwitz A, Birr C et al (1994) Monocyte tissue factor induction by lipopolysaccharide (LPS): dependence on LPS-binding protein and CD14, and inhibition by a recombinant fragment of bactericidal/permeability-increasing protein. Blood 83(9):2516–2525
Mohan Rao LV, Mackman N (2010) Factor VIIa and tissue factor – from cell biology to animal models. Thromb Res 125(S1):S1–S3
Mollnes TE, Lea T, Froland SS, Harboe M (1985) Quantification of the terminal complement complex in human plasma by an enzyme-linked immunosorbent assay based on monoclonal antibodies against a neoantigen of the complex. Scand J Immunol 22(2):197–202
Mollnes TE, Brekke OL, Fung M, Fure H, Christiansen D, Bergseth G et al (2002) Essential role of the C5a receptor in *E. coli*-induced oxidative burst and phagocytosis revealed by a novel lepirudin-based human whole blood model of inflammation. Blood 100(5):1869–1877
Nakae H, Endo S, Inada K, Yoshida M (1996) Chronological changes in the complement system in sepsis. Surg Today 26(4):225–229
Nilsson B, Ekdahl KN, Mollnes TE, Lambris JD (2007) The role of complement in biomaterial-induced inflammation. Mol Immunol 44(1–3):82–94
Oeth PA, Parry GC, Kunsch C, Nantermet P, Rosen CA, Mackman N (1994) Lipopolysaccharide induction of tissue factor gene expression in monocytic cells is mediated by binding of c-Rel/p65 heterodimers to a kappa B-like site. Mol Cell Biol 14(6):3772–3781
Osnes LT, Westvik AB, Joo GB, Okkenhaug C, Kierulf P (1996) Inhibition of IL-1 induced tissue factor (TF) synthesis and procoagulant activity (PCA) in purified human monocytes by IL-4, IL-10 and IL-13. Cytokine 8(11):822–827
Osterud B (1995) Cellular interactions in tissue factor expression by blood monocytes. Blood Coagul Fibrinolysis 6(Suppl 1):S20–S25

Osterud B, Olsen JO, Benjaminsen AW (1984) The role of complement in the induction of thromboplastin synthesis. Haemostasis 14(5):386–392
Osterud B, Breimo ES, Olsen JO (2008) Blood borne tissue factor revisited. Thromb Res 122(3):432–434
Prydz H, Allison AC, Schorlemmer HU (1977) Further link between complement activation and blood coagulation. Nature 270(5633):173–174
Riedemann NC, Guo RF, Ward PA (2003) A key role of C5a/C5aR activation for the development of sepsis. J Leukoc Biol 74(6):966–970
Sahu A, Kay BK, Lambris JD (1996) Inhibition of human complement by a C3-binding peptide isolated from a phage-displayed random peptide library. J Immunol 157(2):884–891
Schumann RR, Leong SR, Flaggs GW, Gray PW, Wright SD, Mathison JC et al (1990) Structure and function of lipopolysaccharide binding protein. Science 249(4975):1429–1431
Siegbahn A (2000) Cellular consequences upon factor VIIa binding to tissue factor. Haemostasis 30(Suppl 2):41–47
Siegbahn A, Johnell M, Sorensen BB, Petersen LC, Heldin CH (2005) Regulation of chemotaxis by the cytoplasmic domain of tissue factor. Thromb Haemost 93(1):27–34
Steinemann S, Ulevitch RJ, Mackman N (1994) Role of the lipopolysaccharide (LPS)-binding protein/CD14 pathway in LPS induction of tissue factor expression in monocytic cells. Arterioscler Thromb 14(7):1202–1209
Strey CW, Markiewski M, Mastellos D, Tudoran R, Spruce LA, Greenbaum LE et al (2003) The proinflammatory mediators C3a and C5a are essential for liver regeneration. J Exp Med 198(6):913–923
Van de Veerdonk FL, Netea MG, Dinarello CA, Joosten LA (2011) Inflammasome activation and IL-1beta and IL-18 processing during infection. Trends Immunol 32(3):110–116
Ward PA (2004) The dark side of C5a in sepsis. Nat Rev Immunol 4(2):133–142

Chapter 9
CR2-Mediated Targeting of Complement Inhibitors: Bench-to-Bedside Using a Novel Strategy for Site-Specific Complement Modulation

V. Michael Holers, Bärbel Rohrer, and Stephen Tomlinson

Abstract Recent approval of the first human complement pathway-directed therapeutics, along with high-profile genetic association studies, has catalyzed renewed biopharmaceutical interest in developing drugs that modulate the complement system. Substantial challenges remain, however, that must be overcome before widespread application of complement inhibitors in inflammatory and autoimmune diseases becomes possible. Among these challenges are the following: (1) defining the complement pathways and effector mechanisms that cause tissue injury in humans and determining whether the relative importance of each varies by disease, (2) blocking or modulating, using traditional small molecule or biologic approaches, the function of complement proteins whose circulating levels are very high and whose turnover rates are relatively rapid, especially in the setting of acute and chronic autoimmune diseases, and (3) avoiding infectious complications or impairment of other important physiological functions of complement when using systemically active complement-blocking agents. This chapter will review data that address these challenges to therapeutic development, with a focus on the development of a novel strategy of blocking specific complement pathways by targeting inhibitors using a recombinant portion of the human complement receptor type 2 (CR2/CD21) which specifically targets to sites of local complement C3 activation where C3 fragments are covalently fixed. Recently, the first of these CR2-targeted proteins has entered human phase I studies in the human disease paroxysmal nocturnal hemoglobinuria. The results of murine translational studies using CR2-targeted inhibitors strongly suggest that a guiding principle going forward in complement therapeutic development may well be to focus on developing strategies to modulate the pathway as precisely as possible by physically localizing therapeutic inhibitory effects.

V.M. Holers (✉)
Departments of Medicine and Immunology, University of Colorado School of Medicine, Aurora, CO, USA
e-mail: michael.holers@ucdenver.edu

B. Rohrer
Department of Ophthalmology, Medical University of South Carolina, Charleston, SC, USA

Ralph H. Johnson Veterans Affairs Medical Center, Charleston, SC, USA

S. Tomlinson
Department of Microbiology and Immunology, Medical University of South Carolina, Charleston, SC, USA

Ralph H. Johnson Veterans Affairs Medical Center, Charleston, SC, USA

J.D. Lambris et al. (eds.), *Complement Therapeutics*, Advances in Experimental Medicine and Biology 735,
DOI 10.1007/978-1-4614-4118-2_9, © Springer Science+Business Media New York 2013

9.1 Overview: Local Complement Activation in Human Disease Pathogenesis

The complement system plays a central role in the pathophysiology of many human diseases [reviewed in Holers (2001, 21.1–21.8; 2008, 223:316)]. This interaction is complex, however, because of the dual roles of complement, being both protective in its ability to clear circulating immune complexes and apoptotic debris as well as damaging when inappropriately activated locally on self-tissues, especially in the disease systemic lupus erythematosus (SLE) but also likely in other conditions (Carroll 1998b, 16:545–568; 2000, 74:61–88). Nevertheless, because complement activation fragments, especially those derived from C3, are present in essentially all autoimmune, inflammatory, and ischemic human diseases, implying a potential pathogenic role, therapeutic manipulation of the system in humans has remained a major goal in drug development (Ricklin and Lambris 2007, 25:1265–1275).

The complement system consists of soluble activation pathway proteins found in the blood and tissues, as well as membrane-bound receptors and both soluble and membrane-bound regulatory proteins (Table 9.1). The biologic roles of the complement system depend on the specific context in which activation occurs and the number and levels of effector molecules that are generated. The beneficial effects of complement activation that have been described are many and include recognition and clearance of foreign and nonself-antigens and pathogens (Fearon and Locksley 1996, 272:50–54), modulation of both humoral (Carroll 1998, 16:545–568) and cellular (Kaya et al. 2001, 2:739–745) immune responses, enhancement of phagocytosis (Brown 1991, 3:76–82), noninflammatory clearance of self-antigens derived from apoptotic processes (Botto and Walport 2002, 205:395–406), transport of immune complexes for disposal (Davies et al. 1992, 90:2075–2083), tissue regeneration following injury (Strey et al. 2003, 198:913–923), responding to endogenous danger and activation signals (Holers and Carroll 2005, 86:137–157), and shaping of the natural antibody repertoire (Holers 2005, 26:405–423).

A key concept for the development of complement therapeutic approaches, however, is that misdirection of the system occurs locally and results in tissue-specific damage and impaired function. Even in diseases in which complement is viewed to be "systemically activated," such as SLE, injury in

Table 9.1 Complement regulatory proteins

Soluble control proteins	Serum concentration (μg/mL)	Molecular weight	Activities
Positive regulation			
Properdin	25	220,000	Stabilize AP C3/C5 convertases
Negative regulation			
C1-INH	200	105,000	Inhibits C1r/C1s, MASPs
C4-bp	250	550,000	Decay-acceleration classical pathway, cofactor activity C4b
Factor H	500	150,000	Decay-acceleration alternative pathway, cofactor activity C3b
Factor I	34	90,000	Cleavage of C3b/C4b
Anaphylatoxin inactivator (carboxypeptidase H)	35	280,000	Generates C3a/C5a desArg
S protein (vitronectin)	500	80,000	Blocks MAC formation
SP-40,40 (clusterin)	60	80,000	Blocks MAC formation
Membrane regulatory proteins			
Decay-accelerating factor (CD55)		70,000	Decay-acceleration classical and alternative pathways
Membrane cofactor protein (CD46)		45,000–70,000	Cofactor activity classical and alternative pathways
CD59		20,000	Blocks C8–C9 and C9
CRIT		30,000	Blocks C2

individual patients is typically focused on a small set of specific tissues such as the kidney (Tsokos 2012, 365:2110–2121). In each of these pathogenic settings, the powerful effector mechanisms of complement that should be focused on foreign pathogens and antigens appear to be instead redirected to self and cause local tissue damage. Importantly, in a disease context when one desires to block complement activation locally, for example, to protect the kidney, it is also highly desirable to maintain positive protective functions at other sites, including the circulation where immune complexes are processed and delivered to the reticuloendothelial system, and apoptotic self-tissues are cleared (Botto and Walport 2002, 205:395–406). This is especially true in autoimmune diseases such as SLE (Davies et al. 1992, 90:2075–2083; Davies et al. 1994, 15:397–416) but is also relevant to hemolytic syndromes such as paroxysmal nocturnal hemoglobinuria (PNH) (Rosse and Nishimura 2003, 77:113–120), chronic local inflammatory diseases such as macular degeneration (Gehrs et al. 2006, 38:450–471), and many other circumstances.

Another important consideration in the complement system is that it is activated by three interacting recognition processes, the classical, alternative, and lectin pathways. The classical pathway is initiated by IgM- or IgG-containing immune complexes, C-reactive protein (CRP), beta amyloid fibrils, serum amyloid P or tissue damage products such as mitochondrial membrane proteins that directly bind C1q (Lachmann and Hughes-Jones 1984, 7:143–162). The alternative pathway undergoes a continuous process of "tickover" that results in activation that is initiated in the fluid phase but is quickly localized onto a surface (Muller-Eberhard 1988, 57:321–347). In this system, certain substances such as repeating polysaccharides, endotoxin, IgA-containing immune complexes, C3 nephritic factor (C3Nef) autoantibody, and some immunoglobulin light chains serve to promote alternative pathway activation. In addition, this pathway also serves as an "amplification loop" when C3b that is deposited by any pathway serves to bind factor B and initiates further cleavage events through the alternative pathway. The lectin pathway is initiated by mannose-binding lectin (MBL) or by ficolins, carbohydrate-binding proteins that are involved in the clearance of foreign organisms such as bacteria and viruses, but it is also activated on local tissues following injury and by agalactosyl (G0) carbohydrates on IgG and IgM in patients such as those with rheumatoid arthritis (Gadjeva et al. 2004, 41:113–121; Reid and Turner 1994, 15:307–325; Runza et al. 2008, 213:297–306), as well as certain pathogenic natural antibodies (Jordan et al. 2001, 1413–1418).

Importantly, through each of the three major processes by which the complement system is initiated, convergence upon the centrally important C3 protein occurs when multi-component C3 convertases (activating enzymes) are formed and C3 is cleaved. The presence of a thioester bond in C3 allows the covalent attachment of this protein in the C3b form through ester or amide linkages to other molecules during the process of activation (Lambris and Sahu 2001, 180:35–48). This irreversibly "marks" the attached targets as immunologically different and allows the subsequent cleavage of C3b to the iC3b/C3dg/C3d forms and binding of the complex to specific C3 fragment receptors (Carroll 1998a, 16:545–568). The activation of C3 is followed by the formation of a C5 convertase, resulting in C5 cleavage and activation, coincident release of soluble anaphylatoxins C3a and C5a, and formation of the pore-like membrane attack complex (MAC) (Wetsel 1995, 7:48–53). It is the deposition of tissue-localized C3 fragments and their subsequent processing to fragments with altered receptor-binding characteristics that underlies the tissue-specific targeting of therapeutics described in this chapter.

9.2 Rationale and Strategies for Development of Targeted Complement Inhibitors

While a great number of different approaches to inhibit complement have been successfully applied in experimental models of disease, few have shown any clinical benefit. Indeed, after more than 40 years of research and development efforts, only two types of complement inhibitors (anti-C5 monoclonal antibody (Rother et al. 2007, 25:1256–1264) and C1-esterase inhibitors (Cicardi and Zanichelli

2012, 46:867–874)) have received FDA approval and only then for very limited applications. There are various reasons why the translation of complement inhibitory strategies to the clinic has failed, but significant contributing factors are the poor bioavailability, the apparent redundant functions of effector pathways (C5aR, C3aR, CR1, CR3/CR4, and MAC) in many disease settings, the large levels of target proteins (e.g., circulating C3 levels are >1.0 mg/mL) with half-lives of a few days, and the need to maintain the protective roles of complement to avoid adverse events. With regard to the latter, it has long been known that complement is an important immune effector mechanism for host defense, but complement can also shape an adaptive immune response, whether directed against a pathogen, an allograft, cancer, or self. As noted above, complement also plays important roles in immune homeostasis mechanisms such as the clearance of immune complexes and apoptotic cells, the effective clearance of which is important for the resolution of inflammation.

Complement also plays important roles in tissue repair and regeneration, not only in terms of removing dead and injured cells, but also through direct signaling mechanisms mediated by complement activation products. There is also cross-talk between the complement, contact, and coagulation systems, and it is important to consider how complement inhibition negatively affects these other systems. Finally, in addition to the above considerations, in many cases, there is an incomplete understanding of the role of complement in the pathogenic and protective mechanisms that are specific for a particular disease; it is clear that the different complement pathways and different complement activation products can contribute differently to a given disease in terms of both pathological injury and protective mechanisms important for a given disease.

Early investigations into complement inhibition focused mainly on systemically blocking the activation of C3, the central step in the complement cascade. However, from the foregoing, it follows that systemic and nonspecific inhibition of complement will not be an optimal therapeutic approach under many circumstances. More recently, two approaches for more specific complement inhibition have been investigated, each with the goal of minimizing the generation of disease-specific pathogenic complement mediators, while minimizing disruption of complement-dependent protective processes. These approaches involve either the targeted inhibition of a particular complement pathway or activation product or the site-specific targeting of a complement inhibitor. With regard to the former approach, the list of molecules that have been investigated is extensive, but examples include monoclonal antibodies to factor D or factor B that block the alternative pathway, anti-C5 monoclonal antibodies and other soluble molecules that block the terminal pathway, and antibodies and antagonists to specific complement receptors [reviewed in Ricklin and Lambris (2007, 25:1265–1275)].

Herein, however, the focus is on site-specific physical targeting of complement inhibition. The basic rationale behind the development of site-targeted complement inhibitors is the concept that localized delivery will not only increase inhibitor bioavailability at sites of disease but also will minimize systemic complement inhibition that may otherwise disrupt the normal physiological and protective mechanisms of complement. In addition, although the benefits of targeted inhibition may be particularly relevant for the treatment of chronic conditions, short-term disruption of homeostatic and host defense mechanisms may not be desirable even during treatment of acute conditions, especially when risk of infection is high, such as in patients that are immunocompromised and in patients undergoing a surgical procedure or suffering from a traumatic injury.

9.3 Complement Receptors and Regulatory Proteins Used to Construct Targeted Inhibitors

One important feature of the use of complement therapeutics targeting to date is the incorporation of fragments of naturally occurring complement inhibitors into the chimera. This section reviews these proteins and the rationale for their use. The core tissue-targeting element for the primary focus of this

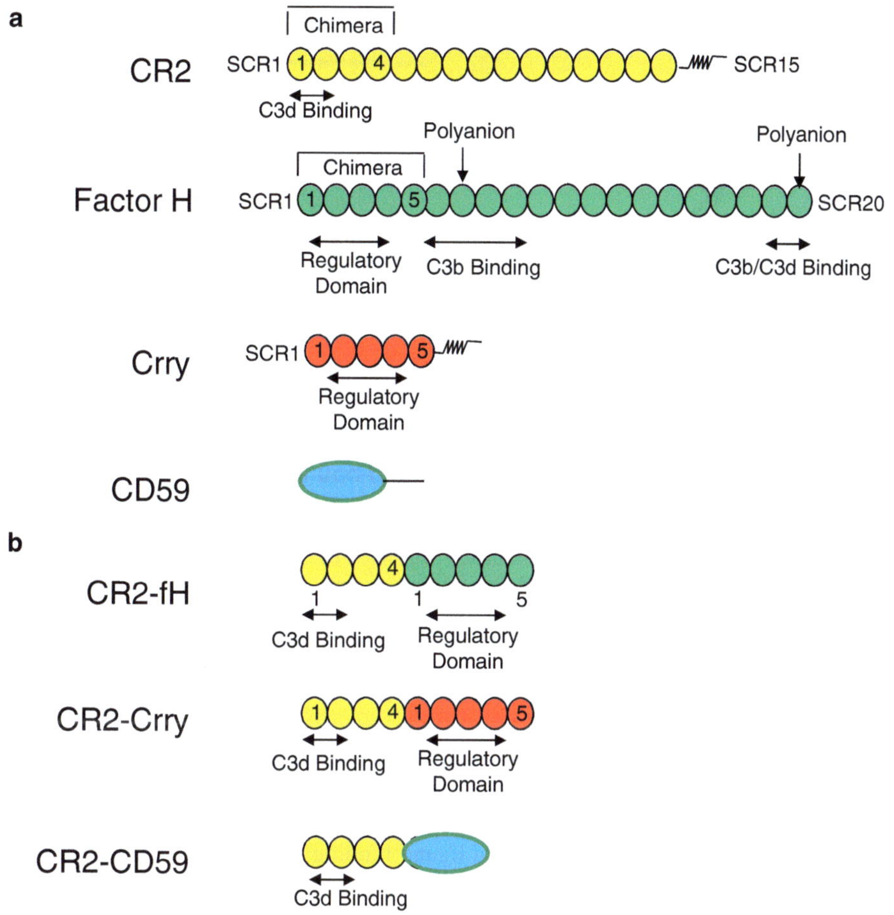

Fig. 9.1 (**a**) Shown are schematic views of the proteins utilized in the targeted inhibitors discussed in this review. (**b**) CR2-targeted inhibitors whose activities are described herein

review is provided by a fragment of CR2, which in its normal role is a multifunctional receptor expressed on B cells, follicular dendritic cells, and a subset of peripheral and thymic T cells [reviewed in Carroll (1998a, 16:545–568); Holers and Boackle (2004, 7:33–48)]. CR2 has several ligands including complement C3 fragments iC3b/C3dg and C3d (Iida et al. 1983, 158:1021–1033; Kalli et al. 1991, 147:590–594), which are covalently attached to target antigens, as well as the gp350/220 viral coat protein of the Epstein-Barr virus (Weis et al. 1988, 167:1047–1066), the immunoregulatory protein CD23 (Aubry et al. 1992, 358:505–507), and interferon-alpha (Asokan et al. 2006, 177:383–394). Notably, the only "tissue-fixed" ligands of CR2 are the C3 fragments iC3b/C3dg/C3d.

CR2 is an ~145-Kd type I transmembrane protein comprised of 15 SCRs followed by a 28-amino acid transmembrane domain and a relatively short 34-amino acid intracytoplasmic tail (Carel et al. 1989, 143:923–930; Weis et al. 1988, 167:1047–1066) (Fig. 9.1). SCRs are 60–70-amino acid long protein modules that comprise the majority of the structures of CR2 and other members of its extended gene family called the regulators of complement activation (RCA) (Hourcade et al. 1989, 45:381–416). CR2 acts as a B cell co-receptor for antigen receptor-mediated signal transduction. On B cell lines or primary B lymphocytes, co-ligation of CR2 with surface IgM, using either mAbs, covalently linked complexes of antigen with C3d ligand, or biotin-conjugated C3dg complexed with biotinylated anti-IgM, results in enhanced intracellular calcium release, proliferation, and/or activation of MAP

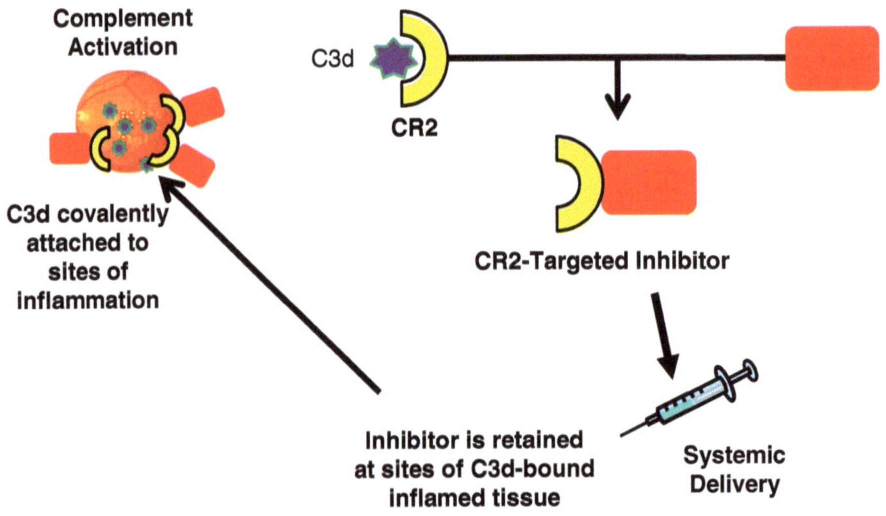

Fig. 9.2 Schematic view of the process of CR2-mediated targeting. The C3 fragments iC3b/C3dg/C3d are covalently attached to sites of complement activation such as the retinal pigment epithelium as indicated. By linking the capability of the CR2 ligand-binding domain that recognizes these fragments, fusion proteins that contain CR2 and the regulatory domain of proteins shown in Fig. 9.1 can be built. When administered systemically, data described herein show that the targeted inhibitor is physically retained at the site of activation and demonstrates enhanced therapeutic characteristics

kinases [reviewed in Carroll (1998a, 10:279–286); Carter et al. (1998, 141:457–463)]. This activity is primarily due to the association of CR2 with CD19 and CD81 in a B cell-specific signal transduction complex.

Structure-function relationships that govern ligand binding to CR2 have been studied by several groups, and what has emerged is a consistent finding that the high-affinity interaction with C3dg and C3d occurs in the SCR 1–2 domain (Lowell et al. 1989, 170:1931–1946; Molina et al. 1995, 154:5426–5435). In contrast to C3dg and C3d, it is not clear whether iC3b binds just to SCR 1–2, as supported by some studies (Kalli et al. 1991, 147:590–594), or to other sites outside of C3dg as supported by others (Esparza et al. 1991, 21:2829–2838; Sarrias et al. 2001, 167:1490–1499). Nevertheless, it is clear that the SCR 1–4 domain alone does interact with iC3b, and recently published studies such as the x-ray crystallography of CR2 with C3d (van den Elsen and Isenman 2011, 332:608–611) support the presence of a high-affinity binding site for all three C3 fragment forms (iC3b, C3dg, C3d) within SCR 1–2. Prior data suggested that SCR 3–4 could help to promote the interaction of CR2 with its ligands in a manner that is not yet understood (Carel et al. 1990, 265:12293–12299).

Another important component of the complement-targeted therapeutic proteins are the complement inhibitory portions, which to date have been derived as recombinant fragments of naturally occurring inhibitors. The inhibitors that have been utilized include the human proteins factor H, decay-accelerating factor (DAF/CD55), and CD59, as well as the murine proteins Crry (complement receptor-related gene/protein y), CD59, and factor H (Table 9.1, Fig. 9.1). With regard to the normal biological roles of these proteins, because of the potent activities manifest by complement, it is a tightly regulated system with inhibitors that act at many steps of the pathway (Liszewski et al. 1996, 61:201–283). These proteins utilize differing mechanisms to block complement activation, but the major activities manifest at the points of C3 and C5 activation acting as a competitive inhibitor, defined as decay-accelerating activity, of multi-component enzymes and/or a cofactor for proteolytic cleavage of C3b and/or C4b. Because of the presence of these many inhibitors at sites in which activators can be found or deposited, the state of the complement pathway at any one point can be considered to reflect the relative balance of these two opposing forces.

In the fluid phase, one major protein that serves to block complement activity at the C3/C5 activation steps is factor H, which serves to accelerate the normally slow spontaneous decay of Bb from C3b in the alternative pathway, thus disabling both the C3 and C5 convertases, and which also exhibits cofactor activity for factor I-mediated cleavage of C3b into the hemolytically inactive form iC3b [reviewed in Zipfel et al. (2002, 30:971–978)]. Factor H is a 150-kDa serum protein with a structure consisting of 20 SCRs. Factor H appears to bind to target cells and tissues primarily through initial contact by SCRs 19–20 with tissue that expresses polyanions and fixed C3b/C3d, followed by the elaboration of protective complement alternative pathway C3 convertase inhibitory function on the tissue itself.

On the cell membrane, other proteins serve similar functions to block complement activation. These proteins include DAF (decay-accelerating factor, CD55), MCP (membrane cofactor protein, CD46), and CD59. DAF is an approximately 70-kDa glycoprotein that, similarly to factor H, acts by binding C3b or C4b on the cell membrane and markedly increasing the spontaneous decay of both the classical pathway C4b2a and alternative pathway C3bBb complexes (Liszewski et al. 1996, 61:201–283). Crry is a rodent-specific complement regulator protein which manifests the activities of DAF and MCP within one widely distributed membrane protein [reviewed in Molina et al. (1992, 175:121–129); Molina (2002, 59:220–229)].

CD59 is a widely distributed, highly disulfide-linked protein with a GPI anchor (Morgan et al. 2005, 12:258–265). CD59 is found on erythrocytes, monocytes, granulocytes, platelets, endothelial cells, and many cells of the nervous and reproductive tissues. With regard to complement regulation, CD59 is able to bind C8 in the C5b-8 complex and to block the effective incorporation of C9. In addition, CD59 is capable of binding C9 already in the MAC and of blocking the subsequent polymerization of C9 and full formation of the transmembrane pore.

9.4 General Principles of Targeted Complement Therapeutics

The first reports on site-specific targeted complement inhibition appeared in 1999 and described two basic approaches. In one approach, cell-specific targeting of CD59 was achieved by linking the inhibitor to an antibody or antibody fragment. The fusion proteins specifically targeted cells expressing cognate antigen and provided targeted, but not untargeted cells, with effective protection from complement-mediated lysis and injury (Zhang et al. 1999, 103:55–61). A similar targeting approach was shown to also significantly enhance the complement inhibitory activity of DAF at a targeted cell surface (Zhang et al. 2001, 276:27270–27275). In the other approach, it was shown that the decoration of soluble CR1 (sCR1) with sialyl Lewis^x (sLE^x) moieties significantly enhanced the protective effect of the inhibitor in a rat model of selectin-dependent lung injury (Mulligan et al. 1999, 162:4952–4959) and in a mouse model of ischemic stroke (Huang et al. 1999, 285:595–599). The sLE^x carbohydrate moiety binds to both P and E selectins, adhesion molecules that are upregulated on activated endothelium. In the lung injury model, the enhanced protective effect of sCR1sLE^x correlated with increased binding of sCR1sLe^x to the lung vasculature when compared to binding of sCR1. sCR1sLE^x also localized to the cerebral vasculature in the stroke model, and in vitro experiments demonstrated that sCR1sLE^x binds to E-selectin and blocks P-selectin-mediated cellular adhesion. Thus, decoration with the sLE^x moiety represents a strategy to increase the efficacy of complement inhibitors by targeting them to selectin-expressing activated endothelial surfaces. Importantly, however, the efficacy of sCR1sLE^x was still dependent on systemic complement inhibition.

A derivative of sCR1 has also been targeted to cell membranes, albeit nonspecifically, by incorporation of a membrane-targeting myristoylated peptide. A membrane-targeted derivative of sCR1 (APT070), consisting of the 3 N-terminal SCRs of CR1, was shown to be 100-fold more active than its parent protein in in vitro complement inhibition assays (Smith 2002, 30:1037–1041); however, the untargeted CR1(SCR 1–3) was itself much less active than sCR1. APT070 was initially characterized in vivo in a rat model of antigen-induced arthritis, where following intra-articular injection, APT070 was retained on cell membranes within the joint and was significantly more protective than its untargeted counterpart (Linton et al. 2012, 43:2590–2597). APT070 has also demonstrated benefit in other models such as intestinal ischemia-reperfusion injury (Souza et al. 2005, 145:1027–1034) and a model of Miller Fisher syndrome (Halstead et al. 2005, 58:203–210). Another application that seems particularly well suited to this strategy of membrane localization is graft protection after cold storage, as perfusion of donor rat kidneys with APT070 reduced tubular injury and increased graft survival after cold ischemia and transplantation (Patel et al. 2006, 17:1102–1111). A similar targeting strategy has also been applied to CD59, and membrane-targeted CD59 has been shown to bind to erythrocytes and provide them with protection from complement-mediated lysis in vitro and in vivo (Fraser et al. 2003, 278:48921–48927).

Another approach to the targeting of complement inhibitors has been developed using single-chain antibody (scFv) fragments as delivery vehicles. Human DAF and mouse Crry were shown to target mouse RBCs when linked to an scFv specific for mouse glycophorin A (Spitzer et al. 2005, 175:7763–7770). Both scFv-DAF and scFv-Crry fusion proteins effectively protected mouse RBCs from complement lysis in vitro, and the scFv-Crry construct restricted complement activation in vivo and reversed a complement sensitive Crry-deficient RBC phenotype. Rat Crry and CD59 have also been linked to an scFv specific for an antigen expressed on rat tubular epithelium (He et al. 2005, 174:5750–5757). The two kidney-targeted inhibitors were equally effective at preventing injury in experimental nephrotic syndrome and were markedly more effective than their untargeted counterparts. In addition, the scFv-inhibitor constructs had very short circulatory half-lives (around 30 min) and provided effective protection against renal injury without systemically inhibiting complement, an important concept in the development of site-specific targeted complement inhibitors.

A somewhat different strategy that has been applied to target complement inhibition involves the use of a prodrug. In this approach, DAF and CD59 were attached to an antibody Fc region by means of a linker incorporating a cleavage site for matrix metalloproteases and/or aggrecanases (Harris et al. 2002, 30:1019–1026). The fusion proteins had limited functional activity (thus limiting their ability to systemically inhibit complement), but an active form of the inhibitor was released upon exposure to certain enzymes that are present in high concentrations at sites of inflammation. Enzymes present in supernatants of cytokine-stimulated chondrocytes, or synovial fluid itself, were both shown to effectively release active DAF from Fc fusion proteins. Finally, complement inhibitors have been targeted to sites of complement activation by means of a CR2-derived delivery vehicle, and that approach is the focus of the remainder of this review.

9.5 In Vitro and In Vivo Studies Validating CR2-Mediated Targeting Concept

CR2-complement inhibitor fusion proteins, specifically CR2-DAF and CR2-CD59, were initially validated using human component proteins in vitro (Song et al. 2003, 111:1875–1885). It was shown that both CR2-DAF and CR2-CD59 bound to C3-opsonized CHO cells and erythrocytes and that they were markedly more effective than the corresponding untargeted inhibitor at providing target cells with protection from complement. The leukocyte receptor CR3 shares the same iC3b ligand with CR2, and it was also shown that the CR2-fusion proteins inhibited CR3-dependent adhesion of U937 cells to C3-opsonized erythrocytes, suggesting an additional potential anti-inflammatory mechanism of these proteins. Fusion proteins with CR2 linked to the C-terminus of each inhibitor were also characterized (i.e., DAF-CR2 and CD59-CR2), but both of the C-terminal fusion proteins were less potent inhibitors than their N-terminal counterparts. Kinetic analysis by surface plasmon resonance revealed that the N-terminal fusion proteins bound C3d with only slightly higher affinity than the C-terminal fusion proteins but that they exhibited much faster association and dissociation rates. This raised the possibility that this fast on-off rate may contribute to efficacy by providing cell surface mobility and that the frequent release of the inhibitor from a fixed orientation may allow for frequent opportunities for a more favorable positioning of the inhibitor relative to its ligand. It is also noteworthy that soluble untargeted CD59 has very poor activity in vitro, even at high concentration, and that for therapeutic application of a CD59 inhibitory strategy, targeting will likely be a requirement.

The in vivo validation of the use of CR2-targeted complement inhibitors in a rodent model required the generation and characterization of a rodent fusion protein. This is because of differences in the relative species-selective activity of complement inhibitors, as well as immunogenicity of foreign proteins. Therefore, a mouse CR2-Crry construct was prepared for characterization of the targeting approach in experimental murine models of disease. CR2-Crry was initially characterized in a mouse model of intestinal ischemia-reperfusion injury, and its efficacy and characteristics were compared with a systemically inhibitory counterpart, Crry-Ig (Atkinson et al. 2005, 115:2444–2453). CR2-Crry was shown to specifically localize to sites of complement activation following intestinal reperfusion and to provide protection from both local (intestine) and remote (lung) complement deposition and injury. Notably, CR2-Crry was 20-fold more effective than non-targeted Crry-Ig on a molar basis in this acute model.

Furthermore, the minimum effective dose of Crry-Ig that provided protection from ischemia-reperfusion injury also markedly increased susceptibility to infection in a model of polymicrobial sepsis, whereas even multiple injections of CR2-Crry had no effect on host defense (Atkinson et al. 2005, 115:2444–2453). In this context, it was also shown that the circulatory half-life of CR2-Crry was only 8.7 h (compared to 40 h for Crry-Ig) and that a therapeutic dose of CR2-Crry had minimal effect on serum complement activity. Because targeted inhibitors are retained and should persist in local targeted tissue, a short circulatory half-life can be considered a positive attribute, since it will minimize systemic complement inhibition. The question of tissue retention at a targeted site was addressed in MRL/lpr mice, a spontaneous model for lupus in which mice develop renal disease with IgG and complement deposition in the kidneys at around 14 weeks of age. Following a single 0.25-mg intravenous injection of CR2-Crry, biodistribution studies demonstrated that the inhibitor localized to the kidneys with a tissue half-life of about 24 h, and CR2-Crry was still detectable in kidneys, co-localized with C3d, at 7 days after injection (Atkinson et al. 2008, 180:1231–1238).

Long-term treatment with CR2-Crry was also investigated in the MRL/lpr model (Atkinson et al. 2008, 180:1231–1238), and the study was instructive in demonstrating other postulated benefits of targeted complement inhibition, in particular with regard to minimizing the effect of complement inhibition on the protective role of complement in disease. Complement plays a dual role in the progression of lupus, having a pathogenic role in driving inflammation, and a protective role in the clearance of immune complexes and apoptotic cells. Treatment of MRL/lpr mice with CR2-Crry after the development of proteinuria provided significant benefits in terms of survival and renal disease (clinical and pathological). CR2-Crry also reduced autoantibody production and did not impact the level of circulating immune complexes as the mice aged. In contrast, the treatment of MRL/lpr mice with Crry-Ig (untargeted counterpart) reduced proteinuria and maintained renal function but did not improve survival or glomerular pathological disease (Bao et al. 2003, 14:670–679). Unlike CR2-Crry, Crry-Ig treatment also significantly increased circulating immune complex levels, and perturbing this protective function of complement likely explains the different efficacies of the two inhibitors. Finally, it is noteworthy that the MRL/lpr mice were treated with a 0.25-mg dose of CR2-Crry once a week, whereas 3 mg of Crry-Ig was administered every other day to maintain systemic complement inhibition and therapeutic effect in this model (Bao et al. 2003, 14:670–679).

CR2-Crry has also shown significant benefit in many other experimental murine models, including models of warm cerebral (Atkinson et al. 2006, 177:7266–7274) and hepatic (Sekine 2009b, 119:2304–2316; Sekine 2009a, 183:4764–4772) ischemia-reperfusion injury, post-transplant cardiac ischemia-reperfusion injury (Atkinson et al. 2010, 185:7007–7013), spinal cord injury (Qiao et al. 2006,

169:1039–1047), rheumatoid arthritis (Song et al. 2007, 179:7860–7869), preeclampsia (Qing et al. 2011, 79:331–339), microvascular injury during acute tracheal rejection (Khan et al. 2011, 109: 1290–1301), paraquat-induced lung injury (Sun et al. 2011b, 45:834–842), and age-related macular degeneration (AMD) (see below) (Schepp-Berglind et al. 2012, 188:6309–6318).

The foregoing highlights the fact that there can be significant benefits to site-specific targeting of a complement inhibitor, even when all complement pathways are centrally blocked at the C3 activation step. However, depending on the role of each complement pathway in pathogenic and protective functions for a given disease, site-specific targeting of a particular pathway has the potential to offer additional therapeutic benefit. In this regard, the alternative pathway is known to play a key role in driving inflammation and propagating injury in many pathological conditions [reviewed in Holers (2008, 223:316); Thurman and Holers (2006, 176:1305–1310)], whereas the classical and lectin pathways are known to play important roles in host defense against some pathogens, as well as provide some important regulatory and homeostatic functions. With this in mind, a CR2-targeted inhibitor specific for the alternative pathway, CR2-fH, was developed. In vitro characterization of mouse CR2-fH (Fig. 9.1) demonstrated specificity for the alternative pathway, as well as a higher specific activity at inhibiting the alternative pathway than CR2-Crry (Huang et al. 2008, 181:8068–8076). In vitro analysis also demonstrated that CR2-fH binding and activity was dependent on CR2- and C3-mediated interactions, and the significance of the CR2-targeting approach was highlighted by the fact that purified endogenous fH failed to provide any significant level of protection against C3 deposition.

CR2-fH was first characterized in vivo in a model of intestinal ischemia-reperfusion injury and its therapeutic effect compared to CR2-Crry (Huang et al. 2008, 181:8068–8076). Although requiring a slightly higher dose than CR2-Crry, CR2-fH was completely protective against both local (intestine) and remote (lung) injury. Earlier studies using complement-deficient mice have demonstrated important roles for both the lectin (Hart et al. 2005, 174:6373–6380) and the alternative (Stahl et al. 2003, 162:449–455) pathways in intestinal ischemia-reperfusion injury, and the efficacy of CR2-fH in this model indicates the presence of alternative pathway dependence in a clinically relevant paradigm. CR2-fH had a circulatory half-life similar to CR2-Crry and was shown to target to the site of local injury. It is interesting that serum factor H fails to provide protection against intestine IRI, even though it can bind to cell surfaces and is present at high concentration; the CR2-targeting mechanism thus significantly enhances the complement inhibitory activity of the N-terminal domain of factor H, the molecular basis for which is not yet understood. Of note, a dimeric factor H construct, CR2-fHfH, displayed higher specific activity than CR2-fH in vitro, but there was no difference between the efficacy of the two inhibitors in the model of intestinal ischemia-reperfusion injury (Huang et al. 2008, 181:8068–8076).

Since its initial characterization in intestinal ischemia-reperfusion injury, CR2-fH has been shown to effectively ameliorate inflammation and injury in many other models of disease, including allergen-induced airway hyperresponsiveness (Takeda et al. 2012, 188:661–667), renal ischemia-reperfusion injury (Renner et al. 2011, 80:165–173), LPS/D-GalN-induced fulminant hepatic failure (Sun et al. 2011a, 6:e26838), paraquat-induced lung injury (Sun et al. 2011b, 45:834–842), spinal cord injury (Qiao et al. 2010, 177:3061–3070), post-transplant cardiac ischemia-reperfusion injury (Atkinson et al. 2010, 185:7007–7013), hepatic ischemia-reperfusion injury (He et al. 2009b, 119:2304–2316), hemorrhage-induced intestinal injury (Hylton et al. 2011, 35:134–140), collagen antibody-induced arthritis (Banda et al. 2009, 183:5928–5937), dextran sodium sulfate-induced colitis (Schepp-Berglind et al. 2012, 188:6309–6318), and osteoarthritis (Wang et al. 2011, 17:1674–1679) age-related macular degeneration (AMD) (see below). Where comparative data exists, CR2-fH and CR2-Crry were similarly effective in treating different acute inflammatory conditions. These data highlight the importance of the alternative pathway in the pathology of inflammation, and since CR2-fH will not directly block any physiologically important functions of the classical or lectin pathway, it follows that CR2-fH could possess superior therapeutic qualities compared to CR2-Crry or its human equivalent.

The arguments above notwithstanding, it has also been shown that for certain disease conditions, specific inhibition of the alternative pathway with CR2-fH can provide significantly better protection than inhibition of all pathways with CR2-Crry. In a side-by-side comparison, CR2-fH provided significantly better protection from glomerulonephritis and renal deposition of IgG than CR2-Crry, in both MRL/lpr (Sekine et al. 2011a, 63:1076–1085) and (NZB/W)F_1 (Sekine et al. 2011b, 49:317–323) models of lupus. Interestingly, the sCR2-targeting vehicle alone significantly reduced autoantibody levels, as well as reduced renal deposition of IgG and complement, indicating that the targeting vehicle is contributing to therapeutic activity by modulating autoimmunity through interruption of endogenous CR2 function on B cells and/or follicular dendritic cells. These results are likely a reflection of the known role of the classical pathway in the catabolism of immune complexes and possibly also the role of the classical and lectin pathways in resolving inflammation via the clearance of apoptotic cells.

Finally, it should be noted that CR2 has also been utilized to target a complement activation signal to a cell surface. CR2 linked to complement-activating Fc domains (CR2-Fc) has been shown to target and amplify complement deposition on tumor cells following treatment with a mAb. By targeting C3d initially deposited by an antitumor mAb, CR2-Fc significantly enhanced monoclonal antibody-dependent ADCC and tumor cell lysis in vitro and significantly enhanced the outcome of mAb therapy in a mouse model of metastatic lymphoma (Imai et al. 2007, 67:9535–9541; Elvington et al. 2012).

9.6 Complement- and CR2-Mediated Targeting in Age-Related Macular Degeneration

AMD is characterized by the progressive loss of central vision due to damage to the photoreceptor cells in this region. Since the loss of photoreceptors appears to occur via two different mechanisms, the terms atrophic (dry) AMD and neovascular or exudative (wet) AMD were coined. Dry AMD, which progresses slowly, is the more prevalent form, affecting up to 90 % of all patients; however, the mechanisms by which photoreceptors degenerate is poorly understood. Wet AMD, which typically progresses very rapidly, causes the more severe vision loss due to its association with choroidal neovascularization (CNV) (Gehrs et al. 2006, 38:450–471). CNV is characterized by newly formed vessels which invade the subretinal space, are inherently leaky, and cause fluid accumulation and retinal detachment that is followed by the loss of the underlying photoreceptors. Because of its importance and the difficult nature of treating vision loss, AMD, especially the dry form, represents one of the most important potential clinical indications for complement inhibitors, is a good example of complex interactions between complement and other pro-inflammatory pathways, and thus is used herein as an example of the preclinical work performed using CR2-targeted complement inhibitors.

Notably, despite the two very different clinical pictures of AMD, there are a number of common features. Structurally, in both forms of AMD, the retinal pigmented epithelium (RPE)-choroid interface is altered. Pathological changes include the deposition of extracellular material between the RPE and Bruch's membrane and the loss of integrity of Bruch's membrane, the basement membrane for the RPE. Etiologically, it is proposed that oxidative stress (Khandhadia and Lotery 2010, 12:e34), complement activation (Gehrs et al. 2006, 38:450–471; Jha et al. 2007, 44:3901–3908), and inflammation (Donoso et al. 2006, 51:137–152) each contribute as mediators in both forms of AMD [reviewed in Gehrs et al. (2006, 38:450–471)]. In addition, with regard to genetic linkage, among the main risk factors are single-nucleotide polymorphisms (SNPs) occurring in complement genes, including genes encoding the alternative pathway inhibitor factor H and the alternative pathway activation proteins factor B and C3, as well as C2 [reviewed in Gorin (2005, 13:793–794)]. It is likely that retinal damage in AMD is at least in part caused by chronic local complement activation, since in addition to SNPs in complement genes, complement components have been found to be associated with the pathological features of AMD such as drusen, Bruch's membrane, and RPE (Gehrs et al. 2006, 38:450–471).

The most studied animal model for wet AMD is the CNV model in rodents. In this model, laser spots are generated by argon laser photocoagulation, rupturing Bruch's membrane, an injury that triggers CNV and complement activation (Nozaki et al. 2006, 103:2328–2333). The involvement of the complement system in mouse CNV is well established. Combined results from several laboratories have shown that complement activation in mouse CNV involves the alternative pathway (Bora et al. 2006, 177:1878; Rohrer et al. 2011, 48:e1–8) that the elimination of the classical or the lectin pathway alone has no protective or deleterious effect. In addition, the alternative pathway alone cannot drive pathology, but it is required in its capacity as the amplification loop (Rohrer et al. 2011, 48:e1–8). In addition to the anaphylatoxins C3a and C5a (Nozaki et al. 2006, 103:2328–2333), the MAC appears to contribute substantially to disease severity, since *CD59*$^{-/-}$ mice develop CNV significantly faster than wild-type mice (Bora et al. 2007, 178:1783–1790); systemic or intravitreal injections of recombinant soluble CD59a-IgG2a fusion-protein (Bora et al. 2010, 285:33826–33833) or gene therapy based on a non-membrane-targeted human soluble CD59 (Cashman et al. 2011, 6:e19078) all reduce CNV. Notably, all CNV experiments have confirmed that an essential downstream molecule required for CNV is VEGF.

In order to assess the role of CR2-targeted inhibitors, we utilized C57BL/6 mice and introduced CNV lesions as described. Following this, murine CR2-factor H (muCR2-fH) (Huang et al. 2008, 181:8068–8076) was administered by tail-vein injections, and the subsequent development of CNV was evaluated (Rohrer et al. 2009, 50:3056–3064). Relative CNV size was determined in flat-mount preparations of RPE-choroid stained with isolectin B examined by confocal microscopy. Retinal function was assessed by electroretinography, taking measurements prior to and after the CNV lesion period. Gene expression in RPE-choroid fractions was performed by quantitative RT-PCR, VEGF measurement by ELISA, and bioavailability of muCR2-fH was determined by immunohistochemistry using a specific mAb against the CR2 portion of the fusion protein. The main results of this study can be summarized as follows: (1) muCR2-fH significantly reduces pathology and improves function in a mouse model of wet AMD, (2) molecular changes known to be causally involved in CNV are normalized by this treatment, and (3) muCR2-fH can be administered systemically rather than requiring local intraocular applications and in this setting is able to localize to the injured tissues.

9.7 Translation of CR2-Mediated Targeting to the Treatment of Human Disease

As noted above, dysregulated and excessive local complement activation, especially of the alternative pathway, is causally associated with numerous human diseases. To further explore the pathogenesis of human alternative pathway-dependent diseases by selectively modulating both C3 and C5 convertase activity at local tissue sites using the CR2-targeting mechanism, TT30, a novel therapeutic fusion protein, was created (Fridkis-Hareli et al. 2011, 118:4705–4713). TT30 was built using the same strategy as muCR2-fH described above, linking the C3frag-binding domain of human CR2 within SCR 1-4 with the alternative pathway inhibitory domain within SCR 1-5 of human factor H.

The activity and mechanism of TT30 prevention of hemolysis was studied using several approaches (Fridkis-Hareli et al. 2011, 118:4705–4713): (1) a model of alternative pathway-mediated hemolysis in which rabbit RBCs are exposed to human serum, (2) flow cytometric phenotyping, using mAbs specific for human iC3b or C3d, of C3frag accumulation on rabbit RBCs exposed to human serum in the presence or absence of TT30, (3) flow cytometric analysis of TT30 binding to rabbit RBCs bearing C3 fragments that interact with CR2, and (4) an ELISA-based alternative pathway-dependent MAC deposition assay. The results demonstrated that TT30 efficiently inhibits alternative pathway-mediated, but not classical pathway-mediated, MAC formation using the ELISA technique, with an IC50 of 3.2 μg/mL. TT30 also inhibited rabbit RBC hemolysis with an IC50 of 50.1 μg/mL. As expected, by using an inhibitory mAb to human CR2 that interrupts its binding to iC3b/C3dg/C3d, both the inhibition and binding activities were found to be dependent upon targeting of TT30 to C3 fragment-bearing surfaces by CR2. Importantly, TT30 was also able to be detected on C3 fragment-bound RBCs during the prevention of rabbit RBC hemolysis and remained on the RBCs for at least 24 h. Thus, these results demonstrate that TT30 displays target cell surface control of alternative pathway-mediated activation and has potential utility for the treatment of complement-mediated diseases in which modulation of alternative pathway activity at both the C3 and C5 convertase points is predicted to be beneficial. Finally, TT30 was found to be bioavailable following subcutaneous injection in cynomolgus monkeys.

Paroxysmal nocturnal hemoglobinuria (PNH) is a human alternative pathway-mediated disease caused by the absence of CD55 on RBCs, which allows alternative pathway-initiated complement activation by C3 and C5 convertases, and the absence of CD59 which allows activated C5 to proceed to formation of the MAC, resulting in intravascular hemolysis (IVH) (Rother et al. 2007, 25:1256–1264). Treatment of patients with the anti-C5 monoclonal antibody eculizumab markedly abrogates IVH and its associated clinical sequelae; however, because of continued apparent C3 convertase activity, covalently bound C3 fragments accumulate on PNH RBCs (Risitano et al. 2009, 113:4094–4100) and likely promote extravascular hemolysis. Notably, TT30 was found to be effective in blocking hemolysis of PNH RBCs as well as in preventing the accumulation of C3 fragments on their surfaces (Pascariello et al. 2012). Finally, TT30, in a manner indistinguishable from muCR2-fH and demonstrating the same phenotypic improvements, has been shown to decrease the development of CNV in the same laser-induced model discussed above (Rohrer et al. 2012).

9.8 Additional Approaches to the Targeting of Complement Inhibitors

There are three major areas of development that can be envisioned with regard to improving the strategy of targeting complement inhibitors to sites of activation. The first is to better understand the biophysical characteristics of the targeted molecules that are important in maximizing the relative localization of inhibitors to sites of complement activation and their durable retention in a functionally active form. There is very little understanding of this issue, highlighted by the finding that adding an additional factor H SCR 1–4 module to muCR2-fH resulted in a molecule that was functionally more active in vitro but did not demonstrate an improved dose-dependent efficacy when tested against muCR2-fH in the murine intestinal ischemia-reperfusion model (Huang et al. 2008, 181:8068–8076).

The second area is to determine whether other types of complement inhibitors with varying functional characteristics could also be effectively targeted. These molecules could include both inhibitors (C4bp, MCP, CR1, etc.) and activators/stabilizers (properdin) of complement activation. One such candidate is CR1, a widely distributed receptor that binds both C4b and C3b, the initial degradation products of C4 and C3, and that also exhibits potent complement regulator functions by acting as a decay-accelerator and cofactor of both the classical and alternative pathway C3 and C5 convertases.

The third area of development involves the method by which the targeting occurs. Several are discussed above, including sCR1sLE^x, APT070, and the use of specific ScFv to target inhibitors to cell surface molecules. In addition, though, Lambris and colleagues have recently developed a novel method to bind factor H from the circulation to a cell surface by initially painting the surface with a peptide that has a high affinity for mid-region SCRs of factor H (Wu et al. 2011, 186:4269–4277).

9.9 Summary

Based on a wide variety of approaches, it appears that substantial therapeutic advantage can be gained by the targeting of complement inhibitors to specific sites. Targeting has been accomplished using sLE^x-binding to selectins, physical membrane insertion, linkage to ScFv that directs inhibitors to specific proteins, and CR2-mediated approach. CR2-mediated targeting, the primary topic of this review, has been shown to demonstrate substantial benefits in preclinical models with regard to the following key determinants: (1) in vitro and in vivo potency as directly compared to non-targeted inhibitors, (2) durability of protective effect in vivo when using intermittent dosing schedules, (3) short circulating half-life in association with a prolonged binding to sites of injury (i.e., dissociating the need for systemic inhibition to generate local protective effects), and (4) decreased risk of infection as manifested in a model of complement-dependent sepsis. Future approaches will likely build upon these initial molecules with a goal of maximizing the efficacy of therapeutics and minimizing safety concerns.

9.10 Acknowledgements

This work discussed herein was supported by National Institutes of Health (NIH) R01 EY019320 (BR), Department of Veterans Affairs merit award RX000444 (BR), an unrestricted grant to MUSC from Research to Prevent Blindness (BR), NIH R01 AR51749 (VMH), NIH R01 HL082485 and HL086576 (ST), and Department of Veterans Affairs merit award BX001218 (ST).

9.11 Disclosures

V.M.H. is a cofounder of Taligen Therapeutics, Inc., formerly of Cambridge, MA, which developed complement inhibitors for therapeutic use, is a consultant to Alexion Pharmaceuticals, Cheshire, CT, and holds licensed patents for CR2-targeted complement inhibitors; B.R. was a consultant to Taligen Therapeutics and holds licensed patents for CR2-targeted complement factor H.S.T. was a consultant to Taligen Therapeutics and holds licensed patents for CR2-targeted complement inhibitors.

References

Asokan R, Hua J, Young KA, Gould HJ, Hannan JP, Kraus DM, Szakonyi G, Grundy GJ, Chen XS, Crow MK, Holers VM (2006) Characterization of human complement receptor type 2 (CR2/CD21) as a receptor for interferon-alpha: a potential role in systemic lupus erythematosus. J Immunol 177:383–394

Atkinson C, Song H, Lu B, Qiao F, Burns TA, Holers VM, Tsokos GC, Tomlinson S (2005) Targeted complement inhibition by C3d recognition ameliorates tissue injury without apparent increase in susceptibility to infection. J Clin Invest 115:2444–2453

Atkinson C, Zhu H, Qiao F, Varela JC, Yu J, Song H, Kindy MS, Tomlinson S (2006) Complement-dependent P-selectin expression and injury following ischemic stroke. J Immunol 177:7266–7274

Atkinson C, Qiao H, Song H, Gilkeson GS, Tomlinson S (2008) Low-dose targeted complement inhibition protects against renal disease and other manifestations of autoimmune disease in MRL/lpr mice. J Immunol 180:1231–1238

Atkinson C, He S, Morris K, Qiao F, Casey S, Goddard M, Tomlinson S (2010) Targeted complement inhibitors protect against posttransplant cardiac ischemia and reperfusion injury and reveal an important role for the alternative pathway of complement activation. J Immunol 185:7007–7013

Aubry JP, Pochon S, Graber P, Jansen KU, Bonnefoy JY (1992) CD21 is a ligand for CD23 and regulates IgE production. Nature 358:505–507

Banda NK, Levitt B, Glogowska M, Thurman JM, Takahashi K, Stahl GL, Tomlinson S, Arend WP, Holers VM (2009) Targeted inhibition of the complement alternative pathway with complement receptor 2 and factor H attenuates collagen antibody-induced arthritis in mice. J Immunol 183:5928–5937

Bao L, Haas M, Kraus DM, Hack BK, Rakstang JK, Holers VM, Quigg RJ (2003) Administration of a soluble recombinant complement C3 inhibitor protects against renal disease in MRL/lpr mice. J Am Soc Nephrol 14:670–679

Bora NS, Kaliappan S, Jha P, Xu Q, Sohn JH, Dhaulakhandi DB, Kaplan HJ, Bora PS (2006) Complement activation via alternative pathway is critical in the development of laser-induced choroidal neovascularization: role of factor B and factor H. J Immunol 177:1872–1878

Bora NS, Kaliappan S, Jha P, Xu Q, Sivasantar B, Harris CL, Morgan BP, Bora PS (2007) CD59, a complement regulatory protein, controls choroidal neovascularization in a mouse model of wet-type age-related macular degeneration. J Immunol 178:1783–1790

Bora NS, Jha P, Lyzogubov VV, Kaliappan S, Liu J, Tytarenko RJ, Fraser DA, Morgan BP, Bora PS (2010) Recombinant membrane-targeted form of CD59 inhibits the growth of choroidal neovascular complex in mice. J Biol Chem 285:33826–33833

Botto M, Walport MJ (2002) C1q, autoimmunity and apoptosis. Immunobiology 205:395–406

Brown EJ (1991) Complement receptors and phagocytosis. Curr Opin Immunol 3:76–82

Carel JC, Frazier B, Ley TJ, Holers VM (1989) Analysis of epitope expression and the functional repertoire of recombinant complement receptor 2 (CR2/CD21) in mouse and human cells. J Immunol 143:923–930

Carel JC, Myones BL, Frazier B, Holers VM (1990) Structural requirements for C3d, g/Epstein-Barr virus receptor (CR2/CD21) ligand binding, internalization, and viral infection. J Biol Chem 265:12293–12299

Carroll MC (1998a) CD21/CD35 in B cell activation. Semin Immunol 10(4):279–286

Carroll MC (1998b) The role of complement and complement receptors in the induction and regulation of immunity. Annu Rev Immunol 16:545–568

Carroll MC (2000) The role of complement in B cell activation and tolerance. Adv Immunol 74:61–88

Carter RH, Spycher MO, Ng YC, Hoffman R, Fearon DT (1998) Synergistic interaction between complement receptor type 2 and membrane IgM on B lymphocytes. J Immunol 141:457–463

Cashman SM, Ramo K, Kumar-Singh R (2011) A non membrane-targeted human soluble CD59 attenuates choroidal neovascularization in a model of age related macular degeneration. PLoS One 6:e19078

Cicardi M, Zanichelli A (2012) Replacement therapy with C1 esterase inhibitors for hereditary angioedema. Drugs Today 46:867–874

Davies KA, Peters AM, Beynon HLC, Walport MJ (1992) Immune complex processing in patients with systemic lupus erythematosus. J Clin Invest 90:2075–2083

Davies KA, Schifferli JA, Walport MJ (1994) Complement deficiency and immune complex disease. Springer Semin Immunopathol 15:397–416

Donoso LA, Kim EE, Frost A, Callahan A, Hageman G (2006) The role of inflammation in the pathogenesis of age-related macular degeneration. Surv Ophthalmol 51:137–152

Elvington M, Huang Y, Morgan BP, Qiao F, van Rooijen N, Atkinson C, Tomlinson S (2012) A targeted complement-dependent strategy to improve the outcome of mAb therapy, and characterization in a murine model of metastatic cancer. Blood Mar 22 [Epub ahead of print]

Esparza I, Becherer JD, Alsenz J, De la Hera A, Lao Z, Tsoukas CD, Lambris JD (1991) Evidence for multiple sites of interaction in C3 for complement receptor type 2 (C3d/EBV receptor, CD21). Eur J Immunol 21:2829–2838

Fearon DT, Locksley RM (1996) The instructive role of innate immunity in the acquired immune response. Science 272:50–54

Fraser DA, Harris CL, Williams AS, Mizuno M, Gallagher S, Smith RA, Morgan BP (2003) Generation of a recombinant, membrane-targeted form of the complement regulator CD59: activity in vitro and in vivo. J Biol Chem 278:48921–48927

Fridkis-Hareli M, Storek M, Maszaroff I, Risitano AM, Lundberg AS, Horvath CJ, Holers VM (2011) Design and development of TT30, a novel C3d-targeted C3/C5 convertase inhibitor for treatment of human complement alternative pathway-mediated diseases. Blood 118:4705–4713

Gadjeva M, Takahashi K, Thiel S (2004) Mannan-binding lectin – a soluble pattern recognition molecule. Mol Immunol 41:113–121

Gehrs KM, Anderson DH, Johnson LV, Hageman GS (2006) Age-related macular degeneration–emerging pathogenetic and therapeutic concepts. Ann Med 38:450–471

Gorin MB (2005) A new vision for age-related macular degeneration. Eur J Hum Genet 13:793–794

Halstead SK, Humphreys PD, Goodfellow JA, Wagner ER, Smith ER, Willison HJ (2005) Complement inhibition abrogates nerve terminal injury in Miller Fisher syndrome. Ann Neurol 58:203–210

Harris CL, Fraser DA, Morgan BP (2002) Tailoring anti-complement therapeutics. Biochem Soc Trans 30:1019–1026

Hart ML, Ceonzo KA, Shaffer LA, Takahashi K, Rother RP, Reenstra WR, Buras JA, Stahl GL (2005) Gastrointestinal ischemia-reperfusion injury is lectin complement pathway dependent without involving C1q. J Immunol 174:6373–6380

He C, Imai M, Song H, Quigg RJ, Tomlinson S (2005) Complement inhibitors targeted to the proximal tubule prevent injury in experimental nephrotic syndrome and demonstrate a key role for C5b-9. J Immunol 174:5750–5757

He S, Atkinson C, Evans Z, Ellett JD, Southwood M, Elvington A, Chavin KD, Tomlinson S (2009a) A role for complement in the enhanced susceptibility of steatotic livers to ischemia and reperfusion injury. J Immunol 183: 4764–4772

He S, Atkinson C, Qiao F, Cianflone K, Chen C, Tomlinson S (2009b) A complement-dependent balance between hepatic ischemia/reperfusion injury and liver regeneration in mice. J Clin Invest 119:2304–2316

Holers VM (2001) Complement. In: Rich R (ed) Principles and practices of clinical immunology. Mosby, St. Louis, MO, pp 21.1–21.8

Holers VM (2005) Complement receptors and the shaping of the natural antibody repertoire. Springer Semin Immunopathol 26:405–423
Holers VM (2008) The spectrum of complement alternative pathway-mediated diseases. Immunol Rev 223:300–316
Holers VM, Boackle SA (2004) Complement receptor 2 and autoimmunity. Curr Dir Autoimmun 7:33–48
Holers VM, Carroll MC (2005) Innate autoimmunity. Adv Immunol 86:137–157
Hourcade D, Holers VM, Atkinson JP (1989) The regulators of complement activation (RCA) gene cluster. Adv Immunol 45:381–416
Huang J, Kim LJ, Mealey R, Marsh HC Jr, Zhang Y, Tenner AJ, Connelly ES Jr, Pinsky DJ (1999) Neuronal protection in stroke by an sLe^x-glycosylated complement inhibitory protein. Science 285:595–599
Huang Y, Qiao F, Atkinson C, Holers VM, Tomlinson S (2008) A novel targeted inhibitor of the alternative pathway and its therapeutic use in ischemia/reperfusion injury. J Immunol 181:8068–8076
Hylton DJ, Hoffman SM, van Rooijen N, Tomlinson S, Fleming SD (2011) Macrophage-produced IL-12p70 mediates hemorrhage-induced damage in a complement-dependent manner. Shock 35:134–140
Iida K, Nadler L, Nussenzweig V (1983) Identification of the membrane receptor for the complement fragment C3d by means of a monoclonal antibody. J Exp Med 158:1021–1033
Imai M, Ohta R, Varela JC, Song H, Tomlinson S (2007) Enhancement of antibody-dependent mechanisms of tumor cell lysis by a targeted activator of complement. Cancer Res 67:9535–9541
Jha P, Bora PS, Bora NS (2007) The role of the complement system in ocular diseases including uveitis and macular degeneration. Mol Immunol 44:3901–3908
Jordan JE, Montalto MC, Stahl GL (2001) Inhibition of mannose-binding lectin reduces postischemic myocardial reperfusion injury. Circulation 104:1413–1418
Kalli KR, Ahearn JM, Fearon DT (1991) Interaction of iC3b with recombinant isotypic and chimeric forms of CR2. J Immunol 147:590–594
Kaya Z, Afanasyeva M, Wang Y, Dohmen KM, Schlichting J, Tretter T, Fairweather D, Holers VM, Rose N (2001) Contribution of the innate immune system to autoimmune myocarditis: a role for complement. Nat Immunol 2:739–745
Khan MA, Jiang X, Dhillon G, Beilke J, Holers VM, Atkinson C, Tomlinson S, Nicolls MR (2011) CD4+ T cells and complement independently mediate graft ischemia in the rejection of mouse orthotopic tracheal transplants. Circ Res 109:1290–1301
Khandhadia S, Lotery A (2010) Oxidation and age-related macular degeneration: insights from molecular biology. Expert Rev Mol Med 12:e34
Lachmann PJ, Hughes-Jones NC (1984) Initiation of complement activation. Springer Semin Immunopathol 7:143–162
Lambris JD, Sahu A (2001) Structure and biology of complement protein C3, a connecting link between innate and acquired immunity. Immunol Rev 180:35–48
Linton SM, Williams AS, Dodd I, Smith RA, Williams BD, Morgan BP (2012) Therapeutic efficacy of a novel membrane-targeted complement regulator in antigen-induced arthritis in the rat. Arthritis Rheum 43:2590–2597
Liszewski MK, Farries TC, Lublin DM, Rooney IA, Atkinson JP (1996) Control of the complement system. Adv Immunol 61:201–283
Lowell CA, Klickstein LB, Carter RH, Mitchell JA, Fearon DT, Ahearn JM (1989) Mapping of the Epstein-Barr virus and C3dg binding sites to a common domain on complement receptor type 2. J Exp Med 170:1931–1946
Molina H (2002) The murine complement regulator Crry: new insights into the immunobiology of complement regulation. Cell Mol Life Sci 59:220–229
Molina H, Wong W, Kinoshita T, Brenner C, Foley S, Holers VM (1992) Distinct receptor and regulatory properties of recombinant mouse Complement Receptor 1 (CR1) and Crry, the two genetic homologs of human CR1. J Exp Med 175:121–129
Molina H, Perkins SJ, Guthridge J, Gorka J, Kinoshita T, Holers VM (1995) Characterization of a complement receptor 2 (CD2, CD21) ligand binding site for C3: an initial model of ligand interaction with two linked SCR modules. J Immunol 154:5426–5435
Morgan BP, Berg CW, Harris CL (2005) "Homologous restriction" in complement lysis: roles of membrane complement regulators. Xenotransplantation 12:258–265
Muller-Eberhard HJ (1988) Molecular organization and function of the complement system. Annu Rev Biochem 57:321–347
Mulligan MS, Warner RL, Rittershaus CW, Thomas LJ, Ryan US, Foreman KE, Crouch LD, Till GO, Ward PA (1999) Endothelial targeting and enhanced antiinflammatory effects of complement inhibitors possessing Sialyl $Lewis^x$ moieties. J Immunol 162:4952–4959
Nozaki M, Raisler BJ, Sakuria E, Sarma JV, Barnum SR, Lambris JD, Chen Y, Zhang K, Ambati BK, Baffi JZ, Ambati J (2006) Drusen complement components C3a and C5a promote choroidal neovascularization. Proc Natl Acad Sci 103:2328–2333

Pascariello C, Selleri C, Del Vecchio L, Horvath CJ, Holers VM, Risitano AM (2012) A candidate complement inhibitor for the treatment of paroxysmal nocturnal hemoglobinuria: preclinical data with TT30, a therapeutic that selectively blocks the complement alternative pathway. Blood May 10 [Epub ahead of print]

Patel H, Smith RAG, Sacks SH, Zhou W (2006) Therapeutic strategy with a membrane-localizing complement regulator to increase the number of usable donor organs after prolonged cold storage. J Am Soc Nephrol 17:1102–1111

Qiao F, Atkinson C, Song H, Pannu R, Singh I, Tomlinson S (2006) Complement plays an important role in spinal cord injury and represents a therapeutic target for improving recovery following trauma. Am J Pathol 169:1039–1047

Qiao F, Atkinson C, Kindy MS, Shunmugavel A, Morgan BP, Song H, Tomlinson S (2010) The alternative and terminal pathways of complement mediate post-traumatic spinal cord inflammation and injury. Am J Pathol 177:3061–3070

Qing X, Redecha PB, Burneister MA, Tomlinson S, D'Agati VD, Davisson RL, Salmon JE (2011) Targeted inhibition of complement activation prevents features of preeclampsia in mice. Kidney Int 79:331–339

Reid KBM, Turner MW (1994) Mammalian lectins in activation and clearance mechanisms involving the complement system. Springer Semin Immunopathol 15:307–325

Renner BR, Ferreira VP, Cortes C, Goldberg R, Ljubanovic D, Pangburn MK, Pickering MC, Tomlinson S, Holland-Neidermyer A, Strassheim D, Holers VM, Thurman JM (2011) Binding of factor H to tubular epithelial cells limits interstitial complement activation in ischemic injury. Kidney Int 80:165–173

Ricklin D, Lambris JD (2007) Complement-targeted therapeutics. Nat Biotechnol 25:1265–1275

Risitano AM, Notaro R, Marando L, Serio B, Ranaldi D, Seneca E, Ricci P, Alfinito F, Camera A, Gianfaldoni G, Amendola A, Boschetti C, Di Bona E, Fratellanza G, Barbano F, Rodeghiero F, Zanella A, Iori AP, Selleri C, Luzzatto L, Rotoli B (2009) Complement fraction 3 binding on erythrocytes as an additional mechanism of disease in paroxysmal nocturnal hemoglobinuria patients treated by eculizumab. Blood 113:4094–4100

Rohrer B, Long Q, Coughlin B, Wilson RB, Huang Y, Qiao F, Tang PH, Kunchithapautham K, Gilkeson GS, Tomlinson S (2009) A targeted inhibitor of the alternative complement pathway reduces angiogenesis in a mouse model of age-related macular degeneration. Invest Opthalmol Vis Sci 50:3056–3064

Rohrer B, Coughlin B, Kunchithapautham K, Long Q, Tomlinson S, Takahashi K, Holers VM (2011) The alternative pathway is required, but not alone sufficient, for retinal pathology in mouse laser-induced choroidal neovascularization. Mol Immunol 48:e1–e8

Rohrer B, Coughlin B, Holers VM (2012) Systemic human CR2-targeted complement alternative pathway inhibitor treatment ameliorates mouse laser-induced choroidal neovascularization. J Ocul Pharmacol Ther Feb 6 [Epub ahead of print]

Rosse WF, Nishimura J (2003) Clinical manifestations of paroxysmal nocturnal hemoglobinuria: present state and future problems. Int J Hematol 77:113–120

Rother RP, Rollins SA, Mojcik CF, Brodsky RA, Bell L (2007) Discovery and development of the complement inhibitor eculizumab for the treatment of paroxysmal nocturnal hemoglobinuria. Nat Biotechnol 25:1256–1264

Runza VL, Schwaeble W, Mannel DN (2008) Ficolins: novel pattern recognition molecules of the innate immune response. Immunobiology 213:297–306

Sarrias MR, Franchini S, Canziani G, Argyropoulos E, Moore WT, Sahu A, Lambris JD (2001) Kinetic analysis of the interactions of complement receptor 2 (CR2, CD21) with its ligands C3d, iC3b, and the EBV glycoprotein gp350/220. J Immunol 167:1490–1499

Schepp-Berglind J, Atkinson C, Elvington M, Qiao F, Mannon P, Tomlinson S (2012) Complement dependent injury and protection in a murine model of acute dextran sulfate sodium-induced colitis. J Immunol Jun 15, 188(12):6309–6318

Sekine H, Kinser TT, Qiao F, Martinez E, Paulling E, Ruiz P, Gilkeson GS, Tomlinson S (2011a) The benefit of targeted and selective inhibition of the alternative complement pathway for modulating autoimmunity and renal disease in MRL/lpr mice. Arthritis Rheum 63:1076–1085

Sekine H, Ruiz P, Gilkeson GS, Tomlinson S (2011b) The dual role of complement in the progression of renal disease in NZB/W F(1) mice and alternative pathway inhibition. Mol Immunol 49:317–323

Smith RA (2002) Targeting anticomplement agents. Biochem Soc Trans 30:1037–1041

Song H, He C, Knaak C, Guthridge JM, Holers VM, Tomlinson S (2003) Complement receptor 2-mediated targeting of complement inhibitors to sites of complement activation. J Clin Invest 111:1875–1885

Song H, Qiao F, Atkinson C, Holers VM, Tomlinson S (2007) A complement C3 inhibitor specifically targeted to sites of complement activation effectively ameliorates collagen-induced arthritis. J Immunol 179:7860–7869

Souza DG, Esser D, Bradford R, Vieira AT, Teixeira MM (2005) APT070 (Mirococept), a membrane-localized complement inhibitor, inhibits inflammatory responses that follow intestinal ischaemia and reperfusion injury. Br J Pharmacol 145:1027–1034

Spitzer D, Unsinger J, Mao D, Wu X, Molina H, Atkinson JP (2005) In vivo correction of complement regulatory protein deficiency with an inhibitor of targeting the red blood cell membrane. J Immunol 175:7763–7770

Stahl GL, Xu Y, Hao L, Miller M, Buras JA, Fung M, Zhao H (2003) Role for the alternative complement pathway in ischemia/reperfusion injury. Am J Pathol 162:449–455

Strey CW, Markiewski M, Matellos D, Tudoran R, Spruce LA, Greenbaum LE, Lambris JD (2003) The proinflammatory mediators C3a and C5a are essential for liver regeneration. J Exp Med 198:913–923
Sun S, Guo Y, Zhao G, Zhou X, Li J, Hu J, Yu H, Chen Y, Song H, Qiao F, Xu G, Yang F, Wu Y, Tomlinson S, Duan Z, Zhou Y (2011a) Complement and the alternative pathway play an important role in LPS/D-GalN-induced fulminant hepatic failure. PLoS One 6:e26838
Sun S, Wang H, Zhao G, An Y, Guo Y, Du L, Song H, Qiao F, Yu H, Wu X, Atkinson C, Jiang S, Tomlinson S, Zhou Y (2011b) Complement inhibition alleviates paraquat-induced acute lung injury. Am J Respir Cell Mol Biol 45:834–842
Takeda K, Thurman JM, Tomlinson S, Okamoto M, Shiraishi Y, Ferreira VP, Cortes C, Pangburn MK, Holers VM, Gelfand EW (2012) The critical role of complement alternative pathway regulator factor H in allergen-induced airway hyperresponsiveness and inflammation. J Immunol 188:661–667
Thurman JM, Holers VM (2006) The central role of the alternative complement pathway in human disease. J Immunol 176:1305–1310
Tsokos GC (2012) Systemic lupus erythematosus. N Engl J Med 365:2110–2121
van den Elsen JM, Isenman DE (2011) A crystal structure of the complex between human complement receptor 2 and its ligand C3d. Science 332:608–611
Wang Q, Rozelle AL, Lepus CM, Scanzello CR, Song JJ, Larsen DM, Crish JF, Bebek G, Ritter SY, Lindstrom TM, Hwang I, Wong HH, Punzi L, Encarnacion A, Shamloo M, Goodman SB, Wyss-Coray T, Goldring SR, Banda NK, Thurman JM, Gobezie R, Crow MK, Holers VM, Lee DM, Robinson WH (2011) Identification of a central role for complement in osteoarthritis. Nat Med 17:1674–1679
Weis JJ, Toothaker LE, Smith JA, Weis JH, Fearon DT (1988) Structure of the human B lymphocyte receptor for C3d and the Epstein-Barr virus and relatedness to other members of the family of C3/C4 binding proteins. J Exp Med 167:1047–1066
Wetsel RA (1995) Structure, function and cellular expression of complement anaphylatoxin receptors. Curr Opin Immunol 7:48–53
Wu YQ, Qu H, Sfyroera G, Tzekou A, Kay BK, Nilsson B, Ekdahl KN, Ricklin D, Lambris JD (2011) Protection of nonself surfaces from complement attack by factor H-binding peptides: implications for therapeutic medicine. J Immunol 186:4269–4277
Zhang H-F, Yu J, Bajwa E, Morrison SL, Tomlinson S (1999) Targeting of functional antibody-CD59 fusion proteins to a cell surface. J Clin Invest 103:55–61
Zhang H, Lu S, Morrison SL, Tomlinson S (2001) Targeting of functional antibody-decay-accelerating factor fusion proteins to a cell surface. J Biol Chem 276:27270–27275
Zipfel PF, Skerka C, Hellwage J, Jokiranta ST, Meri S, Brade V, Kraiczy P, Noris M, Remuzzi G (2002) Factor H family proteins: on complement, microbes and human diseases. Biochem Soc Trans 30:971–978

Chapter 10
Paroxysmal Nocturnal Hemoglobinuria and the Complement System: Recent Insights and Novel Anticomplement Strategies

Antonio M. Risitano

Abstract Paroxysmal nocturnal hemoglobinuria (PNH) is a hematological disorder characterized by complement-mediated hemolytic anemia, thrombophilia, and bone marrow failure. PNH is due to a somatic, acquired mutation in the X-linked *phosphatidylinositol glycan class A (PIG-A)* gene, which impairs the membrane expression on affected blood cells of a number of proteins, including the complement regulators CD55 and CD59. The most evident clinical manifestations of PNH arise from dysregulated complement activation on blood cells; in fact, the hallmark of PNH is chronic, complement-mediated, intravascular hemolysis, which results in anemia, hemoglobinuria, fatigue, and other hemolysis-related disabling symptoms. In addition, the peculiar thromboembolic risk typical of PNH patients is thought as secondary to the complement-mediated hemolysis itself and/or to a complement-mediated activation of platelets. Thus, as a complement-mediated disease, PNH was an appropriate medical condition to develop and to investigate therapeutical complement inhibitors. Indeed, the first complement inhibitor eculizumab, a humanized anti-C5 monoclonal antibody, has been proven safe and effective for the treatment of PNH patients. Chronic treatment with eculizumab results in sustained control of intravascular hemolysis, leading to hemoglobin stabilization and transfusion independence in more than half of the patients. However, recent observations have demonstrated that residual anemia may persist in some patients regardless of sustained fluid-phase terminal complement inhibition. Indeed, persistent dysregulated activation of the early phases of the complement cascade on PNH erythrocytes may lead to progressive C3 deposition on affected cells, which become susceptible to subsequent extravascular hemolysis through the reticuloendothelial system. These findings have renewed the interest for the development of novel complement inhibitors which aim to modulate early phases of complement activation, more specifically at the level of C3 activation. As proof of principle of this concept, an anti-C3 monoclonal antibody has been proven effective in vitro to prevent hemolysis of PNH erythrocytes. More intriguingly, a human fusion protein consisting of the iC3b/C3d-binding region of complement receptor 2 and of the inhibitory domain of the CAP regulator factor H has been recently shown effective in inhibiting, in vitro, both intravascular hemolysis of and surface C3-deposition on PNH erythrocytes, and is now under investigation in phase 1 clinical trials.

A.M. Risitano (✉)
Division of Hematology, Department of Biochemistry and Medical Biotechnologies, Federico II University of Naples, Via Pansini 5, 80131 Naples Italy
e-mail: amrisita@unina.it

J.D. Lambris et al. (eds.), *Complement Therapeutics*, Advances in Experimental Medicine and Biology 735, DOI 10.1007/978-1-4614-4118-2_10,

10.1 Introduction

Paroxysmal nocturnal hemoglobinuria (PNH) is a complex hematological disorder clinically characterized by hemolytic anemia, bone marrow failure, and thrombophilia (Dunn et al. 2000; Parker and Ware 2003; Luzzatto and Notaro 2003). PNH is a rare disease (its estimated prevalence is 1–5 cases per million; Orphanet 2004), but because of its puzzling nature and picturesque presentation, PNH has intrigued generations of investigators since its first extensive description by Dr. Strubing in 1882 (Crosby 1951) and subsequent recognition as a distinct medical entity by Dr. Marchiafava and Dr. Micheli in 1911. Indeed, PNH remained a medical mystery until the 1980s, when most of its pathophysiology was progressively elucidated, first with the description of the molecular defect of PNH cells and then with the identification of the underlying genetic defect. By that time, it was already known that PNH erythrocytes are exquisitely sensitive to lysis upon complement activation, both in vivo and in vitro. Thus, the observation that PNH cells lack from their surface some complement regulators, all included in a specific class of membrane-bound proteins (the so-called glycosyl phosphatidylinositol (GPI)-anchored proteins [GPI-APs]), clearly explained the reason for such sensitivity. Thereafter, the biochemical pathway accounting for GPI-AP surface expression was described, as well as its impairment in PNH cells. Finally, the genetic lesion leading to the aberrant phenotype was identified in distinct mutations in the *phosphatidylinositol glycan class A (PIG-A)* gene. This formally demonstrated that PNH is a clonal hematological disorder characterized by the expansion of abnormal (GPI-AP-deficient, *PIG-A*-mutated) hematopoietic stem cells (HSCs) carrying an intrinsic defect, which accounts for the clinical phenotype of the disease. Nowadays, most pathophysiological events occurring in PNH patients have been extensively described, even if definitive explanations for some disease manifestations (i.e., thromboembolic events) are still elusive. In the last few years, insights into the field pertain to the new treatment strategies that have drastically changed the management and the clinical outcome of PNH patients. In fact, the availability of an inhibitor of the complement cascade – the actual effector mechanism of hemolysis in PNH – has led to the first etiological treatment for PNH, which seems to have a superb impact on the natural history of the disease. Furthermore, this treatment has also allowed a more detailed understanding of biological mechanisms playing a role in disease pathophysiology, making clear that distinct steps of the complement cascade are uncontrolled on PNH cells. As it often occurs in medicine, the observation of achievements and limitations of the current anticomplement treatment in PNH has paved the way for the development of alternative strategies of complement inhibition. Here, we briefly review the pathophysiology of PNH, pointing out the most recent findings which have led to novel candidate agents potentially effective for the treatment of PNH.

10.2 Pathophysiology of PNH

10.2.1 The Genetic Basis of PNH

The peculiar defect of PNH consists in the bizarre lack of several proteins from the blood cell surface (Kunstling and Rosse 1969; Nicholson-Weller et al. 1983; Selvaraj et al. 1988). Notably, in PNH, the abnormal phenotype pertains to erythrocytes, granulocytes, and platelets, but not all blood cells are affected; indeed, PNH patients present a mosaicism of normal and abnormal blood cells. The clonal origin of PNH hematopoiesis was first demonstrated in 1970 by Dr. Luzzatto's group by studying patients heterozygous for the glucose-6-phosphate-dehydrogenase; (Oni et al. 1970) on the other hand, the involvement of all blood lineages suggested that aberrant phenotypes derive from multipotent hematopoietic progenitors or even hematopoietic stem cells (Kinoshita et al. 1985). All the proteins

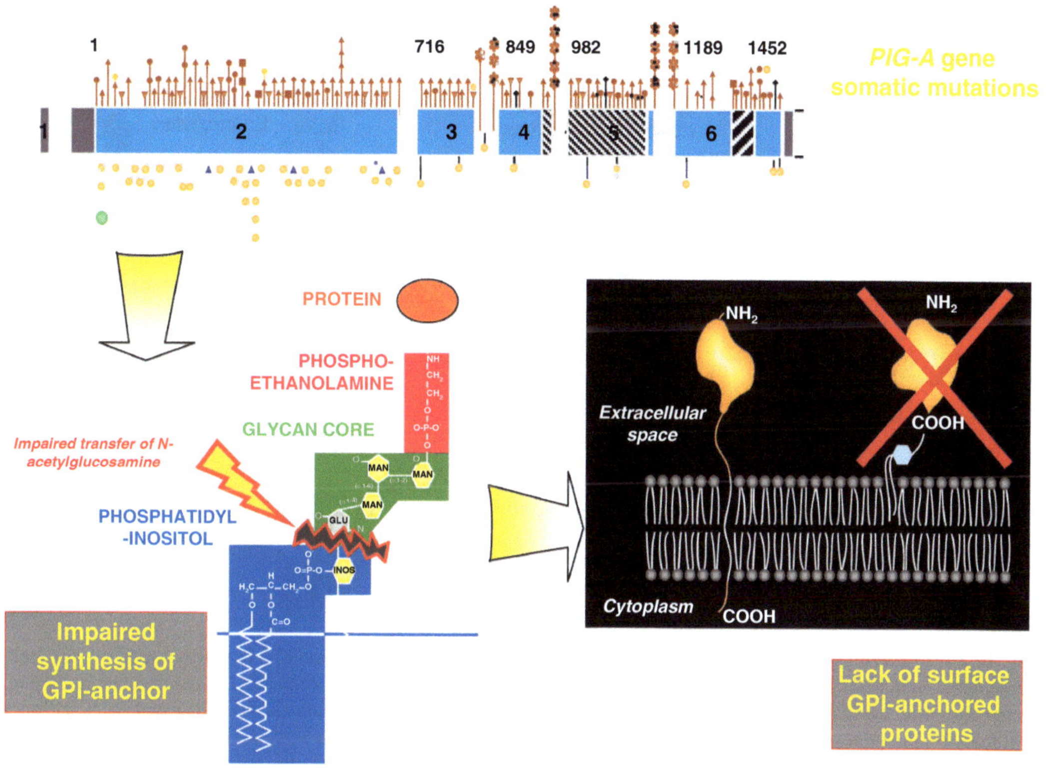

Fig. 10.1 *Genetic basis and biochemical phenotype of PNH cells.* Different acquired, somatic mutations in the *PIG-A* gene (on the *upper* side are depicted as small deletions or insertions, producing frameshift; on the *lower* one, nucleotide substitutions) result in the impairment of the biosynthesis of the GPI anchor. As a result, affected cells loose from their surface all protein linked through the GPI anchor

missing from the PNH cell surface shared a common mechanism responsible for their attachment to the cell membrane (Medof et al. 1987), which is a specific glycolipid structure named glycosyl phosphatidylinositol (GPI) anchor (Mahoney et al. 1992). In 1993, Kinoshita and colleagues demonstrated that PNH is associated with mutations in the *PIG-A* gene (Takeda et al. 1993; Miyata et al. 1993), an X-linked gene which encodes an enzyme essential to transfer N-acetyl glucosamine to phosphatidylinositol. This is the very first step of the GPI-anchor biosynthesis (Armstrong et al. 1992; Hirose et al. 1992; Takahashi et al. 1993), and the impairment of this biochemical pathway precludes the expression of all GPI-linked proteins on the surface of affected cells (Takayashi et al. 1993). PNH develops through an acquired somatic mutation in the *PIG-A* gene occurring in one or a few hematopoietic stem cells, which originate progeny of mature blood cells – erythrocytes, granulocytes, platelets, and possibly lymphocytes – uniquely lacking in all GPI-APs from their surface (Fig. 10.1). PNH is therefore an acquired genetic blood disorder, which cannot be transmitted to the progeny.

10.2.2 The Dual Pathophysiology Theory

A number of observations support the idea that the *PIG-A* mutation itself is necessary but not sufficient to cause PNH. In fact, the expansion of PNH cells over normal hematopoiesis remains a key step to develop the disease phenotype. Several investigators have tried to generate murine models of PNH, with limited success. Indeed, a complete knockout animal could not be produced due to fetal loss (Kawagoe

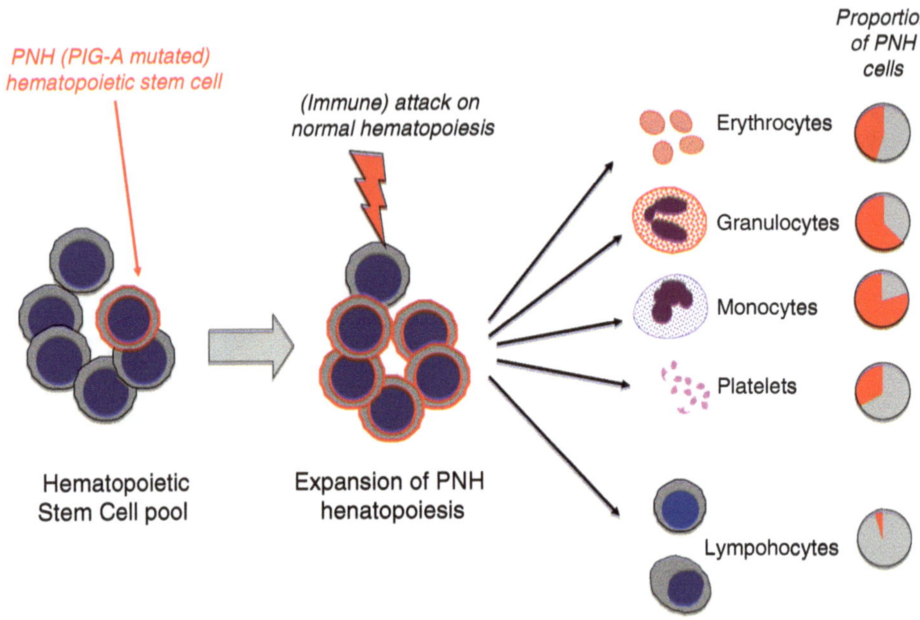

Fig. 10.2 *Mosaicism of hematopoiesis in PNH patients*. Due to the PIG-A mutation of a few hematopoietic stem cells, hematopoiesis in PNH patients is a mixture of normal (PIG-A wild-type) and abnormal (*PIG-A-mutated*) hematopoietic stem cells. Affected (*PIG-A-mutated*) hematopoietic stem cells are able to originate all blood progeny cells, including erythrocytes, granulocytes, monocytes, platelets, and possibly lymphocytes (which however largely derive from homeostatic proliferation of peripheral lymphocyte). As a result, once PNH hematopoietic stem cells expand over normal hematopoiesis (according to the dual pathophysiology theory, see text for details), PNH patients harbor proportion of affected cell within each mature blood lineage

et al. 1996), and even the use of chimeras obtained by inserting *pig-a* knocked-out embryonic stem cells in early embryonic development gave disappointing results (Tremml et al. 1999). Mice having a substantial proportion of PNH cells were subsequently generated by using a conditional inactivation of the murine *pig-a* gene implemented using Cre recombinase specifically targeted to the hematopoietic stem cells. (Keller et al. 2001; Jasinski et al. 2001) However, even these mice do not really mimic the disease phenotype seen in humans, because PNH hematopoiesis tends to decrease over time (Jasinski et al. 2001), leading to the conclusion that the *pig-a* mutation is not sufficient itself to sustain the expansion of PNH-like hematopoiesis over time. On the other hand, even if a *PIG-A* mutation may be identified in all PNH patients, it has been demonstrated that a few PNH-like cells carrying inactivating *PIG-A* mutations may be detected even in normal individuals (without any sign or symptom of PNH), at a frequency of 10–50 cells per million (Araten et al. 1999). This clearly demonstrated that, in vivo, a *PIG-A* mutation was not sufficient to cause the disease; this view is also supported by the observation that, at least in some patients, PNH hematopoiesis may include more than one abnormal clone (Endo et al. 1996; Nishimura et al. 1997), consistent with an expansion based on non-stochastic processes, such as selection, rather than a random process. This background raised the hypothesis of a dual pathophysiology for PNH: the *PIG-A* mutation is not sufficient to cause the disease and requires a second, independent event (Rotoli and Luzzatto 1989). This theory is also known as the "relative advantage" (Luzzatto et al. 1997) or "escape" theory (Young and Maciejewski 2000). According to this view, a mutation in the *PIG-A* gene might be a fairly common phenomenon, with no major biological consequences, because in physiological conditions the mutated cell has no reason for expanding in the presence of a vast majority of normal cells. In fact, no intrinsic proliferative advantage has been demonstrated in PNH hematopoietic progenitors in comparison to normal ones (Araten et al. 2002). However, additional factors may alter this equilibrium, creating the conditions for the expansion of PNH clone(s) (Fig. 10.2) and possibly leading to the occurrence of a single *PIG-A*-mutated stem cell sustaining hematopoiesis even for the rest of the patient's life (Nishimura et al. 2002, 2004). The nature of such second event(s) is thought to be an (auto)immune attack against hematopoiesis, as supported by the well-known clinical overlap between PNH and aplastic anemia (AA, which is in most cases immune-mediated) (Lewis and Dacie 1967), as well as by direct demonstration of immune abnormalities in PNH patients. Indeed, normal and PNH cells are not different in terms of growth or survival, but PNH cells could be spared by immune-mediated damage, finally resulting in a progressive consumption of normal hematopoiesis, with relative expansion of PNH hematopoiesis (Rotoli and Luzzatto 1989; Luzzatto et al. 1997; Young and Maciejewski 2000). This was confirmed by gene expression profiling performed in normal and PNH hematopoietic stem cells obtained from PNH patients: in fact, phenotypically normal (GPI-AP-positive) CD34+ cells harbored diffuse abnormalities of their transcriptome, with overexpression of genes involved in apoptosis and immune activity, paralleling the findings seen in CD34+ cells of AA patients, while phenotypically abnormal PNH CD34+ (GPI-AP negative) showed a gene expression profiling closer to normal (Chen et al. 2005). Notably, normal and PNH CD34+ cells did not show any difference in the genes involved in growth or proliferation, rather suggesting the presence of sublethal extrinsic damage in phenotypically normal HSCs but not in their *PIG-A* mutated counterpart.

10.3 Disease Manifestations and Pathogenic Mechanisms in PNH

According to the dual step theory, the expansion of PNH hematopoiesis may represent a natural escape to prevent a more severe marrow failure; however, on the other hand, the production of substantial proportion of progeny blood cells harboring the PNH phenotype leads to the clinical phenotype of PNH, which is characterized by the typical triad of intravascular hemolysis, thrombophilia, and bone marrow failure (Fig. 10.3; Dunn et al. 2000; Parker and Ware 2003; Luzzatto and Notaro 2003; Parker et al. 2005). The natural history of PNH is quite unpredictable, given the very heterogeneous disease presentation and evolution; indeed, some patients may live with the disease for decades, without major complications, whereas some others present with early life-threatening medical complications. Median survival was estimated above 10 years in the past decade in two independent series (Hillmen et al. 1995; Sociè et al. 1996) and about 22 years in a recent update of the French large database (de Latour et al. 2008), which anyway refers to time period before the introduction of the anticomplement therapy. The leading causes of death are thrombosis, followed by complications of bone marrow failure (infections and hemorrhages).

10.3.1 Intravascular Hemolysis

Intravascular hemolysis is the most typical manifestation of PNH and accounts for the most evident signs and symptoms of PNH (Fig. 10.3). Hemolysis is found in all PNH patients, even if the degree may be variable due to the size of the PNH clone(s), as well as to the type of PIG-A mutation (missense mutations may lead to a partial deficiency of GPI-linked protein, the so-called type II PNH erythrocytes, which have a lower susceptibility to complement-mediated lysis in comparison to the type III PNH erythrocytes; Rosse and Dacie 1966). Typically, the hemolysis is chronic, with possible exacerbations (the paroxysms) which occur due to massive complement activation, often in association with infections or other triggering events. Hemolysis leads to hemoglobinuria, urinary iron loss (seen as hemosiderinuria), and eventually anemia. The extent of anemia is very heterogeneous among patients, being also dependent on other factors such as compensatory erythropoiesis (which may be impaired in patients with more severe marrow failure) or even iron/vitamin deficiency. As a result,

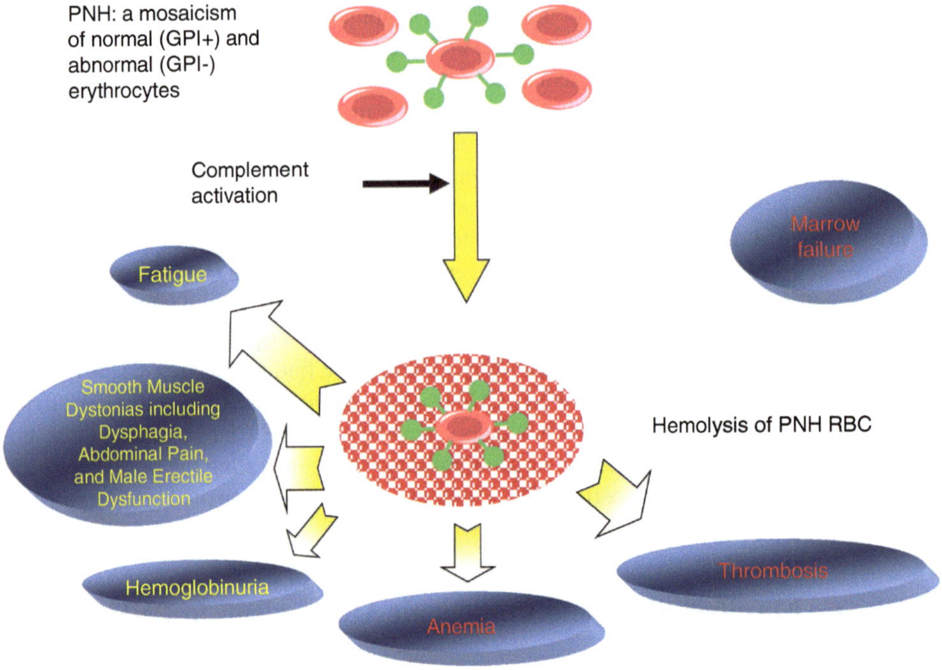

Fig. 10.3 *Pathogenic mechanisms of clinical manifestations in PNH.* Because of the lack of CD55 and CD59 (*green flags*), PNH erythrocytes are particularly susceptible to complement attack and suffer from complement-mediated lysis. Intravascular hemolysis accounts for anemia and other hemolysis-related manifestation; hemolysis (and possibly complement activation itself) may also be involved in the propensity to thrombosis typical of PNH patients. In contrast, bone marrow failure is not related to complement activation or hemolysis and harbors distinct pathogenic mechanisms (see text)

anemia may be severe, requiring frequent transfusions, in some patients, or well compensated in other cases. Besides anemia, it is now accepted that lysis may induce by itself specific disabling symptoms; they include malaise and fatigue (which may exceed than expected based on the low hemoglobin level), painful abdominal crises, dysphagia, and erectile dysfunction (Rother et al. 2005). All these latter symptoms, which usually occur in association with paroxysms of hemolysis, are thought to be due to smooth muscle dystonia secondary to NO consumption (Rother et al. 2005). Hemolysis in PNH mechanistically derives from the aberrant phenotype of affected erythrocytes. In fact, as discussed more in detail later on, the GPI-linked proteins missing on PNH erythrocytes include the two complement regulators CD55 (also known as decay-accelerating factor, DAF; Nicholson-Weller et al. 1982, 1983) and CD59 (or membrane inhibitor of reactive lysis, MIRL; Holguin et al. 1989a, b). As a result, PNH erythrocytes are exquisitely vulnerable to complement activation, resulting in uncontrolled complement activation and subsequent membrane attack complex (MAC) assembly, which leads to chronic intravascular hemolysis. The hierarchical contribution of CD55 and CD59 to hemolysis suggests that CD59 is the key molecule which, if absent, leads to lysis (Wilcox et al. 1991); in contrast, redundant mechanisms (including CD59 itself) usually overcome the isolated deficiency of CD55 (as demonstrated in patients carrying the so-called Inab phenotype, who do not suffer from hemolysis; Holguin et al. 1992). However, it should be reiterated that both complement activation and effector mechanisms are uncontrolled on PNH erythrocytes. Specifically, the lack of CD55 impairs regulation of the C3 convertases (regardless of the triggering pathway – classical or alternative; Mold et al. 1990), leading to increased C3 activation and further progression along the subsequent steps of the cascade. It is also understood that low-grade spontaneous C3 tickover leads to chronic CAP activation on the PNH erythrocyte surface (Pangburn et al. 1981, Pangburn et al. 1983c). Due to the lack of CD55, once activated by the CAP, the complement cascade continues through to MAC assembly, finally coming to lysis due to the lack of CD59 inhibition of C8 and C9 incorporation into the MAC. Thus, given such mechanistic relationship between hemolysis and the complement cascade, the availability of complement inhibitors is a critically important breakthrough for the treatment of PNH patients.

10.3.2 Thrombophilia

Thrombophilia is the second typical manifestation of PNH, with thrombosis developing in about 40% of all patients; accordingly, PNH is the medical condition with the higher risk of thrombosis. Unfortunately, as the underlying pathogenic mechanisms are not fully understood, thromboses are largely unpredictable in PNH patients, even if according to most series they generally develop in patients with large clones and massive hemolysis (Hall et al. 2003; Moyo et al. 2004). The thrombotic risk is peculiar to each patient, possibly as a result of additional independent (environmental or genetic) risk factors that may shape the individual predisposition to thrombosis. Thrombosis of PNH is quite unique because it mostly occurs at venous sites which are unusual for other non-PNH-related thrombosis. Intra-abdominal veins are the most frequent sites, followed by cerebral and limb veins; other possible sites include dermal veins, the lungs – with pulmonary embolus – and the arteries – leading to arterial thrombosis. Thrombotic disease may be life threatening and is the main cause of death for PNH patients (Hillmen et al. 1995; Socie et al. 1996; de Latour et al. 2008). Typical severe presentations of thrombotic PNH include hepatic venous (Budd-Chiari syndrome) (Hoekstra et al. 2009), portal, mesenteric, renal, and cerebral vein thrombosis (Audebert et al. 2005). A number of possible mechanisms accounting for thrombophilia in PNH have been hypothesized. The clinical observation that thrombotic complications are more common in patients with larger PNH clones (Hall et al. 2003; Grünewald et al. 2003; Moyo et al. 2004) and greater hemolysis suggests that the pathogenic mechanism is related to complement activation (Markiewski et al. 2007) and/or hemolysis itself (Fig. 10.3); this is someway confirmed by the reduced thromboembolic risk observed in patients receiving anticomplement therapy (Hillmen et al. 2007). At least four distinct (but not mutually exclusive) mechanisms can be hypothesized. First, uncontrolled complement regulation on platelet surface might lead to platelet activation and aggregation, enhancing clot formation (Gralnick et al. 1995; Louwes et al. 2001). Second, thrombophilia may directly result from hemolysis, due to the buildup of cell-free plasma hemoglobin released by the erythrocytes and subsequent scavenger effect on plasmatic nitric oxide, which might impair its inhibitory action on platelet aggregation and adhesion to endothelium, as well as its regulatory effect on vessel wall tone (Schafer et al. 2004; Rother et al. 2005). A third mechanism could be the generation of procoagulant particles; indeed, microvesicles are known to be released in PNH patients upon hemolysis and complement activation from RBCs (Hugel et al. 1999), WBCs (monocytes) and platelets (Wiedmer et al. 1993; Simak et al. 2004) and even from the endothelium. However, even if their procoagulant action is commonly accepted in other conditions (van Beers et al. 2009), their specific role in the pathophysiology of thromboembolisms in PNH is not yet proven. The fourth possible mechanism of thrombophilia in PNH might be an impairment of the fibrinolytic system, due to the lack of membrane-bound urokinase-type plasminogen activator receptor (uPAR) – which is GPI-linked – and to the excess of its soluble form (Ninomiya et al. 1997; Sloand et al. 2008).

10.3.3 Bone Marrow Failure

The third key clinical feature of PNH is bone marrow failure, which usually becomes evident as neutropenia and thrombocytopenia; however, it may also contribute to anemia in PNH patients, as evident in PNH patients with reticulocyte counts inadequate to the hemoglobin level. The marrow impairment in PNH may range from mild to severe cytopenia, eventually seen as aplastic anemia (AA) (Lewis and Dacie 1967), and with possible modifications during the course of the disease. A substantial proportion of PNH patients (between 30% and 70%) sooner or later develop frank AA (Hillmen et al. 1995; Socie et al. 1996; De Latour et al. 2008); on the other hand, AA patients may also develop PNH. Indeed, up to 40% of AA patients may have detectable PNH clone (Sugimori et al. 2006; Scheinberg et al. 2010), which may increase over time eventually causing hemolysis and clinical PNH. Thus, PNH and AA are closely embedded (Dameshek 1967) and should be considered as different presentations of the same disorder rather than independent conditions; furthermore, according to the dual pathophysiology theory, an underlying bone marrow failure accounts for the expansion of the PNH clone and is not surprising in PNH patients. (Rotoli and Luzzatto 1989). Indeed, regardless of its clinical presentation, PNH is a bone marrow disorder because of both a quantitative (clinical or subclinical) and a qualitative (the aberrant phenotype stems from the *PIG-A*-mutated HSC) impairment of hematopoiesis. Bone marrow impairment in PNH patients has been demonstrated by culture assays, showing a significant reduction in all lineage-committed progenitors (CFU-E, BFU-E, CFU-GM, CFU-GEMM) (Rotoli et al. 1982; Maciejewski et al. 1997), as well as in stem cells/multipotent progenitors (LTC-IC) (Maciejewski et al. 1997). The pathogenic mechanisms accounting for marrow failure in PNH are common to those described in AA and include an immune-mediated damage of the hematopoiesis, with quantitative and functional impairment of the hematopoietic stem cell pool and subsequent pancytopenia (Young and Maciejewski 2000; Young et al. 2006). Several evidences of immune derangement in PNH patients have been produced, including the presence of an oligoclonal T cell repertoire (Karadimitris et al. 2000; Risitano et al. 2004) and of clonal effector T cells (Poggi et al. 2005; Gargiulo et al. 2007), pointing out the pivotal role of cell-mediated immunity.

10.4 Complement Dysregulation in PNH

The complement system is a key component of innate immunity, which has evolved to recognize both exogenous pathogenic microorganisms as well as injured self tissues and to amplify adaptive immunity (Müller-Eberhard 1988; Holers 2008). The complement system encompasses distinct functional pathways with unique mechanisms of activation: the classical, the alternative, and the lectin pathways. These pathways are initiated by specific and distinct triggers, usually infectious agents or injured cells, but finally merge into a common final effector mechanism, namely, the cytolytic MAC. As an exception, the complement alternative pathway (CAP) also exhibits a low-grade continuous activation due to spontaneous hydrolysis of the C3 (the so-called tickover phenomenon) (Pangburn et al. 1981; Pangburn and Müller-Eberhard 1983). The CAP plays a pivotal role in complement functioning also because of its capability of amplifying any initial complement activation, regardless of the specific triggering pathway that initially generates C3b. Fine mechanisms have evolved to prevent possible damage from the complement system; they include membrane-bound proteins (complement receptor 1 [CR1], membrane cofactor protein [MCP], and the membrane proteins CD55 and CD59) as well as fluid-phase components, including complement factor I (FI) and factor H (FH). Among these, CD55 and CD59 are of pivotal importance in PNH, given that they are absent on *PIG-A* mutated cells due to the impaired GPI-anchor biosynthesis. CD55, also known as decay accelerating factor (DAF) is a

70-kd protein which inhibits the formation and the stability of C3 convertase (both C3bBb and C4b2a) (Nicholson-Weller 1992). Historically, CD55 was the first complement regulator reported to be absent on PNH erythrocytes (Pangburn et al. 1983a, b; Nicholson-Weller et al. 1983) possibly accounting for the increased susceptibility of PNH erythrocytes to complement-mediated lysis. However, further studies suggested that factors other than CD55 should also be involved, possibly acting downstream on the complement cascade (Medof et al. 1987; Shin et al. 1986). Subsequently, CD59 (also known as membrane inhibitor of reactive lysis, MIRL) was identified as an additional complement inhibitor which was found deficient on PNH cells (Holguin et al. 1989a). CD59 interferes with the terminal effector complement, blocking the incorporation of C9 onto the C5b-C8 complex, thus preventing MAC formation (Meri et al. 1990). The hierarchical contribution of CD55 and CD59 to hemolysis suggests that CD59 is the key molecule which, if absent, leads to lysis (Wilcox et al. 1991). In contrast, redundant mechanisms (including CD59 itself) usually overcome an isolated deficiency of CD55 (Holguin et al. 1992), as demonstrated in patients carrying the so-called Inab, who do not show any sign of hemolysis (Merry et al. 1989). In contrast, anecdotic cases of inherited CD59 deficiency harbor a clinical phenotype undistinguishable from PNH (Yamashina et al. 1990; Motoyama et al. 1992). It is intelligible that the deficiency of the complement regulatory proteins CD55 and CD59 makes PNH erythrocytes prone to lyse once the complement cascade activates. The in vitro susceptibility of PNH erythrocytes has been demonstrated since the initial descriptions by Dr. Ham, who clearly documented (before being aware of any biochemical and genetic feature of PNH) that erythrocytes from PNH patients lyse in autologous serum upon complement activation by acidification (the so-called acidified serum assay, also known as the Ham test; Ham and Dingle 1939). Several decades later, Rosse and Dacie clearly elucidated that erythrocytes obtained from PNH patients are a mosaicism of two or three different phenotypes, which differ in their sensitivity to complement-mediated lysis in vitro (Rosse and Dacie 1966; Rosse 1971). The first phenotype has a normal (equal to that of erythrocytes from healthy individuals) sensitivity to complement, while abnormal cells may have a dramatic hypersensitivity to complement-mediated lysis (15–25 times the normal one) or just a moderate hypersensitivity (3–5 times the normal). These phenotypes are referred as PNH type I, type III, and type II, respectively (Rosse and Dacie 1966; Rosse 1971); now, we know that they correspond to a normal expression of GPI-APs (type I) or to a complete (type III) or partial (type II) deficiency of GPI-APs, as documented by flow cytometry (van der Schoot et al. 1990). The susceptibility of PNH erythrocytes to complement-mediated lysis has a direct and mechanistic consequence in vivo because the lifespan of PNH erythrocytes is about tenfold reduced in comparison to that of normal red cells. This chronic hemolysis likely results from a steady-state complement activation, due to the low-grade spontaneous C3 tickover leading to continuous CAP activation on PNH erythrocyte surface (Pangburn et al. 1981; Pangburn and Müller-Eberhard 1983). This steady-state chronic hemolysis may undergo tremendous rises in case of specific conditions, such as infections or inflammatory status, which may further activate the complement cascade. It should be reiterated that both the initial complement activation and the downstream effector mechanisms are uncontrolled on PNH erythrocytes; indeed, the lack of CD55 impairs the regulation of C3 convertases (regardless of the triggering pathway – classical or alternative) (Mold et al. 1990), leading to increased C3 activation and further progression along the subsequent steps toward the MAC assembly – which is not prevented due to the lack of CD59 (Fig. 10.4). In this view, PNH can be considered mostly a CAP-mediated disease (Holers 2008), where a low-level chronic CAP-mediated hemolysis is greatly amplified during inflammatory or infectious diseases. Overt hemolysis and paroxysms of PNH patients result from a specific triggering action on the complement system, which may occur along each of the three distinct complement pathways. There are no data demonstrating which complement pathway is activated in specific conditions in vivo (i.e., infections); however, one may speculate that all the three pathways may cooperate, with a possible hierarchical dominance of the CAP, given its capability to amplify any complement activation regardless of the initial triggering pathway (and its uncontrolled activation due to the lack of CD55).

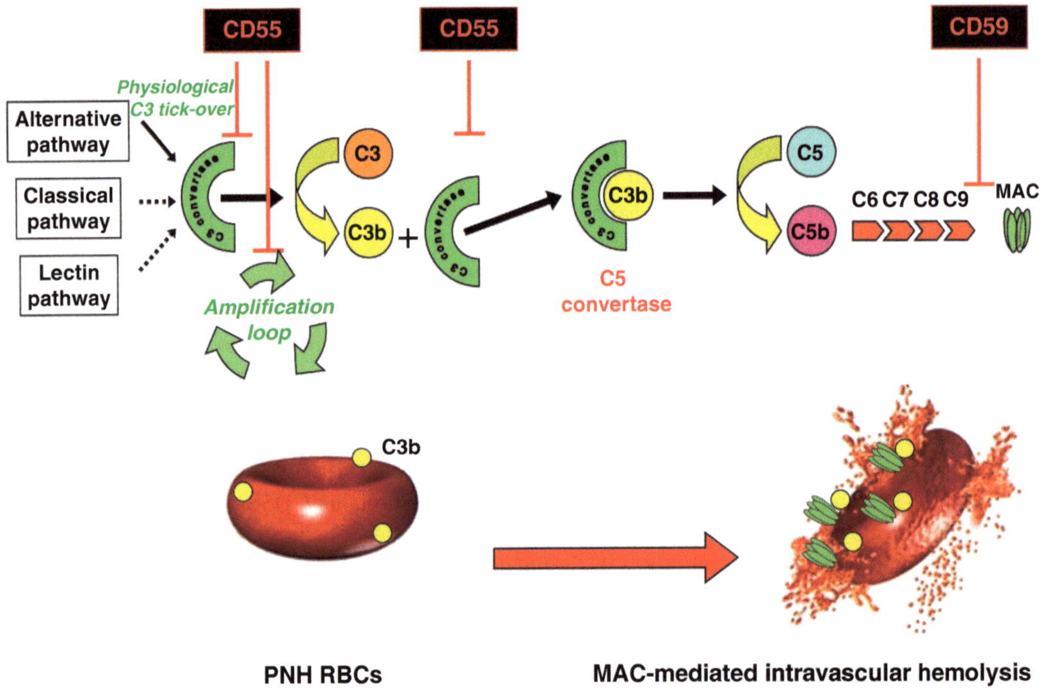

Fig. 10.4 *Complement regulation and intravascular hemolysis of PNH erythrocytes: steady state*. Given their deficiency in surface CD55, PNH erythrocytes are susceptible to complement activation through C3 tickover and the alternative pathway, as well as through any other complement pathway. After initial C3 activation and subsequent C3b binding to erythrocyte surface, the complement cascade may proceed toward the MAV formation, which is not inhibited due to the lack of CD59. As a result, PNH erythrocytes succumb because of MAC-mediated intravascular hemolysis

10.5 Anticomplement Treatment in PNH: The Lesson of Eculizumab

10.5.1 Efficacy and Safety of the Complement Inhibitor Eculizumab

The treatment of PNH has been dramatically changed since the availability of the first complement inhibitor eculizumab. Eculizumab (h5G1.1-mAb, Soliris®, Alexion Pharmaceuticals) is a humanized monoclonal antibody (mAb) (Rother et al. 2007) derived from the murine anti-human C5 mAb; it specifically binds the terminal complement component 5, thereby inhibiting its cleavage to C5a and C5b (Matis and Rollins 1995). Thus, eculizumab blocks the formation of MAC, the terminal effector mechanism accounting for intravascular hemolysis of PNH erythrocytes. Eculizumab was initially investigated in patients suffering from other complement-mediated disorders; however, PNH appeared to be the best candidate disease to benefit from eculizumab treatment. In fact, eculizumab may compensate for the absence of CD59 on PNH erythrocytes, preventing their lysis upon complement activation. The blockade of the complement cascade at the level of C5 does not affect early complement components, preserving pivotal functions such as clearance of immune complexes and microorganisms (Matis and Rollins 1995). In the last few years, eculizumab has been extensively investigated for the treatment of hemolysis in patients with transfusion-dependent PNH. Safety and efficacy of eculizumab were initially established in one phase II pilot study (Hillmen et al. 2004) as well as in two multinational phase III clinical studies [TRIUMPH (Hillmen et al. 2006) and SHEPHERD (Brodsky et al. 2008)] and subsequently were confirmed in a common open-label extension study (Hillmen et al. 2007). Eculizumab was given intravenously dosed at 600 mg weekly for 4 weeks (loading phase), followed 1 week later by 900 mg fortnightly (maintenance phase); all patients were vaccinated against *Neisseria meningitidis* at least 2 weeks before starting the treatment. After the initial pilot study, which provided proof of principle of effective blockade of intravascular hemolysis in 11 heavily transfused PNH patients (Hillmen et al. 2004), eculizumab was tested in a double-blind, placebo-controlled, multinational randomized trial which enrolled 86 transfusion-dependent PNH patients (Hillmen et al. 2006). Treatment with eculizumab resulted in a dramatic reduction of intravascular hemolysis, as measured by LDH, leading to hemoglobin stabilization and transfusion independence in about half of the patients. Control of intravascular hemolysis was found in all patients, and even cases not achieving transfusion independence showed a reduction of their transfusional needs. The effects of eculizumab on hemolysis were evident after the first administration and lasted for the whole study period. In comparison to placebo, eculizumab also resulted in a significant improvement in fatigue and quality of life, as measured by validated questionnaires. The safety profile of eculizumab was excellent, with negligible side effects and incidence of adverse events comparable to that of the placebo. These data were confirmed in the open-label phase III study SHEPHERD, which included a broader PNH population (Brodsky et al. 2008). In the 96 patients enrolled in the study, treatment with eculizumab resulted in an almost complete control of intravascular hemolysis, regardless of the pretreatment transfusion requirement, with transfusion independence achieved in half of the patients, and significant improvement in fatigue and quality of life (Brodsky et al. 2008). The subsequent open-label extension study enrolled a total of 187 patients who have previously completed one of the parent clinical trials (Hillmen et al. 2007). The extension study confirmed the efficacy and the safety of eculizumab with a longer follow-up, confirming that the effects of eculizumab treatment on intravascular hemolysis were retained over time. In addition, by comparing the rate of thrombosis between the pretreatment and treatment periods in the same patients, the extension study allowed an evaluation of the incidence of thromboembolic events in PNH patients on eculizumab, showing an 85% relative reduction of the thromboembolic risk (Hillmen et al. 2007). Whether eculizumab exerts its effect on thrombophilia of PNH directly, or through the blockade of intravascular hemolysis (e.g., by reduction of NO consumption or reduced release of procoagulant microvesicles), is still unknown. However, the effects of eculizumab on thrombophilia of PNH seem quite specific, given that distinct biomarkers of coagulation pathway activation, reactive fibrinolysis, and endothelial cell activation decrease during eculizumab treatment (Helley et al. 2010). All these beneficial effects of eculizumab may potentially result in an improvement of survival of PNH patients, as recently shown in a limited cohort (Kelly et al. 2011).

10.5.2 Biological Insights from Treatment with Eculizumab

The use of eculizumab represented a unique opportunity to improve our understanding of PNH pathophysiology in vivo and in particular to dissect the potential role of early and terminal phases of the complement cascade. Recent observation from our group has led to the identification of a novel mechanism of disease specifically emerging in PNH patients on eculizumab, which could somehow limit the benefit of eculizumab treatment. Indeed, we have demonstrated that PNH erythrocytes of PNH patients on eculizumab escape MAC formation and lysis, thanks to eculizumab, but progressively bind C3 fragments on their surface. In our initial series of 56 PNH patients, we have demonstrated by flow cytometry that all the 41 PNH patients on eculizumab harbored C3 fragments bound to a substantial portion of their PNH erythrocytes (while none of the untreated patients did) (Risitano et al. 2009a). Our data were confirmed in an independent series by another group that exploited a direct antiglobulin test using C3d-specific antisera (Hill et al. 2010). Notably, these biological findings parallel the observation that most PNH patients on eculizumab remain anemic, regardless of an

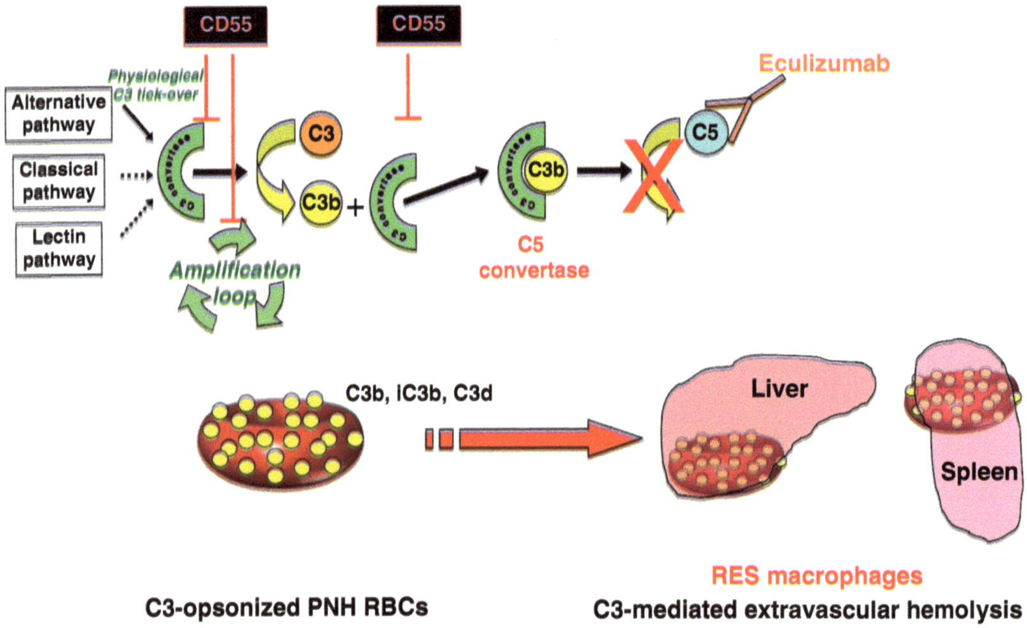

Fig. 10.5 *Complement regulation and intravascular hemolysis of PNH erythrocytes: effects of eculizumab.* Eculizumab binds to C5 in the fluid phase and prevents its cleavage to C5a and C5b. As a result, the progression of the complement cascade toward the MAC assembly is completely inhibited, and PNH erythrocytes result protected from MAC-mediated intravascular hemolysis. However, given the lack of CD55, initial CAP activation through the C3 tickover and the CAP-mediated amplification loop remain uncontrolled; thus, early C3 activation continues, and because of their increased survival, PNH erythrocytes progressively bind substantial amount of C3 fragments on their surface. Subsequently, C3-opsonized PNH erythrocytes become susceptible to C3-receptor-mediated clearance by the hepatosplenic reticuloendothelial system, eventually resulting in C3-mediated extravascular hemolysis

adequate control of intravascular hemolysis (as shown by normal or almost normal LDH levels), and all of them continue to have substantial reticulocytosis (Risitano et al. 2009a). We hypothesized that membrane-bound C3 fragments work as opsonins on PNH erythrocytes, resulting in their entrapment by reticuloendothelial cells through specific C3 receptors and subsequent extravascular hemolysis (Risitano et al. 2009a; Luzzatto et al. 2010). This mechanism is supported by persistent reticulocytosis, hyperbilirubinemia, and anemia in patients on eculizumab and was also confirmed by an in vivo erythrocyte survival study by ^{51}Cr labeling (which showed reduced survival and hepatosplenic ^{51}Cr uptake) (Risitano et al. 2009a). Thus, we have suggested that C3-mediated extravascular hemolysis is a common phenomenon of PNH patients on eculizumab, which may become clinically significant in some patients and largely accounts for residual anemia and heterogeneous hematological benefit of eculizumab treatment (Fig. 10.5; Risitano et al. 2009a). Pathophysiologically, it is clear that such a mechanism becomes evident only when eculizumab prevents MAC-mediated hemolysis, allowing longer survival of PNH erythrocytes, which continue to suffer from uncontrolled C3 convertase activation and C3 fragment deposition due to CD55 deficiency (Luzzatto et al. 2010; Risitano et al. 2010). Indeed, CAP is physiologically in a state of continuous activation because low-grade spontaneous hydrolysis of the internal thioester bond of C3 generates a C3b-like molecule, C3(H_2O); nascent C3(H_2O) is able to recruit factor B in forming (in the fluid phase) an unstable pro-C3 convertase. Once cleaved by factor D (generating C3(H_2O)Bb), this complex will in turn cleave additional C3 molecules to generate C3b, which binds predominantly to glycophorin A and activates (now in a membrane-bound phase) the CAP amplification loop (Parker et al. 1982; Pangburn et al.

1983c; Müller-Eberhard 1988). On PNH erythrocytes, the lack of CD55 will allow this process (which is self-limiting on normal cells) to continue, leading to progressive CAP-mediated amplification, even in the presence of eculizumab (which acts downstream of C3). The reasons why only a fraction of PNH erythrocytes have membrane-bound C3, and why the proportion varies among patients, are not fully understood. Nevertheless, in vitro data support the concept that PNH erythrocytes are all susceptible to C3 deposition once exposed to conditions causing complement activation (Sica et al. 2010). The reasons why a specific erythrocyte may reach the threshold needed to start CAP activation could be simply stochastic or possibly linked to specific microenvironmental conditions generating in particular districts of the circulatory tree. We have also hypothesized that interindividual differences in other physiological inhibitors (such as CR1, complement FH, and complement FI) may modulate the complement activation in a patient-specific fashion, leading to distinct patterns of C3 deposition. In addition, even more complex factors may drive downstream the subsequent fate of C3-bound PNH erythrocytes; in fact, some patients may harbor large proportion of C3-bound PNH erythrocytes, without showing a clinically relevant extravascular hemolysis (Risitano et al. 2010). At the moment, there is yet no ability to predict before starting eculizumab which patients will develop clinically meaningful C3-mediated extravascular hemolysis.

10.6 Novel Strategies of Complement Modulation in PNH

C3 opsonization of PNH erythrocytes is a common phenomenon for PNH patients treated with eculizumab, even if the subsequent extravascular hemolysis may remain limited or well compensated in many cases (Luzzatto et al. 2010). Thus, additional therapeutical strategies to manage clinically meaningful C3-mediated extravascular hemolysis are needed. We reported a patient managed by splenectomy (Risitano et al. 2008), who achieved a substantial improvement of hemoglobin level without any medical complication; however, many physicians raise the concern that this approach may carry an increased lifelong risk of infections, in addition to the risk of intra- or perioperatory complications (especially thrombosis or hemorrhage in thrombocytopenic patients) (Brodsky 2009). Steroids seem to be a poor option because of the well-known side effects of long-term treatment and lack of evidence supporting a clear benefit (Risitano et al. 2010). In some patients, the use of recombinant erythropoietin has proven beneficial by increasing compensatory erythropoiesis (Hill et al. 2007), without any specific effect on residual extravascular hemolysis. Thus, at the moment, there is no established treatment to prevent or to control C3-mediated extravascular hemolysis in PNH.

10.6.1 Targeted Anticomplement Therapies

The emergence of experimental and clinical evidences for CAP-initiated and C3 fragment-mediated extravascular hemolysis suggests that new treatment strategies appropriately targeting the early phases of the complement cascade should be assessed. The ideal agent should prevent the early phase of complement activation on PNH cells and defuse the amplification mechanisms (e.g., the CAP amplification loop). A systemic blockade of C3 activation through all pathways by monoclonal antibodies (similar to the anti-C5 eculizumab) could be considered (e.g., by anti-C3 monoclonal antibodies); however, this approach may carry the risk of infectious and autoimmune complications secondary to a complete switching off of the complement system at this point. Recently, a different type of anti-C3 monoclonal antibody (the murine 3E7 and its chimeric-deimmunized derivative H17) has been developed, which recognizes activated C3 (C3b/iC3b) but not naïve, circulating C3; this antibody seems to affect the activity of the C3/C5 convertases of the CAP only, preserving the classical pathway (Lindorfer et al. 2010). These antibodies have been shown to be effective in inhibiting hemolysis of PNH erythrocytes in vitro; however, it has to be remarked that if C3/C5 convertases are localized on red cell surface, such an antibody may also work as an additional opsonin, possibly further amplifying the clearance of PNH erythrocytes through macrophage Fc receptors. A novel candidate agent has been designed by creating a recombinant fusion protein between two endogenous complement-related proteins, complement FH and complement receptor 2 (CR2). FH is a physiological complement inhibitor that modulates the initial CAP activation in the fluid phase by preventing C3 convertase activity and by promoting C3b inactivation into iC3b (Whaley and Ruddy 1976). Indeed, FH defuses the CAP amplification loop, and it has been demonstrated protective from lysis for PNH erythrocytes in vitro (Ferreira and Pangburn 2007). In the aim to deliver FH activity locally at the site of complement activation, FH was fused with the iC3b/C3d-binding domain of CR2. We have recently investigated the effects of this CR2-FH fusion protein (named TT30) in an in vitro model based on exposure of PNH erythrocytes to CAP-activated serum. In our assay TT30 resulted in a complete inhibition of complement-mediated hemolysis, as well as in a complete prevention of C3 deposition on PNH erythrocytes (Risitano et al. 2009b; Risitano et al. 2012). These in vitro data support the concept that TT30 in vivo could be useful to prevent both intravascular and extravascular hemolysis of PNH erythrocytes. Based on this robust background, a phase I clinical trial has just started to enroll PNH patients in a *first-in-human* study (Alexion Pharmaceuticals, personal communication). A number of additional complement inhibitors or modulators are currently under preclinical development; many of them could be potentially intriguing for the treatment of PNH, especially those targeting the C3 convertase (such as anti-FB antibodies or properdin inhibitors) or C3 itself (such as the small molecule compstatin and its derivatives; Ricklin and Lambris 2008) seem the most appropriate for possible preclinical investigations in PNH. Once these or other next-generation complement inhibitors proceed to clinical development, then we can determine whether such targeted inhibition should be additional or alternative to eculizumab. Indeed, the adequate control of C3 or both C3 and C5 activation on PNH red cells might make the downstream blockade by eculizumab redundant. Of course, any clinical translation should carefully take into account possible detrimental effects of such upstream and broad impairment of the complement cascade, which could expose to potentially dramatic infectious and autoimmune complications, much more than those seen with eculizumab.

10.7 Conclusions

Since its initial description, PNH has been a unique opportunity for physicians and scientist to improve their knowledge in basic science and medicine. Indeed, with its history, PNH paralleled the terrific scientific progress of medical science in the last century; in fact, biochemistry, molecular biology, and genetics all contributed to unravel the mystery of PNH. In some way, together with other hematological disorders, PNH represented one of the best examples of how basic science has changed how we view medicine, with a deep impact on our medical practice. Even in the last decade, PNH continued to be a unique scientific gym: in fact, it represented the opportunity to develop the first complement inhibitor approved for clinical use. In this endless process, PNH has driven a further improvement in our understanding of complement biology, especially in the context of its complex pathophysiology. The observation that novel mechanisms of disease may appear in PNH patients on eculizumab treatment led to a deeper understanding of PNH pathophysiology and, more important, to a growing interest in the development of novel anticomplement strategies. The future years promise to be extremely productive in terms of therapeutical strategies targeting specific pathways and/or steps of the complement cascade; the hope is that they will represent a substantial improvement for the treatment of PNH and other complement-mediated diseases.

References

Araten DJ, Nafa K, Pakdeesuwan K et al (1999) Clonal populations of hematopoietic cells with paroxysmal nocturnal hemoglobinuria genotype and phenotype are present in normal individuals. Proc Natl Acad Sci USA 96:5209–5214
Araten DJ, Bessler M, McKenzie S et al (2002) Dynamics of hematopoiesis in paroxysmal nocturnal hemoglobinuria (PNH): no evidence for intrinsic growth advantage of PNH clones. Leukemia 16:2243–2248
Armstrong C, Schubert J, Ueda E et al (1992) Affected paroxysmal nocturnal hemoglobinuria T lymphocytes harbor a common defect in assembly of N-acetyl-D-glucosamine inositol phospholipid corresponding to that in class A Thy-1-urine lymphoma mutants. J Biol Chem 267:25347–25351
Audebert HJ, Planck J, Eisenburg M et al (2005) Cerebral ischemic infarction in paroxysmal nocturnal hemoglobinuria report of 2 cases and updated review of 7 previously published patients. J Neurol 252:1379–1386
Brodsky RA, Young NS, Antonioli E et al (2008) Multicenter phase III study of the complement inhibitor eculizumab for the treatment of patients with paroxysmal nocturnal hemoglobinuria. Blood 114:1840–1847
Brodsky RA (2009) How I treat paroxysmal nocturnal hemoglobinuria. Blood 113:6522–6527
Chen G, Zeng W, Maciejewski JP et al (2005) Differential gene expression in hematopoietic progenitors from paroxysmal nocturnal hemoglobinuria patients reveals an apoptosis/immune response in 'normal' phenotype cells. Leukemia 19:862–868
Crosby WH (1951) Paroxysmal nocturnal hemoglobinuria; a classic description by Paul Strübling in 1882, and a bibliography of the disease. Blood 63:270–284
Dameshek W (1967) Riddle: what do aplastic anemia, paroxysmal nocturnal hemoglobinuria (PNH) and "hypoplastic" leukemia have in common? Blood 30:251–254
de Latour RP, Mary JY, Salanoubat C et al (2008) French Association of Young Hematologists Paroxysmal nocturnal hemoglobinuria: natural history of disease subcategories. Blood 112:3099–3106
Dunn DE, Liu JM, Young NS (2000) Paroxysmal nocturnal hemoglobinuria. In: Young NS (ed) Bone marrow failure syndromes. W.B. Saunders Company, Philadelphia, pp 99–121
Endo M, Ware RE, Vreeke TM et al (1996) Molecular basis of the heterogeneity of expression of glycosyl phosphatidylinositol anchored proteins in paroxysmal nocturnal hemoglobinuria. Blood 87:2546–2557
Ferreira VP, Pangburn MK (2007) Factor H mediated cell surface protection from complement is critical for the survival of PNH erythrocytes. Blood 15(110):1290–1292
Gargiulo L, Lastraioli S, Cerruti G et al (2007) Highly homologous T-cell receptor beta sequences support a common target for autoreactive T cells in most patients with paroxysmal nocturnal hemoglobinuria. Blood 109:5036–5042
Gralnick HR, Vail M, McKeown LP et al (1995) Activated platelets in paroxysmal nocturnal haemoglobinuria. Br J Haematol 91:697–702
Grünewald M, Siegemund A, Grünewald A et al (2003) Plasmatic coagulation and fibrinolytic system alterations in PNH: relation to clone size. Blood Coagul Fibrinolysis 14:685–695
Hall C, Richards S, Hillmen P (2003) Primary prophylaxis with warfarin prevents thrombosis in paroxysmal nocturnal hemoglobinuria (PNH). Blood 102:3587–3591
Ham TH, Dingle JH (1939) Studies on the destruction of red blood cells. II. Chronic hemolytic anemia with paroxysmal nocturnal hemoglobinuria: certain immunological aspects of the hemolytic mechanism with special reference to serum complement. J Clin Invest 18:657–672
Helley D, de Latour RP, Porcher R et al (2010) French Society of Hematology. Evaluation of hemostasis and endothelial function in patients with paroxysmal nocturnal hemoglobinuria receiving eculizumab. Haematologica 95:574–581
Hill A, Richards SJ, Rother RP et al (2007) Erythropoietin treatment during complement inhibition with eculizumab in a patient with paroxysmal nocturnal hemoglobinuria. Haematologica 92:e31, ECR14
Hill A, Rother RP, Arnold L et al (2010) Eculizumab prevents intravascular hemolysis in patients with paroxysmal nocturnal hemoglobinuria and unmasks low-level extravascular hemolysis occurring through C3 opsonization. Haematologica 95:567–573
Hillmen P, Lewis SM, Bessler M et al (1995) Natural history of paroxysmal nocturnal hemoglobinuria. N Engl J Med 333:1253–1258
Hillmen P, Hall C, Marsh JC et al (2004) Effect of eculizumab on hemolysis and transfusion requirements in patients with paroxysmal nocturnal hemoglobinuria. N Engl J Med 350:552–559
Hillmen P, Young NS, Schubert J et al (2006) The complement inhibitor eculizumab in paroxysmal nocturnal hemoglobinuria. N Engl J Med 355:1233–1243
Hillmen P, Muus P, Duhrsen U et al (2007) Effect of the complement inhibitor eculizumab on thromboembolism in patients with paroxysmal nocturnal hemoglobinuria. Blood 110:4123–4128
Hirose S, Ravi L, Prince GM et al (1992) Synthesis of mannosylglucosaminylinositol phospholipids in normal but not paroxysmal nocturnal hemoglobinuria cells. Proc Natl Acad Sci USA 89:6025–6029
Hoekstra J, Leebeek FW, Plessier A et al (2009) Paroxysmal nocturnal hemoglobinuria in Budd-Chiari syndrome: findings from a cohort study. J Hepatol 51:696–706

Holers VM (2008) The spectrum of complement alternative pathway-mediated diseases. Immunol Rev 223:300–316
Holguin MH, Fredrick LR, Bernshaw NJ et al (1989a) Isolation and characterization of a membrane protein from normal human erythrocytes that inhibits reactive lysis of the erythrocytes of paroxysmal nocturnal hemoglobinuria. J Clin Invest 84:7–17
Holguin MH, Wilcox LA, Bernshaw NJ et al (1989b) Relationship between the membrane inhibitor of reactive lysis and the erythrocyte phenotypes of paroxysmal nocturnal hemoglobinuria. J Clin Invest 84:1387–1394
Holguin MH, Martin CB, Bernshaw NJ et al (1992) Analysis of the effects of activation of the alternative pathway of complement on erythrocytes with an isolated deficiency of decay accelerating factor. J Immunol 148:498–502
Hugel B, Socié G, Vu T et al (1999) Elevated levels of circulating procoagulant microparticles in patients with paroxysmal nocturnal hemoglobinuria and aplastic anemia. Blood 93:3451–3456
Jasinski M, Keller P, Fujiwara Y et al (2001) GATA1-Cre mediates Piga gene inactivation in the erythroid/megakaryocytic lineage and leads to circulating red cells with a partial deficiency in glycosyl phosphatidylinositol-linked proteins (paroxysmal nocturnal hemoglobinuria type II cells). Blood 98:2248–2255
Karadimitris A, Manavalan JS, Thaler HT et al (2000) Abnormal T-cell repertoire is consistent with immune process underlying the pathogenesis of paroxysmal nocturnal hemoglobinuria. Blood 96:2613–2620
Kawagoe K, Kitamura D, Okabe M et al (1996) Glycosylphosphatidylinositol-anchor-deficient mice: implications for clonal dominance of mutant cells in paroxysmal nocturnal hemoglobinuria. Blood 87:3600–3606
Keller P, Payne JL, Tremml G et al (2001) FES-Cre targets phosphatidylinositol glycan class A (PIGA) inactivation to hematopoietic stem cells in the bone marrow. J Exp Med 194:581–589
Kelly RJ, Hill A, Arnold LM et al (2011) Long-term treatment with eculizumab in paroxysmal nocturnal hemoglobinuria: sustained efficacy and improved survival. Blood 117:6786–6792
Kinoshita T, Medof ME, Silber R et al (1985) Distribution of decay-accelerating factor in the peripheral blood of normal individuals and patients with paroxysmal nocturnal hemoglobinuria. J Exp Med 162:75–92
Kunstling TR, Rosse WF (1969) Erythrocyte acetylcholinesterase deficiency in paroxysmal nocturnal hemoglobinuria (PNH). A comparison of the complement-sensitive and insensitive populations. Blood 33:607–616
Lewis SM, Dacie JV (1967) The aplastic anaemia–paroxysmal nocturnal haemoglobinuria syndrome. Br J Haematol 13:236–251
Lindorfer MA, Pawluczkowycz AW, Peek EM et al (2010) A novel approach to preventing the hemolysis of paroxysmal nocturnal hemoglobinuria: both complement-mediated cytolysis and C3 deposition are blocked by a monoclonal antibody specific for the alternative pathway of complement. Blood 115:2283–2291
Louwes H, Vellenga E, de Wolf JT (2001) Abnormal platelet adhesion on abdominal vessels in asymptomatic patients with paroxysmal nocturnal hemoglobinuria. Ann Hematol 80:573–576
Luzzatto L, Notaro R (2003) Paroxysmal nocturnal hemoglobinuria. In: Handin RI, Lux SE, Stossel TP (eds) Blood, principles and practice of hematology, 2nd edn. Lippincott Williams & Wilkins, Philadelphia, pp 319–334
Luzzatto L, Bessler M, Rotoli B (1997) Somatic mutations in paroxysmal nocturnal hemoglobinuria: a blessing in disguise? Cell 88:1–4
Luzzatto L, Risitano AM, Notaro R (2010) Paroxysmal nocturnal hemoglobinuria and eculizumab. Haematologica 95:523–526
Maciejewski JP, Sloand EM, Sato T et al (1997) Impaired hematopoiesis in paroxysmal nocturnal hemoglobinuria/aplastic anemia is not associated with a selective proliferative defect in the glycosylphosphatidylinositol-anchored protein-deficient clone. Blood 89:1173–1181
Mahoney JF, Urakaze M, Hall S et al (1992) Defective glycosylphosphatidylinositol anchor synthesis in paroxysmal nocturnal hemoglobinuria granulocytes. Blood 79:1400–1403
Markiewski MM, Nilsson B, Ekdahl KN et al (2007) Complement and coagulation: strangers or partners in crime? Trends Immunol 28:184–192
Matis LA, Rollins SA (1995) Complement-specific antibodies: designing novel anti-inflammatories. Nat Med 1:839–842
Medof ME, Gottlieb A, Kinoshita T et al (1987) Relationship between decay accelerating factor deficiency, diminished acetylcholinesterase activity, and defective terminal complement pathway restriction in paroxysmal nocturnal hemoglobinuria erythrocytes. J Clin Invest 80:165–174
Meri S, Morgan BP, Davies A et al (1990) Human protectin (CD59), an 18,000–20,000 MW complement lysis restricting factor, inhibits C5b-8 catalysed insertion of C9 into lipid bilayers. Immunology 71:1–9
Merry AH, Rawlinson VI, Uchikawa M et al (1989) Studies on the sensitivity to complement-mediated lysis of erythrocytes (Inab phenotype) with a deficiency of DAF (decay accelerating factor). Br J Haematol 73:248–253
Miyata T, Takeda J, Iida Y et al (1993) The cloning of PIG-A, a component in the early step of GPI-anchor biosynthesis. Science 259:1318–1320
Mold C, Walter EI, Medof ME (1990) The influence of membrane components on regulation of alternative pathway activation by decay-accelerating factor. J Immunol 145:3836–3841
Motoyama N, Okada N, Yamashina M et al (1992) Paroxysmal nocturnal hemoglobinuria due to hereditary nucleotide deletion in the HRF20 (CD59) gene. Eur J Immunol 22:2669–2673

Moyo VM, Mukhina GL, Garrett ES et al (2004) Natural history of paroxysmal nocturnal haemoglobinuria using modern diagnostic assays. Br J Haematol 126:133–138
Müller-Eberhard HJ (1988) Molecular organization and function of the complement system. Annu Rev Biochem 57:321–347
Nicholson-Weller A (1992) Decay accelerating factor (CD55). Curr Top Microbiol Immunol 178:7–30
Nicholson-Weller A, Burge J, Fearon DT et al (1982) Isolation of a human erythrocyte membrane glycoprotein with decay-accelerating activity for C3 convertases of the complement system. J Immunol 129:184–189
Nicholson-Weller A, March JP, Rosenfeld SI et al (1983) Affected erythrocytes of patients with paroxysmal nocturnal hemoglobinuria are deficient in the complement regulatory protein, decay accelerating factor. Proc Natl Acad Sci USA 80:5066–5070
Ninomiya H, Hasegawa Y, Nagasawa T et al (1997) Excess soluble urokinase-type plasminogen activator receptor in the plasma of patients with paroxysmal nocturnal hemoglobinuria inhibits cell-associated fibrinolytic activity. Int J Hematol 65:285–291
Nishimura J, Inoue N, Wada H et al (1997) A patient with paroxysmal nocturnal hemoglobinuria bearing four independent PIG-A mutant clones. Blood 89:3470–3476
Nishimura J Ji, Hirota T, Kanakura Y et al (2002) Long-term support of hematopoiesis by a single stem cell clone in patients with paroxysmal nocturnal hemoglobinuria. Blood 99:2748–2751
Nishimura J, Kanakura Y, Ware RE et al (2004) Clinical course and flow cytometric analysis of paroxysmal nocturnal hemoglobinuria in the United States and Japan. Medicine (Baltimore) 83:193–207
Oni BS, Osunkoya BO, Luzzatto L (1970) Paroxysmal nocturnal hemoglobinuria: evidence for monoclonal origin of abnormal red cells. Blood 36:145–152
Orphanet (2004) Paroxysmal nocturnal hemoglobinuria, 1-12-2004. Ref Type: Internet Communication
Pangburn MK, Müller-Eberhard HJ (1983) Initiation of the alternative complement pathway due to spontaneous hydrolysis of the thioester of C3. Ann N Y Acad Sci 421:291–298
Pangburn MK, Schreiber RD, Müller-Eberhard HJ (1981) Formation of the initial C3 convertase of the alternative complement pathway. Acquisition of C3b-like activities by spontaneous hydrolysis of the putative thioester in native C3. J Exp Med 154:856–867
Pangburn MK, Schreiber RD, Trombold JS et al (1983a) Paroxysmal nocturnal hemoglobinuria: deficiency in factor H-like functions of the abnormal erythrocytes. J Exp Med 157:1971–1980
Pangburn MK, Schreiber RD, Müller-Eberhard HJ (1983b) Deficiency of an erythrocyte membrane protein with complement regulatory activity in paroxysmal nocturnal hemoglobinuria. Proc Natl Acad Sci USA 80:5430–5434
Pangburn MK, Schreiber RD, Müller-Eberhard HJ (1983c) C3b deposition during activation of the alternative complement pathway and the effect of deposition on the activating surface. J Immunol 131:1930–1935
Parker CJ, Ware RE (2003) Paroxysmal nocturnal hemoglobinuria. In: Greer J et al (eds) Wintrobe's clinical hematology, 11th edn. Lippincott Williams & Wilkins, Philadelphia, pp 1203–1221
Parker CJ, Baker PJ, Rosse WF (1982) Increased enzymatic activity of the alternative pathway convertase when bound to the erythrocytes of paroxysmal nocturnal hemoglobinuria. J Clin Invest 69:337–346
Parker C, Omine M, Richards S, International PNH Interest Group et al (2005) Diagnosis and management of paroxysmal nocturnal hemoglobinuria. Blood 106:3699–3709
Poggi A, Negrini S, Zocchi MR et al (2005) Patients with paroxysmal nocturnal hemoglobinuria have a high frequency of peripheral-blood T cells expressing activating isoforms of inhibiting superfamily receptors. Blood 106: 2399–2408
Ricklin D, Lambris JD (2008) Compstatin: a complement inhibitor on its way to clinical application. Adv Exp Med Biol 632:273–292
Risitano AM, Maciejewski JP, Green S et al (2004) In vivo dominant immune responses in aplastic anemia patients: molecular tracking of putatively pathogenic T cells by TCRβ-CDR3 sequencing. Lancet 364:353–363
Risitano AM, Marando L, Seneca E et al (2008) Hemoglobin normalization after splenectomy in a paroxysmal nocturnal hemoglobinuria patient treated by eculizumab. Blood 112:449–451
Risitano AM, Notaro R, Marando L et al (2009a) Complement fraction 3 binding on erythrocytes as additional mechanism of disease in paroxysmal nocturnal hemoglobinuria patients treated by eculizumab. Blood 113:4094–4100
Risitano AM, Holers VM, Rotoli B (2009b) TT30, a novel regulator of the complement alternative pathway (CAP), inhibits hemolysis of paroxysmal nocturnal hemoglobinuria (PNH) erythrocytes and prevents upstream C3 binding on their surface in an in vitro model. Blood 114:158a
Risitano AM, Notaro R, Luzzatto L et al (2010) Paroxysmal nocturnal hemoglobinuria: hemolysis before and after eculizumab. N Engl J Med 363:2270–2272
Risitano AM, Notaro R, Pascariello C et al (2012) The complement receptor 2/factor H fusion protein TT30 protects paroxysmal nocturnal hemoglobinuria erythrocytes from complement mediated hemolysis and C3 fragment. May 2012, [Epub ahead of print]
Rosse WF (1971) The life-span of complement-sensitive and -insensitive red cells in paroxysmal nocturnal hemoglobinuria. Blood 37:556–562

Rosse WF, Dacie JV (1966) Immune lysis of normal human and paroxysmal nocturnal hemoglobinuria (PNH) red blood cells. I. The sensitivity of PNH red cells to lysis by complement and specific antibody. J Clin Invest 45:736–748
Rother RP, Bell L, Hillmen P et al (2005) The clinical sequelae of intravascular hemolysis and extracellular plasma hemoglobin: a novel mechanism of human disease. JAMA 293:1653–1662
Rother RP, Rollins SA, Mojcik CF et al (2007) Discovery and development of the complement inhibitor eculizumab for the treatment of paroxysmal nocturnal hemoglobinuria. Nat Biotechnol 25:1256–1264
Rotoli B, Luzzatto L (1989) Paroxysmal nocturnal haemoglobinuria. Baillieres Clin Haematol 2:113–138
Rotoli B, Robledo R, Luzzatto L (1982) Decreased number of circulating BFU-Es in paroxysmal nocturnal hemoglobinuria. Blood 60:157–159
Schafer A, Wiesmann F, Neubauer S et al (2004) Rapid regulation of platelet activation in vivo by nitric oxide. Circulation 109:1819–1822
Scheinberg P, Marte M, Nunez O et al (2010) Paroxysmal nocturnal hemoglobinuria clones in severe aplastic anemia patients treated with horse anti-thymocyte globulin plus cyclosporine. Haematologica 95:1075–1080
Selvaraj P, Rosse WF, Silber R et al (1988) The major Fc receptor in blood has a phosphatidylinositol anchor and is deficient in paroxysmal nocturnal haemoglobinuria. Nature 333:565–567
Shin ML, Hänsch G, Hu VW et al (1986) Membrane factors responsible for homologous species restriction of complement-mediated lysis: evidence for a factor other than DAF operating at the stage of C8 and C9. J Immunol 136:1777–1782
Sica M, Pascariello C, Rondelli T et al (2010) In vitro complement protein 5 (C5) blockade recapitulates the complement protein 3 (C3) binding to GPI-negative erythrocytes observed in paroxysmal nocturnal hemoglobinuria (PNH) patients on eculizumab. Haematologica 95(s2):196(a)
Simak J, Holada K, Risitano AM et al (2004) Elevated counts of circulating endothelial membrane microparticles in paroxysmal nocturnal hemoglobinuria indicate inflammatory status and ongoing stimulation of vascular endothelium. Br J Hematol 125:804–813
Sloand EM, Pfannes L, Scheinberg P et al (2008) Increased soluble urokinase plasminogen activator receptor (suPAR) is associated with thrombosis and inhibition of plasmin generation in paroxysmal nocturnal hemoglobinuria (PNH) patients. Exp Hematol 36:1616–1624
Socie G, Mary JY, de Gramont A et al (1996) Paroxysmal nocturnal haemoglobinuria: long-term follow-up and prognostic factors. French Society of Haematology. Lancet 348:573–577
Sugimori C, Chuhjo T, Feng X et al (2006) Minor population of CD55-CD59-blood cells predicts response to immunosuppressive therapy and prognosis in patients with aplastic anemia. Blood 107:1308–1314
Takahashi M, Takeda J, Hirose S et al (1993) Deficient biosynthesis of N-acetylglucosaminyl-phosphatidylinositol, the first intermediate of glycosyl phosphatidylinositol anchor biosynthesis, in cell lines established from patients with paroxysmal nocturnal hemoglobinuria. J Exp Med 177:517–521
Takeda J, Miyata T, Kawagoe K et al (1993) Deficiency of the GPI anchor caused by a somatic mutation of the PIG-A gene in paroxysmal nocturnal hemoglobinuria. Cell 73:703–711
Tremml G, Dominguez C, Rosti V et al (1999) Increased sensitivity to complement and a decreased red blood cell life span in mice mosaic for a nonfunctional Piga gene. Blood 94:2945–2954
van Beers EJ, Schaap MC, Berckmans RJ, CURAMA Study Group et al (2009) Circulating erythrocyte-derived microparticles are associated with coagulation activation in sickle cell disease. Haematologica 94:1513–1519
van der Schoot CE, Huizinga TW, van't Veer-Korthof ET et al (1990) Deficiency of glycosyl-phosphatidylinositol-linked membrane glycoproteins of leukocytes in paroxysmal nocturnal hemoglobinuria, description of a new diagnostic cytofluorometric assay. Blood 76:1853–1859
Whaley K, Ruddy S (1976) Modulation of the alternative complement pathway by β1H globulin. J Exp Med 144:1147–1163
Wiedmer T, Hall SE, Ortel TL et al (1993) Complement-induced vesiculation and exposure of membrane prothrombinase sites in platelets of paroxysmal nocturnal hemoglobinuria. Blood 82:1192–1196
Wilcox LA, Ezzell JL, Bernshaw NJ et al (1991) Molecular basis of the enhanced susceptibility of the erythrocytes of paroxysmal nocturnal hemoglobinuria to hemolysis in acidified serum. Blood 78:820–829
Yamashina M, Ueda E, Kinoshita T et al (1990) Inherited complete deficiency of 20-kilodalton homologous restriction factor (CD59) as a cause of paroxysmal nocturnal hemoglobinuria. N Engl J Med 323:1184–1189
Young NS, Maciejewski JP (2000) Genetic and environmental effects in paroxysmal nocturnal hemoglobinuria: this little PIG-A goes "Why? Why? Why?". J Clin Invest 106:637–641
Young NS, Calado RT, Scheinberg P (2006) Current concepts in the pathophysiology and treatment of aplastic anemia. Blood 108:2509–2519

Chapter 11
Defective Complement Action and Control Defines Disease Pathology for Retinal and Renal Disorders and Provides a Basis for New Therapeutic Approaches

Peter F. Zipfel and Nadine Lauer

Abstract The complement system is a central homeostatic system of the vertebrate organism and part of innate immunity. When activated, complement has multiple functions and drives homeostasis and the elimination of infectious microbes (Walport MJ (2001) N Engl J Med 344:1140–1144; Zipfel PF, Skerka C (2009) Nat Rev Immunol 9:729–740). Several inflammatory disorders are caused by defective complement action, and the growing, detailed understanding of the underlying pathophysiological principles translate into therapy with complement inhibitors. As complement inhibitors have been approved for treatment of the complement-mediated disorders hemolytic uremic syndrome (HUS) and paroxysmal nocturnal hemoglobinuria (PNH), there is a growing interest to extended and improve the options for other complement-mediated diseases. Here, we summarize the current understanding and concepts how defective complement action at biological surfaces lead to pathology and disease, and how this understanding can be used for the development of surface targeting complement inhibitors.

Abbreviations

aHUS	Atypical hemolytic uremic syndrome
AMD	Age-related macular degeneration
AP	The alternative pathway
CFHR	Complement Factor H-related protein
CP	The classical pathway
CRP	C-reactive protein
DAMPs	Damage-associated molecular patterns
DEAP-HUS	Deficient for CFHR proteins and Factor H autoantibody positive HUS
FHL1	Factor H-like protein
LP	The lectin pathway
mCRP	Monomeric CRP

P.F. Zipfel (✉)
Department of Infection Biology, Leibniz Institute for Natural Product Research and Infection Biology, Beutenbergstrasse 11a, 07745 Jena, Germany

Friedrich Schiller University, Fürstengraben 1, 07743 Jena, Germany
email: peter.zipfel@hki-jena.de

N. Lauer
Department of Infection Biology, Leibniz Institute for Natural Product Research and Infection Biology, Beutenbergstrasse 11a, 07745 Jena, Germany

J.D. Lambris et al. (eds.), *Complement Therapeutics*, Advances in Experimental Medicine and Biology 735, DOI 10.1007/978-1-4614-4118-2_11,

MDA	Malondialdehyde
MPGN II	Membranoproliferative glomerulonephritis type II
PNH	Paroxysmal nocturnal hemoglobinuria
pCRP	Pentameric CRP
RPE	Retinal pigment epithelial cells
SCR	Short consensus repeat
TCC	Terminal complement complex

11.1 The Human Complement System

The complement system is a central homeostatic system of the vertebrate organism and part of innate immunity. When activated, complement has multiple functions and drives homeostasis by eliminating invading microbes, allowing safe and noninflammatory clearance of damaged or modified self-cells, inducing inflammation, and also by directing and modulating the adaptive immune response (Carroll 2000; Ogden and Elkon 2006; Walport 2001; Zipfel and Skerka 2009). The complement system is activated by three major pathways: the alternative (AP), the classical (CP), and the lectin pathway (LP). Each pathway is initiated by unique activators and all three pathways generate a C3-convertase that cleaves the central complement component C3 into C3a and C3b. In its default setting, C3b is spontaneously and continuously generated via the alternative pathway by the "tick over activation" of native C3 (Pangburn and Muller-Eberhard 1983). In the absence of inhibitors, activated C3b is deposited onto any surface. The newly formed C3b molecule forms the central focus, and activators in combination with regulators determine the fate of the process. If activation proceeds, the reaction is powerful amplified, and many additional C3-convertases are formed (Zipfel et al. 2007). The inflammatory mediator C3a is released, and in addition, more C3b is generated and deposited onto the surface. Such C3b-decorated particles are efficiently phagocytosed and are removed in a silent, noninflammatory manner. In addition, surface-deposited C3b is properly processed, and the major intermediates include iC3b, an important inflammatory surface molecule, as well as C3dg and C3d. Each intermediate of C3 directs to distinct receptors and has unique distinct effector functions (Sahu and Lambris 2001). Further activation then leads to formation of a C5-convertase that by cleaving the plasma protein C5 generates the potent anaphylatoxin C5a, as well as C5b. C5b when stabilized initiates assembly of the terminal complement complex (TCC) that inserts into membranes and causes lysis of a target cell.

11.2 Complement Activation Can be Both, Beneficial and Harmful

From side of the host, this activation, in particular the action of newly generated effector compounds, must be targeted onto the surface of damaged self-cells or infectious microbes. At the same time, activation needs to be restricted on the surface of intact host cells (Zipfel and Skerka 2009). To this end, the vertebrate organism uses a battery of highly specific regulators that discriminate between activator and nonactivator surfaces (also termed self and nonself/modified self) and that – in practice – control, modulate, and also inhibit activation and progression of the cascade. Thus, the complement system, with its diverse effector functions and activation levels, is regulated at each single step to adjust the intensity of the response and also the amount of newly generated effector compounds. The various regulators determine the degree of amplification as well as the intensity of the response, and they thereby block progression of the complement cascade or freeze activation. Multiple redundantly acting regulators with overlapping activities are distributed as soluble proteins in plasma and body fluids or are expressed as integral membrane proteins on nonactivator surfaces.

11.3 From the AP Complement Regulator Factor H and FHL1 to Complement-Related Diseases

The abundant plasma proteins Factor H and the Factor H-like 1 protein (FHL1) represent two major regulators of the complement system (Jozsi and Zipfel 2008). Factor H, as well as FHL1, protect self-surfaces and biomembranes from complement-mediated damage, have anti-inflammatory activities, and display important housekeeping functions. Factor H, a 150-kDa serum protein, is composed of 20 consecutive domains, termed short consensus repeats (SCR). This plasma protein has two central functions: regulation of the AP C3-convertase in fluid phase as well as on self-surfaces via attachment to the surface of host cells. The central complement regulatory region of Factor H is located within the N terminus and is represented by domains SCR1–4. These four SCRs are responsible for controlling the activity and fate of the AP C3-convertase C3bBb. In addition, Factor H uses two surface attachment and recognition regions, which are located in SCR7 and in the C-terminal domains SCR18–20. FHL1 is a second protein encoded by the human *factor H gene*. The FHL1 encoding mRNA is derived by alternative splicing, and this mRNA utilizes exons I–IX of the *factor H gene*, which are shared by FHL1 and Factor H and exon X. Exon X is unique for FHL1 and encodes the four FHL1-specific C-terminal amino acids (SFTL), a specific 3′UTR and polyadenylation signals (Hellwage et al. 1997; Misasi et al. 1989). The mature FHL1 protein is composed of 431 amino acids that form seven SCRs, which correspond to the seven N-terminal SCRs of Factor H. Both FHL1 and Factor H share the N-terminal complement regulatory region in SCR1–4 (Kuhn and Zipfel 1996). However, the two molecules have different surface attachment regions (Fig. 11.1). FHL1 uses SCR7 as the C-terminal surface attachment region which guides this regulator to the surface of damaged self cells, including endothelial and epithelial cells. Factor H however has two surface attachment regions, located in SCR7 and in the C-terminal region SCR18–20 (Blackmore et al. 1996; Hellwage et al. 2002; Rodriguez de Cordoba 2004). Although an important role for SCR7, as a unique and separate surface recognition region is emerging for FHL1, the interplay between the two surface attachment regions within the intact Factor H protein is still unclear. In native Factor H, the C-terminal region represents the first and major surface binding region, which provides the first contact to cells (Oppermann et al. 2006; Kajander et al. 2011). Both attachment regions contact glycosaminoglycans and heparin, which are exposed on the surface of human cells or biomembranes (Meri and Pangburn 1990). This Factor H specific attachment to self-surfaces via SCR7 and via the C terminus directs the regulator to endothelial and epithelial cells, to the surface of modified self-cells, such as apoptotic and necrotic cells as well as to biomembranes, like the glomerular basement membrane of the kidney, or to the Bruch's membrane of the retina (Fritsche et al. 2010; Jokiranta et al. 2005; Lauer et al. 2011; Pangburn 2002).

11.4 Discrimination of Complement Action at Self- and Nonself Surfaces

Proper discrimination between activator and nonactivator surfaces by the plasma proteins FHL1 and Factor H is important for recognition of modified cells, including necrotic, as well as apoptotic cells and also for biomembranes, in particular membranes under complement challenge. The surface of modified self-cells, such as apoptotic and necrotic cells, lack membrane-bound complement regulators such as CR1, CD46, CD55, and CD59. Consequently, complement is activated on these surfaces, and when not properly controlled, inflammation and complement effector functions are induced and initiated (Lauer et al. 2011). However, in this situation, the modified surfaces attach and recruit fluid phase regulators such as FHL1, Factor H, and also C4BP (Mihlan et al. 2011, 2009). A precisely controlled processing of the complement cascade is beneficial and essential on the surface of such modified cells. Controlled activation of complement up to the level of the C3-convertase, and moderate controlled

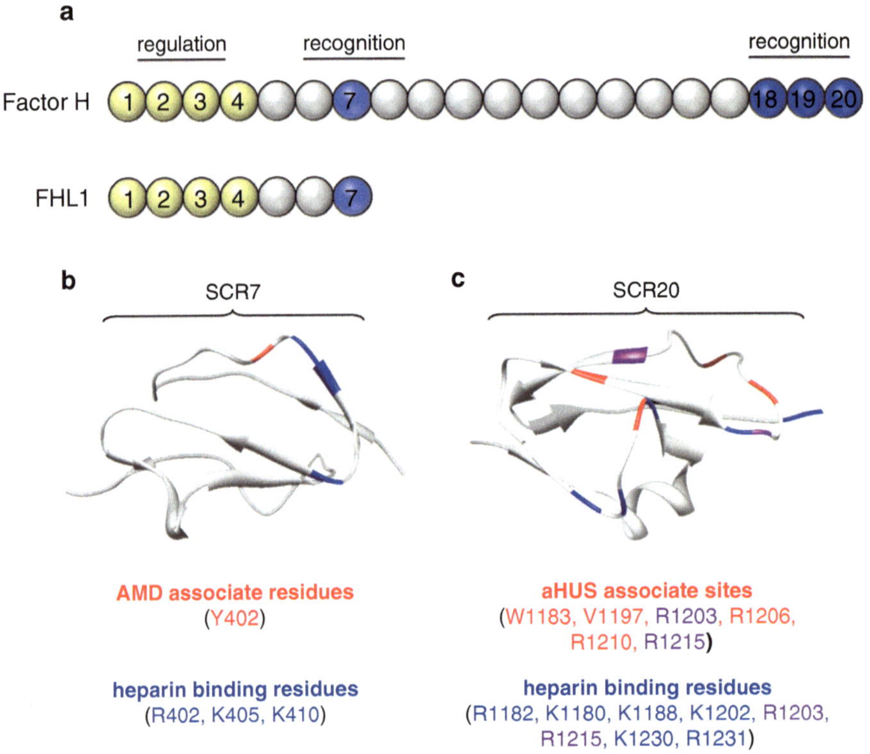

Fig. 11.1 *Structure of Factor H and FHL1*. The mRNAs encoding Factor H and FHL1 are derived from the same gene by means of alternative splicing, and the secreted proteins are composed of short consensus repeats (SCRs). Factor H is composed of 20 SCRs, FHL1 is identical in sequence with the seven N-terminal SCRs of Factor H, but this protein has a C-terminal extension of four unique amino acid residues. The complement regulatory domains are located in the N-terminal part within the first four SCRs (*yellow*), and the surface attachment regions (*blue*) are located in SCR7 (of *FHL1*) and in SCR7 as well as SCR18–20 of Factor H. (**b**/**c**) AMD and aHUS-associated amino acid substitutions are often close to patches of positively charged residues that are responsible for surface binding and consequently can reduced FHL1 or Factor H binding to self- or modified self-surfaces. Amino acids responsible for surface binding are labeled in blue. Disease-associated amino acids, either linked to AMD (*SCR7*, **b**) or aHUS (*SCR20*, **c**) are shown in *red* or *purple* respectively, if they match with amino acids responsible for surface binding. The molecular graphic was created using the UCSF Chimera software (http://www.cgl.ucsf.edu/chimera) and the public crystal structures of SCR7 (2jgx) and SCR20 (2xqw) publish on the protein databank (PDB) (www.rcsb.org)

C3-convertase activity leads to C3b opsonization of the surface. This allows phagocytosis as well as and efficient, silent, and noninflammatory removal (Mihlan et al. 2009; Trouw et al. 2008). However, further progression of the cascade beyond the C3-convertase, to the next level, i.e., the C5-convertase, is blocked, and consequently neither the potent inflammatory marker C5a nor TCC are formed. This scenario defines an important protective and anti-inflammatory role of acquired regulators such as FHL1, Factor H, as well as C4BP on the surface of intact, but in particular on the surface of modified self-surfaces.

11.5 Defective Discrimination Leads to Disease

The relevance of FHL1 and of Factor H bound to the surface of modified self-cells, as well as to biomembranes, is highlighted by disease-associated gene variations in the two surface attachment regions SCR7 and SCR18–20 (Zipfel et al. 2006). The common sequence variation in SCR7, i.e., a Tyrosine (Y) to

Histidine (H) exchange at position 402, increases the risk for the retinal disease age-related macular degeneration (AMD) (Edwards et al. 2005; Haines et al. 2005; Klein et al. 2005). Sequence variants and mutations in the C-terminal surface recognition region are associated with the renal disease atypical hemolytic uremic syndrome (aHUS). The AMD relevant amino acid at position 402 in SCR7 is either protective (Y402) or a risk (H402). The H402 risk variants of both FHL1 and also Factor H bind with reduced intensity to retinal pigment epithelial cells (RPE) and also to glycosaminoglycans of the Bruch's membrane (Clark et al. 2010; Lauer et al. 2011; Skerka et al. 2007). The majorities of aHUS-associated, mostly heterozygous, gene mutations cluster in the C-terminal surface attachment region and reduce surface binding of Factor H due to a reduced binding to glycosaminoglycans and to heparin (Caprioli et al. 2001; Herbert et al. 2006; Manuelian et al. 2003). Furthermore, in the autoimmune form DEAP-HUS disease relevant autoantibodies are present in patient plasma, which specifically bind to the C-terminal region of Factor H and inhibit surface attachment (Zipfel et al. 2010b). In the renal disease membranoproliferative glomerulonephritis type II (MPGN II; also termed dense deposits disease), the known Factor H gene mutations are mostly homozygous and frequently affect disulfide forming cysteine residues, which are essential for the architecture of an individually folded SCR domain (Appel et al. 2005). These mutations in the *factor H gene* affect secretion, while other MPGN II-associated mutations in the N-terminal regulatory domain block the complement regulatory functions. Absence of Factor H in the circulation, as well as inactivation or defective regulatory functions, leads to systemic complement activation, C3 consumption, and to local, uncontrolled complement activation especially at the glomerular basement membrane (Chen et al. 2011; Licht et al. 2006).

Proper complement regulation in fluid phase and efficient attachment of the soluble complement regulators FHL1 and Factor H to biological surfaces is therefore central for the integrity of host tissues. Inappropriate targeting of the regulators causes enhanced local complement activation at cellular or biological surfaces and results in complement progression, inflammation, TCC assembly, and further complement-mediated damage. The relevance of the sequence variation in SCR7 for surface attachment of FHL1 and of Factor H in AMD pathology confirms the important role of the two regulators in the retina. Similarly, the association of *factor H gene* mutations that cluster in the C-terminal recognition region and that are associated with the renal disease aHUS allows to define principles and also the physiological role of FHL1 and of Factor H at cellular surfaces. The growing, detailed insights of these FHL1- and Factor H-associated diseases, which have been obtained over the last decade, allowed to understand the pathophysiological principle for these distinct disorders which manifests in different organs and highlighted the importance of complement and complement regulation for homeostasis and health.

11.6 Age-Related Macular Degeneration: A Complement-Associated Diseases

Age-related macular degeneration (AMD) is the leading cause of blindness in elderly people of Western societies (Jager et al. 2008). This multifactorial disease develops over many years before a definitive diagnosis can be set. Approximately 20 Mio individuals in the United States and Europe suffer from the disease, which is characterized by a gradual degeneration of photoreceptors within the macula and ultimately results in central vision loss (Pascolini et al. 2004). Late disease stages present as two severe forms: geographic atrophy, that is caused by atrophy of photoreceptors and choroidal neovascularization which develop due to the growth of new blood vessels into the retinal layer. Photoreceptors are critically dependent upon functionally intact RPE cells that form a single layer of pigmented cells on the Bruch's membrane and that transfer nutrients to the photoreceptors and remove outer photoreceptor segments via phagocytosis (Strauss 2005). RPE cell damage, cell death, as well as age-related degeneration, are relevant for AMD pathology and can lead to either apoptosis or

Table 11.1 Complement components in AMD – components identified by different assays

Histological analysis of drusen	**C3** (fragments C3a, iC3b, C3d, C3dg), Factor B
	Factor H/FHL1, CFHR1
	C5, TCC
	CD46, Clusterin
	CRP (mCRP, pCRP)
Expression RPE/choroid	**C3**, Factor D, **Factor I**
	Factor H/FHL1
	C5, C7
	CD46, CD55, CD59
Genetic	**C3, Factor I**, Factor B, C2
	Factor H/FHL1, CFHR1, CFHR3, *CFHR5*
	C7

bold association confirmed by genetic and expression studies, in *italic*, not confirmed yet

necrosis. Oxidative stress, for example, induces apoptosis of RPE cells due to lipofuscin formation and accumulation in the mitochondria that consequently affect mitochondrial function, cause leakiness of the inner membrane which in turn releases pro-apoptotic stimuli such as cytochrome C (Cai et al. 1999; Zhou et al. 2006). However, a subpopulation of RPE cells exists that appear rounded and swollen without signs of cell shrinkage or nuclear condensation. These cells are degenerated, most likely due to necrosis (Anderson et al. 2002; Thurman et al. 2009). A hallmark of AMD and usually one of the first clinical symptoms are ocular drusen – which represent extracellular deposits that form between the RPE layer and the Bruch's membrane. Ultrastructural analyses suggest that RPE cells, which overlie such drusen, are degenerated due to necrosis (Hageman et al. 2001). Proteome analyses, as well as histological studies, identified in these deposits a series of inflammatory molecules and complement components and also complement activation fragments, TCC components as well as complement regulators (Crabb et al. 2002; Johnson et al. 2001). Most or all of these proteins are expressed locally, for example, by RPE cells, and are present in the choroid and the Bruch's membrane, thus confirming a local role of complement and inflammation in the pathogenesis of AMD (Table 11.1). This concept is confirmed by genetic studies that revealed a significant associations between AMD and the genes for complement Factor H, complement Factor B, complement component 3 (C3), and complement Factor H-related proteins 1 and 3 (CFHR1/CFHR3) (Fritsche et al. 2010; Gold et al. 2006; Hughes et al. 2006; Yates et al. 2007).

11.7 Role of FHL1 and Factor H in Age-Related Macular Degeneration

Importantly in up to 50% of AMD patients, the genetically defined gene variants are linked to sequence variations in the *factor H* gene. The most relevant AMD-associated single nucleotide variation within the *factor H* gene translates on the protein level into a tyrosine (Y) to histidine (H) exchange at amino acid position 402. Both FHL1 and Factor H with Y402 represent the protective, and the proteins with H402 represent the risk variants. Residue 402 is located within the surface attachment region of SCR7 that is shared by FHL1 and by Factor H. The presence of the H402 risk variant increases the risk for AMD about twofold to fourfold for heterozygote and about threefold to sevenfold for homozygote individuals (Klein et al. 2005). Subsequent studies confirmed the strong genetic association in different ethnic groups (Okamoto et al. 2006; Simonelli et al. 2006; Souied et al. 2005). Functional studies show that the Y402toH402 exchange affects binding of both FHL1 and Factor H to heparin, glycosaminoglycans, cellular surfaces, and in particular to the surface of apoptotic and necrotic cells of the Bruch's membrane (Clark et al. 2010; Prosser et al. 2007; Skerka et al. 2007; Lauer et al. 2011) Reduced binding of either the FHL1 or the Factor H risk variants results consequently in a lower number of surface-bound regulators. This translates into reduced complement inhibition and to subsequent cascade progression on the surface of modified RPE cells and also the exposed Bruch's membrane and provokes local cell damage and inflammation.

11.8 Cell Damage and Lipid Peroxidation in AMD

The eye and also retinal cells have a high level of O_2 consumption which makes this tissue and cell compartment susceptible to oxidative damage (Cai et al. 2000). Local oxidative stress in the retina is caused by different scenarios, including the permanent exposure to light, accumulation of light-sensitive components such as lipofuscin in RPE cells, damage of RPE cells which is followed by the enhanced release of reactive O_2 species from lysosomes, complement activation, as well as physical/chemical agents like cigarette smoking (Fujihara et al. 2008; Thurman et al. 2009; Zhou et al. 2006). Oxidative reactions – in particular – photo oxidative processes are critical for tissue destruction and progression of AMD pathology (Hollyfield et al. 2008). As a consequence endogenous molecules such as lipids, proteins, and DNA are modified which in turn undergo structural changes. This may trigger apoptosis or, after more intensive stress, even necrotic cell death. The oxidative degradation of lipids and proteins create numerous new moieties on the surface of affected cells and generates further reactive decomposition products with oxidative specific immunodominant epitopes (Chou et al. 2008). Such epitopes are termed damage-associated molecular patterns (DAMPs) and are recognized by complement and other components of the innate immune system (Miller et al. 2011). The complement system, as a central element of innate immunity, has a pivotal role in recognizing and responding to DAMPs and when activated complement generates a variety of proinflammatory immune responses. In addition, oxidative stress leads to the activation and deposition of complement components on the surface of damaged cells. Lipid peroxidation is a frequent process and an inducer of inflammation. Malondialdehyde (MDA), a common lipid peroxidation product, is generated on the surface of apoptotic and necrotic cells and is also formed in many other pathophysiological processes (Esterbauer et al. 1991). MDA, as well as other oxidation-specific epitopes, form central ligands for innate immune molecules including the regulator Factor H, TLR4, natural antibodies, and C-reactive protein (Chang et al. 2002). MDA is generated in pathophysiological process of AMD, and both FHL1 and Factor H are two major MDA-binding proteins (Suzuki et al. 2007; Weismann et al. 2011). In addition, the surface of damaged self-cells exposes a variety of Factor H ligands, including monomeric CRP (mCRP), DNA, Annexin-II, phosphatidylserine, and histones (Lauer et al. 2011; Leffler et al. 2010).

FHL1 and Factor H bind to MDA and also to mCRP on the surface-damaged cells, and this interaction is relevant in AMD pathology (Lauer et al. 2011; Weismann et al. 2011). CRP circulates as a 125-kDa pentameric protein (pCRP), binds to the surface of damaged cells, is then destabilized and undergoes conformational changes. Modification of the structure results in the generation of monomeric CRP (mCRP) which has new binding characteristics (Ji et al. 2007). mCRP is exclusively formed on the surface of modified cells, such as apoptotic and necrotic cells (Eisenhardt et al. 2009; Mihlan et al. 2011). Both FHL1 and Factor H bind to MDA and to surface attached mCRP, but not to pCRP. FHL1 uses the recognition region in SCR7 for MDA/mCRP binding and Factor H – the two recognition regions in SCR7 in SCR18–20.

11.9 MDA Targeting of FHL1 and Factor H Is Affected by the Sequence Variation in SCR7

Interestingly, the AMD-associated sequence variation in SCR7 at position 402 affects FHL1 and also Factor H binding to both MDA and to mCRP. The protective Y402 variants of FHL1 and Factor H bind with higher intensity to MDA and also to mCRP as compared to the H402 risk variants. The interaction of the protective variants with both modification products was increased by 40–50%. This enhanced MDA- and mCRP-mediated surface recruitment of regulators to modified cells, for example, necrotic RPE cells, ultimately affects local complement regulation and release of the proinflammatory cytokines (Lauer et al. 2011; Weismann et al. 2011). Furthermore, Factor H bound to MDA blocks the uptake of MDA-modified proteins by macrophages and further inhibits MDA-induced proinflammatory effects in vivo.

FHL1 and Factor H colocalize with MDA and also with mCRP at the surface of modified cells and both modification products – MDA and mCRP – mediate surface attachment of these two innate immune regulators. FHL1 and Factor H complexed to either MDA or mCRP maintain complement regulatory activity and control cofactor activity of Factor I-mediated C3b degradation into iC3b. Surface-deposited C3b/iC3b allows subsequent anti-inflammatory clearance of such marked particles by macrophages and other immune effector cells (Poon et al. 2010; Sjoberg et al. 2009; Trouw et al. 2008). In consequence, the newly formed ligands on the surface of self-cells select for proper effector functions and allow silent noninflammatory removal by immune effector cells. In this scenario, both FHL1 and Factor H equipped with the protective Y402 variant bind to MDA or mCRP with almost 50% higher efficiency, and subsequently, the complement regulatory cofactor activity on the MDA- or mCRP-coated surfaces is enhanced. This resulted in an increased C3b deposition on the surface of about 30%. In presence of the protective FHL/Factor H variants, local generation of these surface bound inflammatory iC3b fragments in the retina is therefore enhanced. Furthermore, surface bound Factor H has an anti-inflammatory activity and reduces release of the inflammatory cytokines TNF-α and IL8 (Mihlan et al. 2009) (Fig. 11.2).

11.10 Surface Targeting of Complement Inhibitors in Inflammation

MDA as well as mCRP are exposed on the surface of modified cells. Cells damaged by oxidative events are continuously generated in the retina and need to be efficiently removed (Cai et al. 2000). One possible scenario for RPE damage and induction of necrosis is that RPE cells are damaged by oxidative stress; the cells undergo necrosis and activate complement, ideally to a moderate degree, to allow opsonization with C3b/iC3b and proper noninflammatory removal. As the risk variants of FHL1 and Factor H bind with reduced intensity to these modified surfaces, complement action is inappropriately and insufficiently controlled, the cascade progresses, particles are less efficiently opsonized, and inflammation is induced. If such a response persits for a long time and becomes chronic, deposits are formed which develop into drusen. Recruitment of the fluid phase regulators FHL1 and Factor H to the surface of modified RPE cells is therefore essential to block complement cascade progression, amplification, and inflammation. This pathological principle is relevant for AMD and also for other chronic inflammatory diseases. Interfering and controlling these processes by complement-based therapeutic will ultimately provide a promising approach for AMD therapy.

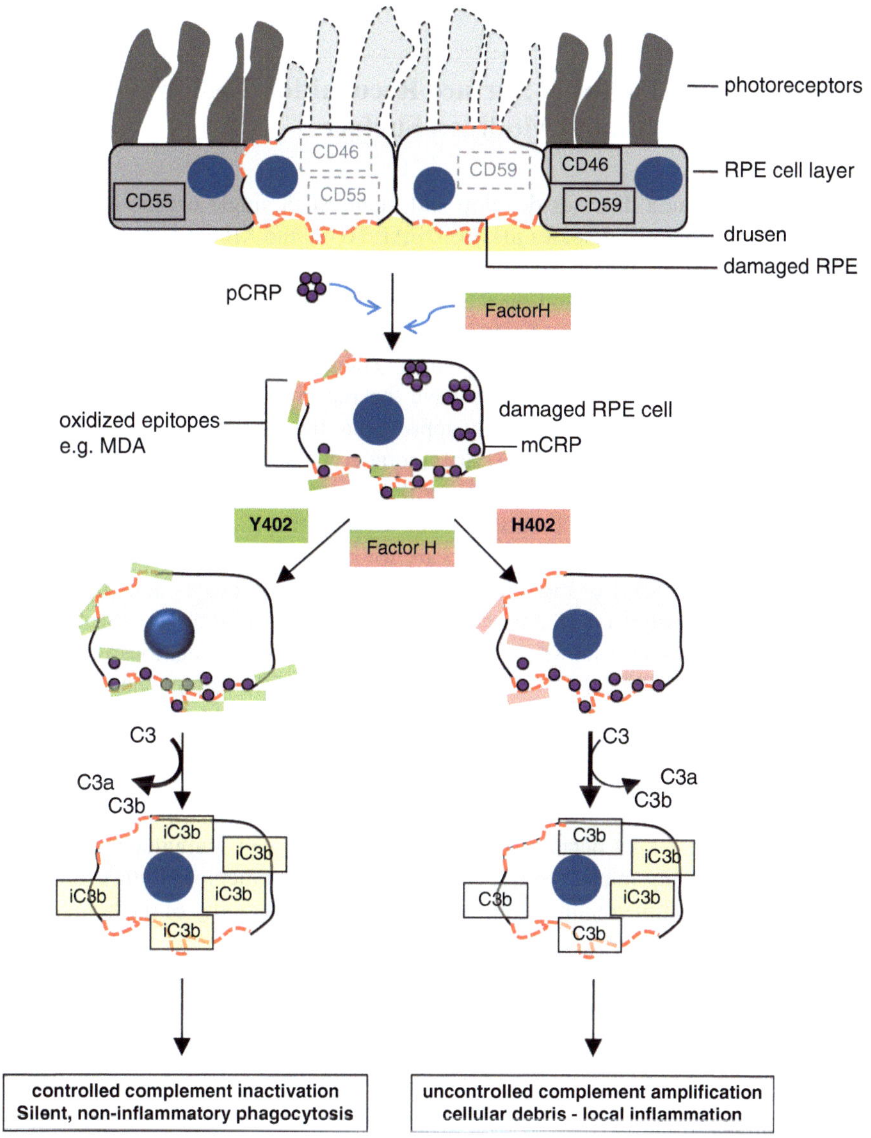

Fig. 11.2 *The Y402 H402 amino acid exchange in SCR7 of FHL1 or Factor H reduces cell recruitment* via *the modification products mCRP and MDA and consequently alters the local complement regulation at the damaged cell surface*. Upon progression of AMD, RPE cells become damaged and degenerate (*dashed orange lines*). Cellular damage in the eye is often caused by local oxidative stress. As consequence endogenous, the membrane components are modified and undergo structural changes that can trigger, for example, necrotic cell death. Alteration of the plasma membrane leads to exposure of altered self-epitopes, such as MDA a ubiquitous lipid peroxidation product. Such damaged RPE cells lack complement regulators on their surface and activate complement and induced inflammation. Consequently, regulation is required to allow silent, noninflammatory removal of these modified particles or debris. Locally expressed Factor H binds to the damaged cellular surfaces, for example, via modified lipids modification products, like MDA. During inflammation and severe tissue damage, pCRP plasma levels increase substantially either locally as well as systemically. The inflammatory mediator pCRP binds to the surface of damaged RPE is modified and dissociated to the monomeric form mCRP. Newly generated mCRP as well as MDA recruit Factor H from the fluid phase to the surface, to the site of severe cell damage, and enhance complement regulation. This step is mandatory because damaged, for example, necrotic, cells lack the membrane bound regulators like CR1, CD46, CD55, and CD59. The protective Factor H Y402 variant (symbolized by *green bars*) binds with higher intensity to surface attached mCRP and MDA, and the surface recruited regulator inactivates complement at the level of C3-convertase. This results in deposition of iC3b on the cellular surface and blocks further progression of the cascade. Consequently, macrophages are attracted that remove the opsonized necrotic debris in a silent, noninflammatory manner. In contrast, the risk Factor H H402 variant (symbolized by *red bars*) binds with lower intensity to mCRP- and MDA-decorated surfaces. Consequently, complement is not as efficiently controlled, and the cascade progresses mare easily to the C5-convertase and to the TCC level

11.11 Exploiting the Factor H Surface Recognition Region to Target Inhibitors to Modified Surfaces

Most complement-associated diseases develop and manifest in single organs and at specific sites. Examples are the severe kidney diseases aHUS, DEAP-HUS, and MPGN where damage occurs primarily in the kidney or AMD, which develops in the retina (Pickering et al. 2007; Zipfel et al. 2006). Thus, defective complement regulation or activation occurs locally, and the systemic action of the complement cascade seems not or less affected. However, most complement therapeutics act systemically and control complement in the whole organism. Therefore, these inhibitors may also block the beneficial effects of complement and thereby have the risk of side effects such as infections. The C5-targeting antibody Eculizumab is currently approved to treat the human diseases PNH (paroxysmal nocturnal hemoglobinuria) and aHUS (Lapeyraque et al. 2011; Rother et al. 2007). In addition, the C3-binding peptide compstatin is an effective fluid phase complement inhibitor that is used in clinical trials for AMD (Ricklin and Lambris 2008). Compstatin is a cyclic 13-residue peptide, initially identified by screening a phage-display library for peptides that bind to native C3. Compstatin inhibits complement at the level of the C3-convertase and thereby blocks activation, as well as the amplification of all three pathways (Sahu et al. 1996). Compstatin binds specifically to human and primate C3, has low toxicity, and shows no adverse effects in humans as demonstrated by in vitro and in vivo studies (Sahu et al. 2003, 2000). The inhibitory activity of the original peptide compstatin was improved more than 260-fold in optimization steps by chemical, biophysical, as well as computational approaches (Qu et al. 2011). In 2009, Potentia Pharmaceuticals successfully completed a phase I clinical trial for the use of compstatin for treatment of AMD (Potentia 2009).

In addition, inappropriate processing of immune complexes and development of an inappropriate humoral immune response may occur during long-term therapy in immunocompromised patients. Therefore, there is a growing interest to target therapeutic complement inhibitors directly to the site of damage and at the same time maintain the systemic action and activity of the homeostatic complement system (Mulligan et al. 1999; Smith and Smith 2001). A rational concept is to direct therapeutic complement inhibitors to local sites where complement is activated in order to allow exact control of the cascade at the side of activation. A benefit of targeted inhibitors that are active at the site of damage is their local action with ideally no or little effects on systemic complement actions. At the same time, systemic action of the complement cascade and activation in other organs remains intact and is not or only mildly affected. The concept of targeted inhibitors was first used for soluble sCR1 which was equipped with a membrane targeting myristoylated peptide (Mulligan et al. 1999). A second successful example is TT30, a chimeric protein, that has the iC3b- and C3d-targeting region of CR2, linked to the complement regulatory region of Factor H, i.e., SCR1–4 in combination with SCR5 (Fridkis-Hareli et al. 2011). This inhibitor TT30 binds to inflamed tissue and when tested in animal models is about 20–30 fold more effective as the nontargeted CR2 homologue. The CR2 region targets the inhibitor to the site of complement activation where iC3b is formed and the N terminus provided by Factor H controls local complement activation (Atkinson et al. 2005; Huang et al. 2008; Rohrer et al. 2009). However, there is continuous interest to develop complement inhibitors with improved local action.

Modified cellular surfaces activate complement and induce inflammation. Under normal conditions, soluble regulators like Factor H attach to these sites and dampen the activation to allow opsonisation with C3b/iC3b and block further downstream C5-convertase-induced inflammation. Inappropriate binding of FHL1 or Factor H to the surface of modified cells or biomembranes triggers the complement cascade and results in inflammation and enhances the damaging effects of activated complement. Such enhanced local complement activation at modified biological surfaces results in immune stress, local inflammation, and cause cell or tissue damage, which is relevant for the retinal disorder AMD and also for other chronic inflammatory diseases (Zipfel et al. 2010a). Several of the rare renal disorders are associated with reduced Factor H surface attachment leading to reduced local complement control. These related disease scenarios demonstrate the relevance of surface attached regulators for homeostasis and define a therapeutic principle for specific targeted complement control at sides that are damaged by complement. These include RPE cells modified by oxidative stress as well as the exposed Bruch's membrane or exposed endothelial kidney cells as well as unprotected glomerular basement membranes. Local control of the overacting complement cascade at such damaged surfaces is an attractive option to treat complement-mediated diseases (Ricklin and Lambris 2007).

11.12 The C terminus of Factor H Recruits Complement Inhibitors to Modified Self-Surfaces

Given the central role of the C-terminal surface attachment region of Factor H for protection of self-surfaces, we aimed to utilize this central sensor as a vehicle to direct synthetic complement inhibitors to modified host surfaces, i.e., cells that are damaged by immune stress (Pangburn 2002; Schmidt et al. 2008). The C-terminal region, SCR18–20 of Factor H, provides the first contact to human surfaces and seems therefore most suitable for proper cellular attachment (Kajander et al. 2011; Leffler et al. 2010; Oppermann et al. 2006). This region binds to the surface of human cells and biomembranes, in particular to necrotic and apoptotic cells via surface ligands such as heparin, glycosaminoglycans, DNA, Annexin, MDA, and mCRP. Consequently, this recognition region of Factor H was used as a tag, to guide the synthetic inhibitor compstatin to sides of immune stress (Lauer et al. 2011). The proposed benefit of such a targeted inhibitor is complement control at local sites, for example, surfaces where complement is activated and where C3 is deposited. The C-terminal Factor H surface attachment region then delivers and targets the attached inhibitor to sites of damage, which under normal conditions bind the endogenous Factor H. Such a targeting chimeric inhibitor should specifically direct the complement inhibitor to local sites that are challenged by immune stress. Consequently we designed and generated a novel chimeric inhibitor which combines the complement inhibitor compstatin with the C-terminal attachment region of Factor H (SCR15–20) (Fig. 11.3). The novel compstatin Factor H SCR15–20 chimera uses the Factor H recognition tag, i.e., SCR15–20 to direct the inhibitor – here compstatin – to self-surfaces, in particular to surfaces damaged or modified by complement activation. Once bound to modified surfaces, the attached compstatin blocks complement at the level of C3-convertase and inhibits C3 activation as well as further downstream effects. The tagged compstatin has two binding sites for C3 and a tenfold higher affinity for C3b as compared to the compstatin peptide alone. Compstatin equipped with the Factor H tag maintains complement regulatory activity in fluid phase binds specifically to the surface of modified cells and regulates complement at such surfaces rather efficiently. When compared at molar levels for their regulatory activity at the surface of modified, nucleated CHO cells-tagged compstatin was almost 100-fold more effective as untagged compstatin.

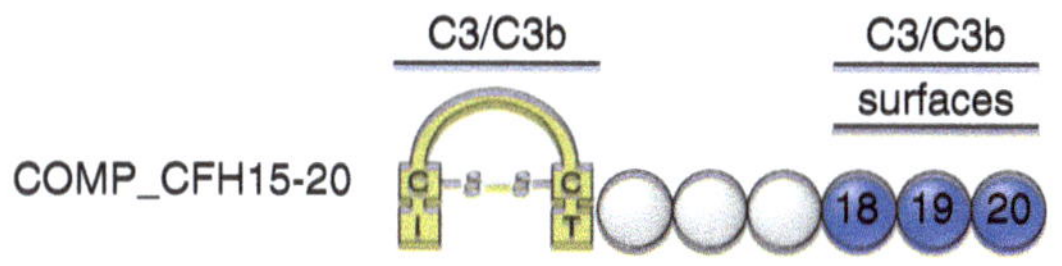

Fig. 11.3 *Structure of the targeted complement inhibitor compstatin – Factor H SCR15–20.* The novel chimeric surface targeting complement inhibitor *COMP_CFH15–20* has potent complement inhibitor compstatin linked the C-terminal surface attachment region of Factor H (domains SCR15–20). The COMP_CFH15–20 chimera has two binding sites for C3b one provided by compstatin itself and the second by the Factor H tag SCR15–20. The Factor H tag includes heparin- and glycosaminoglycan-binding sites

Consequently, the tagged compstatin controls complement locally, at sites of immune stress and such targeted inhibitors can reduce or may even abolish the adverse side effects associated with systemic complement inhibition. Further evaluation of this novel targeted inhibitor in both in vitro as well as in vivo models will define the therapeutic potential for complement-mediated diseases, such as AMD, aHUS, or MPGN-II.

Acknowledgment The work of the authors is supported by the Deutsche Forschungsgemeinschaft (DFG), and NL is supported by a Ph.D. research grant from the German ProRetina Foundation.

References

Anderson DH, Mullins RF, Hageman GS, Johnson LV (2002) A role for local inflammation in the formation of drusen in the aging eye. Am J Ophthalmol 134:411–431

Appel GB, Cook HT, Hageman G, Jennette JC, Kashgarian M, Kirschfink M, Lambris JD, Lanning L, Lutz HU, Meri S, Rose NR, Salant DJ, Sethi S, Smith RJ, Smoyer W, Tully HF, Tully SP, Walker P, Welsh M, Wurzner R, Zipfel PF (2005) Membranoproliferative glomerulonephritis type II (dense deposit disease): an update. J Am Soc Nephrol 16:1392–1403

Atkinson C, Song H, Lu B, Qiao F, Burns TA, Holers VM, Tsokos GC, Tomlinson S (2005) Targeted complement inhibition by C3d recognition ameliorates tissue injury without apparent increase in susceptibility to infection. J Clin Invest 115:2444–2453

Blackmore TK, Sadlon TA, Ward HM, Lublin DM, Gordon DL (1996) Identification of a heparin binding domain in the seventh short consensus repeat of complement factor H. J Immunol 157:5422–5427

Cai J, Wu M, Nelson KC, Sternberg P Jr, Jones DP (1999) Oxidant-induced apoptosis in cultured human retinal pigment epithelial cells. Invest Ophthalmol Vis Sci 40:959–966

Cai J, Nelson KC, Wu M, Sternberg P Jr, Jones DP (2000) Oxidative damage and protection of the RPE. Prog Retin Eye Res 19:205–221

Caprioli J, Bettinaglio P, Zipfel PF, Amadei B, Daina E, Gamba S, Skerka C, Marziliano N, Remuzzi G, Noris M (2001) The molecular basis of familial hemolytic uremic syndrome: mutation analysis of factor H gene reveals a hot spot in short consensus repeat 20. J Am Soc Nephrol 12:297–307

Carroll MC (2000) The role of complement in B cell activation and tolerance. Adv Immunol 74:61–88

Chang MK, Binder CJ, Torzewski M, Witztum JL (2002) C-reactive protein binds to both oxidized LDL and apoptotic cells through recognition of a common ligand: phosphorylcholine of oxidized phospholipids. Proc Natl Acad Sci USA 99:13043–13048

Chen Q, Muller D, Rudolph B, Hartmann A, Kuwertz-Broking E, Wu K, Kirschfink M, Skerka C, Zipfel PF (2011) Combined C3b and factor B autoantibodies and MPGN type II. N Engl J Med 365:2340–2342

Chou MY, Hartvigsen K, Hansen LF, Fogelstrand L, Shaw PX, Boullier A, Binder CJ, Witztum JL (2008) Oxidation-specific epitopes are important targets of innate immunity. J Intern Med 263:479–488

Clark SJ, Perveen R, Hakobyan S, Morgan BP, Sim RB, Bishop PN, Day AJ (2010) Impaired binding of the age-related macular degeneration-associated complement factor H 402H allotype to Bruch's membrane in human retina. J Biol Chem 285:30192–30202

Crabb JW, Miyagi M, Gu X, Shadrach K, West KA, Sakaguchi H, Kamei M, Hasan A, Yan L, Rayborn ME, Salomon RG, Hollyfield JG (2002) Drusen proteome analysis: an approach to the etiology of age-related macular degeneration. Proc Natl Acad Sci USA 99:14682–14687

Edwards AO, Ritter R III, Abel KJ, Manning A, Panhuysen C, Farrer LA (2005) Complement factor H polymorphism and age-related macular degeneration. Science 308:421–424

Eisenhardt SU, Habersberger J, Murphy A, Chen YC, Woollard KJ, Bassler N, Qian H, von Zur Muhlen C, Hagemeyer CE, Ahrens I, Chin-Dusting J, Bobik A, Peter K (2009) Dissociation of pentameric to monomeric C-reactive protein on activated platelets localizes inflammation to atherosclerotic plaques. Circ Res 105:128–137

Esterbauer H, Schaur RJ, Zollner H (1991) Chemistry and biochemistry of 4-hydroxynonenal, malonaldehyde and related aldehydes. Free Radic Biol Med 11:81–128

Fridkis-Hareli M, Storek M, Mazsaroff I, Risitano AM, Lundberg AS, Horvath CJ, Holers VM (2011) Design and development of TT30, a novel C3d-targeted C3/C5 convertase inhibitor for treatment of human complement alternative pathway-mediated diseases. Blood 118:4705–4713

Fritsche LG, Lauer N, Hartmann A, Stippa S, Keilhauer CN, Oppermann M, Pandey MK, Kohl J, Zipfel PF, Weber BH, Skerka C (2010) An imbalance of human complement regulatory proteins CFHR1, CFHR3 and factor H influences risk for age-related macular degeneration (AMD). Hum Mol Genet 23:4694–4704

Fujihara M, Nagai N, Sussan TE, Biswal S, Handa JT (2008) Chronic cigarette smoke causes oxidative damage and apoptosis to retinal pigmented epithelial cells in mice. PLoS One 3:e3119

Gold B, Merriam JE, Zernant J, Hancox LS, Taiber AJ, Gehrs K, Cramer K, Neel J, Bergeron J, Barile GR, Smith RT, Hageman GS, Dean M, Allikmets R (2006) Variation in factor B (BF) and complement component 2 (C2) genes is associated with age-related macular degeneration. Nat Genet 38:458–462

Hageman GS, Luthert PJ, Victor Chong NH, Johnson LV, Anderson DH, Mullins RF (2001) An integrated hypothesis that considers drusen as biomarkers of immune-mediated processes at the RPE-Bruch's membrane interface in aging and age-related macular degeneration. Prog Retin Eye Res 20:705–732

Haines JL, Hauser MA, Schmidt S, Scott WK, Olson LM, Gallins P, Spencer KL, Kwan SY, Noureddine M, Gilbert JR, Schnetz-Boutaud N, Agarwal A, Postel EA, Pericak-Vance MA (2005) Complement factor H variant increases the risk of age-related macular degeneration. Science 308:419–421

Hellwage J, Kuhn S, Zipfel PF (1997) The human complement regulatory factor-H-like protein 1, which represents a truncated form of factor H, displays cell-attachment activity. Biochem J 326:321–327

Hellwage J, Jokiranta TS, Friese MA, Wolk TU, Kampen E, Zipfel PF, Meri S (2002) Complement C3b/C3d and cell surface polyanions are recognized by overlapping binding sites on the most carboxyl-terminal domain of complement factor H. J Immunol 169:6935–6944

Herbert AP, Uhrin D, Lyon M, Pangburn MK, Barlow PN (2006) Disease-associated sequence variations congregate in a polyanion recognition patch on human factor H revealed in three-dimensional structure. J Biol Chem 281:16512–16520

Hollyfield JG, Bonilha VL, Rayborn ME, Yang X, Shadrach KG, Lu L, Ufret RL, Salomon RG, Perez VL (2008) Oxidative damage-induced inflammation initiates age-related macular degeneration. Nat Med 14:194–198

Huang Y, Qiao F, Atkinson C, Holers VM, Tomlinson S (2008) A novel targeted inhibitor of the alternative pathway of complement and its therapeutic application in ischemia/reperfusion injury. J Immunol 181:8068–8076

Hughes AE, Orr N, Esfandiary H, Diaz-Torres M, Goodship T, Chakravarthy U (2006) A common CFH haplotype, with deletion of CFHR1 and CFHR3, is associated with lower risk of age-related macular degeneration. Nat Genet 38:1173–1177

Jager RD, Mieler WF, Miller JW (2008) Age-related macular degeneration. N Engl J Med 358:2606–2617

Ji SR, Wu Y, Zhu L, Potempa LA, Sheng FL, Lu W, Zhao J (2007) Cell membranes and liposomes dissociate C-reactive protein (CRP) to form a new, biologically active structural intermediate: mCRP(m). FASEB J 21:284–294

Johnson LV, Leitner WP, Staples MK, Anderson DH (2001) Complement activation and inflammatory processes in Drusen formation and age related macular degeneration. Exp Eye Res 73:887–896

Jokiranta TS, Cheng ZZ, Seeberger H, Jozsi M, Heinen S, Noris M, Remuzzi G, Ormsby R, Gordon DL, Meri S, Hellwage J, Zipfel PF (2005) Binding of complement factor H to endothelial cells is mediated by the carboxy-terminal glycosaminoglycan binding site. Am J Pathol 167:1173–1181

Józsi M, Zipfel PF (2008) Factor H family proteins and human diseases. Trends Immunol 29:380–387

Kajander T, Lehtinen MJ, Hyvarinen S, Bhattacharjee A, Leung E, Isenman DE, Meri S, Goldman A, Jokiranta TS (2011) Dual interaction of factor H with C3d and glycosaminoglycans in host-nonhost discrimination by complement. Proc Natl Acad Sci USA 108:2897–2902

Klein RJ, Zeiss C, Chew EY, Tsai JY, Sackler RS, Haynes C, Henning AK, SanGiovanni JP, Mane SM, Mayne ST, Bracken MB, Ferris FL, Ott J, Barnstable C, Hoh J (2005) Complement factor H polymorphism in age-related macular degeneration. Science 308:385–389

Kuhn S, Zipfel PF (1996) Mapping of the domains required for decay acceleration activity of the human factor H-like protein 1 and factor H. Eur J Immunol 26:2383–2387

Lapeyraque AL, Fremeaux-Bacchi V, Robitaille P (2011) Efficacy of eculizumab in a patient with factor-H-associated atypical hemolytic uremic syndrome. Pediatr Nephrol 26:621–624

Lauer N, Mihlan M, Hartmann A, Schlotzer-Schrehardt U, Keilhauer C, Scholl HP, Charbel IP, Holz F, Weber BH, Skerka C, Zipfel PF (2011) Complement regulation at necrotic cell lesions is impaired by the age-related macular degeneration-associated factor-H His402 risk variant. J Immunol 187:4374–4383

Leffler J, Herbert AP, Norstrom E, Schmidt CQ, Barlow PN, Blom AM, Martin M (2010) Annexin-II, DNA, and histones serve as factor H ligands on the surface of apoptotic cells. J Biol Chem 285:3766–3776

Licht C, Heinen S, Jozsi M, Loschmann I, Saunders RE, Perkins SJ, Waldherr R, Skerka C, Kirschfink M, Hoppe B, Zipfel PF (2006) Deletion of Lys224 in regulatory domain 4 of Factor H reveals a novel pathomechanism for dense deposit disease (MPGN II). Kidney Int 70:42–50

Manuelian T, Hellwage J, Meri S, Caprioli J, Noris M, Heinen S, Jozsi M, Neumann HP, Remuzzi G, Zipfel PF (2003) Mutations in factor H reduce binding affinity to C3b and heparin and surface attachment to endothelial cells in hemolytic uremic syndrome. J Clin Invest 111:1181–1190

Meri S, Pangburn MK (1990) Discrimination between activators and nonactivators of the alternative pathway of complement: regulation via a sialic acid/polyanion binding site on factor H. Proc Natl Acad Sci USA 87: 3982–3986

Mihlan M, Stippa S, Jozsi M, Zipfel PF (2009) Monomeric CRP contributes to complement control in fluid phase and on cellular surfaces and increases phagocytosis by recruiting factor H. Cell Death Differ 16:1630–1640
Mihlan M, Blom AM, Kupreishvili K, Lauer N, Stelzner K, Bergstrom F, Niessen HW, Zipfel PF (2011) Monomeric C-reactive protein modulates classic complement activation on necrotic cells. FASEB J 25:4198–4210
Miller YI, Choi SH, Wiesner P, Fang L, Harkewicz R, Hartvigsen K, Boullier A, Gonen A, Diehl CJ, Que X, Montano E, Shaw PX, Tsimikas S, Binder CJ, Witztum JL (2011) Oxidation-specific epitopes are danger-associated molecular patterns recognized by pattern recognition receptors of innate immunity. Circ Res 108:235–248
Misasi R, Huemer HP, Schwaeble W, Solder E, Larcher C, Dierich MP (1989) Human complement factor H: an additional gene product of 43 kDa isolated from human plasma shows cofactor activity for the cleavage of the third component of complement. Eur J Immunol 19:1765–1768
Mulligan MS, Warner RL, Rittershaus CW, Thomas LJ, Ryan US, Foreman KE, Crouch LD, Till GO, Ward PA (1999) Endothelial targeting and enhanced antiinflammatory effects of complement inhibitors possessing sialyl Lewisx moieties. J Immunol 162:4952–4959
Ogden CA, Elkon KB (2006) Role of complement and other innate immune mechanisms in the removal of apoptotic cells. Curr Dir Autoimmun 9:120–142
Okamoto H, Umeda S, Obazawa M, Minami M, Noda T, Mizota A, Honda M, Tanaka M, Koyama R, Takagi I, Sakamoto Y, Saito Y, Miyake Y, Iwata T (2006) Complement factor H polymorphisms in Japanese population with age-related macular degeneration. Mol Vis 12:156–158
Oppermann M, Manuelian T, Jozsi M, Brandt E, Jokiranta TS, Heinen S, Meri S, Skerka C, Götze O, Zipfel PF (2006) Recognition mechanisms of the complement regulator factor H: Target binding via the C-terminus precedes complement regulation by the N-terminus. Clin Exp Immunol 144:342–352
Pangburn MK (2002) Cutting edge: localization of the host recognition functions of complement factor H at the carboxyl-terminal: implications for hemolytic uremic syndrome. J Immunol 169:4702–4706
Pangburn MK, Muller-Eberhard HJ (1983) Initiation of the alternative complement pathway due to spontaneous hydrolysis of the thioester of C3. Ann N Y Acad Sci 421:291–298
Pascolini D, Mariotti SP, Pokharel GP, Pararajasegaram R, Etya'ale D, Negrel AD, Resnikoff S (2004) Global update of available data on visual impairment: a compilation of population-based prevalence studies. Ophthalmic Epidemiol 11:67–115
Pickering MC, de Jorge EG, Martinez-Barricarte R, Recalde S, Garcia-Layana A, Rose KL, Moss J, Walport MJ, Cook HT, de Cordoba SR, Botto M (2007) Spontaneous hemolytic uremic syndrome triggered by complement factor H lacking surface recognition domains. J Exp Med 204:1249–1256
Poon IK, Hulett MD, Parish CR (2010) Molecular mechanisms of late apoptotic/necrotic cell clearance. Cell Death Differ 17:381–397
Potentia P (2009) POT-4 drug candidate for age-related macular degeneration successfully completes phase I clinical trial – prepares for phase II. Press Release 2009
Prosser BE, Johnson S, Roversi P, Herbert AP, Blaum BS, Tyrrell J, Jowitt TA, Clark SJ, Tarelli E, Uhrin D, Barlow PN, Sim RB, Day AJ, Lea SM (2007) Structural basis for complement factor H linked age-related macular degeneration. J Exp Med 204:2277–2283
Qu H, Magotti P, Ricklin D, Wu EL, Kourtzelis I, Wu YQ, Kaznessis YN, Lambris JD (2011) Novel analogues of the therapeutic complement inhibitor compstatin with significantly improved affinity and potency. Mol Immunol 48:481–489
Ricklin D, Lambris JD (2007) Complement-targeted therapeutics. Nat Biotechnol 25:1265–1275
Ricklin D, Lambris JD (2008) Compstatin: a complement inhibitor on its way to clinical application. Adv Exp Med Biol 632:273–292
Rodriguez de Cordoba S, Esparza-Gordillo J, Goicoechea de Jorge E, Lopez-Trascasa M, Sanchez-Corral P (2004) The human complement factor H: functional roles, genetic variations and disease associations. Mol Immunol 41:355–367
Rohrer B, Long Q, Coughlin B, Wilson RB, Huang Y, Qiao F, Tang PH, Kunchithapautham K, Gilkeson GS, Tomlinson S (2009) A targeted inhibitor of the alternative complement pathway reduces angiogenesis in a mouse model of age-related macular degeneration. Invest Ophthalmol Vis Sci 50:3056–3064
Rother RP, Rollins SA, Mojcik CF, Brodsky RA, Bell L (2007) Discovery and development of the complement inhibitor eculizumab for the treatment of paroxysmal nocturnal hemoglobinuria. Nat Biotechnol 25:1256–1264
Sahu A, Lambris JD (2001) Structure and biology of complement protein C3, a connecting link between innate and acquired immunity. Immunol Rev 180:35–48
Sahu A, Kay BK, Lambris JD (1996) Inhibition of human complement by a C3-binding peptide isolated from a phage-displayed random peptide library. J Immunol 157:884–891
Sahu A, Soulika AM, Morikis D, Spruce L, Moore WT, Lambris JD (2000) Binding kinetics, structure-activity relationship, and biotransformation of the complement inhibitor compstatin. J Immunol 165:2491–2499
Sahu A, Morikis D, Lambris JD (2003) Compstatin, a peptide inhibitor of complement, exhibits species-specific binding to complement component C3. Mol Immunol 39:557–566

Schmidt CQ, Herbert AP, Hocking HG, Uhrin D, Barlow PN (2008) Translational mini-review series on complement factor H: structural and functional correlations for factor H. Clin Exp Immunol 151:14–24
Simonelli F, Frisso G, Testa F, di Fiore R, Vitale DF, Manitto MP, Brancato R, Rinaldi E, Sacchetti L (2006) Polymorphism p.402Y>H in the complement factor H protein is a risk factor for age related macular degeneration in an Italian population. Br J Ophthalmol 90:1142–1145
Sjoberg AP, Trouw LA, Blom AM (2009) Complement activation and inhibition: a delicate balance. Trends Immunol 30:83–90
Skerka C, Lauer N, Weinberger AA, Keilhauer CN, Suhnel J, Smith R, Schlotzer-Schrehardt U, Fritsche L, Heinen S, Hartmann A, Weber BH, Zipfel PF (2007) Defective complement control of factor H (Y402H) and FHL-1 in age-related macular degeneration. Mol Immunol 44:3398–3406
Smith GP, Smith RA (2001) Membrane-targeted complement inhibitors. Mol Immunol 38:249–255
Souied EH, Leveziel N, Richard F, Dragon-Durey MA, Coscas G, Soubrane G, Benlian P, Fremeaux-Bacchi V (2005) Y402H complement factor H polymorphism associated with exudative age-related macular degeneration in the French population. Mol Vis 11:1135–1140
Strauss O (2005) The retinal pigment epithelium in visual function. Physiol Rev 85:845–881
Suzuki M, Kamei M, Itabe H, Yoneda K, Bando H, Kume N, Tano Y (2007) Oxidized phospholipids in the macula increase with age and in eyes with age-related macular degeneration. Mol Vis 13:772–778
Thurman JM, Renner B, Kunchithapautham K, Ferreira VP, Pangburn MK, Ablonczy Z, Tomlinson S, Holers VM, Rohrer B (2009) Oxidative stress renders retinal pigment epithelial cells susceptible to complement-mediated injury. J Biol Chem 284:16939–16947
Trouw LA, Blom AM, Gasque P (2008) Role of complement and complement regulators in the removal of apoptotic cells. Mol Immunol 45:1199–1207
Walport MJ (2001) Complement. First of two parts. N Engl J Med 344:1058–1066
Weismann D, Hartvigsen K, Lauer N, Bennett KL, Scholl HP, Charbel IP, Cano M, Brandstatter H, Tsimikas S, Skerka C, Superti-Furga G, Handa JT, Zipfel PF, Witztum JL, Binder CJ (2011) Complement factor H binds malondialdehyde epitopes and protects from oxidative stress. Nature 478:76–81
Yates JR, Sepp T, Matharu BK, Khan JC, Thurlby DA, Shahid H, Clayton DG, Hayward C, Morgan J, Wright AF, Armbrecht AM, Dhillon B, Deary IJ, Redmond E, Bird AC, Moore AT (2007) Complement C3 variant and the risk of age-related macular degeneration. N Engl J Med 357:553–561
Zhou J, Jang YP, Kim SR, Sparrow JR (2006) Complement activation by photooxidation products of A2E, a lipofuscin constituent of the retinal pigment epithelium. Proc Natl Acad Sci USA 103:16182–16187
Zipfel PF, Skerka C (2009) Complement regulators and inhibitory proteins. Nat Rev Immunol 9:729–740
Zipfel PF, Heinen S, Jozsi M, Skerka C (2006) Complement and diseases: defective alternative pathway control results in kidney and eye diseases. Mol Immunol 43:97–106
Zipfel PF, Mihlan M, Skerka C (2007) The alternative pathway of complement: a pattern recognition system. Adv Exp Med Biol 598:80–92
Zipfel PF, Lauer N, Skerka C (2010a) The role of complement in AMD. Adv Exp Med Biol 703:9–24
Zipfel PF, Mache C, Muller D, Licht C, Wigger M, Skerka C (2010b) DEAP-HUS: deficiency of CFHR plasma proteins and autoantibody-positive form of hemolytic uremic syndrome. Pediatr Nephrol 10:2009–2019

Chapter 12
C3 Glomerulonephritis/CFHR5 Nephropathy Is an Endemic Disease in Cyprus: Clinical and Molecular Findings in 21 Families

Constantinos Deltas, Daniel Gale, Terence Cook, Konstantinos Voskarides, Yiannis Athanasiou, and Alkis Pierides

Abstract Microscopic haematuria is the presenting symptom of several conditions, either heritable or acquired. A well-recognized familial condition is Alport syndrome, either of X-linked or autosomal recessive inheritance, as well as thin basement membrane nephropathy (TBMN) because of heterozygous collagen IV mutations. Even though microscopic haematuria of TBMN was long considered as a benign disease with excellent prognosis, more recent data suggest that development of chronic kidney disease (CKD) and even end-stage kidney disease (ESKD) is not a rare finding, perhaps owing to the cofounding role of modifier genes and other factors. Recent investigations in London and Cyprus culminated in the identification of another autosomal dominant condition that presents with microscopic haematuria because of heterozygous mutations in the *CFHR5* gene, which apparently plays a pivotal role in the regulation of the alterative pathway of complement system, which constitutes a significant part of innate immunity in humans. Histologically, the hallmark observation is the isolated glomerular deposition of C3 complement in the absence of immune complexes. It is considered one of the C3 glomerulopathies, and it may or may not be accompanied by mild membranoproliferative glomerulonephritis. Interestingly, a single mutation has been identified so far, a duplication of exons 2–3 of the *CFHR5* gene, and it has been described in patients of Greek-Cypriot descend only, perhaps originating on the Troodos mountains of Cyprus. Thus far, no patient with a mutation in this gene has been diagnosed in any other population. In Cyprus, it has been found in clusters of families in neighbouring villages in a total of 136 patients, and it constitutes a strong founder phenomenon. About 50% of patients over 50 years have progressed to CKD, and 14% of all patients progressed to ESKD. It is not quite well understood why males run a much higher risk to progress to CKD, compared to women.

C. Deltas (✉) • K. Voskarides
Molecular Medicine Research Center and Laboratory of Molecular and Medical Genetics, Department of Biological Sciences, University of Cyprus, Kallipoleos 75, 1678 Nicosia, Cyprus
e-mail: deltas@ucy.ac.cy

D. Gale • T. Cook
Imperial College Kidney and Transplant Institute, London, UK

Y. Athanasiou
Department of Nephrology, Nicosia General Hospital, Nicosia, Cyprus

A. Pierides
Department of Nephrology, Hippocrateon Hospital, Nicosia, Cyprus

J.D. Lambris et al. (eds.), *Complement Therapeutics*, Advances in Experimental Medicine and Biology 735, DOI 10.1007/978-1-4614-4118-2_12,

Microscopic haematuria is presently recognized as the initial presenting symptom in patients with one of several familial diseases. These diseases are linked to mutations in genes which encode either chains of type IV collagen (a major structural component of the glomerular basement membrane) or the complement regulator complement factor H-related protein 5 (CFHR5) or in other yet unknown genes (Kashtan 2005; Deltas et al. 2012).

COL4A3, *COL4A4* and *COL4A5* are three collagen IV genes that encode for proteins participating in the formation of trimers making up a meshwork of matrix within the lamina densa of the glomerular basement membrane. Mutations in these genes can present in a number of clinical contexts: (a) young males with X-linked Alport syndrome and mutations in the *COL4A5* gene, (b) female heterozygous carriers of the X-linked Alport syndrome, (c) male and female patients with the autosomal recessive form of Alport syndrome and homozygous mutations in the *COL4A3/COL4A4* genes or (d) autosomal dominant thin basement membrane nephropathy (TBMN), with or without progressive focal segmental glomerular sclerosis and renal impairment in adulthood, associated with heterozygous mutations in one of the *COL4A3/COL4A4* genes (Kashtan 2005; Deltas et al. 2012).

CFHR5 is a relatively newly described gene encoding for a regulator of the alternative complement pathway. The protein was co-localized in the glomerulus with complement under pathological conditions of the kidney (Murphy et al. 2002). A mutation in this gene leads to C3 glomerulonephritis (C3GN, glomerular inflammation with complement C3 but not immunoglobulin or C1q deposited in the kidney) following an autosomal dominant mode of inheritance and presenting with isolated microscopic haematuria with or without intermittent macroscopic haematuria and progressive renal impairment, especially in males. Here, we will concentrate on this disease, which is referred to as CFHR5 nephropathy.

As a first line of defence against pathogens, the complement system constitutes a significant part of innate immunity in humans. While the classical complement system requires immune complexes which act as the triggers for its activation, the alternative complement pathway requires no immune complexes, is independent of the presence of antibodies and may be active on a permanent basis. Recognition that, in CFHR5 nephropathy, a mutation in the *CFHR5* gene is responsible for familial C3 glomerulonephritis has suggested that this gene plays an important role in regulation of the alternative complement pathway in the kidney. This story was based on two parallel events. The first was that one of us, DG, embarked on the study of a few patients of Cypriot origin living in London whom he encountered in clinical practice and who were observed by the histopathologist (TC) to manifest a highly unusual form of familial glomerulonephritis. The second event has been the ongoing collection of samples and preparation of a DNA biobank from Cypriot families with inherited kidney disorders, at the laboratory of Molecular and Medical Genetics at the University of Cyprus, by CD, AP and colleagues (Gale et al. 2010; Athanasiou et al. 2011).

The index patient and affected relatives with the disease presented with microscopic haematuria as well as episodes of macroscopic haematuria following upper respiratory tract infections (a pattern termed 'synpharyngitic macroscopic haematuria'). Based on this presentation, it was reasonable to suspect IgA nephropathy which however was rejected based on renal biopsy results which did not show glomerular deposition of IgA. Additionally, IgA nephropathy is sporadic in the overwhelming majority of cases. Some familial cases have been reported recently, and a locus has been mapped, although no gene has been cloned as yet.

The histopathologist (TC) noted that the biopsies were highly unusual: there was a mild membranoproliferative glomerulonephritis (MPGN), also referred to as mesangiocapillary glomerulonephritis, with slight increase in mesangial cells and matrix. Some cells had slight capillary wall thickening. The EM showed subendothelial glomerular basement membrane electron dense C3 deposition. Importantly, there was no deposition of immune complexes. It is worth mentioning that not all the biopsies in CFHR5 nephropathy actually show MPGN.

These appearances, which are now termed C3 glomerulonephritis, implicated dysregulation of the complement alternative pathway (Gale et al. 2010). Molecular investigation of this potentially monogenic disorder was commenced in the initial two families in whom there was apparently autosomal dominant inheritance of the disease (Fig. 12.1).

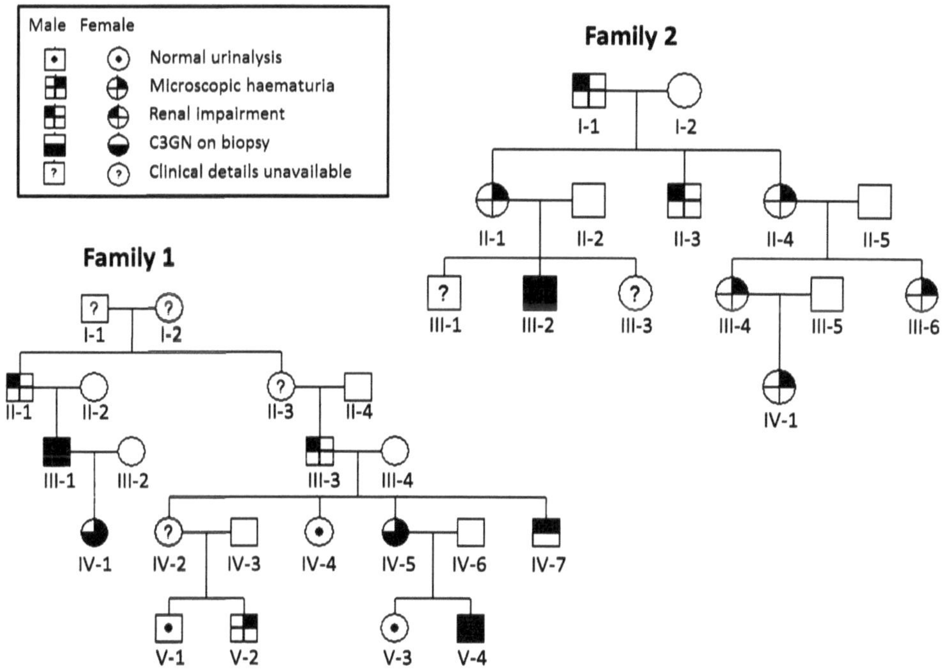

Fig. 12.1 Pedigrees of the two original Greek-Cypriot families that were described in London

12.1 Molecular Studies and More Families

DNA linkage analysis in the initial two families using 6,000 SNPs across the genome established linkage to locus 1q31, where the family of complement factor H-related genes map. This locus includes the original complement factor H (*CFH*) gene, which encodes the major alternative complement pathway regulator, along with five CFH related genes, *CFHR1–5*. Initial DNA resequencing of candidate genes, including *CFH* and *CFHR5*, revealed no likely causative sequence variants. A multiplex ligation-dependent probe amplification technique (Schouten et al. 2002) revealed evidence of a duplication of exons 2–3 in the *CFHR5* gene which was confirmed by Southern blotting, which also revealed the genomic location and size of the duplication. This duplication results in the synthesis of a larger protein which is immunologically detectable in the serum of patients, in addition to the normal protein. A polymerase chain reaction amplification test was developed to allow rapid detection of the mutant allele, and all affected individuals in the two initial families were found to harbour this mutation; surprisingly, almost all the patients with severe disease, including end-stage kidney disease (ESKD), were males (Gale et al. 2010; Athanasiou et al. 2011). Both families were of Cypriot origin and specifically from two villages in the Troodos mountains of Cyprus, namely, Kalopanayiotis and Gerakies. In addition to sharing the same mutation, all patients also shared an extended common haplotype of 8.74 cM, thereby indicating a founder mutation. Molecular dating showed that the most recent common affected ancestor of both families probably lived 11–16 generations ago and we have several examples of spreading of the mutation elsewhere on the island, in addition to London. So far, no patients of another ethnic background have been identified with this exon 2–3 duplication in the *CFHR5* gene. It is reasonable, however, to propose that similar forms of glomerular disease might be caused by other mutations of *CHFR5* in unrelated individuals since patients with similar clinicopathological presentations have been reported in the past in several populations including Japanese and Caucasians (Grekas et al. 1984; Servais et al. 2007), although these earlier studies have not highlighted a familial pattern of inheritance.

Fig. 12.2 Genetic map of Cyprus showing the CFHR5 clusters identified thus far

CFHR5 nephropathy appears to be remarkably common in Cyprus, as there are 136 affected individuals identified so far, within a total Greek-Cypriot population of 659,350, according to the 2011 census. This results in a prevalence of 1/4,848 in the Greek-Cypriot population, rendering it one of the more common rare diseases.

This *CFHR5* gene mutation represents one more founder phenomenon, among many in Cyprus. While this high frequency could result from the effects of genetic drift (where the frequency of rare alleles becomes amplified within an isolated population), the possibility remains that the *CFHR5* mutation associated with CFHR5 nephropathy also provides some selective advantage—perhaps by enhancing innate immune responses to an infectious disease which was endemic in Cyprus.

While DG and his colleagues were evaluating their finding on C3GN in Cypriot families living in the West London, they contacted A. Pierides and C. Deltas in Cyprus, asking for our contribution in expanding the work in the Cypriot population. Immediately after they communicated to us the nature of the mutation found in the *CFHR5* gene, we went into our freezers and dug out a number of samples from our renal DNA biobank that had been diagnosed with sporadic or familial glomerulopathy of unknown aetiology. During the past 20 years, we have been engaged in investigating inherited renal disorders among the Cypriot population, and we had indeed archived high numbers of samples that had escaped a clear diagnosis until then. To our surprise, a number of those samples scored positive for the *CFHR5* gene exon 2–3 duplication. The finding of the mutation in some sporadic samples enabled us to enter the families and offer a molecular diagnosis to additional at-risk members. In the meanwhile, all nephrologists in Cyprus started searching more carefully and wittingly for new suspects with this condition. Soon, we were all surprised by the number of families and patients that proved positive and segregated this one common mutation. Of great interest was also the fact that many families, although reportedly unrelated, have their origin in Troodos mountains and same or neighbouring villages as the original two families that were diagnosed in London. It seemed likely that the other families actually represented descendants of migrants from Kalopanayiotis or Gerakies to other cities around the island (Fig. 12.2).

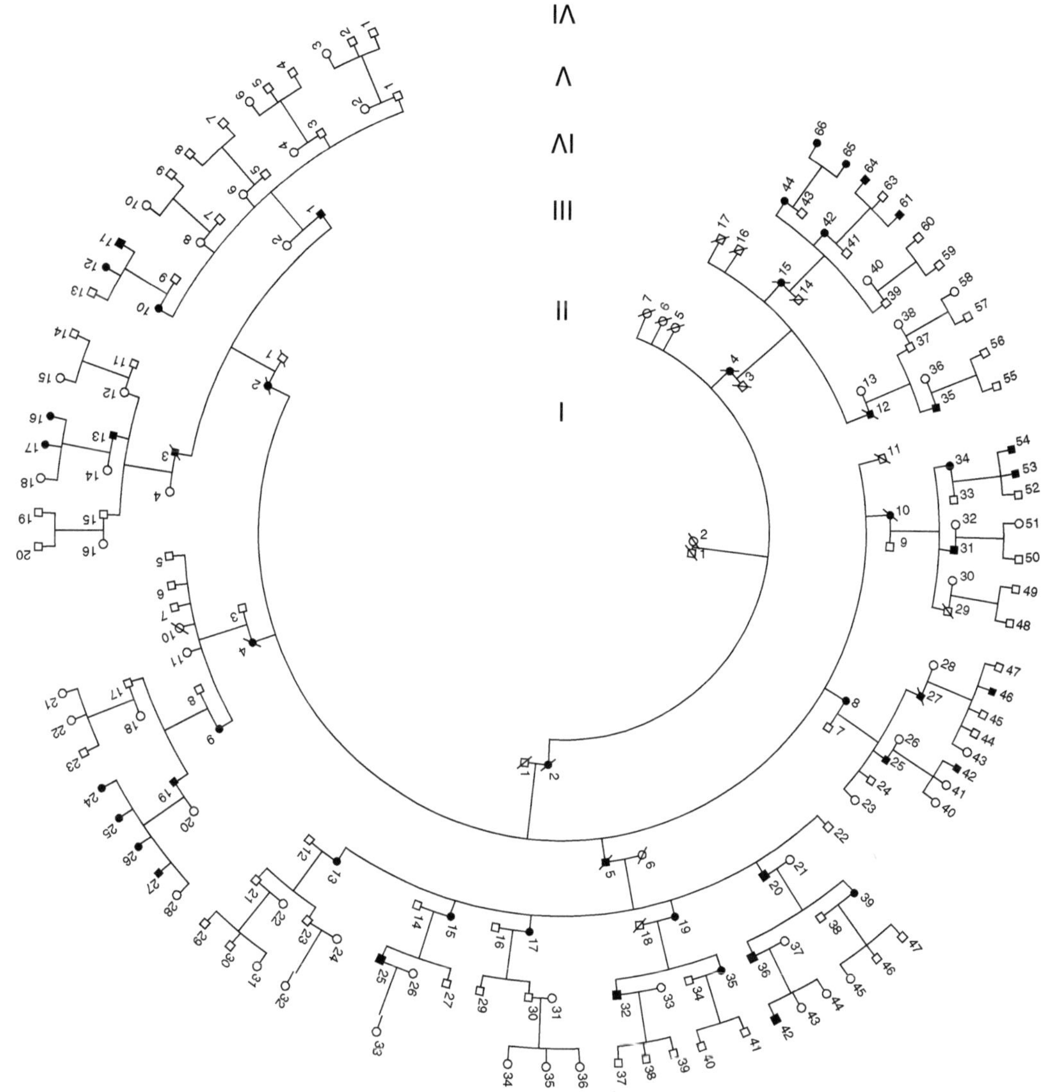

Fig. 12.3 Pedigree CY5308, of the largest family studied in Cyprus, extending in six generations

In the following months after the initial identification in 2009, we engaged in to more scrupulously investigating for and including families with patients who presented with microscopic or macroscopic episodes of haematuria and/or chronic kidney disease (CKD) with or without a clear cut histological diagnosis. In fact, we introduced a policy according to which every patient that is under investigation for a familial form of microscopic haematuria is also screened for the exon 2–3 duplication in *CFHR5*. So far, we have identified 136 patients who carry the mutation, belonging to 21 separate families, the largest family being CY5308 with 37 mutation carriers (Fig. 12.3). There is clear evidence for reduced penetrance, 90%, as 14 of 136 patients tested are negative on urine findings. 59% of the patients have haematuria only, 4% haematuria and proteinuria but no CKD and 50% (30/61, 24 M) of patients over 50 years have progressed to CKD. In total, 19 patients reached ESKD (14%), 13 of them, all males, after the age of 50 years. It is of great interest that of all 19 patients who progressed to ESKD, only 3 (16%) are females. In 11 of the patients who reached ESKD, there have been 13 renal transplants that have a superb survival ranging from 1 to 23 years. It is also interesting that all males who progressed to ESKD had demonstrated episodes of macroscopic haematuria during childhood and adolescence after upper respiratory tract infections. This was indeed a finding that complicated the clinical diagnosis as it was occasionally confused with IgA nephropathy that also presents with episodes of synpharyngitic macroscopic haematuria, associated with infection and pyrexia.

As is the case with thin basement membrane nephropathy and similarly to a Cypriot cohort that we presented previously, patients who carry this mutation in *CFHR5* are nearly asymptomatic until the age of about 30 years, with only isolated microscopic haematuria or negative urine findings. After the age of 30, proteinuria develops in most patients, and once this occurs, patients may progress to chronic renal failure (Voskarides et al. 2007; Pierides et al. 2009). It is presently unknown what protects women from reaching a more severe phenotype. This great phenotypic heterogeneity among patients of even same families is hypothesized to be attributed to unknown modified genes and perhaps environmental factors. A couple of candidate modifier genes are under investigation in our laboratory in Cyprus at the Molecular Medicine Research Centre. We recently reported our evidence that the glutamine variant of the R229Q podocin (*NPHS2*) mutation may act as a high-risk genetic factor predisposing patients with familial haematuria (TBMN or CFHR5 nephropathy) to a more severe disease (Voskarides et al. 2012). We had similar findings for another genetic modifier, this time a polymorphism in the target of miRNA gene hsa-miR-1207-5p, in the 3′UTR of *HBEGF* (C1936T, rs13385) (Papagregoriou et al. 2012).

12.2 Founder Mutations and Geographic Clustering

Over the past 20 years, we have identified and characterized a number of founder mutations in kidney-related genes that are responsible for monogenic disorders (see Fig. 3 of Voskarides et al. 2008 and Deltas 2004). The mutation in the *CFHR5* gene, exon 2–3 duplication, thus far has only been identified in patients with Cypriot origin. We have not been able to identify another mutation. It is presently unknown what the prevalence of this disease is in other populations. However, it is more than certain that other mutations are to be found soon by other investigators.

Detailed haplotype analysis has shown, in addition to a common 1 cM haplotype flanking the mutation in all affected families studied, that several seemingly unrelated families from the region around Nicosia, the capital, also share a larger (~20–30 cM) haplotype flanking in this region which is not present in the Troodos pedigrees. This large common haplotype may indicate a rather recent event which did not have adequate time to recombine. It is also probable that the affected populations expanded from a common origin in two separate areas. Whether the mutation originally arose in the Troodos mountains and subsequently spread to and became established in Nicosia, or vice versa, is not known, although the spreading from Troodos seems more likely. As regards the prevalence of the disease in London, it is worth mentioning that London is said to include the largest number of Greek-Cypriots of any city, where thousands of people migrated for economic reasons mostly during the decades of 1940s–1960s and then immediately after the military intervention by Turkey in July 1974.

12.3 Algorithm for Molecular Testing

As the differential diagnosis for familial microscopic haematuria is expanding fast, it is becoming equally important to introduce molecular genetics as a frontline tool for the diagnosis of suspect mutations. Perhaps it is worth performing at least one renal biopsy to the index patient if there is clear

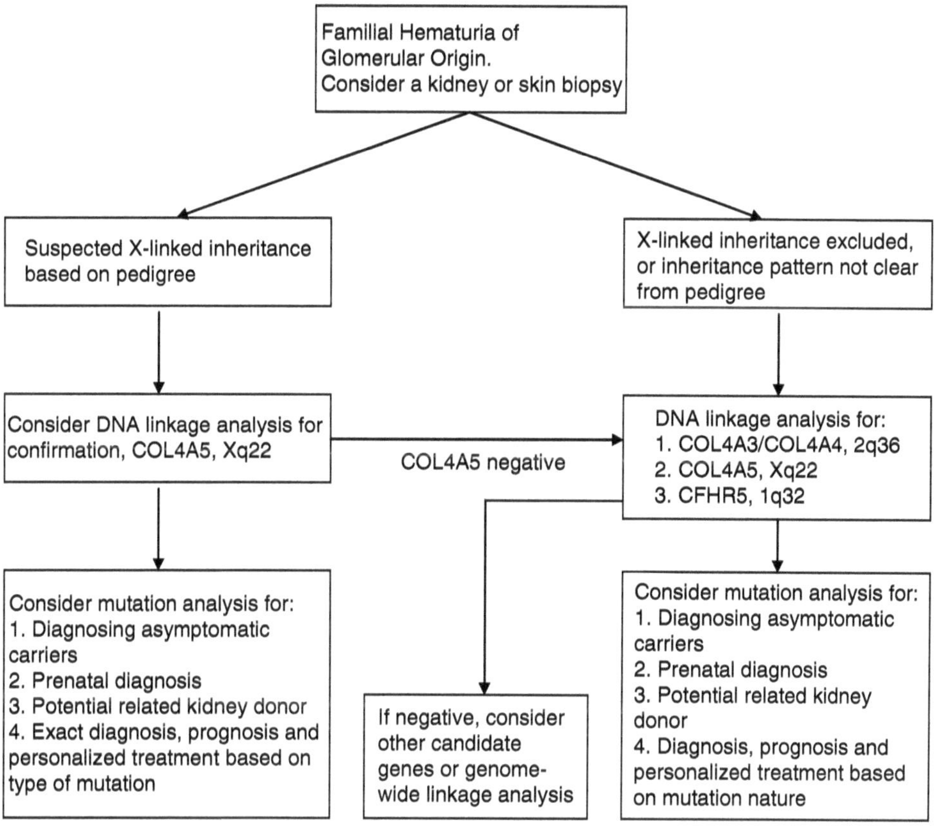

Fig. 12.4 Algorithm for studying families segregating glomerular microscopic haematuria. For patients with familial microscopic haematuria, a kidney or a skin biopsy may be performed before or after the molecular analysis depending on the clinical status or disease progression, for histological evaluation. The results of the biopsy may direct molecular testing and save time. In most, but certainly not all, families where a diagnosis has been established by molecular means, a simple molecular testing of the subject under study, for detecting the responsible affected haplotype or germinal mutation, might be adequate. A molecular diagnosis will acquire a superb position in CFHR5 and collagen IV nephropathies diagnosis and treatment, if early intervention proves decisive for disease prevention (From Deltas et al. (2011))

evidence for a familial segregation of isolated microscopic haematuria. Skin biopsy, although not always diagnostic, may also be useful. If the pedigree structure clearly excludes X-linked inheritance, autosomal collagen IV genes and the *CHFR5* gene are the immediate candidates. If analysis of these candidate loci is negative and the pedigree is large enough, then a genome-wide DNA linkage analysis is advised (Deltas et al. 2012). An algorithm for studying families segregating microscopic haematuria of glomerular origin is suggested in Fig. 12.4. One expects that more genes will be found to be responsible for familial forms of microscopic haematuria, as in our biobank in Cyprus, we have families that do not map to the above loci.

Acknowledgements The authors thank all patients and relatives who participated in this work, in London and Cyprus. The work was partly funded by grants to CD, specifically grant PENEK ENIΣX/0505/02 from the Cyprus Research Promotion Foundation, a grant co-funded by the European Regional Development Fund and the Republic of Cyprus through the Research Promotion Foundation (Strategic Infrastructure Project NEW INFRASTRUCTURE/STRATEGIC/0308/24, and through the University of Cyprus Articles 3/311 and 3/346.

References

Athanasiou Y, Voskarides K, Gale D, Damianou L, Patsias C, Zavros M et al (2011) Familial C3 glomerulopathy associated with *CFHR5* mutations: clinical characteristics of 91 patients in 16 pedigrees. Clin J Am Soc Nephrol 6:1436–1446

Deltas C (2004) Inherited diseases and Cyprus reality. A historico-genetic approach. In: The 30th annals of the Cyprus Research Center (Epetirida), Ministry of Education and Culture, Nicosia, Cyprus (Greek), pp 457–489

Deltas C, Pierides A, Voskarides K (2012) The role of molecular genetics in diagnosing familial hematuria(s). Pediatr Nephrol, DOI: 10.1007/s00467-011-1935-5

Gale DP, de Jorge EG, Cook HT, Martinez-Barricarte R, Hadjisavvas A, Adam G et al (2010) Identification of a mutation in complement factor H-related protein 5 in patients of Cypriot origin with glomerulonephritis. Lancet 376:794–801

Grekas D, Morley AR, Wilkinson R, Kerr DN (1984) Isolated C3 deposition in patients without systemic disease. Clin Nephrol 21:270–274

Murphy B, Georgiou T, Machet D, Hill P, McRae J (2002) Factor H-related protein-5: a novel component of human glomerular immune deposits. Am J Kidney Dis 39:24–27

Kashtan CE (2005) Familial hematurias: what we know and what we don't. Pediatr Nephrol 20:1027–1035

Papagregoriou G, Erguler K, Dweep H, Voskarides K, Koupepidou P, Athanasiou A et al (2012) A miR-1207-5p binding site polymorphism abolishes regulation of *HBEGF* and is associated with disease severity in CFHR5 nephropathy. PLoS ONE 7(2):e31021 [Epub 2012 Feb 2]

Pierides A, Voskarides K, Athanasiou Y, Ioannou K, Damianou L, Arsali M et al (2009) Clinico-pathological correlations in 127 patients in 11 large pedigrees, segregating one of three heterozygous mutations in the *COL4A3/COL4A4* genes associated with familial haematuria and significant late progression to proteinuria and chronic kidney disease from focal segmental glomerulosclerosis. Nephrol Dial Transplant 24:2721–2729

Schouten JP, McElgunn CJ, Waaijer R, Zwijnenburg D, Diepvens F, Pals G (2002) Relative quantification of 40 nucleic acid sequences by multiplex ligation-dependent probe amplification. Nucleic Acids Res 30:e57

Servais A, Frémeaux-Bacchi V, Lequintrec M, Salomon R, Blouin J, Knebelmann B et al (2007) Primary glomerulonephritis with isolated C3 deposits: a new entity which shares common genetic risk factors with haemolytic uraemic syndrome. J Med Genet 44:193–199

Voskarides K, Damianou L, Neocleous V, Zouvani I, Christodoulidou S, Hadjiconstantinou V et al (2007) *COL4A3/COL4A4* mutations producing focal segmental glomerulosclerosis and renal failure in thin basement membrane nephropathy. J Am Soc Nephrol 18:3004–3016

Voskarides K, Patsias C, Pierides A, Deltas C (2008) *COL4A3* Founder mutations in Greek-Cypriot families with thin basement membrane nephropathy and focal segmental glomerulosclerosis dating from around 18th century. Genet Test 12(2):273–278

Voskarides K, Arsali M, Athanasiou Y, Elia A, Pierides A, Deltas C (2012) Evidence that *NPHS2*-R229Q predisposes to proteinuria and renal failure in familial hematuria. Pediatr Nephrol, DOI: 10.1007/s00467-011-2084-6

Chapter 13
Complement-Targeted Therapeutics in Periodontitis

George Hajishengallis and John D. Lambris

Abstract Periodontitis is a prevalent oral chronic inflammatory disease which, in severe forms, may exert a major impact on systemic health. Clinical and histological observations, as well as experimental animal studies, suggest involvement of the complement system in periodontitis. However, the precise roles of the various complement components and pathways in periodontitis have only recently started to be elucidated. In this chapter, we review recent progress in the field and discuss the potential of complement-targeted therapeutics in the treatment of periodontitis.

13.1 Introduction

Periodontal disease is a prevalent chronic condition that causes inflammatory destruction of the tooth-supporting tissues (Pihlstrom et al. 2005). Severe periodontitis exerts a systemic impact on health, and the patients run increased risk for systemic diseases, such as atherosclerosis, diabetes, aspiration pneumonia, averse pregnancy outcomes, and perhaps rheumatoid arthritis (Tonetti et al. 2007; de Pablo et al. 2009; Kebschull et al. 2010; Pihlstrom et al. 2005; Scannapieco et al. 2010; Genco and Van Dyke 2010). Treated periodontal patients often develop recurrent disease for reasons that are not clear, thus necessitating better understanding of the underlying immunopathology (Armitage 2002; Hajishengallis 2009b). The annual cost of periodontal therapy in the USA exceeds $14 billion (Brown et al. 2002), and the suspected association of periodontitis with systemic conditions underscores the importance of implementing new and effective treatment options.

Although a group of tooth-associated subgingival anaerobic bacteria is strongly associated with periodontitis (Socransky et al. 1998), it is the host inflammatory response to uncontrolled bacterial challenge, rather than direct bacterial toxic effects, that primarily mediates periodontal tissue destruction (Gaffen and Hajishengallis 2008; Graves 2008). In this context, periodontal health represents a dynamic state where proinflammatory and antimicrobial activities are optimally regulated to prevent unwarranted host reactions (Gaffen and Hajishengallis 2008). This homeostatic balance may be disrupted, however, either by genetic immunoregulatory defects or by pathogens that subvert the host

G. Hajishengallis (✉)
Department of Microbiology, School of Dental Medicine, University of Pennsylvania, Philadelphia, PA 19104, USA
e-mail: geoh@upenn.edu

J.D. Lambris
Department of Pathology and Laboratory Medicine, School of Medicine, University of Pennsylvania, Philadelphia, PA 19104, USA

J.D. Lambris et al. (eds.), *Complement Therapeutics*, Advances in Experimental Medicine and Biology 735, DOI 10.1007/978-1-4614-4118-2_13, © Springer Science+Business Media New York 2013

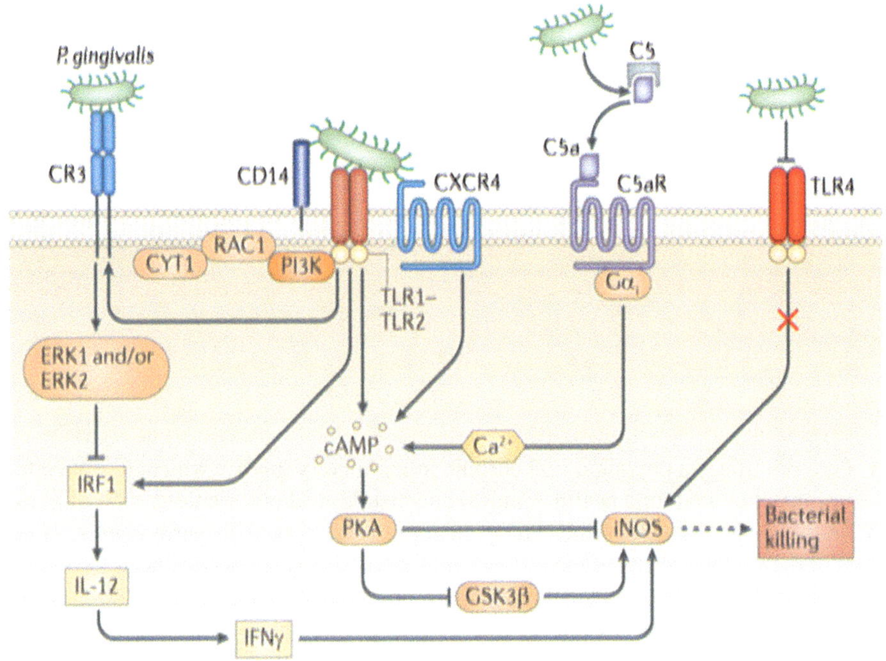

Fig. 13.1 *Exploitation of C5aR and other innate immune receptors by P. gingivalis to undermine host immunity. P. gingivalis* has surface structures that interact with Toll-like receptor (TLR)-2 (specifically with the CD14–TLR2–TLR1 signaling complex) and with TLR4. The activation of TLR4, however, is blocked by the bacterium's atypical lipopolysaccharide which acts as an antagonist; therefore, TLR4 is unlikely to induce protective responses. The TLR2 response is proactively modified through crosstalk with other receptors that are under *P. gingivalis* control. *P. gingivalis* regulates C5a receptor (C5aR) by virtue of Arg-specific cysteine proteinases, which attack C5 and release biologically active C5a. C5a stimulates intracellular Ca^{2+} signaling which synergistically enhances the otherwise weak cAMP responses induced by TLR2 activation alone. Maximal cAMP induction requires the participation of CXC-chemokine receptor 4 (CXCR4), which is activated directly by the bacterium's fimbriae. The resulting activation of the cAMP-dependent protein kinase A (PKA) inactivates glycogen synthase kinase-3β (GSK3β) and inhibits the inducible nitric oxide synthase (iNOS)-dependent killing of the pathogen in macrophages. An additional pathway-induced downstream of TLR2 is an inside-out signaling pathway, mediated by RAC1, phosphatidylinositol-3 kinase (PI3K), and cytohesin 1 (CYT1), which transactivates complement receptor-3 (CR3). Activated CR3 binds *P. gingivalis* and induces extracellular signal-regulated kinase-1/ERK2 signaling, which in turn selectively downregulates IL-12 p35 and p40 mRNA expression through suppression of interferon regulatory factor 1 (IRF1). Inhibition of bioactive IL-12, and secondarily IFNγ, leads to impaired immune clearance of *P. gingivalis* [From Hajishengallis and Lambris (2011) Nat Rev Immunol 11:187–200 (used by permission)]

response, thereby leading to nonprotective and nonresolving chronic inflammation (Gaffen and Hajishengallis 2008; Kinane et al. 2006; Kumpf and Schumann 2008). Available evidence implicates the periodontal pathogen *Porphyromonas gingivalis* as a master of immune subversion (Hajishengallis 2009a) (Fig. 13.1). Indeed, *P. gingivalis* inhibits critical antimicrobial responses that could eliminate it, while on the other hand, stimulates local inflammation, which may facilitate nutrient acquisition (e.g., gingival crevicular fluid-derived hemin) and additionally cause collateral tissue damage (Hajishengallis et al. 2008; Coats et al. 2009; Wang et al. 2007; Burns et al. 2006; Hajishengallis et al. 2007; Slaney and Curtis 2008; Potempa and Pike 2009; Krauss et al. 2010). Recently, *P. gingivalis* was shown to act as a keystone pathogen which promotes the survival and virulence of the entire microbial community (Hajishengallis et al. 2011).

At least in principle, periodontitis could be inhibited by interventions aiming to control inflammation and counteract microbial subversion of the host response. This concept is discussed here in the context of the complement system, which is now recognized as a central network that orchestrates the host response (Ricklin et al. 2010). Specifically, besides its classic antimicrobial functions (*see below*), complement crosstalks with and regulates other signaling systems, including Toll-like receptor (TLR) pathways (Hajishengallis and Lambris 2010). Despite its potentially host-protective role, however, complement forms a major link between infection and inflammatory pathology if overactivated or deregulated (Ricklin and Lambris 2007; Wagner and Frank 2010; Holers 2008; Ricklin et al. 2010).

13.2 Complement and Periodontal Disease

Besides the characteristic group of serum proteins (C1–9), the integrated complement system includes pattern-recognition molecules, convertases and other proteases, regulators, and receptors for interactions with immune mediators (Ricklin et al. 2010). The complement cascade can be triggered via distinct pathways (classical, lectin, or alternative) which converge at the third complement component (C3). The activation of the classical pathway is initiated by antigen-antibody complexes recognized by the C1q subunit of C1. The lectin pathway is triggered through interaction of the mannose-binding lectin, a secreted pattern-recognition molecule, with specific carbohydrate groups on microbial surfaces. Both the classical and lectin pathways proceed through C4 and C2 cleavage for the generation of the classical/lectin C3 convertase. The alternative pathway is initiated by low-level, spontaneous hydrolysis of C3 to C3[H_2O], which forms the initial alternative pathway C3 convertase in the presence of factors B (fB) and D (fD). As long as there is no sufficient negative regulation (as in the case of bacteria or other nonself surfaces), this initiation is followed by rapid propagation of the alternative pathway involving an amplification loop (Dunkelberger and Song 2010; Ricklin et al. 2010). The alternative pathway can also be triggered by lipopolysaccharide and lipooligosaccharide in a way dependent upon the plasma protein properdin attached to bacterial surfaces (Spitzer et al. 2007; Kimura et al. 2008) and may potentially contribute to ≥80% of total complement activation (Harboe and Mollnes 2008). C3 activation by pathway-specific C3 convertases leads to the generation of effector molecules involved in (a) the recruitment and activation of inflammatory cells (e.g., the C3a and C5a anaphylatoxins that activate specific G-protein–coupled receptors, C3aR and C5aR [CD88], respectively), (b) microbial opsonization and phagocytosis (e.g., through the C3b opsonin), and (c) direct lysis of targeted pathogens (by means of the C5b-9 membrane attack complex [MAC]) (Ricklin et al. 2010). An alternative receptor for C5a is the C5a-like receptor 2 (C5L2; GPR77), which has been assigned both regulatory and proinflammatory roles (Ward 2009; Hajishengallis and Lambris 2010; Zhang et al. 2010; Bamberg et al. 2010).

At least in principle, local complement activation could promote periodontal inflammation through multiple pathways, including C5a-induced vasodilation, increased vascular permeability, and chemotactic recruitment of inflammatory cells, including neutrophils. In addition to their role in acute inflammation, neutrophils have been implicated in chronic inflammatory diseases (e.g., rheumatoid arthritis, inflammatory bowel disease, and chronic obstructive pulmonary disease) (Kasama et al. 2005; Kanazawa and Furukawa 2007; Simpson et al. 2009; Kitsis and Weissmann 1991). Neutrophils are also key effectors of inflammatory tissue injury in periodontitis (Serhan et al. 2008; Nussbaum and Shapira 2011) and can be found in great numbers in the gingival crevice (≥95% of total leukocytes) (Delima and Van Dyke 2003). Although gingival crevicular neutrophils form what looks like a "defense wall" against the periodontal bacteria, they largely fail to control the bacteria despite maintaining viability and capacity to elicit inflammatory responses (Delima and Van Dyke 2003; Lange and Schroeder 1971; Newman 1980; Schroeder and Listgarten 1997, Vitkov et al. 2010). The underlying reasons are largely unexplored.

Clinical and histological observations suggest that complement is indeed involved in periodontitis (Hajishengallis 2010). Chronically inflamed gingiva or gingival crevicular fluid from periodontitis patients displays increased levels of activated complement fragments relative to control samples from healthy individuals (Patters et al. 1989; Nikolopoulou-Papaconstantinou et al. 1987; Rautemaa and Meri 1996; Beikler et al. 2008; Courts et al. 1977; Schenkein and Genco 1977; Niekrash and Patters 1986). Importantly, induction of experimental gingival inflammation in human volunteers causes progressive elevation of complement cleavage products correlating with increased clinical indices of inflammation (Patters et al. 1989).

Interestingly, a single nucleotide polymorphism of C5 (rs17611), which is associated with increased serum C5 levels and susceptibility to liver fibrosis and rheumatoid arthritis (Hillebrandt et al. 2005; Chang et al. 2008), was shown to be more prevalent in periodontitis patients than in healthy controls (Chai et al. 2010). Moreover, an immunohistochemical study showed weaker expression of CD59 in the gingiva of periodontitis patients compared to healthy controls, suggesting reduced protection of diseased tissues against autologous MAC-mediated tissue damage (Rautemaa and Meri 1996). A case of aggressive periodontitis accompanied by severe gingival angioedema was linked to dysregulated complement function, specifically C1INH deficiency (Roberts et al. 2003).

These studies suggest complement involvement in periodontal inflammation and pathogenesis. However, their correlative nature does not allow reliable conclusions as to the precise role(s) of the various complement pathways, nor do these observations necessarily imply that all complement pathways mediate destructive inflammation. In this regard, partial *C4* gene deficiencies are significantly more frequent in periodontal patients relative to healthy controls (Seppanen et al. 2007), therefore suggesting involvement of the classical and/or lectin pathway in a protective function. For instance, C3b generation via the C4-dependent classical and/or lectin pathways could promote opsonophagocytosis of periodontal bacteria, secondarily contributing to control of infection-induced inflammation. In conclusion, it has been uncertain which specific complement pathways need to be blocked to attenuate inflammatory pathology or kept intact to promote host defense. However, considerable insights have been gained by studies in preclinical models. At this point, there is sufficient evidence to implicate the C5a–C5aR axis in the pathogenesis of periodontitis (below).

13.3 Involvement of the C5a–C5aR Pathway in Periodontitis

The C5a anaphylatoxin is perhaps the most powerful effector molecule of the complement cascade, as it mediates chemotactic recruitment and activation of neutrophils and other inflammatory cells and is involved in synergistic complement interactions with Toll-like receptors (Guo and Ward 2005; Zhang et al. 2007). These immunostimulatory effects of C5a can potentially protect the host against microbial pathogens. In this regard, a major medical pathogen, *Staphylococcus aureus*, has evolved a strategy that allows it to block C5a binding and C5aR activation, via a secreted chemotaxis inhibitory protein (de Haas et al. 2004). Nevertheless, C5aR signaling can contribute to the pathogenesis of a number of acute or chronic inflammatory diseases, such as sepsis, acute lung injury, ischemia-reperfusion injury, and rheumatoid arthritis (Guo and Ward 2005; Okroj et al. 2007).

Intriguingly, in contrast to the *S. aureus* strategy, *P. gingivalis* is proactively involved in C5aR activation. Specifically, *P. gingivalis* employs its Arg-specific gingpains to generate biologically active C5a through limited degradation of C5, whereas the C5b remnant is proteolytically destroyed, ostensibly to prevent activation of the terminal complement pathway (Wingrove et al. 1992; Popadiak et al. 2007; Liang et al. 2011; Wang et al. 2010). *P. gingivalis* in fact can generate high levels of C5a (>30 nM) after a 30-min incubation in heat-inactivated human serum (Wang et al. 2010). This activity may appear to be counterproductive for the pathogen, given the important role of C5a in host defense. Strikingly, however, *P. gingivalis* was shown to exploit C5a to impair the killing function of macrophages via manipulation of specific signaling events in the absence of generalized immune suppression (Wang et al. 2010).

The mechanism involves synergistic production of high and sustained cAMP levels, which inhibit nitric oxide-dependent killing of *P. gingivalis* (Wang et al. 2010). This synergism requires a crosstalk between C5a-activated C5aR and *P. gingivalis*-activated TLR2, whereas downstream players include cAMP-dependent protein kinase A and glycogen synthase kinase-3β, the interplay of which inhibits the inducible nitric oxide synthase (Wang et al. 2010) (Fig. 13.1).

Moreover, the *P. gingivalis*-induced C5aR-TLR2 crosstalk regulates cytokine production that favors the pathogen (Liang et al. 2011). In this regard, the C5aR-TLR2 crosstalk inhibits TLR2-induced interleukin (IL)-12p70 which promotes immune clearance of *P. gingivalis* (Liang et al. 2011). In contrast, the same C5aR-TLR2 crosstalk upregulates inflammatory and bone-resorptive cytokines (IL-1β, IL-6, and TNF-α) which do not seem to harm *P. gingivalis*. These effects collectively lead to enhanced persistence of *P. gingivalis* in the host and set the stage for inflammatory tissue damage. This notion is supported by experimental periodontitis studies in the mouse model: Mice deficient in either C5aR or TLR2 were protected against *P. gingivalis*-induced inflammatory periodontal bone loss, whereas wild-type controls developed inflammation and suffer serious bone loss (Liang et al. 2011). In this inflammatory context, the proactive release of C5a by *P. gingivalis* could contribute to stimulation of inflammatory exudate for acquisition of nutrients like hemin and tissue breakdown products (peptides) which are essential for *P. gingivalis* and other asaccharolytic periodontal bacteria (Krauss et al. 2010). On the other hand, an isogenic mutant of *P. gingivalis*, which is deficient in all gingipain genes (KDP128), failed to persist in vivo and could not cause periodontitis (Liang et al. 2011; Hajishengallis et al. 2011). This difference in survival capacity may be related, at least in part, to the inability of KDP128 to generate C5a (Liang et al. 2011). These studies collectively indicate that the C5a–C5aR axis exerts a destructive role in periodontitis through a dual mechanism: (a) It is exploited by *P. gingivalis* to escape host defense, and (b) it mediates inflammatory periodontal bone loss.

13.4 C5aR-Targeted Intervention in Periodontitis

The above discussed findings provided a solid basis for rational C5aR-targeted intervention against *P. gingivalis* and periodontitis. This notion is supported by experimental evidence. Indeed, blockade of C5aR with a specific antagonist (Ac-F[OP(D)Cha-WR]; also known as PMX-53) abrogated the C5a-dependent subversive strategy of *P. gingivalis* and facilitated its immune clearance in vitro and in vivo (Wang et al. 2010; Liang et al. 2011). This effect may not necessarily imply protection against periodontal infection and inflammation in general, given the polymicrobial nature of periodontitis. However, it was recently shown that complement subversion by *P. gingivalis* can additionally benefit bystander periodontal bacteria in the same biofilm, which thereby displays quantitative and qualitative changes that lead to complement-dependent periodontitis at least in a mouse model (Hajishengallis et al. 2011) (Fig. 13.2).

Specifically, oral inoculation of mice with *P. gingivalis* exerts growth-enhancing effects and compositional changes in the oral microbiota, despite the very low colonization levels of *P. gingivalis* (<0.01% of the total bacterial counts) (Hajishengallis et al. 2011). Collectively, these actions lead to destructive inflammatory disease that requires the presence of the commensal microbiota and intact complement pathways since *P. gingivalis* fails to cause periodontitis in germ-free mice or conventionally raised mice deficient in C3aR or C5aR (Hajishengallis et al. 2011) (Fig. 13.3). Moreover, *P. gingivalis* fails to cause changes to the oral commensal microbiota of *C3aR*$^{-/-}$ or *C5aR*$^{-/-}$ mice, in contrast to normal specific pathogen-free (SPF) mice (Fig. 13.3).

The inability of *P. gingivalis* to alter the oral microbiota in *C5aR*$^{-/-}$ mice can be explained by the lack of C5aR, which is required by *P. gingivalis* to inhibit the killing capacity of leukocytes (Wang et al. 2010; Liang et al. 2011). *P. gingivalis*-affected leukocytes with impaired killing capacity would likely allow uncontrolled growth of other bacterial species in the same biofilm, accounting for the

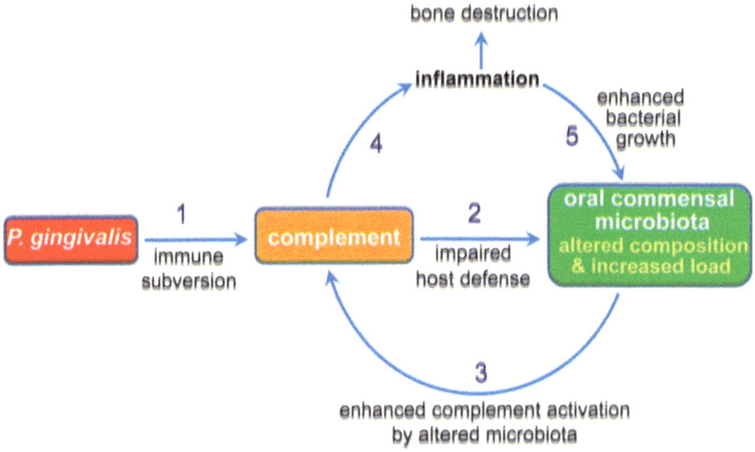

Fig. 13.2 *P. gingivalis* subverts complement leading to alterations in the oral microbiota and development of periodontitis. *P. gingivalis* impairs innate immunity through complement subversion leading to quantitative and qualitative changes to the oral commensal microbiota. Through complement, the altered microbiota causes inflammatory bone loss, whereas tissue breakdown products may further stimulate oral bacterial growth [From Hajishengallis et al. (2011) Cell Host Microbe 10:497–506 (used by permission)]

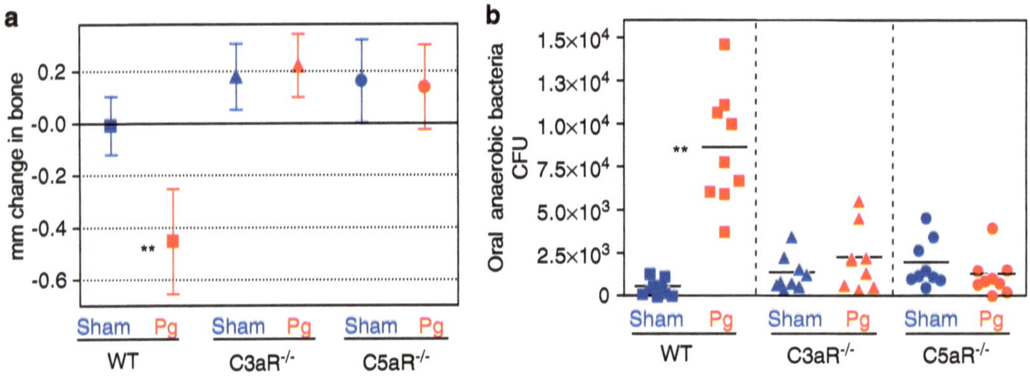

Fig. 13.3 *Complement-dependent elevation of the oral bacterial load and induction of bone loss.* BALB/c mice, either wild type (WT) or deficient in C3aR (*C3aR*$^{-/-}$) or C5aR (*C5aR*$^{-/-}$), were orally inoculated with *P. gingivalis* (Pg) or vehicle only (Sham) and assessed for bone loss (**a**) and levels of oral anaerobic bacteria (**b**). *Negative values* indicate bone loss relative to bone levels of the indicated controls (zero baseline), whereas *positive values* indicate increased bone levels. CFU counts are shown for each individual mouse with horizontal lines denoting mean values. $^{**}p<0.01$ versus corresponding control [From Hajishengallis et al. (2011) Cell Host Microbe 10:497–506 (used by permission)]

elevation of the total bacterial numbers. Consistent with the importance of gingipains in C5aR-dependent subversion of leukocytes (Liang et al. 2011), a gingipain-deficient isogenic mutant (KDP128) of *P. gingivalis* failed to elevate the oral bacterial load even in normal SPF mice. The C3aR requirement for the *P. gingivalis* effect on the oral microbiota and bone loss may be related to its synergistic interactions with C5aR that include reciprocal augmentation of receptor expression (Ricklin et al. 2010). In this regard, *P. gingivalis*-inoculated *C3aR*$^{-/-}$ mice displayed significantly reduced expression of C5aR (Hajishengallis et al. 2011).

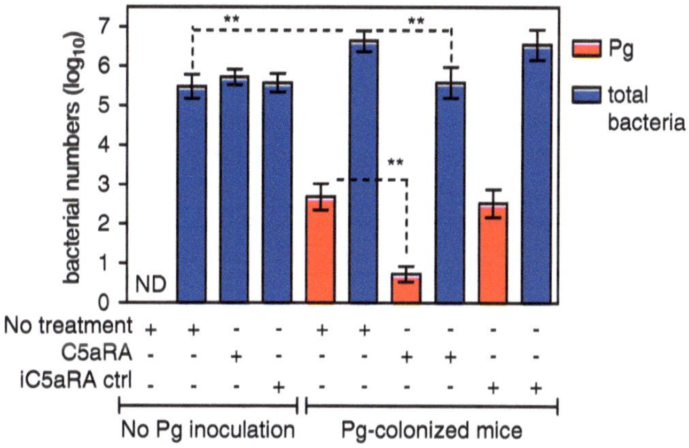

Fig. 13.4 Effects of C5aR antagonist (C5aRA), or inactive analog control (iC5aRA ctrl), on the numbers of *P. gingivalis* or total bacteria in the periodontal tissue of mice with or without previous inoculation with *P. gingivalis* Groups of mice were orally inoculated or not with *P. gingivalis* (as described above) and 7 days later, were injected with C5aRA or iC5aRA control, into the palatal gingiva, on the mesial of the first molar and in the papillae between first and second and third molars on both sides of the maxilla (1 μL of 1 μg per site; total of six sites). Two days later (day 9), the mice were sacrificed, and maxillary periodontal tissue was harvested to determine the levels of *P. gingivalis* colonization and the number of total oral bacteria, using quantitative real-time PCR of the *ISPg1* gene (*P. gingivalis*) or the 16 S rRNA gene (total oral bacteria). Two groups of mice, one of which was inoculated with *P. gingivalis*, were not treated with C5aRA or iC5aRA at day 7 and were sacrificed the same day for determining the levels of *P. gingivalis* and of total oral bacteria prior to the C5aRA or iC5aRA interventions. ND, Pg not detected. $^{**}p<0.01$ between indicated groups [From Hajishengallis et al. (2011) Cell Host Microbe 10:497–506 (used by permission)]

Since C5aR is required for the in vivo survival of *P. gingivalis* (Wang et al. 2010; Liang et al. 2011), local administration of a C5aR antagonist (PMX-53) could block the persistence of *P. gingivalis*, leading to its removal from the periodontal tissue and perhaps adversely affecting the total microbiota. Indeed, local administration of PMX-53 resulted in almost complete elimination of *P. gingivalis*, accompanied by a >10-fold reduction in the total numbers of oral anaerobic bacteria, which returned to their original lower levels (Hajishengallis et al. 2011) (Fig. 13.4). This reduction in the total microbiota was not a direct effect on the commensal bacteria by C5aRA since the antagonist failed to reduce the total oral bacterial numbers in mice not colonized with *P. gingivalis* (Fig. 13.4). These data clearly indicated that the experimental removal of *P. gingivalis* from the periodontal tissue, similar to its introduction, exerts a major influence on the oral microbiota. By analogy to the "keystone species" concept in macroecology, i.e., low-abundance species with a major supporting role for an entire ecological community (Power et al. 1996; Brown et al. 2001; Ebenman and Jonsson 2005), *P. gingivalis* may be regarded as such a species by fulfilling the criteria of low abundance and major influence on the microbial community. In fact, *P. gingivalis* could be characterized as a keystone pathogen, defined as "a keystone species which supports and shapes a microbial community in ways that also promote disease pathogenesis" (Hajishengallis et al. 2011).

13.5 Conclusions and Future Directions

Despite being a very minor constituent of the total oral microbiota, *P. gingivalis* can alter the numbers and community organization of the commensal bacteria, the presence of which is essential for inflammatory periodontal bone loss. Although PMX-53 has not been shown yet to inhibit periodontal bone loss after local administration, its capacity to block *P. gingivalis* persistence in the periodontal tissue and to inhibit the overgrowth of the oral microbiota strongly suggests that it is a promising therapeutic against periodontitis. Moreover, C5aR inhibitors may have important therapeutic implications also in other infections or inflammatory diseases where *P. gingivalis* is thought to be implicated, such as oral aspiration pneumonia and atherosclerosis (Gibson et al. 2006; Okuda et al. 2005).

Despite this progress, additional systematic approaches are required to comprehensively identify the precise roles of the various complement pathways in the context of periodontal pathogenesis. Such information could reveal which specific pathways need to be blocked to reverse inflammatory pathology or, conversely, be enhanced (or left intact) to promote host defense. This would greatly facilitate complement-targeted therapeutic intervention against periodontitis.

Importantly, a number of complement-specific drugs are already in clinical trials for other inflammatory diseases (Ricklin and Lambris 2007; Wagner and Frank 2010). Among them, PMX-53, which blocks both mouse and human C5aR, has a good safety record when given orally to humans (Ricklin and Lambris 2007; Wagner and Frank 2010). This, and potentially other complement-targeted drugs, may prove effective as adjunctive treatments for periodontitis in the near future.

Acknowledgments The authors acknowledge support by U.S. Public Health Service Grants DE015254, DE018292, DE021685 (GH) and AI068730, AI030040, AI072106, and GM062134 (JDL).

References

Armitage GC (2002) Classifying periodontal diseases – a long-standing dilemma. Periodontol 2000 30:9–23

Bamberg CE, Mackay CR, Lee H, Zahra D, Jackson J, Lim YS, Whitfeld PL, Craig S, Corsini E, Lu B, Gerard C, Gerard NP (2010) The C5a receptor (C5aR) C5L2 is a modulator of C5aR-mediated signal transduction. J Biol Chem 285:7633–7644

Beikler T, Peters U, Prior K, Eisenacher M, Flemmig TF (2008) Gene expression in periodontal tissues following treatment. BMC Med Genomics 1:30

Brown LJ, Johns BA, Wall TP (2002) The economics of periodontal diseases. Periodontol 2000 29:223–234

Brown JH, Whitham TG, Morgan Ernest SK, Gehring CA (2001) Complex species interactions and the dynamics of ecological systems: long-term experiments. Science 293:643–650

Burns E, Bachrach G, Shapira L, Nussbaum G (2006) Cutting edge: TLR2 is required for the innate response to *Porphyromonas gingivalis*: activation leads to bacterial persistence and TLR2 deficiency attenuates induced alveolar bone resorption. J Immunol 177:8296–8300

Chai L, Song Y-Q, Zee K-Y, Leung WK (2010) Single nucleotide polymorphisms of complement component 5 and periodontitis. J Periodont Res 45:301–308

Chang M, Rowland CM, Garcia VE, Schrodi SJ, Catanese JJ, van der Helm-van Mil AH, Ardlie KG, Amos CI, Criswell LA, Kastner DL, Gregersen PK, Kurreeman FA, Toes RE, Huizinga TW, Seldin MF, Begovich AB (2008) A large-scale rheumatoid arthritis genetic study identifies association at chromosome 9q33.2. PLoS Genet 4:e1000107

Coats SR, Jones JW, Do CT, Braham PH, Bainbridge BW, To TT, Goodlett DR, Ernst RK, Darveau RP (2009) Human toll-like receptor 4 responses to *P. gingivalis* are regulated by lipid A 1- and 4′-phosphatase activities. Cell Microbiol 11:1587–1599

Courts FJ, Boackle RJ, Fudenberg HH, Silverman MS (1977) Detection of functional complement components in gingival crevicular fluid from humans with periodontal diseases. J Dent Res 56:327–331

de Haas CJC, Veldkamp KE, Peschel A, Weerkamp F, Van Wamel WJB, Heezius ECJM, Poppelier MJJG, Van Kessel KPM, van Strijp JAG (2004) Chemotaxis inhibitory protein of *Staphylococcus aureus*, a bacterial antiinflammatory agent. J Exp Med 199:687–695

de Pablo P, Chapple IL, Buckley CD, Dietrich T (2009) Periodontitis in systemic rheumatic diseases. Nat Rev Rheumatol 5:218–224

Delima AJ, Van Dyke TE (2003) Origin and function of the cellular components in gingival crevice fluid. Periodontol 2000 31:55–76

Dunkelberger JR, Song WC (2010) Complement and its role in innate and adaptive immune responses. Cell Res 20:34–50

Ebenman B, Jonsson T (2005) Using community viability analysis to identify fragile systems and keystone species. Trends Ecol Evol 20:568–575

Gaffen SL, Hajishengallis G (2008) A new inflammatory cytokine on the block: re-thinking periodontal disease and the Th1/Th2 paradigm in the context of Th17 cells and IL-17. J Dent Res 87:817–828
Genco RJ, Van Dyke TE (2010) Prevention: reducing the risk of CVD in patients with periodontitis. Nat Rev Cardiol 7:479–480
Gibson FC III, Yumoto H, Takahashi Y, Chou HH, Genco CA (2006) Innate immune signaling and *Porphyromonas gingivalis*-accelerated atherosclerosis. J Dent Res 85:106–121
Graves D (2008) Cytokines that promote periodontal tissue destruction. J Periodontol 79:1585–1591
Guo RF, Ward PA (2005) Role of C5a in inflammatory responses. Annu Rev Immunol 23:821–852
Hajishengallis G (2009a) *Porphyromonas gingivalis*-host interactions: open war or intelligent guerilla tactics? Microbes Infect 11:637–645
Hajishengallis G (2009b) Toll gates to periodontal host modulation and vaccine therapy. Periodontol 2000 51:181–207
Hajishengallis G (2010) Complement and periodontitis. Biochem Pharmacol 80:1992–2001
Hajishengallis G, Lambris JD (2010) Crosstalk pathways between Toll-like receptors and the complement system. Trends Immunol 31:154–163
Hajishengallis G, Lambris JD (2011) Microbial manipulation of receptor crosstalk in innate immunity. Nat Rev Immunol 11:187–200
Hajishengallis G, Shakhatreh M-AK, Wang M, Liang S (2007) Complement receptor 3 blockade promotes IL-12-mediated clearance of *Porphyromonas gingivalis* and negates its virulence in vivo. J Immunol 179:2359–2367
Hajishengallis G, Wang M, Liang S, Triantafilou M, Triantafilou K (2008) Pathogen induction of CXCR4/TLR2 cross-talk impairs host defense function. Proc Natl Acad Sci USA 105:13532–13537
Hajishengallis G, Liang S, Payne MA, Hashim A, Jotwani R, Eskan MA, McIntosh ML, Alsam A, Kirkwood KL, Lambris JD, Darveau RP, Curtis MA (2011) Low-abundance biofilm species orchestrates inflammatory periodontal disease through the commensal microbiota and complement. Cell Host Microbe 10:497–506
Harboe M, Mollnes TE (2008) The alternative complement pathway revisited. J Cell Mol Med 12:1074–1084
Hillebrandt S, Wasmuth HE, Weiskirchen R, Hellerbrand C, Keppeler H, Werth A, Schirin-Sokhan R, Wilkens G, Geier A, Lorenzen J, Kohl J, Gressner AM, Matern S, Lammert F (2005) Complement factor 5 is a quantitative trait gene that modifies liver fibrogenesis in mice and humans. Nat Genet 37:835–843
Holers VM (2008) The spectrum of complement alternative pathway-mediated diseases. Immunol Rev 223:300–316
Kanazawa N, Furukawa F (2007) Autoinflammatory syndromes with a dermatological perspective. J Dermatol 34:601–618
Kasama T, Miwa Y, Isozaki T, Odai T, Adachi M, Kunkel SL (2005) Neutrophil-derived cytokines: potential therapeutic targets in inflammation. Curr Drug Targets Inflamm Allergy 4:273–279
Kebschull M, Demmer RT, Papapanou PN (2010) "Gum bug leave my heart alone" – epidemiologic and mechanistic evidence linking periodontal infections and atherosclerosis. J Dent Res 89:879–902
Kimura Y, Miwa T, Zhou L, Song W-C (2008) Activator-specific requirement of properdin in the initiation and amplification of the alternative pathway complement. Blood 111:732–740
Kinane DF, Peterson M, Stathopoulou PG (2006) Environmental and other modifying factors of the periodontal diseases. Periodontol 2000 40:107–119
Kitsis E, Weissmann G (1991) The role of the neutrophil in rheumatoid arthritis. Clin Orthop Relat Res 265:63–72
Krauss JL, Potempa J, Lambris JD, Hajishengallis G (2010) Complementary tolls in the periodontium: how periodontal bacteria modify complement and Toll-like receptor responses to prevail in the host. Periodontol 2000 52:141–162
Kumpf O, Schumann RR (2008) Genetic influence on bloodstream infections and sepsis. Int J Antimicrob Agents 32(Suppl 1):S44–S50
Lange D, Schroeder HE (1971) Cytochemistry and ultrastructure of gingival sulcus cells. Helv Odontol Acta 15:65–86
Liang S, Krauss JL, Domon H, McIntosh ML, Hosur KB, Qu H, Li F, Tzekou A, Lambris JD, Hajishengallis G (2011) The C5a receptor impairs IL-12-dependent clearance of *Porphyromonas gingivalis* and is required for induction of periodontal bone loss. J Immunol 186:869–877
Newman HN (1980) Neutrophils and IgG at the host-plaque interface on children's teeth. J Periodontol 51:642–651
Niekrash CE, Patters MR (1986) Assessment of complement cleavage in gingival fluid in humans with and without periodontal disease. J Periodont Res 21:233–242
Nikolopoulou-Papaconstantinou AA, Johannessen AC, Kristoffersen T (1987) Deposits of immunoglobulins, complement, and immune complexes in inflamed human gingiva. Acta Odontol Scand 45:187–193
Nussbaum G, Shapira L (2011) How has neutrophil research improved our understanding of periodontal pathogenesis? J Clin Periodontol 38:49–59
Okroj M, Heinegard D, Holmdahl R, Blom AM (2007) Rheumatoid arthritis and the complement system. Ann Med 39:517–530
Okuda K, Kimizuka R, Abe S, Kato T, Ishihara K (2005) Involvement of periodontopathic anaerobes in aspiration pneumonia. J Periodontol 76:2154–2160

Patters MR, Niekrash CE, Lang NP (1989) Assessment of complement cleavage in gingival fluid during experimental gingivitis in man. J Clin Periodontol 16:33–37
Pihlstrom BL, Michalowicz BS, Johnson NW (2005) Periodontal diseases. Lancet 366:1809–1820
Popadiak K, Potempa J, Riesbeck K, Blom AM (2007) Biphasic effect of gingipains from *Porphyromonas gingivalis* on the human complement system. J Immunol 178:7242–7250
Potempa J, Pike RN (2009) Corruption of innate immunity by bacterial proteases. J Innate Immun 1:70–87
Power ME, Tilman D, Estes JA, Menge BA, Bond WJ, Mills LS, Daily G, Castilla JC, Lubchenco J, Paine RT (1996) Challenges in the quest for keystones. Bioscience 46:609–620
Rautemaa R, Meri S (1996) Protection of gingival epithelium against complement-mediated damage by strong expression of the membrane attack complex inhibitor protectin (CD59). J Dent Res 75:568–574
Ricklin D, Lambris JD (2007) Complement-targeted therapeutics. Nat Biotechnol 25:1265–1275
Ricklin D, Hajishengallis G, Yang K, Lambris JD (2010) Complement: a key system for immune surveillance and homeostasis. Nat Immunol 11:785–797
Roberts A, Shah M, Chapple IL (2003) C-1 esterase inhibitor dysfunction localised to the periodontal tissues: clues to the role of stress in the pathogenesis of chronic periodontitis? J Clin Periodontol 30:271–277
Scannapieco FA, Dasanayake AP, Chhun N (2010) Does periodontal therapy reduce the risk for systemic diseases? Dent Clin North Am 54:163–181
Schenkein HA, Genco RJ (1977) Gingival fluid and serum in periodontal diseases. II. Evidence for cleavage of complement components C3, C3 proactivator (factor B) and C4 in gingival fluid. J Periodontol 48:778–784
Schroeder HE, Listgarten MA (1997) The gingival tissues: the architecture of periodontal protection. Periodontol 2000 13:91–120
Seppanen M, Lokki ML, Notkola IL, Mattila K, Valtonen V, Nieminen A, Vesanen M, Asikainen S, Meri S (2007) Complement and c4 null alleles in severe chronic adult periodontitis. Scand J Immunol 65:176–181
Serhan CN, Chiang N, Van Dyke TE (2008) Resolving inflammation: dual anti-inflammatory and pro-resolution lipid mediators. Nat Rev Immunol 8:349–361
Simpson JL, Phipps S, Gibson PG (2009) Inflammatory mechanisms and treatment of obstructive airway diseases with neutrophilic bronchitis. Pharmacol Ther 124:86–95
Slaney JM, Curtis MA (2008) Mechanisms of evasion of complement by *Porphyromonas gingivalis*. Front Biosci 13:188–196
Socransky SS, Haffajee AD, Cugini MA, Smith C, Kent RL Jr (1998) Microbial complexes in subgingival plaque. J Clin Periodontol 25:134–144
Spitzer D, Mitchell LM, Atkinson JP, Hourcade DE (2007) Properdin can initiate complement activation by binding specific target surfaces and providing a platform for de novo convertase assembly. J Immunol 179:2600–2608
Tonetti MS, D'Aiuto F, Nibali L, Donald A, Storry C, Parkar M, Suvan J, Hingorani AD, Vallance P, Deanfield J (2007) Treatment of periodontitis and endothelial function. N Engl J Med 356:911–920
Vitkov L, Klappacher M, Hannig M, Krautgartner WD (2010) Neutrophil fate in gingival crevicular fluid. Ultrastruct Pathol 34:25–30
Wagner E, Frank MM (2010) Therapeutic potential of complement modulation. Nat Rev Drug Discov 9:43–56
Wang M, Shakhatreh M-AK, James D, Liang S, S-i N, Yoshimura F, Demuth DR, Hajishengallis G (2007) Fimbrial proteins of *Porphyromonas gingivalis* mediate in vivo virulence and exploit TLR2 and complement receptor 3 to persist in macrophages. J Immunol 179:2349–2358
Wang M, Krauss JL, Domon H, Hosur KB, Liang S, Magotti P, Triantafilou M, Triantafilou K, Lambris JD, Hajishengallis G (2010) Microbial hijacking of complement-toll-like receptor crosstalk. Sci Signal 3:ra11
Ward PA (2009) Functions of C5a receptors. J Mol Med 87:375–378
Wingrove JA, DiScipio RG, Chen Z, Potempa J, Travis J, Hugli TE (1992) Activation of complement components C3 and C5 by a cysteine proteinase (gingipain-1) from *Porphyromonas (Bacteroides) gingivalis*. J Biol Chem 267:18902–18907
Zhang X, Kimura Y, Fang C, Zhou L, Sfyroera G, Lambris JD, Wetsel RA, Miwa T, Song WC (2007) Regulation of Toll-like receptor-mediated inflammatory response by complement in vivo. Blood 110:228–236
Zhang X, Schmudde I, Laumonnier Y, Pandey MK, Clark JR, Konig P, Gerard NP, Gerard C, Wills-Karp M, Kohl J (2010) A critical role for C5L2 in the pathogenesis of experimental allergic asthma. J Immunol 185:6741–6752

Chapter 14
Complement System Activation in Cardiac and Skeletal Muscle Pathology: Friend or Foe?

Maro Syriga and Manolis Mavroidis

Abstract A major goal in current cardiology practice is to determine optimal strategies for minimizing myocardial necrosis and optimizing cardiac repair following an acute myocardial infarction. Temporally regulated activation and suppression of innate immunity may be critical for achieving this goal. Extensive experimental data in various animal models have indicated that inhibiting complement activation offers protection to cardiac tissue after ischemia/reperfusion. However, the results of clinical studies using complement inhibitors (mainly at the C5 level) in patients with acute myocardial infarction have largely been disappointing.

In cases in which complement activation participates in the initial events of muscle cell destruction, as in autoimmune myocarditis or autoimmune muscle disorders, inhibition of complement activation is expected to prove a successful treatment. In other pathologic conditions in which complement is recruited by degenerating or dying muscle cells, as in ischemia, the ideal approach is probably to modulate rather than abruptly blunt complement activation. Beneficial effects of complement action with regard to waste disposal, recruitment of stem cells, regeneration, angiogenesis, and better utilization of energy sources under hypoxic conditions may also prove important for successful disease treatment. Patient outcome after myocardial infarction almost certainly depend upon the combined activation of several distinct but potentially interrelated signaling pathways, suggesting that a combination of treatments targeted to different pathways should be the therapy of choice, and modulation of complement could be one of them.

14.1 Complement Activation in Cardiac Muscle

14.1.1 Myocardial Infarction and Innate Immunity

Myocardial infarction is the most frequent cardiovascular event in the western world and is responsible for a large fraction of all cardiovascular deaths. Improved clinical management of acute myocardial infarction has significantly lowered immediate mortality over the past two decades (Lloyd-Jones et al. 2009), but surviving patients still face another major complication: the development of heart failure resulting in part from inadequate healing of their original infarct. Heart failure is a major health problem that has reached almost epidemic proportions in the U.S., in that it affects 2% of the American

---

M. Syriga • M. Mavroidis, Ph.D. (✉)
Center of Basic Research, Biomedical Research Foundation, Academy of Athens, Athens 11527, Greece
e-mail: emavroeid@bioacademy.gr

J.D. Lambris et al. (eds.), *Complement Therapeutics*, Advances in Experimental Medicine and Biology 735,
DOI 10.1007/978-1-4614-4118-2_14,

population. It is also an economic problem: The cost of hospitalizations related to heart failure is currently twice that for all forms of cancer and myocardial infarctions combined. Therefore, a major goal in current cardiology practice is to determine optimal strategies for minimizing myocardial necrosis and optimizing cardiac repair following acute myocardial infarctions.

Myocardial infarction results in decreased oxygen tension within cardiomyocytes and a subsequent loss of oxidative phosphorylation and decreased generation of high-energy phosphates. The mammalian heart cannot produce enough energy under anaerobic conditions to sustain cardiac function and viability, and therefore, irreversible cardiomyocyte injury develops after 40–60 min of sustained severe ischemia (Jennings et al. 1990). The predominant mechanism of cardiomyocyte death in the infarcted heart is coagulation necrosis, although apoptosis is also likely to contribute to cardiomyocyte loss. Dying cells activate innate immune mechanisms, initiating an inflammatory reaction that has a dual effect, primarily repairing the trauma but also producing adverse ventricular remodeling, with detrimental consequences if activation is enhanced or uncontrolled. Innate immunity is a primordial system that was evolved in the lower phyla as a host reaction system primarily for repairing cutaneous trauma. The critical goals of this reaction system were to prevent an infection; dispose of apoptotic, necrotic cells and damaged tissue; and, in these and other ways, facilitate wound repair. The size of the eventual scar was not the key issue. On the other hand, scar tissue is largely acellular and lacks the normal biochemical properties of the host cells, and in cardiac tissue, this situation leads to electrical uncoupling, mechanical dysfunction, and loss of structural integrity. A compensatory hypertrophy of the noninfarcted area, accompanied by chamber dilation and upregulation of fetal gene expression, also develops at a later point, and ultimately, these changes lead to the development of heart failure (Jennings et al. 1995). We have put forward the hypothesis that temporally regulated activation and suppression of innate immunity may be critical for minimizing cardiac tissue injury and achieving effective cardiac repair and regeneration (Jiang and Liao 2010).

14.1.2 Complement System Activation in Cardiac Ischemia/Reperfusion

The complement system is an important arm of innate immunity whose essential role in ischemia/reperfusion damage was first described by Hill and Ward in the early 1970s (Hill and Ward 1971). After ischemia and reperfusion, myocardial levels of C3 and C9 have been shown to exceed the levels of production of these complement components in the liver in a rabbit model (Yasojima et al. 1998). Extensive experimental data from a variety of animal models have indicated that inhibiting complement activation offers protection to cardiac tissue after ischemia/reperfusion [for a review, see Diepenhorst et al. (2009)]. In patients with acute MI, however, the results of clinical studies using complement inhibitors (four studies of more than 10,500 patients using pexelizumab, an anti-C5 antibody, and studies using C1 esterase inhibitor or soluble complement receptor 1) have largely been disappointing [for a review, see Oksjoki et al. (2007) and Diepenhorst et al. (2009)]. Over the past 30 years, the use of specific strategies that modulate the inflammatory response have led to dramatic reductions in infarct size and attenuation of adverse remodeling in numerous experimental animal models [reviewed in Frangogiannis (2008)]. Nevertheless, attempts to mitigate the inflammatory reaction in human patients after MI have yielded disappointing results. At present, no specific immunomodulatory approach is used in patients after MI. Established therapeutic strategies such as β-adrenergic blockade, ACE inhibition, and anticoagulation treatment may exert their beneficial effects in part by interfering with the inflammatory cascade (Kilgore et al. 1994; Chen et al. 1998).

Why have anti-inflammatory strategies that yield very promising results in animal models of cardiac tissue injury been unsuccessful in human patients? Although fundamental differences between animal models and their respective human diseases may provide an explanation in some cases, what is becoming evident from the studies conducted thus far is that a deeper understanding of the biology of this complex network of molecular and cellular interactions is necessary before any specific interventions can be pursued in therapeutic trials. In addition, in the clinical reality, variables such as age;

the presence of comorbid conditions such as diabetes, obesity, and hyperlipidemia; the timing of reperfusion; and genetic variations between individuals greatly complicate the prediction of the potential effects of a therapeutic intervention. Finally, in the various animal models of MI, the outcome of the cardiac injury may also differ according to various conditions, such as whether the MI is small or large, whether a small or large amount of myocardial regeneration occurs, and whether the ischemia is permanent or transient. If the duration of the ischemia is limited (e.g., 30 min, as in many of the MI/reperfusion animal models used to assess the effects of complement system inhibition), the main determinants in the outcome of the cardiac injury are the extent of the initial injury of the vessel endothelium after the reperfusion and the degree of activation of the complement system. Complement activation stimulates neutrophils by inducing chemotactic migration, aggregation (Crawford et al. 1988), and the release of cytotoxic products such as proteases, elastases, and reactive oxygen species that can compromise tissue integrity (Engler 1987; Hori and Nishida 2009). Neutrophils activated by C5a or by iC3b deposited within vessel walls (Hirahashi et al. 2006) can also contribute to microvascular obstruction by promoting fibrin deposition and thrombosis; thus, the complement system can contribute to a vicious cycle of vasoconstriction, microvascular hypoperfusion, and cell death (Saraste et al. 1997; Rezkalla and Kloner 2002) that further compromises cardiac tissue integrity. Thus, in the case of ischemia of limited duration with subsequent reperfusion, there is wide evidence that oxygen-derived free radicals, neutrophils, and activated complement components are crucial players in the destruction of cardiac tissue (Kilgore et al. 1994; Diepenhorst et al. 2009). If it were possible to restore the oxygen supply to cardiomyocytes without inciting the harmful acute inflammatory reaction, then the restoration of cardiac function could be maximized.

In longer-lasting ischemia, when irreversible cardiomyocyte injury develops as a direct result of oxygen deprivation, the extent of cardiomyocyte death determines the outcome of the cardiac dysfunction. Damaged and destroyed cardiomyocytes in the ischemic area can be roughly divided into two categories: (a) cardiomyocytes under severe and extended stress whose death is inevitable and (b) those that, while under stress, will eventually manage to survive through the upregulation of cytoprotective mechanisms. The intrinsic cytoprotective mechanisms of cardiomyocytes can involve the upregulation of small heat shock proteins, antioxidant enzymes, antiapoptotic molecules, and/or a fetal gene expression program. As previously mentioned, dying cardiomyocytes can trigger an inflammatory reaction, activating reparative pathways that ultimately result in the formation of a scar (Frangogiannis 2008). Enhanced or uncontrolled activation can lead to adverse ventricular remodeling, with detrimental consequences. Timely resolution of the inflammatory infiltrate and spatial containment of the inflammatory and reparative response to the infarcted area are essential for optimal infarct healing.

What is the role of complement activation in this milieu of interactions among secreted factors, resident cells, and cells that have infiltrated into the site of injury? Although the complement system often contributes to the tissue damage described in the previous paragraph, it is not easy to separate this process from the beneficial effects of complement with regard to waste disposal, recruitment of stem cells (Ratajczak et al. 2006), regeneration (Markiewski et al. 2009), and angiogenesis (Nozaki et al. 2006), as has been recognized in other tissue injury models. The ideal approach is likely to be modulating rather than abruptly blunting complement activation.

14.1.3 Complement Activation as a "Friend" in Cardiac Tissue Injury

Complement activation leads to the deposition of iC3b at the site of necrotic cardiomyocytes; the iC3b then binds to complement receptors CR3 and CR4 and mediates phagocytosis, thereby participating in the removal of cell debris and containment of the inflammatory reaction. CR3/4-mediated phagocytosis, per se, does not elicit proinflammatory signals in phagocytes nor does it provoke a respiratory

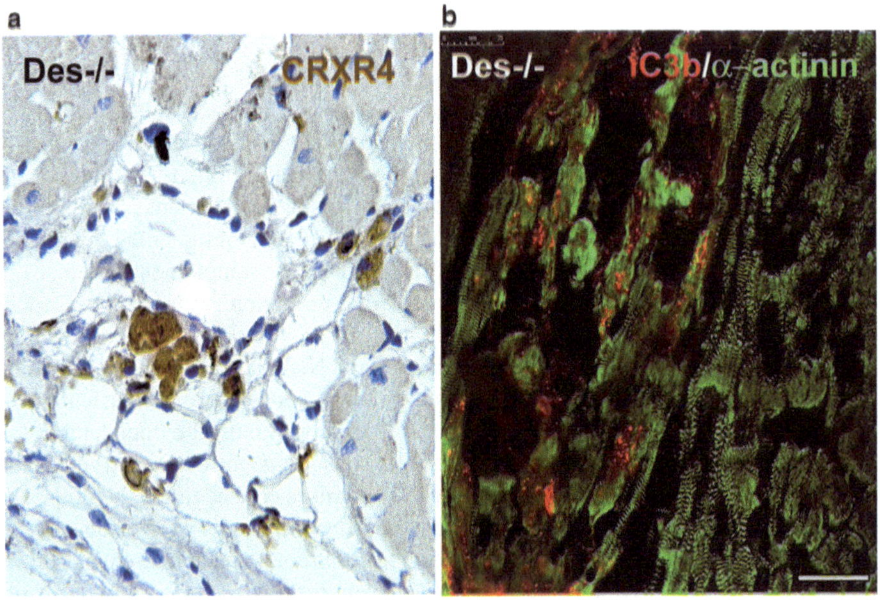

Fig. 14.1 Activated complement components and CRXR4-positive cells are present in the myocardium of the desmin-null cardiac tissue injury/heart failure model. (**a**) CRXR4-positive cells are present in areas of cardiac tissue injury and neovascularization in the desmin-null myocardium. (**b**) Extended iC3b deposition (*red*) is observed in the desmin null cardiac tissue, in areas of cardiomyocyte degeneration. Actinin (*green*) is a cardiomyocyte marker (bar = 25 μm). Extended cardiomyocyte injury and acute inflammatory reaction are observed in the desmin-null mice (Mavroidis and Capetanaki 2002), together with a 7.5-fold upregulation of CR3 RNA levels (Psarras et al. 2011)

burst. Considering that CR3 has primarily evolved to support the physiologic functions necessary to balance tissue homeostasis, i.e., through clearance of apoptotic cells, the lack of inflammation is not surprising. In order to promote leukocyte adhesion and transmigration through the endothelium as well as inflammatory responses, C3 needs to be preactivated to a high-affinity state in order for ligation to result in an inflammatory response. Signaling pathways that induce this high-affinity status include those downstream of activating immunoglobulin G receptors (FcγRs) and the chemoattractant G-protein-coupled receptors (GPCR).

Secreted inflammatory mediators play an important role in mobilizing progenitor cells and may regulate their homing to the infarcted myocardium (Vandervelde et al. 2005). Most of the positive effects reported thus far for cell-based therapy have been attributed to paracrine effects rather than to the direct differentiation of engrafted cells to cardiomyocytes (Choi et al. 2011). Thus, although it has not yet been clearly established for cardiac tissue injury, C3 cleavage fragments may exert a beneficial effect by participating in the chemoattraction and tethering of circulating CXCR4+ stem cells to injured cardiac tissue, as has been reported in other tissue injury models (Ratajczak et al. 2006). All of the players (activated complement components, CRXR4-positive cells, SDF1) in this interaction are present in the injured myocardium ((Askari et al. 2003; Abbott et al. 2004); see also Fig. 14.1).

The prorepair role of complement is apparent in liver regeneration, in which anaphylatoxin-induced IL-6 and TNF prosurvival signaling promote hepatocyte growth and proliferation, which are dramatically impaired in C3–/–, C3ar–/–, and C5ar–/– mice (Markiewski et al. 2009). The role of TNF-α in myocardial ischemia/reperfusion injury is ambivalent. Excessive TNF-α expression and subsequent stimulation of cardiomyocyte TNF receptor type 1 induce contractile dysfunction, hypertrophy, fibrosis, and cell death, whereas lower TNF-α concentrations and subsequent stimulation of cardiomyocyte TNF receptor type 2 are protective (Schulz and Heusch 2009). TNF-α can increase protective signals such as the wingless-type integration site family member 1 (WNT1) in cardiomyocytes and cardiac

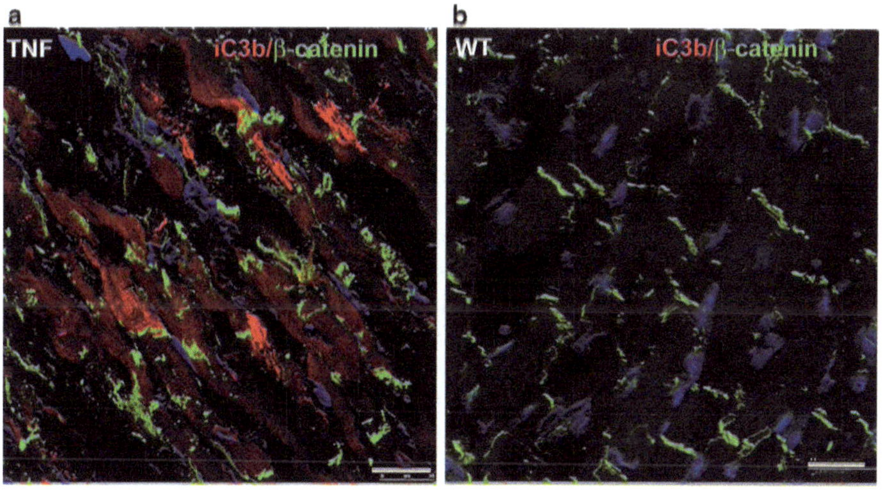

Fig. 14.2 (**a**) Activated C3 (*red*) is seen in aggregated form and perinuclearly in the cardiomyocytes of cardiac-specific tumor necrosis factor (TNF-α)-overexpressing transgenic mice. In TNF-α-overexpressing mice there is a mislocalization of the intercalated disc components to the lateral sarcolemma of the cardiomyocytes, as shown by β-catenin (*green*) staining. (**b**) A cardiac tissue section of wild-type mice is negative for activated C3 staining, and β-catenin is mainly localized to the bipolar ends of the rod-shaped cardiomyocytes. Frozen cardiac tissue sections; *blue* denotes nuclear staining with DAPI, and the bar = 25 μm. TNF-α-overexpressing mice are from Li et al. (2000)

fibroblasts (Venkatachalam et al. 2009). It has been shown in in vitro experiments that complement activation can induce TNF-α synthesis in cardiac myocytes (Zwaka et al. 2002), so a reasonable hypothesis could be advanced that, depending on the timing, the concentration, and the cardiac injury model chosen, induction of TNF-α by C3 activation could be protective. In liver injury, it has been shown that there is a threshold of complement activation for optimal liver regeneration (He et al. 2009), and this threshold is associated with increased early hepatic production of IL-6 and TNF-α and diminished systemic levels of the inflammatory cytokines by 48 hours after partial hepatectomy.

On the other hand, in TNF-α-overexpressing mice, analysis of gene expression profiling during the transition to heart failure (Tang et al. 2004) has demonstrated the development of autoimmune myocarditis, with C3 and MHC class II antigen expression being strongly upregulated (see Fig. 14.2). Given that complement activation is critical for the induction of experimental autoimmune myocarditis (Kaya et al. 2001; Eriksson et al. 2003), depletion of C3 may be essential to preventing TNF-α-induced heart failure.

14.1.4 A Potential Role for the Connection of Complement to Fatty Acid and Glucose Metabolism in the Adaptation of the Heart to Stress

A common feature of a variety of cardiac pathophysiological conditions, including hypoxia, ischemia, and hypertrophy, is a return to a pattern of fetal metabolism. This adaptation is associated with a whole program of cell survival under stress, and there is evidence that, at the same level of stress, survival is greater in the fetal-adapted heart than in the adult heart (Rajabi et al. 2007). A hallmark of fetal metabolism is the predominance of carbohydrates as substrates for energy provision. Under normal conditions, the preferred substrate for the heart is fatty acids (80–100%). During ischemia/reperfusion, the diminished O_2 supply for respiration and oxidative phosphorylation leads to a rapid decline in ATP levels and stimulates anaerobic ATP generation, with an increase in glycolysis and lactate production (Oram et al. 1973; Neely and Morgan 1974). Although the fetal-adapted heart can survive better than the adult under hypoxic stress, this adaptation cannot meet the high cardiac pumping workload (and, consequently, high energy demand) of an adult organism. At a certain point, the fetal gene program is no longer sufficient to support cardiac structure and function, and the heart succumbs to maladaptation, cell death, and ultimately organ failure (Rajabi et al. 2007).

An emerging role of complement system in the regulation of lipid metabolism has become apparent in the last decade. Acylation-stimulating protein (ASP), which has been identified as C3adesArg (produced after cleavage of the COOH-terminal arginine of C3a by carboxypeptidase B), has lipogenic activity. In adipocytes, ASP interacts with its cell-surface receptor C5L2, resulting in increased nonesterified fatty acid (NEFA) uptake and triglyceride synthesis (MacLaren et al. 2008). Conversely, in muscle, ASP has the opposite effect and increases lipolysis, indicating a reduction in NEFA trapping within muscle (Faraj and Cianflone 2004). ASP also increases glucose uptake in a number of cell models, including human adipocytes and L6 myotubes, via the translocation to the cell surface of the glucose transporters GLUT4 and GLUT3 (in myotubes), independently of, but additively with, insulin. Thus, the generation of C3adesArg at the site of cardiac tissue injury through complement activation could offer a potential means of improving the energy utilization of glucose and lipids when under hypoxic stress.

14.2 The Role of Complement in Skeletal Muscle

14.2.1 Skeletal Muscle Regeneration

Skeletal muscle is the most abundant tissue in the human body, contributing to motion, protein and carbohydrate storage, and heat production. Given its superficial body distribution, it is constantly exposed to mechanical insults caused by muscle contraction and to external physical trauma. As a consequence of the significance of its role and its continuous exposure to mechanical stress, skeletal muscle has developed the ability to regenerate and restore its original tissue architecture and functional integrity after damage. Acute muscle injury and genetic deficiencies both lead to compromised muscle function, along with rupture, myofibrillar necrosis, and inflammation. The degenerative phase is followed by tissue recovery and repair, which are characterized by the removal of necrotized tissue and regeneration of the damaged myofibers (Tidball and Villalta 2010). This phase depends on a population of progenitor myogenic cells, the satellite cells, which are located in the basal lamina of the muscle fibers. A complex repertoire of cytokines, growth factors, and extracellular molecules (ECM) triggers the activation, proliferation, and differentiation of satellite cells that result in the subsequent fusion and formation of new, intact myofibers (Corti et al. 2001). A key role in this process is performed by the tightly regulated interplay between the injured skeletal muscle and the immune system (Tidball 2005).

Immediately after damage, neutrophils invade the muscle and promote the progress of the inflammatory reaction in the damaged area. The subsequent accumulation of macrophages continues the proinflammatory phase of muscle injury (Nguyen and Tidball 2003). The M1 phenotype dominates the initial macrophage population, with the release of proinflammatory cytokines such as TNF-α and IL-6 (Collins and Grounds 2001). TNF-α and the rest of the Th1 cytokines produced in this phase contribute to muscle damage, but at the same time they also promote regeneration and repair by triggering the activation, proliferation, and early differentiation stages of satellite muscle cells (Collins and Grounds 2001; Villalta et al. 2009). The M1 phenotype is replaced by infiltrating M2 macrophages, which attenuate the previous proinflammatory reaction and release anti-inflammatory cytokines such as IL-10, IL-4, and IL-13, thus promoting tissue growth and regeneration (Tidball and Wehling-Henricks 2007).

14.2.2 Complement Activation in Skeletal Muscle

Initial studies have revealed the biosynthesis of complement by skeletal muscle cells in vitro. The alternative pathway proteins C3, factor B, and factor H as well as the classical pathway components C1, C2, and C4 have been shown to be produced by human myoblast cell lines (Legoedec et al. 1995, 1997). Despite the activation of autologous complement, muscle fibers are protected from destruction and cell lysis by their abundant expression of the regulatory molecules MCP, CD59, and C4BP (Gasque et al. 1996). Complement synthesis by muscle cells is upregulated by inflammatory cytokines, including INF-γ and IL-1 (Legoedec et al. 1995), suggesting a potent role for the local production and activation of complement under inflammatory pathophysiological conditions in skeletal muscle. A series of studies have pinpointed the involvement of the complement cascade in the complete reconstruction and regeneration of body parts such as limbs and eye tissue in several species of urodele amphibians (Kimura et al. 2003). The central proteins C3 and C5 of the complement cascade are specifically expressed at both the transcriptional and translational (protein) levels in the areas responsible for the repopulation and reformation of the old, amputated (or injured) structures (i.e., the urodele blastema) into the appropriate differentiated, newly reconstructed body parts.

An important question is how complement proteins are related to the molecules responsible for the establishment of cellular adhesion and communication. C3 interacts with molecules of the extracellular matrix, such as laminin and fibronectin (Leivo and Engvall 1986), and with receptors of the integrin family (Law et al. 1987); factor B and C2 share homologous domains with cartilage matrix protein, von Willebrand factor, and the collagen-binding domain of alkaline phosphatase. Given the upregulation of these molecules during the synthesis of new cell membranes in the regeneration phase (Tsonis et al. 1996), these interactions strongly suggest that complement participates in the progress of remodeling and the differentiation of the regenerating parts. Recent reports have highlighted the important role of complement in removing cellular debris during successful skeletal muscle remodeling after chemical or ischemic injury in stem cell antigen 1 (Sca1)-deficient mice (Long et al. 2011). Recruitment of IgM and complement C3 has been shown to be essential for tissue regeneration since the inability of damaged cells to be cleared from C3-deficient mice leads to a profound fibrotic condition (Mevorach et al. 1998; Markiewski et al. 2004).

14.2.3 Complement-Mediated Inflammatory Muscle Diseases

Inflammatory myopathies are a heterogeneous class of autoimmune muscle disorders that are characterized by the production of specific autoantibodies and immune attack against muscle antigens, with subsequent tissue malfunction. As an important link between innate and adaptive immunity, complement plays a key role in the pathogenesis of serious autoimmune diseases such as myasthenia gravis (Vincent and Drachman 2002) and the inflammatory myopathies dermatomyositis and juvenile dermatomyositis (Dalakas 2010b). In addition, complement contributes to the pathogenesis of a class of inflammation-associated muscle disorders, the dysferlinopathies (Liu et al. 1998), in which a defective repair of the plasma membrane produces constant leakage of cytoplasmic contents to the extracellular environment.

The immune system recruits complement in the course of many other pathologic muscle conditions, thus preserving the inflammatory "waves" in the local area of the affected skeletal muscle and leading to either progressive or acute tissue destruction, as in the case of muscle ischemia/reperfusion injury.

14.2.3.1 Complement in Myasthenia Gravis

Myasthenia gravis (from Greek and Latin, literally meaning "grave muscle weakness") is a chronic autoimmune disease that is characterized by the production of autoantibodies that target the acetylcholine receptor (AchR) at the neuromuscular junction (NMJ) (Vincent and Drachman 2002). Studies conducted over the past few decades have shown that complement is a significant participant in the destructive pathway that leads to neuromuscular disruption (Lennon et al. 1978; Tuzun et al. 2003). Initial evidence has revealed a reduction in circulating serum levels of several complement proteins in myasthenia gravis patients, in parallel with the deposition and colocalization of IgG, C3, and C9 on the NMJ membrane (Sahashi et al. 1980). In agreement with the clinical evidence, in mice with experimental autoimmune myasthenia gravis (EAMG), IgG and complement proteins are detected at the degenerating NMJs (Tuzun et al. 2004). Studies including the manipulation of complement activation in EAMG animals, either through a genetic deficiency or pharmacologic blockade, have confirmed a role for complement in the disease mechanism. C3 and C4 knockout mice immunized with the AchR show an improved phenotype when compared to wild-type controls (Tuzun et al. 2007; Zhou et al. 2007; Soltys et al. 2009). However, examination of the disease progress in C5aR knockout mice has revealed that the incidence in EAMG mice is similar to that of wild-type mice, suggesting that the anaphylatoxin C5a is not involved in the pathogenesis of myasthenia gravis; instead, C5 plays a role in pathogenesis in the form of C5b, which contributes to the subsequent formation of the membrane attack complex (MAC) (Chamberlain-Banoub et al. 2006; Qi et al. 2008). Novel therapeutic reagents specifically targeting the activation of the classical pathway and preventing MAC formation at the NMJs, such as monoclonal antibodies or even complement inhibitors conjugated to anti-AchR antibody fragments (Spitzer et al. 2004), could potentially eliminate myasthenia gravis symptoms by preventing complement activation specifically at the site of pathology rather than systemically.

14.2.3.2 Complement in Dermatomyositis (DM)

Inflammatory myopathies comprise a heterogeneous group of degenerative muscle diseases with the common characteristics of mild to severe muscle weakness and the presence of inflammatory infiltrates in muscle biopsies. This group consists of four clinically distinct conditions: dermatomyositis, polymyositis, necrotizing autoimmune myositis, and sporadic inclusion body myositis (Dalakas 2010b). Although their general phenotypic characteristics are similar, the pathophysiological mechanisms of these diseases differ and involve the recruitment of different branches of the immune system. Complement plays a significant role in the pathogenesis of dermatomyositis, and therefore complement activity is considered a useful diagnostic measure and a strong candidate as a future therapeutic target.

Dermatomyositis occurs in both adults and children (juvenile dermatomyositis) and predominantly affects the skin and muscles. It is characterized by a humorally mediated autoimmune attack against the vascular endothelium of the capillaries. Deposited autoantibodies activate the complement cascade, with the subsequent formation of the MAC complex (Kissel et al. 1986; Emslie-Smith and Engel 1990). Immune attack and complement activation lead to extended capillary damage and necrosis, resulting in an insufficient blood supply to the muscle. The detection of MAC offers high sensitivity and specificity for discriminating dermatomyositis from other inflammatory myopathies at both the early and advanced stages (Jain et al. 2011). The future unraveling of complement-mediated destruction and the related cytokine milieu can potentially give us the ability to design targeted therapeutic strategies. Monoclonal antibodies against immune components, including complement (Basta and Dalakas 1994), cytokines, T- and B-cell activating factors, and adhesion molecules, are promising candidates and have proven effective in some cases of dermatomyositis (Dalakas 2010a).

14.2.3.3 Complement in Dysferlin Myopathy

Dysferlinopathies comprise a large group of autosomal recessive myopathies caused by the complete or partial absence of dysferlin. Dysferlin is a 230-kDa protein that acts as a key player in the membrane repair system (Cenacchi et al. 2005). Membrane tears cause an increase in Ca^{2+} influx, which triggers the accumulation of dysferlin-containing vesicles and fusion to the plasma membrane at the site of the membrane disruption (Lennon et al. 2003). Dysferlin mutations give rise to three types of autosomal recessive muscle wasting diseases: limb-girdle muscular dystrophy type 2B, Miyoshi myopathy, and a distal anterior compartment myopathy (Liu et al. 1998).

The immune system plays a complex role in the pathogenesis of dysferlinopathy. Muscle biopsies of dysferlin-deficient patients exhibit a prominent inflammatory infiltration (Gallardo et al. 2001), while the lack of dysferlin in the SJL/J mouse strain results in an altered behavior of immune cells, with an increase in phagocytic activity and upregulation of the inflammasome molecular pathway (Rawat et al. 2010). Complement activation with subsequent MAC deposition on the surface of non-necrotic muscle fibers has been detected on both human and murine dysferlin-deficient skeletal muscles, suggesting a potential role in the pathology (Spuler and Engel 1998). Microarray data have indicated the upregulation of complement proteins and downregulation of complement regulators (Suzuki et al. 2005). In accordance with these microarray results, decay-accelerating factor (DAF) has been found to be eliminated at both the RNA and protein levels in skeletal muscle from dysferlin-deficient mice and patients diagnosed with dysferlinopathy (Wenzel et al. 2005), and this loss of DAF has been shown to result in an increased susceptibility to complement attack. Further investigation has revealed that disruption of the C3 gene in dysferlin-deficient skeletal muscle attenuates muscle pathology by reducing inflammatory infiltration, central nucleation, fibrosis, and fat replacement (Han et al. 2010). Interestingly, ablation of the terminal complement component C5 has only a minimal effect on muscle pathology in this mouse model, suggesting that it is not the terminal activation of the complement system but the activation of C3 that accelerates muscle injury.

In summary, dysferlinopathy causes a strong activation of the complement cascade with a simultaneous attenuation of regulatory control, making the muscle fibers more susceptible to complement attack. The additional involvement of dysferlin in cytokine release is another aspect of the compromised muscle remodeling and healing. Based on these findings, an inhibition of the complement cascade at the C3 level could offer potential therapeutic options for this currently untreatable muscle wasting disorder.

Acknowledgments We thank Dr. Yassemi Capetanaki for the continuous support, Dr. Douglas L. Mann for providing the TNF-α overexpressing mice, and Dr. Deborah McClellan for the editorial assistance.

References

Abbott JD, Huang Y et al (2004) Stromal cell-derived factor-1alpha plays a critical role in stem cell recruitment to the heart after myocardial infarction but is not sufficient to induce homing in the absence of injury. Circulation 110(21):3300–3305

Askari AT, Unzek S et al (2003) Effect of stromal-cell-derived factor 1 on stem-cell homing and tissue regeneration in ischaemic cardiomyopathy. Lancet 362(9385):697–703

Basta M, Dalakas MC (1994) High-dose intravenous immunoglobulin exerts its beneficial effect in patients with dermatomyositis by blocking endomysial deposition of activated complement fragments. J Clin Invest 94(5): 1729–1735

Cenacchi G, Fanin M et al (2005) Ultrastructural changes in dysferlinopathy support defective membrane repair mechanism. J Clin Pathol 58(2):190–195

Chamberlain-Banoub J, Neal JW et al (2006) Complement membrane attack is required for endplate damage and clinical disease in passive experimental myasthenia gravis in Lewis rats. Clin Exp Immunol 146(2):278–286
Chen XL, Tummala PE et al (1998) Angiotensin II induces monocyte chemoattractant protein-1 gene expression in rat vascular smooth muscle cells. Circ Res 83(9):952–959
Choi YH, Kurtz A et al (2011) Mesenchymal stem cells for cardiac cell therapy. Hum Gene Ther 22(1):3–17
Collins RA, Grounds MD (2001) The role of tumor necrosis factor-alpha (TNF-alpha) in skeletal muscle regeneration. Studies in TNF-alpha(−/−) and TNF-alpha(−/−)/LT-alpha(−/−) mice. J Histochem Cytochem 49(8):989–1001
Corti S, Salani S et al (2001) Chemotactic factors enhance myogenic cell migration across an endothelial monolayer. Exp Cell Res 268(1):36–44
Crawford MH, Grover FL et al (1988) Complement and neutrophil activation in the pathogenesis of ischemic myocardial injury. Circulation 78(6):1449–1458
Dalakas MC (2010a) Immunotherapy of myositis: issues, concerns and future prospects. Nat Rev Rheumatol 6(3):129–137
Dalakas MC (2010b) Inflammatory muscle diseases: a critical review on pathogenesis and therapies. Curr Opin Pharmacol 10(3):346–352
Diepenhorst GM, van Gulik TM et al (2009) Complement-mediated ischemia-reperfusion injury: lessons learned from animal and clinical studies. Ann Surg 249(6):889–899
Emslie-Smith AM, Engel AG (1990) Microvascular changes in early and advanced dermatomyositis: a quantitative study. Ann Neurol 27(4):343–356
Engler R (1987) Consequences of activation and adenosine-mediated inhibition of granulocytes during myocardial ischemia. Fed Proc 46(7):2407–2412
Eriksson U, Kurrer MO et al (2003) Interleukin-6-deficient mice resist development of autoimmune myocarditis associated with impaired upregulation of complement C3. Circulation 107(2):320–325
Faraj M, Cianflone K (2004) Differential regulation of fatty acid trapping in mouse adipose tissue and muscle by ASP. Am J Physiol Endocrinol Metab 287(1):E150–E159
Frangogiannis NG (2008) The immune system and cardiac repair. Pharmacol Res 58(2):88–111
Gallardo E, Rojas-Garcia R et al (2001) Inflammation in dysferlin myopathy: immunohistochemical characterization of 13 patients. Neurology 57(11):2136–2138
Gasque P, Morgan BP et al (1996) Human skeletal myoblasts spontaneously activate allogeneic complement but are resistant to killing. J Immunol 156(9):3402–3411
Han R, Frett EM et al (2010) Genetic ablation of complement C3 attenuates muscle pathology in dysferlin-deficient mice. J Clin Invest 120(12):4366–4374
He S, Atkinson C et al (2009) A complement-dependent balance between hepatic ischemia/reperfusion injury and liver regeneration in mice. J Clin Invest 119(8):2304–2316
Hill JH, Ward PA (1971) The phlogistic role of C3 leukotactic fragments in myocardial infarcts of rats. J Exp Med 133(4):885–900
Hirahashi J, Mekala D et al (2006) Mac-1 signaling via Src-family and Syk kinases results in elastase-dependent thrombohemorrhagic vasculopathy. Immunity 25(2):271–283
Hori M, Nishida K (2009) Oxidative stress and left ventricular remodelling after myocardial infarction. Cardiovasc Res 81(3):457–464
Jain A, Sharma MC et al (2011) Detection of the membrane attack complex as a diagnostic tool in dermatomyositis. Acta Neurol Scand 123(2):122–129
Jennings RB, Murry CE et al (1990) Development of cell injury in sustained acute ischemia. Circulation 82(3 Suppl):II2–II12
Jennings RB, Steenbergen C Jr et al (1995) Myocardial ischemia and reperfusion. Monogr Pathol 37:47–80
Jiang B, Liao R (2010) The paradoxical role of inflammation in cardiac repair and regeneration. J Cardiovasc Transl Res 3(4):410–416
Kaya Z, Afanasyeva M et al (2001) Contribution of the innate immune system to autoimmune myocarditis: a role for complement. Nat Immunol 2(8):739–745
Kilgore KS, Homeister JW et al (1994) Sulfhydryl compounds, captopril, and MPG inhibit complement-mediated myocardial injury. Am J Physiol 266(1 Pt 2):H28–H35
Kimura Y, Madhavan M et al (2003) Expression of complement 3 and complement 5 in newt limb and lens regeneration. J Immunol 170(5):2331–2339
Kissel JT, Mendell JR et al (1986) Microvascular deposition of complement membrane attack complex in dermatomyositis. N Engl J Med 314(6):329–334
Law SK, Gagnon J et al (1987) The primary structure of the beta-subunit of the cell surface adhesion glycoproteins LFA-1, CR3 and p150,95 and its relationship to the fibronectin receptor. EMBO J 6(4):915–919
Legoedec J, Gasque P et al (1995) Expression of the complement alternative pathway by human myoblasts in vitro: biosynthesis of C3, factor B, factor H and factor I. Eur J Immunol 25(12):3460–3466

Legoedec J, Gasque P et al (1997) Complement classical pathway expression by human skeletal myoblasts in vitro. Mol Immunol 34(10):735–741

Leivo I, Engvall E (1986) C3d fragment of complement interacts with laminin and binds to basement membranes of glomerulus and trophoblast. J Cell Biol 103(3):1091–1100

Lennon VA, Seybold ME et al (1978) Role of complement in the pathogenesis of experimental autoimmune myasthenia gravis. J Exp Med 147(4):973–983

Lennon NJ, Kho A et al (2003) Dysferlin interacts with annexins A1 and A2 and mediates sarcolemmal wound-healing. J Biol Chem 278(50):50466–50473

Li X, Moody MR et al (2000) Cardiac-specific overexpression of tumor necrosis factor-alpha causes oxidative stress and contractile dysfunction in mouse diaphragm. Circulation 102(14):1690–1696

Liu J, Aoki M et al (1998) Dysferlin, a novel skeletal muscle gene, is mutated in Miyoshi myopathy and limb girdle muscular dystrophy. Nat Genet 20(1):31–36

Lloyd-Jones D, Adams R et al (2009) Heart disease and stroke statistics–2009 update: a report from the American Heart Association Statistics Committee and Stroke Statistics Subcommittee. Circulation 119(3):480–486

Long KK, Pavlath GK et al (2011) Sca-1 influences the innate immune response during skeletal muscle regeneration. Am J Physiol Cell Physiol 300(2):C287–C294

MacLaren R, Cui W et al (2008) Adipokines and the immune system: an adipocentric view. Adv Exp Med Biol 632:1–21

Markiewski MM, Mastellos D et al (2004) C3a and C3b activation products of the third component of complement (C3) are critical for normal liver recovery after toxic injury. J Immunol 173(2):747–754

Markiewski MM, DeAngelis RA et al (2009) The regulation of liver cell survival by complement. J Immunol 182(9):5412–5418

Mavroidis M, Capetanaki Y (2002) Extensive induction of important mediators of fibrosis and dystrophic calcification in desmin-deficient cardiomyopathy. Am J Pathol 160(3):943–952

Mevorach D, Mascarenhas JO et al (1998) Complement-dependent clearance of apoptotic cells by human macrophages. J Exp Med 188(12):2313–2320

Neely JR, Morgan HE (1974) Relationship between carbohydrate and lipid metabolism and the energy balance of heart muscle. Annu Rev Physiol 36:413–459

Nguyen HX, Tidball JG (2003) Interactions between neutrophils and macrophages promote macrophage killing of rat muscle cells in vitro. J Physiol 547(Pt 1):125–132

Nozaki M, Raisler BJ et al (2006) Drusen complement components C3a and C5a promote choroidal neovascularization. Proc Natl Acad Sci USA 103(7):2328–2333

Oksjoki R, Kovanen PT et al (2007) Function and regulation of the complement system in cardiovascular diseases. Front Biosci 12:4696–4708

Oram JF, Bennetch SL et al (1973) Regulation of fatty acid utilization in isolated perfused rat hearts. J Biol Chem 248(15):5299–5309

Psarras S, Mavroidis M et al (2011) Regulation of adverse remodelling by osteopontin in a genetic heart failure model. Eur Heart J [Epub ahead of print]. doi:10.1093/eurheart/ehr119

Qi H, Tuzun E et al (2008) C5a is not involved in experimental autoimmune myasthenia gravis pathogenesis. J Neuroimmunol 196(1–2):101–106

Rajabi M, Kassiotis C et al (2007) Return to the fetal gene program protects the stressed heart: a strong hypothesis. Heart Fail Rev 12(3–4):331–343

Ratajczak MZ, Reca R et al (2006) Modulation of the SDF-1-CXCR4 axis by the third complement component (C3) – implications for trafficking of CXCR4+ stem cells. Exp Hematol 34(8):986–995

Rawat R, Cohen TV et al (2010) Inflammasome up-regulation and activation in dysferlin-deficient skeletal muscle. Am J Pathol 176(6):2891–2900

Rezkalla SH, Kloner RA (2002) No-reflow phenomenon. Circulation 105(5):656–662

Sahashi K, Engel AG et al (1980) Ultrastructural localization of the terminal and lytic ninth complement component (C9) at the motor end-plate in myasthenia gravis. J Neuropathol Exp Neurol 39(2):160–172

Saraste A, Pulkki K et al (1997) Apoptosis in human acute myocardial infarction. Circulation 95(2):320–323

Schulz R, Heusch G (2009) Tumor necrosis factor-alpha and its receptors 1 and 2: Yin and Yang in myocardial infarction? Circulation 119(10):1355–1357

Soltys J, Kusner LL et al (2009) Novel complement inhibitor limits severity of experimentally myasthenia gravis. Ann Neurol 65(1):67–75

Spitzer D, Unsinger J et al (2004) ScFv-mediated in vivo targeting of DAF to erythrocytes inhibits lysis by complement. Mol Immunol 40(13):911–919

Spuler S, Engel AG (1998) Unexpected sarcolemmal complement membrane attack complex deposits on nonnecrotic muscle fibers in muscular dystrophies. Neurology 50(1):41–46

Suzuki N, Aoki M et al (2005) Expression profiling with progression of dystrophic change in dysferlin-deficient mice (SJL). Neurosci Res 52(1):47–60

Tang Z, McGowan BS et al (2004) Gene expression profiling during the transition to failure in TNF-alpha over-expressing mice demonstrates the development of autoimmune myocarditis. J Mol Cell Cardiol 36(4):515–530
Tidball JG (2005) Inflammatory processes in muscle injury and repair. Am J Physiol Regul Integr Comp Physiol 288(2):R345–R353
Tidball JG, Villalta SA (2010) Regulatory interactions between muscle and the immune system during muscle regeneration. Am J Physiol Regul Integr Comp Physiol 298(5):R1173–R1187
Tidball JG, Wehling-Henricks M (2007) Macrophages promote muscle membrane repair and muscle fibre growth and regeneration during modified muscle loading in mice in vivo. J Physiol 578(Pt 1):327–336
Tsonis PA, Del Rio-Tsonis K et al (1996) Can insights into urodele limb regeneration be achieved with cell cultures and retroviruses? Int J Dev Biol 40(4):813–816
Tuzun E, Scott BG et al (2003) Genetic evidence for involvement of classical complement pathway in induction of experimental autoimmune myasthenia gravis. J Immunol 171(7):3847–3854
Tuzun E, Scott BG et al (2004) Circulating immune complexes augment severity of antibody-mediated myasthenia gravis in hypogammaglobulinemic RIIIS/J mice. J Immunol 172(9):5743–5752
Tuzun E, Li J et al (2007) Pros and cons of treating murine myasthenia gravis with anti-C1q antibody. J Neuroimmunol 182(1–2):167–176
Vandervelde S, van Luyn MJ et al (2005) Signaling factors in stem cell-mediated repair of infarcted myocardium. J Mol Cell Cardiol 39(2):363–376
Venkatachalam K, Venkatesan B et al (2009) WISP1, a pro-mitogenic, pro-survival factor, mediates tumor necrosis factor-alpha (TNF-alpha)-stimulated cardiac fibroblast proliferation but inhibits TNF-alpha-induced cardiomyocyte death. J Biol Chem 284(21):14414–14427
Villalta SA, Nguyen HX et al (2009) Shifts in macrophage phenotypes and macrophage competition for arginine metabolism affect the severity of muscle pathology in muscular dystrophy. Hum Mol Genet 18(3):482–496
Vincent A, Drachman DB (2002) Myasthenia gravis. Adv Neurol 88:159–188
Wenzel K, Zabojszcza J et al (2005) Increased susceptibility to complement attack due to down-regulation of decay-accelerating factor/CD55 in dysferlin-deficient muscular dystrophy. J Immunol 175(9):6219–6225
Yasojima K, Kilgore KS et al (1998) Complement gene expression by rabbit heart: upregulation by ischemia and reperfusion. Circ Res 82(11):1224–1230
Zhou Y, Gong B et al (2007) Anti-C5 antibody treatment ameliorates weakness in experimentally acquired myasthenia gravis. J Immunol 179(12):8562–8567
Zwaka TP, Manolov D et al (2002) Complement and dilated cardiomyopathy: a role of sublytic terminal complement complex-induced tumor necrosis factor-alpha synthesis in cardiac myocytes. Am J Pathol 161(2):449–457

Chapter 15
Innate Immunity as Orchestrator of Bone Marrow Homing for Hematopoietic Stem/Progenitor Cells

Mariusz Z. Ratajczak, ChiHwa Kim, Janina Ratajczak, and Anna Janowska-Wieczorek

Abstract The first step that precedes hematopoietic transplantation is elimination of pathological hematopoiesis by administration of myeloablative doses of radiochemotherapy. This eliminates hematolymphopoietic cells and at the same time damages hematopoietic microenvironment in bone marrow (BM). The damage of BM tissue leads to activation of complement cascade (CC), and bioactive CC cleavage fragments modulate several steps of BM recovery after transplantation of hematopoietic stem progenitor cells (HSPCs). Accordingly, C3 cleavage fragments (soluble $C3a/_{desArg}C3a$ and solid phase iC3b) and generation of soluble form of C5b-C9 also known as membrane attack complex (MAC) as well as release of antimicrobial cationic peptides from stromal cells (cathelicidin or LL-37 and β-2 defensin) promote homing of HSPCs. To support this, C3 cleavage fragments and antimicrobial cationic peptides increase homing responsiveness of transplanted HSPCs to stroma-derived factor-1 (SDF-1) gradient. Furthermore, damaged BM cells release several other chemoattractants for HSPCs such as bioactive lipids sphingosine-1-phosphate (S1P) and ceramide-1-phosphate (C1P) and chemotactic purines (ATP and UTP). In this chapter, we will discuss the current view on homing of transplanted HSPCs into BM that in addition to SDF-1 is orchestrated by CC, antimicrobial cationic peptides, and several other prohoming factors. We also propose modulation of CC as a novel strategy to optimize/accelerate homing of HSPCs.

15.1 Introduction

Transplantations of hematopoietic stem progenitor cells (HSPCs) harvested from bone marrow (BM), mobilized peripheral blood (mPB), or umbilical cord blood (UCB) are well-established therapeutic strategy to treat patients with leukemias, lymphomas, inborn defects of hematopoiesis, and some immunological disorders. Hematopoietic transplantation is based on intravenous infusion of histocompatible HSPCs. These cells could be derived from the BM, mPB, or UCB of histocompatible donor (allotransplant) or could be isolated before high-dose chemotherapy from the patient and then infused after treatment to facilitate recovery (autotransplant).

M.Z. Ratajczak, M.D., Ph.D. (✉) • C. Kim • J. Ratajczak
Stem Cell Biology Program at the James Graham Brown Cancer Center,
University of Louisville, 500 S. Floyd Street, Rm. 107, Louisville, KY, 40202, USA
e-mail: mzrata01@louisville.edu

A. Janowska-Wieczorek
Department of Medicine, University of Alberta,
Edmonton, Canada

J.D. Lambris et al. (eds.), *Complement Therapeutics*, Advances in Experimental Medicine and Biology 735,
DOI 10.1007/978-1-4614-4118-2_15,

An important preceding step that facilitates homing and engraftment of HSPCs is myeloablative conditioning of the recipient by radiochemotherapy. Myeloablative conditioning for transplantation is required to destroy old pathological hematopoiesis and to empty stem cell niches to accommodate newly transplanted HSPCs. The myeloablative procedure leads to extensive damage of cells in BM microenvironment, and thus, it is a strong activator of complement cascade (CC). The importance of activation of the CC in hematopoietic transplants of HSPCs has been demonstrated in CC component-deficient mice. As reported, while mice deficient in C3 and C5 components of CC engraft less successfully with HSPCs from wild-type (WT) animals (Ratajczak et al. 2004a, b, 2010a; Pitchford et al. 2009; Kim et al. 2011a), HSPCs obtained from C3a receptor (C3aR)-deficient mice show defective engraftment in WT littermates (Ratajczak et al. 2004b).

The hematopoietic transplantation after administration of HSPCs consists of several steps. In the first step, HSPCs infused into patient's peripheral blood (PB) have to home to the BM stem cell niches. The homing according to the definition is the process in which HSPCs infused into PB lodge into their niches in BM microenvironment. This process is directed by chemoattractants secreted from BM microenvironment and depends on adhesive interactions of HSPCs with endothelium in BM sinusoids. For many years, an α-chemokine stromal-derived factor-1 (SDF-1), a ligand for seven-transmembrane-span $G\alpha_{\iota}$-protein-coupled receptor CXCR4, was considered as the only chemoattractant for HSPCs. Recent evidence, however, reveals that in addition to SDF-1 also other factors such as bioactive lipids sphingosine-1-phosphate (S1P) and ceramide-1-phosphate (C1P) (Ratajczak et al. 2010a; Kim et al. 2011a, b) as well as some purines (ATP and UTP) (Rossi et al. 2007) play a pivotal role in this process. In the next step, HSPCs, being chemoattracted to BM endothelium in order to home/lodge into BM, must attach to endothelial cells in BM sinusoids and subsequently cross the BM-PB barrier (Lee et al. 2009; Jalili et al. 2010a). This process involves metalloproteinases (MMPs) that are secreted by HSPCs. As reported, several MMPs are involved in this process including MT1-MMP and MMP-9 (Pelus et al. 2004; Jalili et al. 2009; Vagima et al. 2009; Shirvaikar et al. 2011). The final destination for transplanted HSPCs ate hematopoietic niches inside BM microenvironment (Taichman 2005; Levesque et al. 2010; Doan and Chute 2011). There are two important anatomical sites identified where HSPCs reside in BM – osteoblastic niches lining trabecular bones and endothelial niches around BM sinusoids (Levesque et al. 2010). Recent stereomorphological evidence suggests that very often both niches in fact overlap (Bengtsson et al. 2011). The homing of HSPCs after transplantation and their lodging in osteoblastic and endothelial stem cells niches lead to engraftment of HSPCs, a process that reestablishes new graft-derived hematopoiesis.

The HSPCs are subsequently retained in BM osteoblastic and endothelial niches by SDF-1 – CXCR4 receptor and in very late antigen-4 (VLA-4, also known as $\alpha_4\beta_1$ integrin) and its ligand vascular adhesion molecule-1 (VCAM-1, also known as CD106) axes. While HSPCs express CXCR4 and VLA-4, their corresponding ligands, SDF-1 and VCAM-1, are expressed by cells in the BM microenvironment (e.g., osteoblasts, stroma fibroblasts, and endothelial cells). SDF-1 is also an important chemoattractant for HSPCs in contrast to VCMA-1 that does not have any chemotactic activity (Peled et al. 1999; Levesque et al. 2001; Rettig et al. 2011).

In this chapter, we will focus on accumulating evidence that several elements of innate immunity including CC cleavage fragments and secreted by damaged BM stroma antimicrobial cationic peptides (cathelicidin [LL-37] and β-2 defensin) play an important role in BM homing of HSPCs after transplantation (Lee et al. 2009; Wu et al. 2011). Elements of innate immunity facilitate homing responsiveness of HSPCs to SDF-1 gradient and in addition evidence accumulate that may play a role in increasing the level of HSPCs homing factors (SDF-1, S1P, C1P, ATP, and UTP) within BM microenvironment.

15.2 Conditioning for Hematopoietic Transplantation Activates Complement Cascade

Innate immunity plays an important role as a senior mechanism that guards tissue/organ integrity. This explains why BM damage by radiochemotherapy activates CC after conditioning for transplantation (Ratajczak et al. 2004a; Kim et al. 2011a, b). Accordingly, CC may become activated in BM microenvironment by several mechanisms that are initiated by both classical (immunoglobulin-dependent) and alternative (immunoglobulin-independent) pathway of CC activation (Ratajczak et al. 2010b). In classical immunoglobulin (Ig)-dependent pathway of activation, BM tissue damaged by radiochemotherapy exposes so-called neoepitope, an antigen that is recognized by circulating in PB naturally occurring antibodies (Ratajczak et al. 2006). Binding of naturally occurring IgM antibodies to neoepitope triggers activation of the C1 complex (composed of one molecule of C1q, two molecules of C1r, and two molecules of C1s, thus forming $C1qr^2s^2$) that leads to activation of classical CC pathway (Danet et al. 2002; Lee and Ratajczak 2009; Jalili et al. 2010b). In addition to neoepitope that marks damaged tissues, also microvesicles, apoptotic bodies, DNA fragments, and proteolytic enzymes released from damaged cells in BM microenvironment all together may activate directly or indirectly both pathways of CC. To support this, activation of CC during conditioning for transplantation in mice after lethal irradiation or administration of cyclophosphamide was confirmed by ELISA to detect C3a and C5a cleavage fragments in both PB plasma and in BM extracts and in addition by histochemical detection of membrane attack complex (MAC) deposits directly in BM sections (Lee et al. 2010; Kim et al. 2011a).

Thus, the myeloablative conditioning for hematopoietic transplantation induces prohoming microenvironment for HSPCs circulating in PB, and SDF-1 was envisioned for many years as a factor responsible for chemoattraction and homing of transplanted HSPCs (Lapidot et al. 2005; Dar et al. 2006). However, the role of SDF-1-CXCR4 axis in homing of HSPCs has been challenged by several observations supporting existence of SDF-1–CXCR4-independent homing mechanisms. In particular, (1) $CXCR4^{-/-}$ fetal liver HSPCs may home to BM in an SDF-1-independent manner (Ma et al. 1999; Levesque et al. 2003); (2) homing of murine HSPCs made refractory to SDF-1 by incubation, and coinjection with a CXCR4 receptor antagonist (AMD3100) is normal or only mildly reduced (Christopherson et al. 2004); and (3) HSPCs in which CXCR4 has been knocked down by means of an SDF-1 intrakine strategy are able to engraft, even in lethally irradiated recipients (Onai et al. 2000). The most recent cumulative evidence shows that besides SDF-1, the chemotactic gradients of S1P and C1P which are products of membrane lipid metabolism and ATP and UTP nucleotides released from the damaged cells are all together involved in homing of HSPCs (Rossi et al. 2007; Granado et al. 2009; Kronlage et al. 2010; Ratajczak et al. 2010a; Kim et al. 2011a). Thus, it is obvious that all these factors may efficiently replace SDF-1 chemotactic gradient when HSPCs lack functional CXCR4 and are rendered insensitive to this chemokine.

Another important problem with SDF-1 chemotactic gradient is a fact that, as it has been recently documented, myeloablative conditioning for transplantation induces a highly proteolytic microenvironment in BM that leads to proteolytic degradation of SDF-1 (Kim et al. 2011a). This potentially impairs chemotactic SDF-1 gradient in BM at time when this gradient is needed to facilitate homing and engraftment of HSPCs. However, as recently reported in order to retain chemotactic power of SDF-1 in damaged BM microenvironment, some of the CC cleavage fragments (e.g., C3a and $_{desArg}$C3a) as well as stroma-derived antimicrobial cationic peptides (e.g., cathelicidin or LL-37 and β2-defensin) become upregulated and increase responsiveness of HSPCs to SDF-1 gradient (Lee et al. 2009; Wu et al. 2011). This phenomenon called "priming effect" (Fig. 15.1) facilitates homing of HSPCs to decreasing SDF-1 gradient in proteolytic microenvironment of BM conditioned by myeloablative therapy for transplantation (Ratajczak et al. 2004a; Reca et al. 2007; Lee et al. 2009).

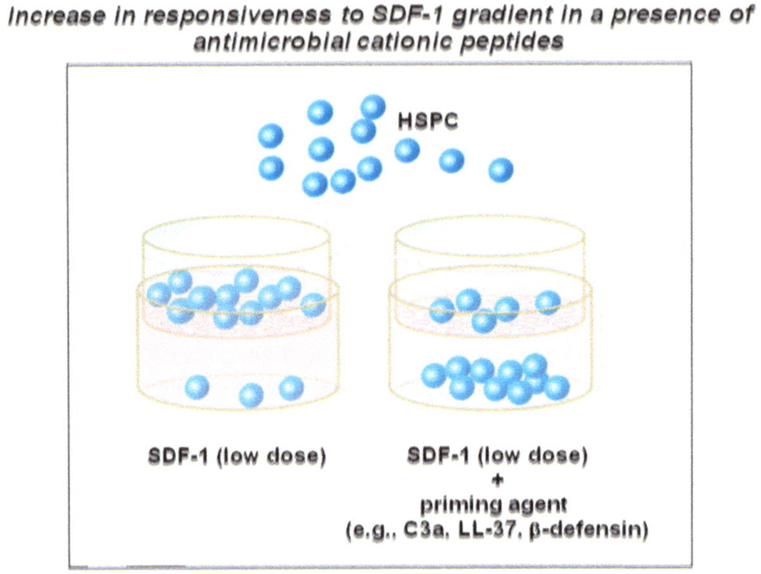

Fig. 15.1 *A priming effect increases the responsiveness of HSPCs to low SDF-1 gradients.* The overall scheme of chemotactic assays performed in the Transwell system to evaluate the HSPC priming phenomenon. In the presence of a priming agent (e.g., antimicrobial cationic peptides such as C3a or cathelicidin [LL-37] or β2-defensin), HSPCs respond more robustly to low doses of SDF-1. This phenomenon is currently being tested in the clinic, where UCB is exposed *ex vivo* to a priming agent (C3a antimicrobial cationic peptide) before transplantation

The protein components of CC are activated through proteolysis in a cascade-like fashion leading to the generation of activated/cleaved protein fragments that bind to the CC-activating surface and small liquid phase activation peptides termed anaphylatoxins such as C3 (C3a, ${}_{desArg}$C3a) and C5 (C5a and ${}_{desArg}$C5a) cleavage fragments. Overall data from our laboratories indicate that activation of CC has important impact on homing of HSPCs, and we will discuss separately consequences of activation of proximal (C3) and distal part (C5) of CC and their involvement in this process (Lee et al. 2009; Reca et al. 2007; Kim et al. 2011a).

15.3 Prohoming Mechanisms Related to Activation of Proximal Part of CC

Pathways that lead to cleavage of C3 that is an abundant protein in PB plasma (1 mg/mL) are considered as activation of proximal part of CC. An activation of both classical and alternative pathway of CC during conditioning for transplants leads in a first step to cleavage of C3, and two groups of C3 cleavage fragments are distinguished – fluid phase anaphylatoxins (C3a, ${}_{des\text{-}Arg}$C3a) and cell- or extracellular matrix-bound (C3b, iC3b, C3dg, C3d) fragments. Liquid phase anaphylatoxin C3a and solid phase C3b are the first cleavage products of C3, and each has a short half-life in plasma. In the next step, C3a is processed by serum carboxypeptidase N to $C3a_{des\text{-}Arg}$ (long half-life cleavage product), and C3b is cleaved into iC3b (long half-life cleavage product) by factor I (Reca et al. 2003, 2007; Ratajczak et al. 2004a, 2010b).

To address the role of activation of proximal CC in homing of HSPCs, we focused on mice deficient in complement C3 ($C3^{-/-}$). These mice are hematologically normal under steady state conditions but displayed a significant delay in hematopoietic recovery from either irradiation or transplantation of wild-type (WT) HSPC (Ratajczak et al. 2004a, b). Transplantation of histocompatible WT $Sca\text{-}1^{+}$

cells into $C3^{-/-}$ mice resulted in delayed hematopoietic recovery after transplantation. Accordingly, we observed a (1) decrease in day-12 colony-forming units in spleen (CFU-S) of transplanted $C3^{-/-}$ mice, (2) 5–7-day delay in platelet and leukocyte recovery, and (3) reduced number of BM hematopoietic clonogenic progenitors at day 16 after transplantation. The fact that HSPC from $C3^{-/-}$ mice engrafted normally into irradiated WT mice suggests that there was a defect in the hematopoietic environment of $C3^{-/-}$ mice and no intrinsic defect of $C3^{-/-}$ mice-derived HSPCs.

Since $C3^{-/-}$ mice cannot activate/cleave C3, the C3 fragments C3a, $C3a_{des\text{-}Arg}$, and iC3b were examined for a role in HSPC engraftment (Ratajczak et al. 2006; Wysoczynski et al. 2009). We found that liquid phase C3a and $C3a_{des\text{-}Arg}$ increased CXCR4 incorporation into membrane lipid rafts (thus potentiating HSPCs responses to SDF-1 gradients), whereas iC3b was deposited onto irradiated BM endothelial and stromal cells and via its receptor CR3 (CD11b/CD18) expressed on HSPCs functioned as ligand to tether HSPCs. To support further the involvement of CR3-iC3b interaction in homing of HSPCs, we demonstrated that HSPCs from $CR3^{-/-}$ mice have defective adhesion to iC3b deposited after activation of CC on BM stromal cells.

To explain further why $C3^{-/-}$ mice poorly engraft with WT HSPCs, we also demonstrated that soluble CC cleavage fragments C3a and ${}_{desArg}C3a$ increase/prime responsiveness of HSPCs to the low SDF-1 gradients (Reca et al. 2003; Ratajczak et al. 2004a, b, 2006; Wysoczynski et al. 2005, 2009). It is very important because as stated above SDF-1 level decreases in BM proteolytic microenvironment after lethal irradiation. We recently provided a molecular explanation for this intriguing phenomenon called "priming effect" based on the observation that actively signaling SDF-1 binding CXCR4 receptor is associated with lipid rafts (Wysoczynski et al. 2005; Lee et al. 2009; Wu et al. 2011). Lipid rafts are membrane domains rich in sphingolipids and cholesterol, which form a lateral assembly in a saturated glycerophospholipid environment. The raft domains are known to serve as moving platforms on the cell surface and are more ordered and resistant to nonionic detergents than other areas of the membrane. These domains are also good sites for cross talk between various cellular signaling proteins. For example, it has been recently reported that small guanine nucleotide triphosphatases (GTPases) such as Rac-1 and Rac-2, which are crucial for engraftment of hematopoietic cells after transplantation, are associated with lipid rafts on migrating HSPCs (Gu et al. 2003; Filippi et al. 2004; Cancelas et al. 2005). Therefore, since the CXCR4 receptor is a lipid raft-associated protein, its signaling ability is enhanced if CXCR4 is incorporated into membrane lipid rafts and where it may better interact with several signaling molecules, including the small GTPase Rac-1 (Yang et al. 2001; Nguyen and Taub 2002; Gu et al. 2003; Gómez-Moutón et al. 2004; Guan 2004). This colocalization of CXCR4 and Rac-1 in lipid rafts facilitates GTP binding and activation of Rac-1. Thus, generation of C3 cleavage fragments in the BM microenvironment may somehow act as a mechanism aimed at increasing responsiveness of HSPCs to the degraded proteolytic microenvironment SDF-1 gradient. In C3-deficient mice, this phenomenon is attenuated, explaining why these animals are showing delayed engraftment. In this context, increase in C3a and $C3a_{desArg}$ level in BM after myeloablative conditioning could be envisioned as one of the mechanisms that promote homing of HSPCs (Ratajczak et al. 2004a).

Furthermore, as demonstrated the priming effect of C3 cleavage fragments to enhance responsiveness of HSPCs to SDF-1 gradient does not depend as reported on potential interaction of C3a and ${}_{desArg}C3a$ with their specific receptors, C3aR and C5L2, respectively. However, we noticed that HSPCs from $C3aR^{-/-}$ mice have a defective homing/engraftment in WT normal littermates (Wysoczynski et al. 2009). Accordingly, transplantation of HSPCs from $C3aR^{-/-}$ mice into lethally irradiated WT recipients resulted in (1) 5–7-day delay in recovery of platelets and leukocytes counts, (2) decrease in formation of day-12 CFU-S, and (3) decrease in the number of donor-derived clonogenic progenitors detectable in the BM cavities at day 16 after transplantation. In agreement with the murine data, blockage of C3aR on human umbilical cord blood $CD34^{+}$ cells by C3aR antagonist SB290157 also impaired their engraftment in nonobese diabetic/severe combined immunodeficient (NOD/SCID) mice (Wysoczynski et al. 2009). Since as mentioned above priming effect of C3a does not depend on interaction of C3a with C3aR, the defective engraftment of C3aR$^{-/-}$ HSPCs in WT animals was unclear and suggested involvement of some other "nonpriming effects" in homing of these cells in WT animals. This has been recently resolved by demonstrating that C3a-C3aR interaction is playing an important role in induction of secretion of matrix metalloprotease-9 (MMP-9) and increasing adhesion of HSPCs to stromal cells (Wysoczynski et al. 2009). Thus, we conclude that C3a, in addition to enhancing responsiveness of HSPCs to SDF-1 gradient in a C3aR-independent manner (priming effect), may also directly modulate HSPC homing by augmenting C3aR-mediated secretion of MMP-9 and cell adhesion.

Based on this, activation of the proximal part of CC and release of C3 cleavage fragments play an important role in homing and engraftment of HSPCs after transplantation by membrane lipid raft-dependent and C3aR-mediated mechanisms.

15.4 Prohoming Mechanism Triggered by Activation of Distal Part of CC

Activation of distal part of CC leads to C5 cleavage and release of C5a anaphylatoxin and C5b fragment (Ratajczak et al. 2010b; Tegla et al. 2011). Liquid phase anaphylatoxin C5a that has a short half-life in plasma is processed by serum carboxypeptidase N to $C5a_{des\text{-}Arg}$ (long half-life cleavage product). The another cleavage fragment C5b then recruits and assembles C6, C7, C7, C8, and C9 molecules to assemble the C5b-C9 membrane attack complex (MAC) (Tegla et al. 2011). MAC is present in biological fluids in two forms – lytic and sublytic. The lytic MAC forms transmembrane channels. These channels disrupt the phospholipid bilayer of target cells, leading to cell lysis and death. The sublytic (soluble) MAC in contrast may bind to cell membranes, independent of any receptor, does not lyse cells, activates multiple signaling pathways, and has wide-range effects on many cell types leading to cellular responses, such as secretion, adherence, aggregation, chemotaxis, and even cell division (Rus et al. 1996; Niculescu et al. 1999a, b; Dashiell et al. 2000; Badea et al. 2002; Tegla et al. 2011). We envision that this soluble form of MAC is also involved in homing of HSPCs (Kim et al. 2011a).

To support this, our recent transplant experiments in C5-deficient (C5$^{-/-}$) mice revealed an important and unrecognized before role of activation of distal part of CC in homing of HSPCs. We noticed that C5$^{-/-}$ animals similarly as C3$^{-/-}$ engraft poorly with WT HSPCs (Lee et al. 2009). At the same time, HSPCs from C5$^{-/-}$ mice engrafted properly in WT recipients. This observation indicated that C5 cleavage fragments (C5a and $_{desArg}$C5a) as well as C5b-C9 (MAC) could be also involved in homing of HSPCs.

To explain involvement of C5a and $_{desArg}$C5a in HSPCs homing, both these soluble C5 cleavage fragments activate bone marrow stroma fibroblasts and enhance secretion of cathelicidin [LL-37] and β2-defensin, two important antimicrobial cationic peptides that are potent priming factors increasing similarly as C3a and $_{desArg}$C3a homing responsiveness of HSPCs to an SDF-1 gradient (Wu et al. 2011). This has again an important implication because of the mentioned-above fact that SDF-1 level paradoxically decreases in proteolytic microenvironment of BM after conditioning for transplantation by radiochemotherapy. C5a as reported also enhances the expression of MMPs (MT-MMP1 and MMP-9) important in BM-PB transendothelial barrier migration and homing of HSPCs (DiScipio et al. 2006; Jalili et al. 2010a; Speidl et al. 2011).

In support of the role for soluble C5b-C9 (MAC) in homing and engraftment, we found that soluble MAC (sMAC) enhances in a CR3 (CD11b/CD18)-dependent manner adhesion of HSPCs to iC3b deposits on BM endothelial and stromal cells and increases the secretion of SDF-1 by BM stromal cells (Kim et al. 2011a). In addition, soluble MAC also enhances in BM similarly as C5a secretion of important priming factors that are cathelicidin (LL-37) and β2-defensin.

Thus, activation of distal part of CC similarly as activation of proximal part plays an important role in modulating homing of HSPCs into BM.

15.5 Cationic Antimicrobial Peptides Cathelicidin (LL-37) and β2-Defensin Important Modulators of Responsiveness of HSPCs to SDF-1 Gradient

Cathelicidin (LL-37) and β2-defensin similarly as C3a belong to the family of antimicrobial cationic peptides that as mentioned above increase (positively prime) responsiveness of HSPCs to SDF-1 homing gradient (Lee et al. 2009). Antimicrobial peptides are cationic proteins and host defense peptides, an evolutionarily conserved component of the innate immune response (Ganz 2003; Ciornei et al. 2005; Bucki et al. 2010; Zughaier et al. 2010). They have been demonstrated to kill bacteria, enveloped viruses, fungi, and even transformed or cancerous cells but do not affect viability of eukaryotic cells. Selective effects of these peptides on prokaryotic cells killing are known to be dependent on the characteristics of prokaryote cell membranes that are susceptible to strong electrostatic and hydrophobic interactions with these "natural antibiotics." In contrast, cell membranes of eukaryotic cells, because of high cholesterol content and weak hydrophobic interaction with cationic peptides, are more resistant to potential toxic effects of these peptides.

As mentioned above, the responsiveness of HSPCs to SDF-1 could be enhanced by antimicrobial cationic peptides (C3a and $_{desArg}$C3a, cathelicidin or LL-37, and β2-defensin) that are released by activated BM stromal cells and granulocytes in C5a- and C5b–C9 (MAC)-dependent manner (Kim et al. 2011a, b; Shirvaikar et al. 2011). This priming phenomenon as mentioned depends on promoting the incorporation of CXCR4 into membrane lipid rafts. Since membrane lipid rafts are enriched for several signaling molecules, incorporation of CXCR4 into lipid raft facilitates signaling, and thus, CXCR4 is activated more efficiently in the presence of low doses of SDF-1 (Lee et al. 2009; Wu et al. 2011).

Priming effect could be easily evaluated in vitro in the Transwell migration assay where two chambers (an upper chamber containing tested cells and a lower chamber containing chemoattractant) are separated by a porous membrane that allows transmigration of cells in response to the chemotactic gradient (Fig. 15.1). Cells that respond to this gradient migrate and accumulate in the lower chamber. Figure 15.2 shows that the chemotaxis of HSCPs to low SDF-1 gradient may be significantly enhanced in the presence of antimicrobial cationic peptides.

As discussed above, biological activity of SDF-1 decreases in BM due to the induction of proteolytic microenvironment after conditioning for transplantation by lethal irradiation (Kim et al. 2011a). Accordingly, a few amino acids located at the N-terminus of SDF-1 are crucial for the biological activity of this peptide. We observed that removal of this peptide fragment, for example, by metalloproteinase-2 (MMP-2) or MMP-9 inhibits completely SDF-1 chemotactic activity. However, at the same time exposure of SDF-1 to MMPs does not affect detection of the SDF-1 protein in tissues by employing antibodies (e.g., in an ELISA assay or histochemistry) targeted to other fragments of the SDF-1 peptide (Kim et al. 2011a). This indicates that antibody-based SDF-1 detection does not correlate with the chemotactic activity of SDF-1, unless antibodies are specifically directed to its N-terminus and do not interact with inactive forms of SDF-1.

Further studies are needed to see whether, in addition to CXCR4, receptors for other chemoattractants of HSPCs such as S1P, C1P, ATP, and UTP are also lipid raft regulated and antimicrobial peptides enhance their incorporation into membrane lipid rafts. Of note, it has been reported that stimulation of the S1P receptor type 1 (S1PR1) by its agonist, FY720, may increase the responsiveness of HSPCs to an SDF-1 gradient (Sugita et al. 2010). However, this probably occurs due to intercellular cross talk between CXCR4 and S1PR1. Since a receptor for another bioactive lipid, C1P, has not yet been identified, it is not clear whether C1P signaling is also lipid raft regulated. However, our data indicate that this receptor is expressed on HSPCs and is sensitive to pertussis toxin, which suggests that, like S1P, it is a $G_{\alpha I}$-protein-coupled receptor (Kim et al. 2011a). Also $G_{\alpha I}$-protein-coupled receptors are purinergic receptors for ATP and UTP (Junger 2008; Kronlage et al. 2010). The possibility of modulation of activity of these receptors by C3a, LL-37, and β2-defesin requires further studies.

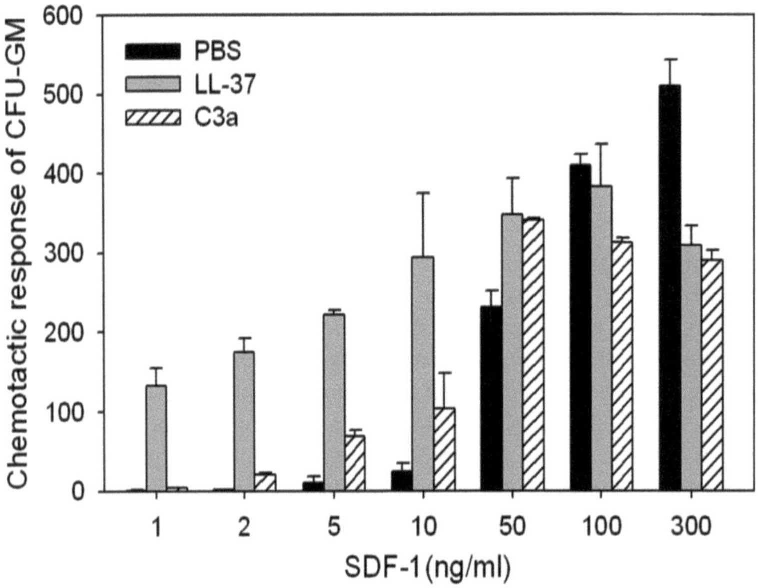

Fig. 15.2 *Antimicrobial cationic peptides C3a and LL-37 enhance responsiveness of murine BM- and human UCB-derived HSPCs to an SDF-1 gradient.* Chemotaxis of murine BM CFU-GM in response to different concentrations of SDF-1, with and without C3a or LL-37. Values are the fold increase of the number of migrated cells compared to the number of migrated cells in medium alone. *Gray bars* indicate the presence of LL-37 (2.5 μg/mL) and C3a (1 μg/mL) in the lower Transwell chambers, and *black bars* indicate its absence. The data represent the combined results from three independent experiments performed in duplicate per group (n = 6)

In addition to cationic peptides, some other small molecules (e.g., prostaglandin E2 [PGE2] or hyaluronic acid) have also been purported to increase responsiveness of HSPCs to an SDF-1 gradient (Hoggatt et al. 2009; Hoggatt and Pelus 2010; Shirvaikar et al. 2011). Of note, PGE2 is also a bioactive lipid derivative and, as previously reported, plays an important role in homing of HSPCs by upregulating expression of CXCR4 on HSPCs (Hoggatt et al. 2009), and this mechanism seems to be responsible for increasing chemotaxis in response to an SDF-1 gradient after pretreatment of HSPCs by PGE2. Interestingly, it has been reported that C5a modulates the activity of coxygenase-2 (COX2) and thus affects synthesis of PGE2 in BM, which explains why C5a increases PGE2 activity in BM stromal cells and why an elevated PGE2 level is detectable in conditioned media harvested from irradiated BM cells (Ohinata et al. 2009). Thus, some of the effects of PGE2 in homing may be also related to activation of CC. This, however, requires further studies.

15.6 Bioactive Lipids in Homing of Transplanted HSPCs

As mentioned above, activation of the CC in BM induces a highly proteolytic microenvironment that degrades SDF-1, which has been accepted for many years as the only major homing factor for HSPCs (Lapidot et al. 2005; Kim et al. 2011a). However, as discussed earlier in this chapter, several doubts have accumulated about whether SDF-1 is the only homing factor responsible for HSPC lodgment/homing into BM.

In further support of these doubts, we found that media conditioned by cells recovered from murine long bones 24 h after lethal irradiation strongly chemoattract HSPCs in an SDF-1-independent manner (Ratajczak et al. 2010a; Kim et al. 2011a). In particular, we observed that (1) chemotaxis occurred

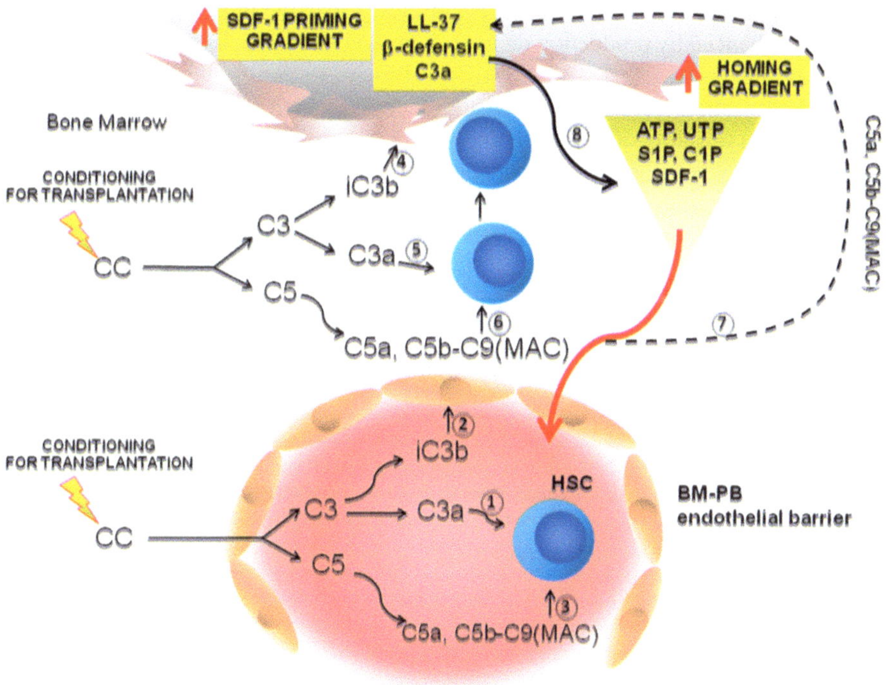

Fig. 15.3 *The involvement of elements of innate immunity in homing and engraftment of HSPCs.* Conditioning for transplantation by radiochemotherapy induces a proteolytic microenvironment in BM and SDF-1 level due to the induction of proteolytic microenvironment decreases. However, at the same time, BM cells damaged by conditioning for transplantation by lethal irradiation release bioactive lipids (S1P and C1P) that are potent chemoattractants for HSPCs. In addition to S1P and C1P, there are also purines released from damaged cells (ATP, UTP) that are endowed with chemotactic activity against HSPCs. Induced by myeloablative treatment, BM damage activation of CC leads to release of C3 and C5 cleavage fragments, C3a (*1, 5*) and C5a, respectively, and generation of soluble C5ab-C9 (MAC) (*3, 6*). While C3a enhances responsiveness of HSPCs to SDF-1 gradient (*1, 5*), iC3b deposits on BM endothelium (*2*), stromal cells, and osteoblasts (*4*) tether HSPCs in CR3-dependent manner. C5a and soluble MAC (C5b-C9) enhance secretion of SDF-1 by stromal cells and secretion of two important antimicrobial cationic peptides by BM stromal cells (cathelicidin [LL-37] and β2-defensin) (7) that similarly as C3a (*1, 5*) enhance responsiveness of HSPCs to SDF-1 gradient (*8*). This increase in SDF-1 secretion and increase in responsiveness of HSPCs to SDF-1 gradient ameliorate the drop in SDF-1 level that occurs in highly proteolytic microenvironment of BM conditioned by radiochemotherapy. Soluble MAC (C5b-C9), in addition to antimicrobial cationic peptides (C3a, LL-37, and β2-defensin), may also enhance responsiveness of HSPCs to SDF-1 gradient (*3, 6*). In addition to SDF-1, homing of HSPCs is mediated by bioactive lipids (S1P and C1P) and some nucleotides (ATP and UTP) that are released from damaged cells. The potential involvement of CC cleavage fragments in modulating BM level of S1P, C1P, ATP, and UTP requires further studies

in the presence of the CXCR4 antagonist AMD3100 and (2) it was resistant to heat inactivation. Based on these findings and data from the literature, we became interested in potential involvement of bioactive lipids and found that S1P and C1P are upregulated in BM conditioned for transplantation and are present at biologically relevant concentrations in conditioned media harvested from irradiated BM that chemoattracts HSPCs.

Based on these findings, we proposed that S1P and C1P are able to support the SDF-1 homing gradient, which decreases after induction of a proteolytic microenvironment by conditioning for transplantation (Kim et al. 2011a). Thus, involvement of S1P and C1P explains occurrence of SDF-1-independent homing of transplanted HSPCs (Fig. 15.3).

On open question that requires further studies is a potential relationship between CC activation and S1P and C1P level in BM conditioned for transplantation. It is important to study if increase of S1P and C1P in irradiated BM is the result of release of these bioactive lipids by leaky damaged cells or perhaps could be enhanced by production de novo from BM stromal cells and osteoblasts in response to CC cleavage fragments.

To address this question, several enzymes are identified which are involved in biosynthesis and degradation of bioactive sphingolipids. S1P and C1P which are important components of cell membranes are derived from the aliphatic amino alcohol sphingosine (Mitsutake et al. 2004; Lamour et al. 2007; Ohkawa et al. 2008; Fyrst and Saba 2010; Ratajczak et al. 2011). S1P is a product of two sphingosine phosphatases (SK1 and SK2) and degraded by S1P lyase (SPL), lipid phosphate phosphatases (LPP1–3), and S1P-specific phosphatases (SPP1 and SPP2). The structurally related lipid C1P is a product of ceramide (N-acyl sphingosine) phosphorylation by ceramide kinase (CERK) and is degraded by LPP1–3. Unlike ceramide (which is often proapoptotic), C1P has been reported to promote cell growth, survival, and migration through an unknown receptor-initiated signaling pathway that is pertussis toxin sensitive and therefore likely to involve $G_{\alpha I}$-protein-coupled seven-transmembrane-spanning receptors (Arana et al. 2010).

Based on fact that C5a, for example, modulates synthesis of some other bioactive lipids (e.g., PGE2), further studies are required if these two very evolutionary conserved biological systems CC and bioactive lipids signaling molecules (S1P and C1P) are closely related and if CC cleavage fragments may be involved in the metabolism (synthesis/degradation) of these important chemoattractants for HSPCs.

15.7 Conclusions

Augmenting evidence demonstrates that innate immunity is playing an important role in homing/engraftment into BM of transplanted HSPCs. In this chapter, we have discussed the involvement of C3 and C5 cleavage fragments as well as a novel role of soluble MAC in this process. Recent evidence also indicates that in addition to C3 and C5 cleavage fragments, also other components of CC may play an important role in homing (Kim et al. 2011a). For example, our recent work shows that C1q which is involved in initial steps of activation of CC by classical Ig-dependent manner may also prime/enhance the chemotactic response of HSPCs to a low SDF-1 gradient and increase secretion of MMP-9 by these cells (Jalili et al. 2010b; Marquez-Curtis et al. 2011). This effect however is mediated directly by C1q receptor (C1qRp) expressed on surface of HSPCs. This suggests that many other components of CC may directly or indirectly affect homing of HSPCs to BM microenvironment. The potential protein that requires further studies on its role in stem cell trafficking is, for example, C4a anaphylatoxin.

We also postulate that these innate immunity-based mechanisms play a more universal role and are involved in regulating migration of other types of stem cells, such as circulating mesenchymal stem cells (MSCs), endothelial progenitor cells (EPCs), and very small embryonic-like (VSEL) stem cells (Ratajczak 2010; Ratajczak et al. 2012; Kim et al. 2011b). Similar mechanisms of homing probably play a role in recruitment of stem cells in other types of organ injury, for example, heart infarct, stroke, and damaged liver or kidney (Wojakowski et al. 2011; Borlongan 2011). This may support involvement of innate immunity in regeneration of damaged tissues.

These observations also open a new area for optimizing trafficking of stem cells by modulating their responsiveness to homing signals by elements of innate immunity (e.g., CC cleavage fragments or antimicrobial cationic peptides). As example based on the observation that the priming strategy of short *ex vivo* exposure of HSPCs to C3a or cathelicidin (LL-37) before transplantation may accelerate homing and engraftment of HSPCs, this strategy is currently under clinical translational evaluation by hematopoietic transplantation centers in Charlottesville, Virginia, USA, and Minneapolis, Minnesota,

USA. In this trial, umbilical cord blood (UCB)-derived HSPCs are ex vivo primed for 30 min before infusion to the patients with recombinant C3a. We envision that this will improve homing of transplanted HSPCs.

Acknowledgments This work was supported by NIH R01 DK074720 and Stella and Henry Hoenig Endowment to MZR.

References

Arana L, Gangoiti P, Ouro A, Trueba M, Gómez-Muñoz A (2010) Ceramide and ceramide 1-phosphate in health and disease. Lipids Health Dis 9:15–26

Badea T, Niculescu F, Soane L, Fosbrink M, Sorana H, Rus V, Shin ML, Rus H (2002) RGC-32 increases p34CDC2 kinase activity and entry of aortic smooth muscle cells into S-phase. J Biol Chem 277:502–508

Bengtsson NE, Kim S, Lin L, Walter GA, Scott EW (2011) Ultra-high-field MRI real-time imaging of HSC engraftment of the bone marrow niche. Leukemia 25:1223–1231

Borlongan CV (2011) Bone marrow stem cell mobilization in stroke: a 'bonehead' may be good after all! Leukemia 25:1674–1686

Bucki R, Leszczyn'ska K, Namiot A, Sokolowski W (2010) Cathelicidin LL-37: a multitask antimicrobial peptide. Arch Immunol Ther Exp 58:15–25

Cancelas JA, Lee AW, Prabhakar R, Stringer KF, Zheng Y, Williams DA (2005) Rac GTPases differentially integrate signals regulating hematopoietic stem cell localization. Nat Med 11:886–891

Christopherson KW II, Hangoc G, Mantel CR, Broxmeyer HE (2004) Modulation of hematopoietic stem cell homing and engraftment by CD26. Science 305:1000–1003

Ciornei CD, Sigurdardottir T, Schmidtchen A, Bodelsson M (2005) Antimicrobial and chemoattractant activity, lipopolysaccharide neutralization, cytotoxicity, and inhibition by serum of analogs of human cathelicidin LL-37. Antimicrob Agents Chemother 49:2845–2850

Danet GH, Luongo JL, Butler G, Lu MM, Tenner AJ, Simon MC, Bonnet DA (2002) C1qRp defines a new human stem cell population with hematopoietic and hepatic potential. Proc Natl Acad Sci 99:10441–10445

Dar A, Kollet O, Lapidot T (2006) Mutual, reciprocal SDF-1/CXCR4 interactions between hematopoietic and bone marrow stromal cells regulate human stem cell migration and development in NOD/SCID chimeric mice. Exp Hematol 34:967–975

Dashiell SM, Rus H, Koski CL (2000) Terminal complement complexes concomitantly stimulate proliferation and rescue of Schwann cells from apoptosis. Glia 30:187–198

DiScipio RG, Schraufstatter IU, Sikora L, Zuraw BL, Sriramarao P (2006) C5a mediates secretion and activation of matrix metalloproteinase 9 from human eosinophils and neutrophils. Int Immunopharmacol 6:1109–1118

Doan PL, Chute JP (2011) The vascular niche: home for normal and malignant hematopoietic stem cells. Leukemia. doi:10.1038/leu.2011.236

Filippi MD, Harris CE, Meller J, Gu Y, Zheng Y, Williams DA (2004) Localization of Rac2 via the C terminus and aspartic acid 150 specifies superoxide generation, actin polarity and chemotaxis in neutrophils. Nat Immunol 5:744–751

Fyrst H, Saba JD (2010) An update on sphingosine-1-phosphate and other sphingolipid mediators. Nat Chem Biol 6:489–497

Ganz T (2003) Defensins: antimicrobial peptides of innate immunity. Nat Rev Immunol 3:710–720

Gómez-Moutón C, Lacalle RA, Mira E, Jiménez-Baranda S, Barber DF, Carrera AC, Martínez-A C, Mañes S (2004) Dynamic redistribution of raft domains as an organizing platform for signaling during cell chemotaxis. J Cell Biol 164:759–768

Granado MH, Gangoiti P, Ouro A, Arana L, Gonzalez M, Trueba M, Gómez-Muñoz A (2009) Ceramide 1-phosphate (C1P) promotes cell migration involvement of a specific C1P receptor. Cell Signal 21:405–412

Gu Y, Filippi MD, Cancelas JA, Siefring JE, Williams EP, Jasti AC, Harris CE, Lee AW, Prabhakar R, Atkinson SJ, Kwiatkowski DJ, Williams DA (2003) Hematopoietic cell regulation by Rac1 and Rac2 guanosine triphosphatases. Science 302:445–449

Guan JL (2004) Cell biology: integrins, rafts, Rac, and Rho. Science 302:773–774

Hoggatt J, Pelus LM (2010) Eicosanoid regulation of hematopoiesis and hematopoietic stem and progenitor trafficking. Leukemia 24:1993–2002

Hoggatt J, Singh P, Sampath J, Pelus LM (2009) Prostaglandin E2 enhances hematopoietic stem cell homing, survival, and proliferation. Blood 113:5444–5455

Jalili A, Shirvaikar N, Marquez-Curtis LA, Turner AR, Janowska-Wieczorek A (2009) The HGF/c-Met axis synergizes with G-CSF in the mobilization of hematopoietic stem/progenitor cells. Stem Cells Dev 19:1143–1151
Jalili A, Shirvaikar N, Marquez-Curtis L, Qui Y, Korol C, Lee H, Turner AR, Ratajczak MZ, Janowska-Wieczorek A (2010a) Fifth complement cascade protein (C5) cleavage fragments disrupt the SDF-1/CXCR4 axis: further evidence that innate immunity orchestrates the mobilization of hematopoietic stem/progenitor cells. Exp Hematol 38:321–332
Jalili A, Marquez-Curtis L, Shirvaikar N, Wysoczynski M, Ratajczak MZ, Janowska-Wieczorek A (2010b) Complement C1q enhances homing-related responses of hematopoietic stem/progenitor cells. Transfusion 50:2002–2010
Junger WG (2008) Purinergic regulation of neutrophil chemotaxis. Cell Mol Life Sci 65:2528–2540
Kim CH, Wu W, Wysoczynski M, Abdel-Latif A, Sunkara M, Morris A, Kucia M, Ratajczak J, Ratajczak MZ (2011a) Conditioning for hematopoietic transplantation activates the complement cascade and induces a proteolytic environment in bone marrow: a novel role for bioactive lipids and soluble C5b-C9 as homing factors. Leukemia. doi:10.1038/leu.2011.185
Kim CH, Liu R, Kucia M, Ratajczak MZ (2011b) New evidence that the bioactive lipid ceramide-1-phosphate (C1P) is a potent chemoattractant for mesenchymal stromal cells (MSC), endothelial progenitor cells (EPCs) and very small embryonic-like stem cells (VSELs), demonstrating its potential involvement in tissue/organ repair and angiogenesis. 2011 American Society of Hematology meeting. Abstract # 2387
Kronlage M, Song J, Sorokin L, Isfort K, Schwerdtle T, Leipziger J, Robaye B, Conley PB, Kim HC, Sargin S, Schön P, Schwab A, Hanley PJ (2010) Autocrine purinergic receptor signaling is essential for macrophage chemotaxis. Sci Signal 3:ra55
Lamour NF, Stahelin RV, Wijesinghe DS, Maceyka M, Wang E, Allegood JC, Merrill AH Jr, Cho W, Chalfant CE (2007) Ceramide kinase uses ceramide provided by ceramide transport protein: localization to organelles of eicosanoid synthesis. J Lipid Res 48:1293–1304
Lapidot T, Dar A, Kollet O (2005) How do stem cells find their way home? Blood 106:1901–1910
Lee H, Ratajczak MZ (2009) Innate immunity: a key player in the mobilization of hematopoietic stem/progenitor cells. Arch Immunol Ther Exp (Warsz) 57:269–278
Lee HM, Wu W, Wysoczynski M, Liu R, Zuba-Surma EK, Kucia M, Ratajczak J, Ratajczak MZ (2009) Impaired mobilization of hematopoietic stem/progenitor cells in C5-deficient mice supports the pivotal involvement of innate immunity in this process and reveals novel promobilization effects of granulocytes. Leukemia 23:2052–2062
Lee HM, Wysoczynski M, Liu R, Shin DM, Kucia M, Botto M, Ratajczak J, Ratajczak MZ (2010) Mobilization studies in complement-deficient mice reveal that optimal AMD3100 mobilization of hematopoietic stem cells depends on complement cascade activation by AMD3100-stimulated granulocytes. Leukemia 24:573–582
Levesque JP, Takamatsu Y, Nilsson SK, Haylock DN, Simmons PJ (2001) Vascular cell adhesion molecule-1 (CD106) is cleaved by neutrophil proteases in the bone marrow following hematopoietic progenitor cell mobilization by granulocyte colony-stimulating factor. Blood 98:1289–1297
Levesque JP, Hendy J, Takamatsu Y, Simmons PJ, Bendall LJ (2003) Disruption of the CXCR4/CXCL12 chemotactic interaction during hematopoietic stem cell mobilization induced by GCSF or cyclophosphamide. J Clin Invest 111:187–196
Levesque JP, Helwani FM, Winkler IG (2010) The endosteal 'osteoblastic' niche and its role in hematopoietic stem cell homing and mobilization. Leukemia 24:1979–1992
Ma Q, Jones D, Springer TA (1999) The chemokine receptor CXCR4 is required for the retention of B lineage and granulocytic precursors within the bone marrow microenvironment. Immunity 10:463–471
Marquez-Curtis LA, Turner AR, Sridharan S, Ratajczak MZ, Janowska-Wieczorek A (2011) The ins and outs of hematopoietic stem cells: studies to improve transplantation outcomes. Stem Cell Rev 7:590–607
Mitsutake S, Kim TJ, Inagaki Y, Kato M, Yamashita T, Igarashi Y (2004) Ceramide kinase is a mediator of calcium-dependent degranulation in mast cells. J Biol Chem 279:17570–17577
Nguyen DH, Taub D (2002) CXCR4 function requires membrane cholesterol: implications for HIV infection. J Immunol 168:4121–4126
Niculescu F, Soane L, Badea T, Shin M, Rus H (1999a) Tyrosine phosphorylation and activation of Janus kinase 1 and STAT3 by sublytic C5b-9 complement complex in aortic endothelial cells. Immunopharmacology 42:187–193
Niculescu F, Badea T, Rus H (1999b) Sublytic C5b-9 induces proliferation of human aortic smooth muscle cells: role of mitogen activated protein kinase and phosphatidylinositol 3-kinase. Atherosclerosis 142:47–56
Ohinata K, Takagi K, Biyajima K, Kaneko K, Miyamoto C, Asakawa A, Eguchi N, Urade Y, Inui A, Yoshikawa M (2009) Complement C5a stimulates food intake via a prostaglandin D(2)- and neuropeptide Y-dependent mechanism in mice. Prostaglandins Other Lipid Mediat 90:81–84
Ohkawa R, Nakamura K, Okubo S, Hosogaya S, Ozaki Y, Tozuka M, Osima N, Yokota H, Ikeda H, Yatomi Y (2008) Plasma sphingosine-1-phosphate measurement in healthy subjects: close correlation with red blood cell parameters. Ann Clin Biochem 45:356–363
Onai N, Zhang YY, Yoneyama H, Kitamura T, Ishikawa S, Matsushima K (2000) Impairment of lymphopoiesis and myelopoiesis in mice reconstituted with bone marrow-hematopoietic progenitor cells expressing SDF-1-intrakine. Blood 96:2074–2080

Peled A, Grabovsky V, Habler L, Sandbank J, Arenzana-Seisdedos F, Petit I, Ben-Hur H, Lapidot T, Alon R (1999) The chemokine SDF-1 stimulates integrin-mediated arrest of CD34+ cells on vascular endothelium under shear flow. J Clin Invest 104:1199–1211
Pelus LM, Bian H, King AG, Fukuda S (2004) Neutrophil-derived MMP-9 mediates synergistic mobilization of hematopoietic stem and progenitor cells by the combination of G-CSF and the chemokines GRObeta/CXCL2 and GRObetaT/CXCL2delta4. Blood 103:110–119
Pitchford SC, Furze RC, Jones CP, Wengner AM, Rankin SM (2009) Differential mobilization of subsets of progenitor cells from the bone marrow. Cell Stem Cell 4:62–72
Ratajczak MZ (2010) Spotlight series on stem cell mobilization: many hands on the ball, but who is the quarterback? Leukemia 24:1665–1666
Ratajczak MZ, Reca R, Wysoczynski M, Kucia M, Baran JT, Allendorf DJ, Ratajczak J, Ross GD (2004a) Transplantation studies in C3-deficient animals reveal a novel role of the third complement component (C3) in engraftment of bone marrow cells. Leukemia 18:1482–1490
Ratajczak J, Reca R, Kucia M, Majka M, Allendorf DJ, Baran JT, Janowska-Wieczorek A, Wetsel RA, Ross GD, Ratajczak MZ (2004b) Mobilization studies in mice deficient in either C3 or C3a receptor (C3aR) reveal a novel role for complement in retention of hematopoietic stem/progenitor cells in bone marrow. Blood 103:2071–2078
Ratajczak MZ, Reca R, Wysoczynski M, Yan J, Ratajczak J (2006) Modulation of the SDF-1-CXCR4 axis by the third complement component (C3) – implications for trafficking of CXCR4+ stem cells. Exp Hematol 34:986–995
Ratajczak MZ, Lee H, Wysoczynski M, Wan W, Marlicz W, Laughlin MJ, Kucia M, Janowska-Wieczorek A, Ratajczak J (2010a) Novel insight into stem cell mobilization-plasma sphingosine-1-phosphate is a major chemoattractant that directs the egress of hematopoietic stem progenitor cells from the bone marrow and its level in peripheral blood increases during mobilization due to activation of complement cascade/membrane attack complex. Leukemia 24:976–985
Ratajczak MZ, Kim CH, Wojakowski W, Janowska-Wieczorek A, Kucia M, Ratajczak J (2010b) Innate immunity as orchestrator of stem cell mobilization. Leukemia 24:1667–1675
Ratajczak MZ, Kim CH, Abdel-Latif A, Schneider G, Kucia M, Morris AJ, Laughlin MJ, Ratajczak J (2011) A novel perspective on stem cell homing and mobilization: review on bioactive lipids as potent chemoattractants and cationic peptides as underappreciated modulators of responsiveness to SDF-1 gradients. Leukemia. doi:10.1038/leu.2011.242
Ratajczak MZ, Kim CH, Wu W, Shin DM, Bryndza E, Kucia M, Ratajczak J (2012) The role of innate immunity in trafficking of hematopoietic stem cells – an emerging link between activation of complement cascade and chemotactic gradients of bioactive sphingolipids. Adv Exp Med Biol 946:37–54
Reca R, Mastellos D, Majka M, Marquez L, Ratajczak J, Franchini S, Glodek A, Honczarenko M, Spruce LA, Janowska-Wieczorek A, Lambris JD, Ratajczak MZ (2003) Functional receptor for C3a anaphylatoxin is expressed by normal hematopoietic stem/progenitor cells, and C3a enhances their homing-related responses to SDF-1. Blood 101:3784–3793
Reca R, Cramer D, Yan J, Laughlin MJ, Janowska-Wieczorek A, Ratajczak J, Ratajczak MZ (2007) A novel role of complement in mobilization: immunodeficient mice are poor granulocyte-colony stimulating factor mobilizers because they lack complement-activating immunoglobulins. Stem Cells 25:3093–3100
Rettig MP, Ansstas G, Dipersio JF (2011) Mobilization of hematopoietic stem and progenitor cells using inhibitors of CXCR4 and VLA-4. Leukemia. doi:10.1038/leu.2011.197
Rossi L, Manfredini R, Bertolini F, Ferrari D, Fogli M, Zini R, Salati S, Salvestrini V, Gulinelli S, Adinolfi E, Ferrari S, Di Virgilio F, Baccarani M, Lemoli RM (2007) The extracellular nucleotide UTP is a potent inducer of hematopoietic stem cell migration. Blood 109:533–542
Rus HG, Niculescu F, Shin ML (1996) Sublytic complement attack induces cell cycle in oligodendrocytes. J Immunol 156:4892–4900
Shirvaikar N, Marquez-Curtis LA, Ratajczak MZ, Janowska-Wieczorek A (2011) Hyaluronic acid and thrombin upregulate MT1-MMP through PI3K and Rac-1 signaling and prime the homing-related responses of cord blood hematopoietic stem/progenitor cells. Stem Cells Dev 20:19–30
Speidl WS, Kastl SP, Hutter R, Katsaros KM, Kaun C, Bauriedel G, Maurer G, Huber K, Badimon JJ, Wojta J (2011) The complement component C5a is present in human coronary lesions in vivo and induces the expression of MMP-1 and MMP-9 in human macrophages in vitro. FASEB J 25:35–44
Sugita K, Kabashima K, Sakabe J, Yoshiki R, Tanizaki H, Tokura Y (2010) FTY720 regulates bone marrow egress of eosinophils and modulates late-phase skin reaction in mice. Am J Pathol 177:1881–1887
Taichman RS (2005) Blood and bone: two tissues whose fates are intertwined to create the hematopoietic stem-cell niche. Blood 105:2631–2639
Tegla CA, Cudrici C, Patel S, Trippe R 3rd, Rus V, Niculescu F, Rus H (2011) Membrane attack by complement: the assembly and biology of terminal complement complexes. Immunol Res 51:45–60
Vagima Y, Avigdor A, Goichberg P, Shivtiel S, Tesio M, Kalinkovich A, Golan K, Dar A, Kollet O, Petit I, Perl O, Rosenthal E, Resnick I, Hardan I, Gellman YN, Naor D, Nagler A, Lapidot T (2009) MT1-MMP and RECK are involved in human CD34+ progenitor cell retention, egress, and mobilization. J Clin Invest 119:492–503

Wojakowski W, Landmesser U, Bachowski R, Jadczyk T, Tendera M (2011) Mobilization of stem and progenitor cells in cardiovascular diseases. Leukemia. doi:10.1038/leu.2011.184

Wu W, Kim CH, Liu R, Kucia M, Marlicz W, Greco N, Ratajczak J, Laughlin MJ, Ratajczak MZ (2011) The bone marrow-expressed antimicrobial cationic peptide LL-37 enhances the responsiveness of hematopoietic stem progenitor cells to an SDF-1 gradient and accelerates their engraftment after transplantation. Leukemia. doi:10.1038/leu.2011.252

Wysoczynski M, Reca R, Ratajczak J, Kucia M, Shirvaikar N, Honczarenko M, Mills M, Wanzeck J, Janowska-Wieczorek A, Ratajczak MZ (2005) Incorporation of CXCR4 into membrane lipid rafts primes homing-related responses of hematopoietic stem/progenitor cells to an SDF-1 gradient. Blood 105:40–48

Wysoczynski M, Reca R, Lee H, Wu W, Ratajczak J, Ratajczak MZ (2009) Defective engraftment of C3aR–/– hematopoietic stem progenitor cells shows a novel role of the C3a-C3aR axis in bone marrow homing. Leukemia 23:1455–1461

Yang FC, Atkinson SJ, Gu Y, Borneo JB, Roberts AW, Zheng Y, Pennington J, Williams DA (2001) GTPases control hematopoietic stem cell shape, adhesion, migration, and mobilization. Proc Natl Acad Sci 98:5614–5618

Zughaier SM, Svoboda P, Pohl J, Stephens DS, Shafer WM (2010) The human host defense peptide LL-37 interacts with *Neisseria meningitidis* capsular polysaccharides and inhibits inflammatory mediators release. PLoS One 5:e13627

Chapter 16
Complement-Mediated Microvascular Injury Leads to Chronic Rejection

Mohammad A. Khan and Mark R. Nicolls

Abstract Microvascular loss may be an unappreciated root cause of chronic rejection for all solid organ transplants. As the only solid organ transplant that does not undergo primary systemic arterial revascularization at the time of surgery, lung transplants rely on the establishment of a microcirculation and are especially vulnerable to the effects of microvascular loss. Microangiopathy, with its attendant ischemia, can lead to tissue infarction and airway fibrosis. Maintaining healthy vasculature in lung allografts may be critical for preventing terminal airway fibrosis, also known as the bronchiolitis obliterans syndrome (BOS). BOS is the major obstacle to lung transplant success and affects up to 60% of patients surviving 5 years. The role of complement in causing acute microvascular loss and ischemia during rejection has recently been examined using the mouse orthotopic tracheal transplantation; this is an ideal model for parsing the role of airway vasculature in rejection. Prior to the development of airway fibrosis in rejecting tracheal allografts, C3 deposits on the vascular endothelium just as tissue hypoxia is first detected. With the eventual destruction of vessels, microvascular blood flow to the graft stops altogether for several days. Complement deficiency and complement inhibition lead to markedly improved tissue oxygenation in transplants, diminished airway remodeling, and accelerated vascular repair. $CD4^+$ T cells and antibody-dependent complement activity independently mediate vascular destruction and sustained tissue ischemia during acute rejection. Consequently, interceding against complement-mediated microvascular injury with adjunctive therapy during acute rejection episodes, in addition to standard immunosuppression which targets $CD4^+$ T cells, may help prevent the subsequent development of chronic rejection.

16.1 Introduction

Chronic rejection after transplantation is the primary cause of long-term morbidity and mortality in solid organ transplant recipients (Libby and Pober 2001). Although not widely studied at this time, emerging clues from preclinical models and clinical studies suggest that the maintenance of a functional microvasculature is required for immunosuppression to be effective (Ozdemir et al. 2004; Babu et al. 2007). Chronic rejection of solid organ transplants develops in close association with microvascular attrition (Luckraz et al. 2004, 2006; Bishop et al. 1989; Matsumoto et al. 1993). In lung transplantation,

M.A. Khan • M.R. Nicolls (✉)
Division of Pulmonary and Critical Care Medicine, Department of Medicine,
VA Palo Alto/Stanford University School of Medicine,
300 Pasteur Dr A283 MC 5351, Stanford, CA 94305, USA
e-mail: mnicolls@stanford.edu

J.D. Lambris et al. (eds.), *Complement Therapeutics*, Advances in Experimental Medicine and Biology 735,
DOI 10.1007/978-1-4614-4118-2_16, © Springer Science+Business Media New York 2013

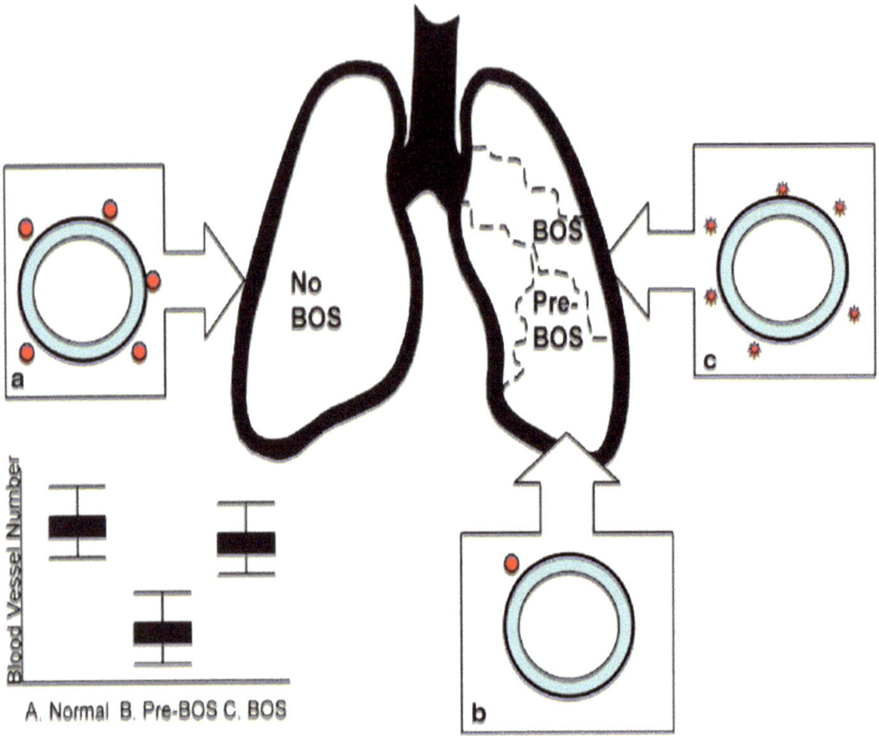

Fig. 16.1 *Microvascular dropout prior to BOS development.* An autopsy study (Luckraz et al. 2004, 2006) has shown that lung transplant patients who die without BOS (A) have normal numbers of blood vessels around airways. In those patients dying with BOS, otherwise normal airways adjacent to BOS airways (i.e., pre-BOS) exhibit diminished microvasculature (B), whereas in adjacent lung with BOS, there are increased numbers of small-caliber blood vessels (C)

chronic rejection is manifested by BOS (Trulock et al. 2006; Yousem et al. 1985). Microvascular loss results in local tissue ischemia and may be an important cause of fibrotic wound healing (Babu et al. 2007; Luckraz et al. 2004, 2006; Minami et al. 2006; Platt et al. 1991). While ischemia-reperfusion injury due to the sudden recirculation of devitalized tissue following transplantation surgery is well recognized, microvascular-injury-associated ischemia, which occurs because of acute rejection, was only recently described by our group (Babu et al. 2007; Jiang et al. 2011). Therapeutics targeting critical pathways involved in microvascular injury are expected to improve clinical outcomes in transplantation (Contreras and Briscoe 2007), but information is lacking about what immune factors are directly responsible for tissue ischemia during acute rejection. This chapter mainly focuses on the role of complement in vascular destruction in transplanted lungs, a phenomenon that is presumably at play in other solid organ transplants.

To study this issue in a model relevant to lung transplantation, our group has utilized mouse orthotopic tracheal transplants (OTTs) (Babu et al. 2007; Jiang et al. 2011; Khan et al. 2011). Grafted trachea is functional transplants through which mice breathe and, in rejection, the airways pathologically replicate lymphocytic bronchiolitis (a large airway precursor of BOS) (Sato and Keshavjee 2008). Findings about fibrosis development in large airways from OTT research can, with appropriate caveats, be extrapolated to fibrogenesis in terminal bronchioles (Babu et al. 2007; Jiang et al. 2011; Murakawa et al. 2005; Kuo et al. 2006). The OTT model is useful because the well-organized planar anatomy of airway microvasculature supports the study of relatively long segments of microvessels (Babu et al. 2007). Recently, the Papworth autopsy study demonstrated a marked loss of microvessels in preobliterative bronchiolitis (OB) foci of human lung transplants which suggested that a loss of microcirculation and airway ischemia precede the onset of OB (Luckraz et al. 2004, 2006) (Fig. 16.1).

These preclinical and clinical studies cumulatively suggest that preserving normal airway circulation is of likely benefit to the overall health (and patency) of the respiratory tree.

During acute rejection, profound physiologic events are occurring in the transplanted tissue beyond inflammation: most notably significant tissue hypoxia due to microvascular injury (Babu et al. 2007; Jiang et al. 2011). Typical histological techniques do not capture this information, and subsequently, the dynamic changes of microvascular blood flow and tissue oxygenation remain unknown for solid organ transplants undergoing acute rejection. Endothelial cells are a well-established target for both adaptive and innate immune responses in allograft rejection (Al-Lamki et al. 2008). Prior studies have determined that effector $CD4^+$ T cells and complement appear to be directly injurious to allogeneic endothelial cells (Minami et al. 2006; Platt et al. 1991; Shiao et al. 2007; Baldwin et al. 2000; Choi et al. 2007). As described below, we recently determined that antibody-dependent complement activity, working independently of $CD4^+$ cells, is sufficient to induce microvascular injury and tissue ischemia observed during acute rejection. This chapter will review the special role that complement-mediated vascular injury plays in airway transplant damage and chronic rejection.

16.2 Lung Transplantation and the Vasculature

BOS is characterized by a decline in ventilatory function and fibro-obliteration of small airways and is the major obstacle to survival following lung transplantation (Trulock et al. 2006). Despite identification of risk factors for the development of BOS, such as acute rejection and cytomegalovirus infection, the etiology of the fibroproliferative changes associated with BOS remains unknown. Recent autopsy studies from Luckraz and colleagues at Papworth Hospital, Cambridge, demonstrated a significant loss of microvasculature in nonobliterated small airways from BOS lungs, suggesting that airway ischemia is a preceding condition to airway fibrosis (Luckraz et al. 2004, 2006). In this study, it was shown that lung transplant patients who die without BOS have normal numbers of blood vessels around their airways. In those patients dying with BOS, otherwise normal airways adjacent to BOS airways (i.e., pre-BOS airways) exhibit significantly diminished microvasculature, whereas in adjacent lung with BOS, there are increased numbers of small-caliber blood vessels (Fig. 16.1).

Lung transplants are the only solid organ allografts that do not routinely undergo direct systemic arterial reconnection at the time of surgery. Airways are normally supplied by a dual circulation derived from the bronchial arteries and the pulmonary artery, and the source of microvessels around airways following transplantation is not fully known. Only the pulmonary artery circulation is restored at the time of transplantation rejection. Because the bronchial anastomosis heals well with the current surgical procedure, reconnecting the bronchial arteries has been deemed unnecessary (Patterson 1993). Therefore, following lung transplantation, the low-O_2 pulmonary artery circulation is the major source of blood and microvasculature for transplanted lungs. It is important to note that, prior to transplantation, about 50% of blood flow to airways normally comes from the bronchial arteries and 50% from the poorly oxygenated pulmonary artery circulation (Barman et al. 1988).

Bronchial artery revascularization at the time of lung transplantation is feasible and was originally performed in several hundred patients (mainly in Copenhagen) in the early 1990s. However, this procedure was no longer utilized when it was noted that tracheal anastomoses healed well without bronchial artery reconnection. For this reason, the highly oxygenated bronchial artery circulation is now sacrificed in all lung transplant recipients. Omitting this rearterialization step may have more distant effects not evident in the early postoperative clinical course. Preclinical and preliminary clinical studies demonstrate that bronchial artery revascularization is durable (Norgaard et al. 1997), improves tissue perfusion with more highly oxygenated blood (Sundset et al. 1997; Kamler et al. 2004), is associated with less epithelial metaplasia (Norgaard et al. 1999), is protective of pulmonary endothelium and type II pneumocytes (Nowak et al. 2002), and may delay the development of BOS while improving patient survival (Norgaard et al. 1998). Without this native circulation, lung transplants rely on the establishment of a microcirculation to provide perfusion to the airways. However, even if bronchial artery revascularization were routinely performed, donor vasculature expressing foreign MHC antigens will always remain a target for alloimmune rejection responses.

16.3 Hypoxia, Ischemia, and Fibrosis

In airway transplantation, there is evidence that the constellation of hypoxia, alloimmune inflammation, and ultimately ischemia culminates in irreversible fibrosis (Babu et al. 2007). The mechanisms by which hypoxia and ischemia contribute to postinflammatory fibrosis are not established but are observed in several clinical scenarios such as normal skin wound healing (Gurtner et al. 2008) and chronic kidney diseases (Fine and Norman 2008). In pulmonary fibrosis studies in humans and rodents, microarray data sets demonstrate that hypoxia signaling is a prominently dysregulated pathway (Cosgrove et al. 2004; Kaminski and Rosas 2006; Zuo et al. 2002). In vitro studies have shown profibrotic phenotypic change of fibroblasts in response to hypoxia (Cool et al. 2006; Karakiulakis et al. 2007). Epithelial and endothelial cells can undergo mesenchymal transition (EMT) under ischemia to become another source of activated fibroblasts (Manotham et al. 2004). Hypoxia likely directly contributes to the progression of fibrosis by increasing the release of major extracellular matrix proteins (Distler et al. 2007). Transforming growth factor-β2-induced fibrosis is associated with remarkable vasoconstriction and tissue hypoxia (Ledbetter et al. 2000). Which of the above phenomena (i.e., activated fibroblasts, EMT, release of matrix proteins) contributes the most to airway fibrosis is not established, but it is clear that inflamed tissue subject to low pO_2 and ischemia is at considerable risk for fibrotic remodeling.

16.4 Experimental Models of Lung Transplantation

There are currently no completely satisfactory models of experimental lung transplantation which replicate the fibrotic BOS lesion in terminal bronchioles. While orthotopic lung transplantation in mice is a recently developed technique (Okazaki et al. 2007), the procedure itself is technically challenging with three surgical anastomoses and has neither the throughput nor the planar anatomy to be conducive to large-scale studies of the microvasculature. Additionally, this model does not develop airway fibrosis. Heterotopic tracheal transplantation has been employed in mice and does develop an occlusive airways disease akin to BOS (Hertz et al. 1993). While this has been used to dissect out certain mechanisms of fibrotic airway occlusion, its nonnative and nonaerated position also makes it somewhat removed from clinical BOS. The use of OTT as a model for alloimmune airway rejection is now well established (Murakawa et al. 2005; Genden et al. 2002; Minamoto and Pinsky 2002). Airway remodeling as a result of chronic injury is manifested by subepithelial fibrosis, akin to the large airways in human lung transplantation (Paradis et al. 1993). The OTT model is not an obstructive airways disease model, like heterotopic transplants, but rather reproduces the fibrosis that follows lymphocytic bronchitis (i.e., large airway inflammation that can precede BOS) (Ross et al. 1997; Glanville et al. 2008). In clinical BOS, which is characterized by terminal bronchiole luminal fibrosis, large airways correspondingly show subepithelial collagen deposition (Zheng et al. 1997) similar to chronically rejecting OTTs. Therefore, the OTT model is useful for studying airway fibrosis but only allows inferences about the fibrotic process that takes place in terminal bronchioles. The benefit over heterotopic transplants is that the OTT model is functional and has, as its microenvironment, normal superior mediastinal structures. Therefore, dendritic cell clearance of shed antigens to draining lymph nodes for antigen presentation to T cells is reproduced in its native position. Transplanting airways into their normal anatomic sites may preserve the regular dendritic cell and lymphocyte trafficking pathways critical for the study of normal airway immunity. OTTs are exposed to ambient air, and so the unique respiratory interface between host and environment is maintained. Further, the inferior pole of the trachea is normally supplied by the bronchial artery circulation and is interrupted at the time of surgery. A significant advantage of the OTT model is that the microvasculature is densely arrayed in a single tissue plane allowing relatively easy quantitation and description. This is much less readily studied in lung parenchyma or terminal bronchioles where the anatomy is not so concentrated in a defined area. These favorable features have led our group to favor the OTT model for airway vascular research.

16.5 Complement, Antibodies, and Vascular Endothelium

As an important component of the innate immune system, the complement system represents a biochemical cascade consisting of a number of small proteins that circulate as inactive zymogens. When stimulated by one of several triggers such as antibody deposition on vascular endothelium, proteases in the system cleave specific proteins to release cytokines and initiate an amplifying cascade of further cleavages. The complement system is subdivided into three cascades: the classical, alternate, and lectin pathways. All three pathways converge on the C3 protein. This process culminates in the activation of the cell-killing membrane attack complex (MAC). Since the 1990s, the major laboratories examining complement and the microvasculature in transplantation and ischemia-reperfusion injury have been under the direction of Jeffrey Platt, George Tsokos, Stephen Tomlinson, Michael Carroll, Greg Stahl, and Michael Holers. Platt's group has long focused on the deleterious role that complement plays and the protective role of complement regulatory proteins (CRPs) on the vascular endothelium of rejecting xenografts (Dalmasso et al. 1992; Byrne et al. 1995; Bustos et al. 1997; Daggett et al. 1997; Gaca et al. 2006). Endothelial injury in Platt's lung xenotransplantation studies leads to vascular leakiness, macrovascular congestion and thrombus formation, pulmonary edema, and neutrophilic invasion of the microvasculature (Gaca et al. 2006; Yeatman et al. 1999). Like all lung transplantation studies to date, this pioneering work did not assess tissue perfusion or oxygenation but rather globally assesses lung tissue for injury. George Tsokos, Michael Carroll, Mike Holers, and Stephen Tomlinson's groups have primarily focused on ischemia-reperfusion injury. These elegant studies have variously demonstrated the ameliorating impact of targeted complement inhibition and the role of natural antibody in complement-mediated injury. A critical distinction in these ischemia-reperfusion studies is that ischemia happens *first* by manual manipulation and that complement-mediated injury ensues. The OTT model suggests that the converse process occurs in vivo; complement activity *leads* to microvascular destruction and ischemia, and tissue injury follows.

Figure 16.2 illustrates the fate of microvessels following rejection in OTT (Babu et al. 2007; Jiang et al. 2011; Khan et al. 2011). Following transplantation, donor and recipient microvessels reconnect through angiogenesis and fusion at the surgical anastomosis site (Babu et al. 2007). Alloimmune injury from antibody-dependent complement activity and $CD4^{+}$ T cells leads to a loss of connection between the vascular systems of the recipient and the donor; the graft becomes ischemic and not rescuable with immunosuppression (Babu et al. 2007; Khan et al. 2011). Thus, the health of the transplant is closely tied to a functional microvasculature in the transplant. Figures 16.3 and 16.4 illustrate what happens to airway perfusion and oxygenation with rejection and a loss of the functional microvasculature. Figure 16.4 is the whole-mount appearance of animals treated with fluorescein isothiocyanate conjugated to lectin (FITC-lectin) i.v. just prior to sacrifice. All vessels that are connected to the systemic circulation will fluoresce, and when the connection is lost, FITC-lectin staining is absent.

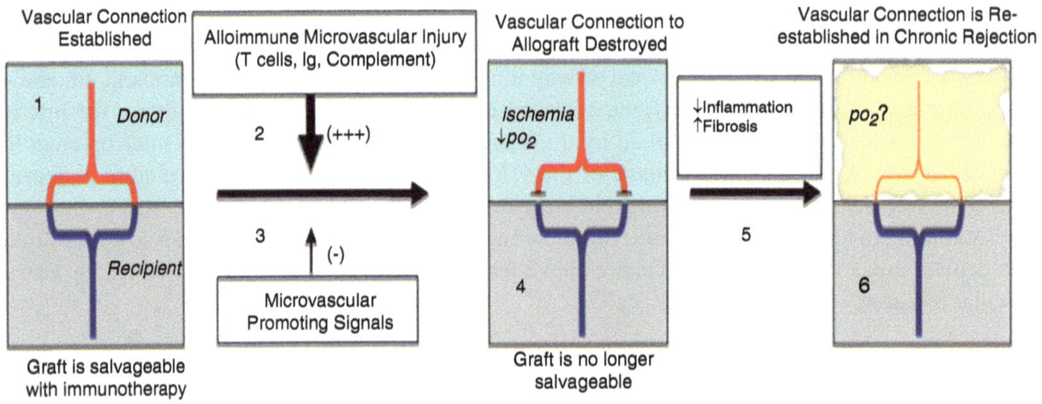

Fig. 16.2 *Model of microvasculature destruction in airway allografts.* (*1*) Within days of transplantation, recipient vessels (*blue*) fuse with donor blood vessels (*red*) at the anastomosis site, and blood flow to the graft is transiently restored (Luckraz et al. 2006). (*2*) With uncontrolled rejection, the tissue then becomes progressively hypoxic, and microvasculature is increasingly injured by alloimmune attack; notably T cells, Ig, and complement. (*3*) Hypoxia induces signals that promote microvascular integrity and growth. HIF-1α is a "master regulator" of this hypoxic response. However, despite multiple proangiogenic signals, the microvascular connection is eventually destroyed resulting in ischemia, decreased tissue pO2, and a loss of paracrine signaling between the vessels and neighboring tissues. (In diagram, this is indicated by alloimmune injury represented by a *larger arrow* than the microvascular promoting signals). (*4*) At this point, immunotherapy will not rescue the graft tissue from chronic rejection (Luckraz et al. 2006). (*5*) Endogenous anti-inflammatory molecules subsequently promote the clearance of invading leukocytes and early fibrosis occurs. (*6*) With fibroproliferation, the vascular connection is reestablished, but the fibrotic graft can no longer be restored to normal cellular architecture with immunotherapy. Therefore, the key to understanding and treating chronic rejection may require increased understanding of factors causing microvascular injury and, conversely, the pathways promoting microvascular stabilization

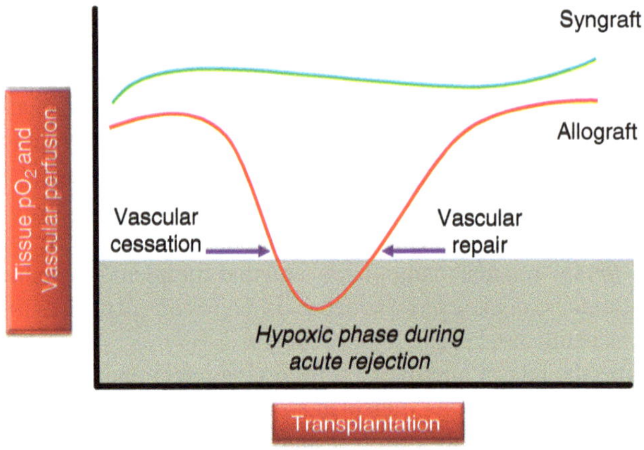

Fig. 16.3 Graphic representation of allograft tissue pO_2 and blood vascular perfusion trends in syngeneic and allogeneic transplants during course of acute rejection in mouse OTT model

Following the loss of microvessels, there is a late return of perfusion as recipient-derived vasculature pierces the ischemic transplant. We have shown that this return of microvessels (vascular repair) occurs through a hypoxia-inducible factor-1α (HIF-1α)-mediated process (Jiang et al. 2011); augmenting the HIF-1α response decreases the hypoxic burden of rejection by promoting microvessel

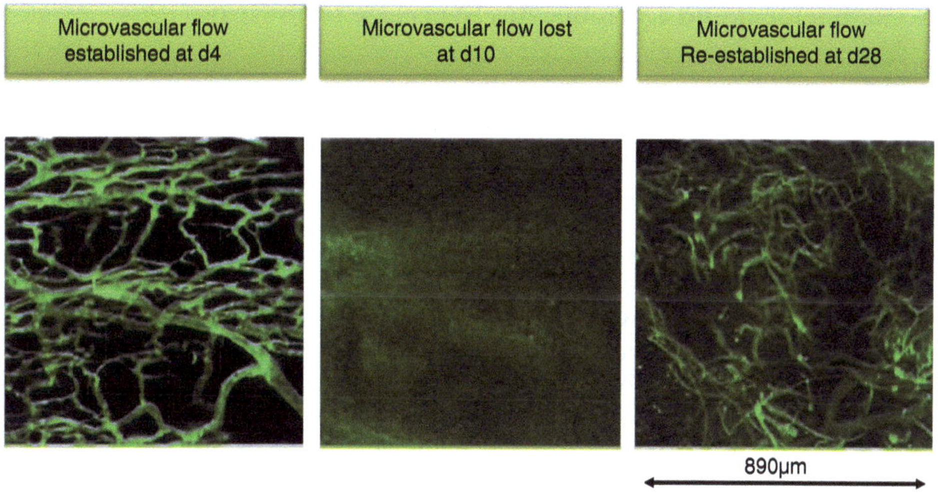

Fig. 16.4 FITC-lectin perfusion shows microvascular flow detected in rejecting allograft at different time period during rejection in mouse orthotopic tracheal transplantation model

integrity through the recruitment of recipient-derived endothelial cells. Thus, a major idea evolving from these animal studies is that rejection is associated with significant tissue hypoxia and limiting the hypoxic burden reduces the extent of fibrotic wound healing and chronic rejection.

16.6 Complement and Rejection in Lung Transplantation

While established in heart and kidney transplantation, complement-mediated injury is less well understood in lung transplantation. In the last several years, there have been diverging reports about the extent and implication of complement deposition in lung transplant recipients (Magro et al. 2003a, b, 2006; Wallace et al. 2005; Ionescu et al. 2005). Baldwin and colleagues demonstrated that rat orthotopic lung transplants performed at different academic centers indeed demonstrated complement deposition on vascular endothelium similar to rejecting heart and kidney transplants (Murata et al. 2008). In heterotopic tracheal transplantation, blockade of complement with soluble complement receptor 1 (sCR1) inhibits airway occlusive disease (Kallio et al. 2000). While Baldwin's group has shown that complement deficiency results in less vascular injury to lung allografts (Nakashima et al. 2002), no study prior to our recent OTT study (Khan et al. 2011) had previously explored how complement-mediated injury to the vascular endothelium affected the functionality of the vessels and whether complement-mediated injury to the vasculature could lead to progressive tissue hypoxia and frank ischemia. A randomized, placebo-controlled clinical study demonstrated that complement inhibition led to improved early outcomes following lung transplantation (as reflected in earlier extubations) (Keshavjee et al. 2005). This therapy was targeting the ischemia-reperfusion injury which occurs at the time of lung transplantation and may have protected the airway microvasculature. In summary, the study of complement in lung transplant rejection is relatively new and, although contentious, shows significant promise for providing insights into disease pathobiology and therapeutic approaches. While generalized complement activation as a result of ischemia perfusion injury at the time of lung transplantation has been investigated, no study in lung transplantation (or any other form of transplantation) has systematically investigated the impact that complement activation has on vascular flow during rejection.

16.7 Complement and Lymphocytes Synergize to Destroy Microvessels in Rejecting Airways

As noted, microvascular loss has been strongly associated with chronic rejection (Ozdemir et al. 2004; Babu et al. 2007; Luckraz et al. 2004, 2006; Jiang et al. 2011; Ishii et al. 2005; Labarrere et al. 2001). An important task was to try to attribute physiologic derangements to individual immune components. By focusing on the functionality of graft microvasculature, we were recently able to demonstrate that CD4^{+} T cells and antibody-dependent complement activity independently mediated graft ischemia during acute rejection (Khan et al. 2011). C3 deposition is first noted in microvessels approximately after 1 week of rejection, several days prior to the loss of microvessels (Fig. 16.5). C3-mediated vascular injury appears to be mediated by the formation of MAC in airway microvessels (unpublished observations, Fig. 16.6). CD8^{+} T cells, in isolation, do not induce microvascular-injury-associated ischemia. Airway ischemia during acute rejection was closely associated with the presence of subepithelial fibrosis. The duration of airway ischemia closely correlated with the appearance of the overlying epithelium with a brief cessation of airway perfusion during rejection still allowing the preservation of an intact overlying columnar lining but prolonged ischemia being tightly associated with a flattened dysplastic-appearing overlying epithelium. Thus, as our other studies have recently demonstrated (Babu et al. 2007; Jiang et al. 2011), microvascular-injury-associated ischemia in airways, during acute rejection episodes, appears to be causally linked to the airway remodeling observed in chronic rejection.

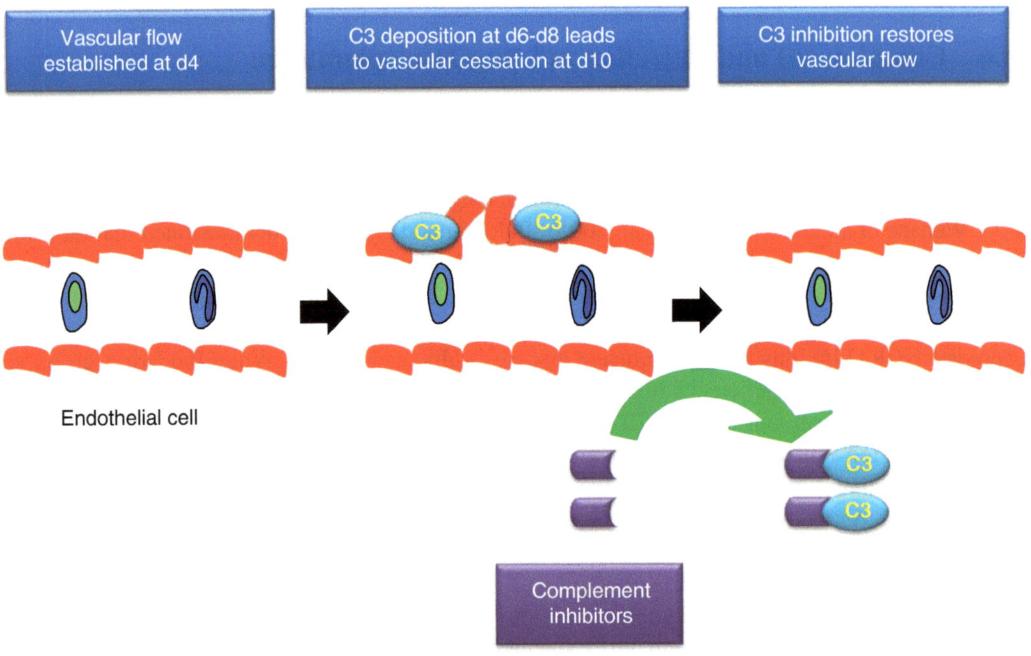

Fig. 16.5 Schematic representation of complement-mediated microvascular loss and a novel approach to optimize microvascular flow by specific complement inhibitors in lung transplantation

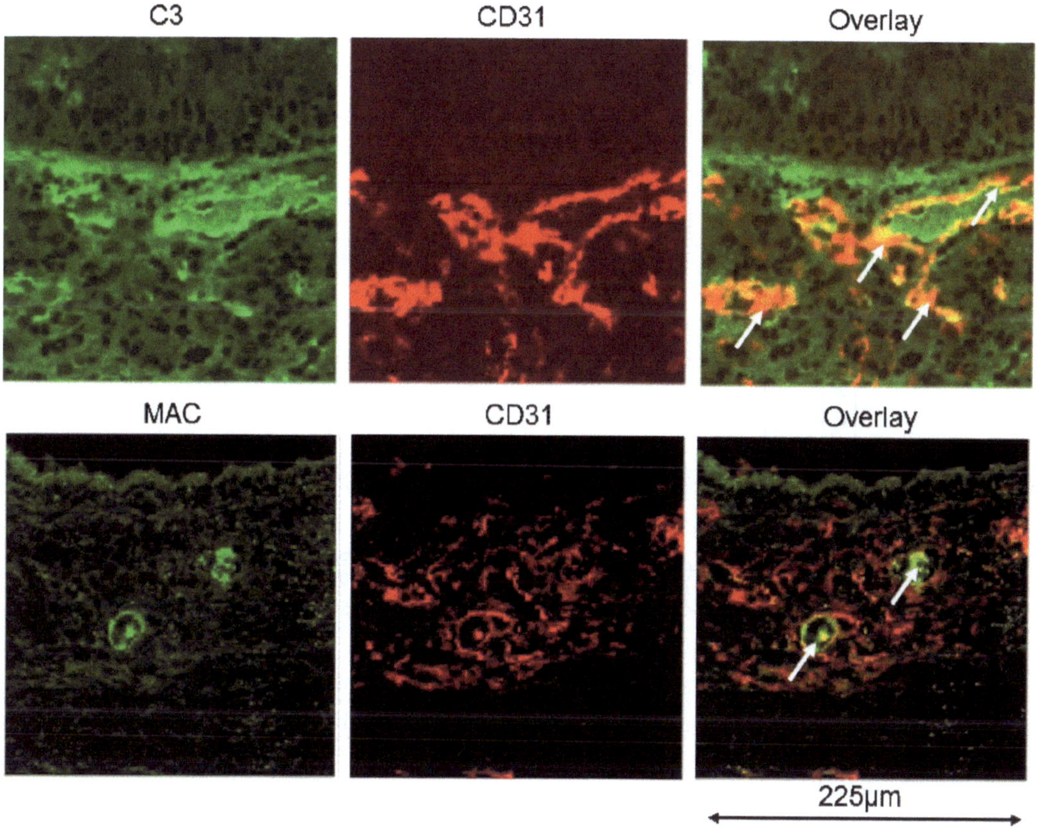

Fig. 16.6 Activated complement (C3) and membrane attack complex (MAC) deposition during acute rejection. *Arrows* indicate colocalization of C3 and MAC on CD31$^+$ vascular endothelial cells

In our recent study (Khan et al. 2011), the following sequence of key results (gleaned from multiple experimental groups) subsequently led to the discovery of two independent effector pathways for microvascular-injury-associated ischemia: (1) Allografts in unreconstituted complement-replete RAG1$^{-/-}$ mice (which lack B and T cells) do not become ischemic. (2) CD4-reconstitution of RAG1$^{-/-}$ recipients results in transplant ischemia. (3) CD4-replete/C3-inhibited recipients demonstrate graft ischemia. (4) Adoptive transfer of donor-specific MHC class II antibodies restores graft ischemia in B and T cell-deficient/complement-replete RAG1$^{-/-}$ recipients. (5) C3-deficient/antibody-replete/CD4-depleted WT recipients do not develop ischemia. The latter two results indicated that in the absence of CD4 cells, C3 and antibody are both required for graft ischemia to occur. The cumulative results indicate that CD4$^+$ T cells and antibody-dependent complement activation independently mediate microvascular-injury-associated ischemia. These results are summarized in Table 16.1. Because C3-deficient/CD4-depleted WT recipients did not become ischemic in this OTT study, additional effector pathways were not required to explain the loss of perfusion in airway transplants. Thus, the possibility was raised that increased complement activity during acute rejection episodes could cause allograft damage that is not addressed with the standard approach of escalating steroids. An unexpected finding in this study, pointing to how adjunctive complement inhibitors may be helpful, was that an administered complement blocker (CR2-Crry) limited graft ischemia by promoting enhanced recipient-derived angiogenesis (i.e., treatment with a complement inhibitor leads to a rapid restoration of graft vascularity by promoting recipient vessels to invade the transplant).

Table 16.1 Role of complement, antibodies and lymphocytes in maintaining microvascular flow in tracheal transplants

Experimental groups			Recipient immune characteristics				Allograft characteristics			
OTT donor	OTT recipient	Treatment	C3	B cells/Ig	CD4 cells	CD8 cells	Acute rejection[a]	Graft hypoxia[b]	Graft ischemia[c]	Chronic rejection
B6	B6	None	Present	Present	Present	Present	No	No	Absent	No
BALB/c	B6	None	Present	Present	Present	Present	Yes	Yes	Moderate	Yes
BALB/c	B6 RAG1$^{-/-}$	None	Present	Absent	Absent	Absent	No	No	Absent	No
BALB/c	B6 RAG1$^{-/-}$	CD4-reconstitution	Present	Absent	Present	Absent	Yes	Yes	Severe	Yes
BALB/c	B6	Anti-CD8	Present	Present	Present	Absent	Yes	Yes	Severe	Yes
BALB/c	B6 RAG1$^{-/-}$	CD8-reconstitution	Present	Absent	Absent	Present	Yes	No	Absent	(ND)[d]
BALB/c	B6	Anti-CD4	Present	Present	Absent	Present	Yes	Yes	Mild	No
BALB/c	B6	Anti-CD4+anti-CD8	Present	Present	Absent	Absent	No	Yes	Mild	No
BALB/c	B6 RAG1$^{-/-}$	None	Present	Absent	Absent	Absent	Yes	No	Absent	No
BALB/c	B6 RAG1$^{-/-}$	B4 IgM	Present	Anti-Annexin IV IgM	Absent	Absent	No	No	Absent	(ND)
BALB/c	B6 RAG1$^{-/-}$	Anti-IAd MHC II IgG	Present	Donor-Specific IgG	Absent	Absent	No	Yes	Present[e]	(ND)
BALB/c	B6 C3$^{-/-}$	Anti-CD4	Absent	Present	Absent	Present	No	No	Absent	No
BALB/c	B6 C3$^{-/-}$	Anti-CD8	Absent	Present	Present	Absent	Yes	Yes	Severe	(ND)
BALB/c	B6 C3$^{-/-}$	Anti-CD4+ anti-CD8	Absent	Present	Absent	Absent	No	No	Absent	No
BALB/c	B6	CR2-Crry	Inhibited	Present	Present	Present	Yes	Yes	Mild	No

Adapted from Khan et al. (2011)

[a]Acute rejection is defined mononuclear infiltration of the transplant

[b]Graft hypoxia defined as significantly different that syngraft oxygenation at any respective time point

[c]Graft ischemia defined as "absent" when none detected, "mild" when the ischemic period was <2 days, "moderate" when perfusion took >2 and <18 days to be restored, and "severe" when reperfusion was not detected during a 28 day period

[d]*ND* not determined)

[e]Only 1 time point assessed (d10) and, therefore, period of ischemia is not described

16.8 Complement Activation and Dysregulated Angiogenesis

Complement activation causes a dysregulation of normal angiogenesis required for healthy fetal development, and C3 inhibition with Crry-Ig blocks the pathological increase in soluble VEGFR-1, a potent inhibitor of VEGF activity, and rescues pregnancies in mice (Girardi et al. 2006). As just noted, in our recent study, CR2-Crry treatment promoted enhanced neovascularization of rejecting transplants in associated with upregulated proangiogenic factors within the transplant; these factors may have caused the rapid reinvestment of recipient-derived microvessels into these rejected airways (Khan et al. 2011). $C3^{-/-}$ recipients had a similar profile of upregulated proangiogenic factors associated with a faster revascularization period. The loss of donor-derived microvasculature is evident in $C3^{-/-}$ recipients; their rapid replacement with nontransgenic vessels is consistent with recipient vessels growing into the transplant as an accelerated response to increased graft-derived proangiogenic factors. So, while antibody-dependent complement activity was implicated in the destruction of microvessels during rejection, antagonism of C3 activity (when unopposed $CD4^{+}$ cells are still present and capable of independently destroying the vessels) promoted rapid vascular repair and limited graft ischemia. Thus, there appears to be a dual rationale for complement inhibitor treatment during rejection episodes: (1) preventing vascular destruction and (2) promoting vascular repair.

16.9 Complement-Mediated Microvascular Injury in Transplanted Airways and the Development of Chronic Rejection

Our recent studies have strongly suggested that the duration of hypoxia and ischemia in allograft rejection is relevant to the subsequent development of significant airway remodeling as manifested by a loss of normal epithelium and increased subepithelial fibrosis (Jiang et al. 2011; Khan et al. 2011). Lung transplant recipients are particularly vulnerable to the deleterious effects of airway hypoxia and ischemia. The lung is unique among solid organ transplants because of the lack of the surgical restoration of a vascular connection to the systemic circulation. Blood supply to the airways in lung transplant recipients, in contrast to the normal dual circulation, presumably comes from the deoxygenated pulmonary artery circulation (Nicolls and Zamora 2010). Therefore, from the outset, lung transplant airways have an impaired microcirculation due to the lack of bronchial artery restoration, which, we have recently demonstrated, results in relative airway tissue hypoxia in lung transplant patients (Dhillon et al. 2010). We have hypothesized that this baseline airway hypoxia may be a diathesis for chronic rejection in lung transplant recipients (Dhillon et al. 2010) and therapies which preserve microvascular integrity may be especially relevant.

Profound tissue ischemia, undetectable by histology, may "silently" occur during acute rejection in transplant recipients. Understanding how complement activation can cause microvascular-injury-associated ischemia should help to more logically target therapeutics designed to preserve microvessel integrity. Recently obtained physiologic information suggests that targeted complement inhibitors could safely synergize with conventional therapy during rejection episodes to prevent sustained tissue ischemia and limit the development of chronic rejection.

References

Al-Lamki RS, Bradley JR, Pober JS (2008) Endothelial cells in allograft rejection. Transplantation 86(10):1340–1348

Babu AN, Murakawa T, Thurman JM, Miller EJ, Henson PM, Zamora MR et al (2007) Microvascular destruction identifies murine allografts that cannot be rescued from airway fibrosis. J Clin Invest 117(12):3774–3785

Baldwin WM III, Qian Z, Ota H, Samaniego M, Wasowska B, Sanfilippo F et al (2000) Complement as a mediator of vascular inflammation and activation in allografts. J Heart Lung Transplant 19(8):723–730
Barman SA, Ardell JL, Parker JC, Perry ML, Taylor AE (1988) Pulmonary and systemic blood flow contributions to upper airways in canine lung. Am J Physiol 255(5 Pt 2):H1130–H1135
Bishop GA, Waugh JA, Landers DV, Krensky AM, Hall BM (1989) Microvascular destruction in renal transplant rejection. Transplantation 48(3):408–414
Bustos M, Coffman TM, Saadi S, Platt JL (1997) Modulation of eicosanoid metabolism in endothelial cells in a xenograft model. Role of cyclooxygenase-2. J Clin Invest 100(5):1150–1158
Byrne GW, McCurry KR, Kagan D, Quinn C, Martin MJ, Platt JL et al (1995) Protection of xenogeneic cardiac endothelium from human complement by expression of CD59 or DAF in transgenic mice. Transplantation 60(10):1149–1156
Choi J, Cho YM, Yang WS, Park TJ, Chang JW, Park SK (2007) Peritubular capillary C4d deposition and renal outcome in post-transplant IgA nephropathy. Clin Transplant 21(2):159–165
Contreras AG, Briscoe DM (2007) Every allograft needs a silver lining. J Clin Invest 117(12):3645–3648
Cool CD, Groshong SD, Rai PR, Henson PM, Stewart JS, Brown KK (2006) Fibroblast foci are not discrete sites of lung injury or repair: the fibroblast reticulum. Am J Respir Crit Care Med 174(6):654–658
Cosgrove GP, Brown KK, Schiemann WP, Serls AE, Parr JE, Geraci MW et al (2004) Pigment epithelium-derived factor in idiopathic pulmonary fibrosis: a role in aberrant angiogenesis. Am J Respir Crit Care Med 170(3):242–251
Daggett CW, Yeatman M, Lodge AJ, Chen EP, Van Trigt P, Byrne GW et al (1997) Swine lungs expressing human complement-regulatory proteins are protected against acute pulmonary dysfunction in a human plasma perfusion model. J Thorac Cardiovasc Surg 113(2):390–398
Dalmasso AP, Vercellotti GM, Fischel RJ, Bolman RM, Bach FH, Platt JL (1992) Mechanism of complement activation in the hyperacute rejection of porcine organs transplanted into primate recipients. Am J Pathol 140(5):1157–1166
Dhillon GS, Zamora MR, Roos JE, Sheahan D, Sista RR, Van der Starre P et al (2010) Lung transplant airway hypoxia: a diathesis to fibrosis? Am J Respir Crit Care Med 182(2):230–236
Distler JH, Jungel A, Pileckyte M, Zwerina J, Michel BA, Gay RE et al (2007) Hypoxia-induced increase in the production of extracellular matrix proteins in systemic sclerosis. Arthritis Rheum 56(12):4203–4215
Fine LG, Norman JT (2008) Chronic hypoxia as a mechanism of progression of chronic kidney diseases: from hypothesis to novel therapeutics. Kidney Int 74(7):867–872
Gaca JG, Appel JZ III, Lukes JG, Gonzalez-Stawinski GV, Lesher A, Palestrant D et al (2006) Effect of an anti-C5a monoclonal antibody indicates a prominent role for anaphylatoxin in pulmonary xenograft dysfunction. Transplantation 81(12):1686–1694
Genden EM, Boros P, Liu J, Bromberg JS, Mayer L (2002) Orthotopic tracheal transplantation in the murine model. Transplantation 73(9):1420–1425
Girardi G, Yarilin D, Thurman JM, Holers VM, Salmon JE (2006) Complement activation induces dysregulation of angiogenic factors and causes fetal rejection and growth restriction. J Exp Med 203(9):2165–2175
Glanville AR, Aboyoun CL, Havryk A, Plit M, Rainer S, Malouf MA (2008) Severity of lymphocytic bronchiolitis predicts long-term outcome after lung transplantation. Am J Respir Crit Care Med 177(9):1033–1040
Gurtner GC, Werner S, Barrandon Y, Longaker MT (2008) Wound repair and regeneration. Nature 453(7193): 314–321
Hertz MI, Jessurun J, King MB, Savik SK, Murray JJ (1993) Reproduction of the obliterative bronchiolitis lesion after heterotopic transplantation of mouse airways. Am J Pathol 142(6):1945–1951
Ionescu DN, Girnita AL, Zeevi A, Duquesnoy R, Pilewski J, Johnson B et al (2005) C4d deposition in lung allografts is associated with circulating anti-HLA alloantibody. Transpl Immunol 15(1):63–68
Ishii Y, Sawada T, Kubota K, Fuchinoue S, Teraoka S, Shimizu A (2005) Injury and progressive loss of peritubular capillaries in the development of chronic allograft nephropathy. Kidney Int 67(1):321–332
Jiang X, Khan MA, Tian W, Beilke J, Natarajan R, Kosek J et al (2011) Adenovirus-mediated HIF-1alpha gene transfer promotes repair of mouse airway allograft microvasculature and attenuates chronic rejection. J Clin Invest 121(6):2336–2349
Kallio EA, Lemstrom KB, Hayry PJ, Ryan US, Koskinen PK (2000) Blockade of complement inhibits obliterative bronchiolitis in rat tracheal allografts. Am J Respir Crit Care Med 161(4 Pt 1):1332–1339
Kaminski N, Rosas IO (2006) Gene expression profiling as a window into idiopathic pulmonary fibrosis pathogenesis: can we identify the right target genes? Proc Am Thorac Soc 3(4):339–344
Kamler M, Nowak K, Bock M, Herold U, Motsch J, Hagl S et al (2004) Bronchial artery revascularization restores peribronchial tissue oxygenation after lung transplantation. J Heart Lung Transplant 23(6):763–766
Karakiulakis G, Papakonstantinou E, Aletras AJ, Tamm M, Roth M (2007) Cell type-specific effect of hypoxia and platelet-derived growth factor-BB on extracellular matrix turnover and its consequences for lung remodeling. J Biol Chem 282(2):908–915
Keshavjee S, Davis RD, Zamora MR, de Perrot M, Patterson GA (2005) A randomized, placebo-controlled trial of complement inhibition in ischemia-reperfusion injury after lung transplantation in human beings. J Thorac Cardiovasc Surg 129(2):423–428

Khan MA, Jiang X, Dhillon G, Beilke J, Holers VM, Atkinson C et al (2011) CD4+ T cells and complement independently mediate graft ischemia in the rejection of mouse orthotopic tracheal transplants. Circ Res 109(11):1290–1301

Kuo E, Bharat A, Goers T, Chapman W, Yan L, Street T et al (2006) Respiratory viral infection in obliterative airway disease after orthotopic tracheal transplantation. Ann Thorac Surg 82(3):1043–1050

Labarrere CA, Nelson DR, Park JW (2001) Pathologic markers of allograft arteriopathy: insight into the pathophysiology of cardiac allograft chronic rejection. Curr Opin Cardiol 16(2):110–117

Ledbetter S, Kurtzberg L, Doyle S, Pratt BM (2000) Renal fibrosis in mice treated with human recombinant transforming growth factor-beta2. Kidney Int 58(6):2367–2376

Libby P, Pober JS (2001) Chronic rejection. Immunity 14(4):387–397

Luckraz H, Goddard M, McNeil K, Atkinson C, Charman SC, Stewart S et al (2004) Microvascular changes in small airways predispose to obliterative bronchiolitis after lung transplantation. J Heart Lung Transplant 23(5):527–531

Luckraz H, Goddard M, McNeil K, Atkinson C, Sharples LD, Wallwork J (2006) Is obliterative bronchiolitis in lung transplantation associated with microvascular damage to small airways? Ann Thorac Surg 82(4):1212–1218

Magro CM, Klinger DM, Adams PW, Orosz CG, Pope-Harman AL, Waldman WJ et al (2003a) Evidence that humoral allograft rejection in lung transplant patients is not histocompatibility antigen-related. Am J Transplant 3(10):1264–1272

Magro CM, Ross P Jr, Kelsey M, Waldman WJ, Pope-Harman A (2003b) Association of humoral immunity and bronchiolitis obliterans syndrome. Am J Transplant 3(9):1155–1166

Magro CM, Abbas AE, Seilstad K, Pope-Harman AL, Nadasdy T, Ross P Jr (2006) C3d and the septal microvasculature as a predictor of chronic lung allograft dysfunction. Hum Immunol 67(4–5):274–283

Manotham K, Tanaka T, Matsumoto M, Ohse T, Inagi R, Miyata T et al (2004) Transdifferentiation of cultured tubular cells induced by hypoxia. Kidney Int 65(3):871–880

Matsumoto Y, McCaughan GW, Painter DM, Bishop GA (1993) Evidence that portal tract microvascular destruction precedes bile duct loss in human liver allograft rejection. Transplantation 56(1):69–75

Minami K, Murata K, Lee CY, Fox-Talbot K, Wasowska BA, Pescovitz MD et al (2006) C4d deposition and clearance in cardiac transplants correlates with alloantibody levels and rejection in rats. Am J Transplant 6(5 Pt 1):923–932

Minamoto K, Pinsky DJ (2002) Recipient iNOS but not eNOS deficiency reduces luminal narrowing in tracheal allografts. J Exp Med 196(10):1321–1333

Murakawa T, Kerklo MM, Zamora MR, Wei Y, Gill RG, Henson PM et al (2005) Simultaneous LFA-1 and CD40 ligand antagonism prevents airway remodeling in orthotopic airway transplantation: implications for the role of respiratory epithelium as a modulator of fibrosis. J Immunol 174(7):3869–3879

Murata K, Iwata T, Nakashima S, Fox-Talbot K, Qian Z, Wilkes DS et al (2008) C4d deposition and cellular infiltrates as markers of acute rejection in rat models of orthotopic lung transplantation. Transplantation 86(1):123–129

Nakashima S, Qian Z, Rahimi S, Wasowska BA, Baldwin WM III (2002) Membrane attack complex contributes to destruction of vascular integrity in acute lung allograft rejection. J Immunol 169(8):4620–4627

Nicolls MR, Zamora MR (2010) Bronchial blood supply after lung transplantation without bronchial artery revascularization. Curr Opin Organ Transplant 15(5):563–567

Norgaard MA, Efsen F, Andersen CB, Svendsen UG, Pettersson G (1997) Medium-term patency and anatomic changes after direct bronchial artery revascularization in lung and heart-lung transplantation with the internal thoracic artery conduit. J Thorac Cardiovasc Surg 114(3):326–331

Norgaard MA, Andersen CB, Pettersson G (1998) Does bronchial artery revascularization influence results concerning bronchiolitis obliterans syndrome and/or obliterative bronchiolitis after lung transplantation? Eur J Cardiothorac Surg 14(3):311–318

Norgaard MA, Andersen CB, Pettersson G (1999) Airway epithelium of transplanted lungs with and without direct bronchial artery revascularization. Eur J Cardiothorac Surg 15(1):37–44

Nowak K, Kamler M, Bock M, Motsch J, Hagl S, Jakob H et al (2002) Bronchial artery revascularization affects graft recovery after lung transplantation. Am J Respir Crit Care Med 165(2):216–220

Okazaki M, Krupnick AS, Kornfeld CG, Lai JM, Ritter JH, Richardson SB et al (2007) A mouse model of orthotopic vascularized aerated lung transplantation. Am J Transplant 7(6):1672–1679

Ozdemir BH, Demirhan B, Ozdemir FN, Dalgic A, Haberal M (2004) The role of microvascular injury on steroid and OKT3 response in renal allograft rejection. Transplantation 78(5):734–740

Paradis I, Yousem S, Griffith B (1993) Airway obstruction and bronchiolitis obliterans after lung transplantation. Clin Chest Med 14(4):751–763

Patterson GA (1993) Airway revascularization: is it necessary? Ann Thorac Surg 56(4):807–808

Platt JL, Fischel RJ, Matas AJ, Reif SA, Bolman RM, Bach FH (1991) Immunopathology of hyperacute xenograft rejection in a swine-to-primate model. Transplantation 52(2):214–220

Ross DJ, Marchevsky A, Kramer M, Kass RM (1997) "Refractoriness" of airflow obstruction associated with isolated lymphocytic bronchiolitis/bronchitis in pulmonary allografts. J Heart Lung Transplant 16(8):832–838

Sato M, Keshavjee S (2008) Bronchiolitis obliterans syndrome: alloimmune-dependent and -independent injury with aberrant tissue remodeling. Semin Thorac Cardiovasc Surg 20(2):173–182, Summer

Shiao SL, Kirkiles-Smith NC, Shepherd BR, McNiff JM, Carr EJ, Pober JS (2007) Human effector memory CD4+ T cells directly recognize allogeneic endothelial cells in vitro and in vivo. J Immunol 179(7):4397–4404
Sundset A, Tadjkarimi S, Khaghani A, Kvernebo K, Yacoub MH (1997) Human en bloc double-lung transplantation: bronchial artery revascularization improves airway perfusion. Ann Thorac Surg 63(3):790–795
Trulock EP, Edwards LB, Taylor DO, Boucek MM, Keck BM, Hertz MI (2006) Registry of the International Society for Heart and Lung Transplantation: twenty-third official adult lung and heart-lung transplantation report–2006. J Heart Lung Transplant 25(8):880–892
Wallace WD, Reed EF, Ross D, Lassman CR, Fishbein MC (2005) C4d staining of pulmonary allograft biopsies: an immunoperoxidase study. J Heart Lung Transplant 24(10):1565–1570
Yeatman M, Daggett CW, Lau CL, Byrne GW, Logan JS, Platt JL et al (1999) Human complement regulatory proteins protect swine lungs from xenogeneic injury. Ann Thorac Surg 67(3):769–775
Yousem SA, Burke CM, Billingham ME (1985) Pathologic pulmonary alterations in long-term human heart-lung transplantation. Hum Pathol 16(9):911–923
Zheng L, Ward C, Snell GI, Orsida BE, Li X, Wilson JW et al (1997) Scar collagen deposition in the airways of allografts of lung transplant recipients. Am J Respir Crit Care Med 155(6):2072–2077
Zuo F, Kaminski N, Eugui E, Allard J, Yakhini Z, Ben-Dor A et al (2002) Gene expression analysis reveals matrilysin as a key regulator of pulmonary fibrosis in mice and humans. Proc Natl Acad Sci USA 99(9):6292–6297

Chapter 17
Targeting Complement at the Time of Transplantation

Steven Sacks, Julia Karegli, Conrad A. Farrar, Elham Asgari, Wilhelm Schwaeble, Wuding Zhou, and Richard A. Smith

Abstract Complement activation occurs in at least two phases when an organ is transplanted into a naive recipient: during reperfusion with recipient blood particularly when the donor organ has undergone a significant period of ischaemia and then during acute rejection once the recipient immune system has recognised the donor tissue as non-self. Both of these reactions are most obvious in the extravascular compartment of the transplanted organ and involve local synthesis of some of the key complement components as well as loss of controls that limit the activation of the pivotal component C3. In contrast, sensitised individuals with pre-existing circulating antibodies have an immediate reaction against the transplant organ that is also complement dependent but is enacted in the intravascular space. All three types of injury (ischaemia-reperfusion, acute rejection, hyperacute rejection) have a critical effect on transplant outcome. Here we discuss therapeutic strategies that are designed to overcome the impact of these factors at the start of transplantation with the aim of improving long-term transplant outcomes. These include the concept of treating the donor organ with modified therapeutic regulators that are engineered to be retained by the donor organ after transplantation and prevent inflammatory injury during the critical early period. By targeting the donor organ with anchored therapeutic proteins, the systemic functions of complement including host defence remain intact. The control of complement activation during the first stages of transplantation, including the possibility that this will reduce the capacity of the graft for stimulating the adaptive immune system, offers an important prospect for increasing the longevity of the transplant and offsetting demand on the limited supply of donor organs. It also provides a model in which the benefits and indications for localised therapy to maximise therapeutic efficiency and minimise the systemic disturbance may be instructive in other complement-related disorders.

S. Sacks (✉) • J. Karegli • C.A. Farrar • E. Asgari • W. Zhou • R.A. Smith
Division of Transplantation Immunology and Mucosal Biology, MRC Centre for Transplantation, 5th Floor, Tower Wing, Great Maze Pond, Guy's Hospital, London SE1 9RT, UK

NIHR Comprehensive Biomedical Research Centre, Guy's & St Thomas' NHS Foundation Trust and King's College London, 16th Floor, Tower Wing, Guy's Hospital, Great Maze Pond, London SE1 9RT, UK
e-mail: steven.sacks@kcl.ac.uk

W. Schwaeble
Department of Infection, Immunity and Inflammation, University of Leicester, Leicester, UK

J.D. Lambris et al. (eds.), *Complement Therapeutics*, Advances in Experimental Medicine and Biology 735,
DOI 10.1007/978-1-4614-4118-2_17,

17.1 Basic Concepts

There are three major shortcomings of modern transplantation which affect most organ and tissue transplants, including kidney, liver, pancreas, heart, lung and pancreatic islets (Lechler et al. 2005). These limitations are: the shortage of donor organs, the impact of the inflammatory response on graft survival and the dangers of systemic treatment given to prevent rejection. Inflammation is initiated non-specifically at the time of transplantation by metabolic or physical stress and is then perpetuated by the adaptive immune response when it recognises alloantigen on the donor tissue. These inflammatory and adaptive mechanisms combine to provide a potent effector response that reduces the chance of long-term survival of the graft and applies further pressure on the limited pool of donor organs. Furthermore, as a result of broadly acting immunosuppressive drugs given to prevent rejection, there is a significant risk of developing infection and cancer.

Several aspects of this early inflammatory and alloimmune response are dependent on complement, and this includes both the afferent and efferent limbs of the immune reaction (Li et al. 2007). Building a therapeutic strategy not only depends on a body of critical knowledge of the relevant biochemical targets in the complement cascade at different stages of transplant injury but also on understanding the timing and localisation of these events. Thus, it is critical to understand precisely when complement is triggered, where it causes maximum tissue damage, how and when complement augments the alloimmune response and at which tissue sites the complement components that mediate these deleterious effects are produced.

The strategy described here involves the construction of therapeutic proteins that are retained in the donor organ after pretreatment and remain active for a critical period following transplantation. Thus, using non-genetic means to modify the biological properties of the donor organ, we have been able to examine the effects on recovery of graft function after transplantation and investigate the impact on the potential for long-term graft survival. The purpose of this article is to reflect on the principles underpinning this approach and examine the initial results in preclinical and early clinical development. We will briefly touch upon other approaches being taken that build on similar principles.

17.2 Identification of Therapeutic Targets in Transplanted Kidney

Complement activation occurs during the early course of the transplant procedure, initially upon reperfusion of the ischaemic organ and subsequently at the onset of transplant rejection, which in untreated animals is marked by cellular infiltration of the donor organ within a few days of transplant surgery.

Ischaemia-reperfusion injury is the product of tissue hypoxia, mitochondrial damage and ATP depletion, followed by the generation of free radicals and tissue inflammation once reperfusion with recipient blood has started. In kidney, the focus of this injury is the renal tubule, where complement is despotised on the outer (basolateral) surface of the proximal tubules (Zhou et al. 2000). In controlled experiments, complement deposition mainly occurs at the time of reperfusion of the ischaemic organ. But in the clinical setting, complement deposition can already be present before the organ is connected to the recipient circulation, presumably as a result of complement activation before the flushed organ is removed from a deceased donor (Naesens et al. 2009; Damman et al. 2011a). Denudation of membrane-bound regulators of complement activation has also been described in association with ischaemia-reperfusion injury of the kidney (Thurman et al. 2006). A dual mechanism, namely, complement triggering and loss of control of complement activation on the cell surface, may thus explain the vulnerability of the kidney (and possibly other organs) to complement-mediated organ damage following ischaemia.

Two of the main effectors contributing to post-ischaemic damage are C5a and C5b-9. The interaction of C5a with its specific receptor on the tubule epithelium and on polymorphs contributes to post-ischaemia acute renal failure in experimental animals, as shown by pharmacological blockade (de Vries et al. 2003; Lewis et al. 2008). The formation of C5b-9 on the renal tubule following reperfusion injury also has a substantial effect, as shown in mice with genetic defects in C5b-9 formation (Zhou et al. 2000). Since the cleavage of C5 into C5a and C5b-9 is dependent on C3 cleavage and since renal tissue has significant capacity to synthesise C3 (Tang et al. 1999a), the local production of complement by intrarenal synthesis has a more critical effect on ischaemia-reperfusion injury than does circulating C3 (produced mainly by hepatic synthesis) (Farrar et al. 2006). Limitation of C3 synthesis and/or cleavage therefore offers a candidate therapeutic approach to the control of ischaemia-reperfusion damage. The rationale for targeting C3 is extended by renal transplant data in patients showing that slow decline of organ function after transplantation correlates with C3 gene expression in the pretransplant biopsy (Naesens et al. 2009).

A second bout of renal transplant injury occurs during the onset of cell-mediated rejection and coincides with a massive increase in complement activation (Pratt et al. 2002). This is detected on the tubule surface and in the surrounding interstitium, associated with an infiltrate of T cells and macrophages. Interestingly, several types of migratory leucocytes including macrophages, dendritic cells, T cells and polymorphs have the capability for complement synthesis (Peng et al. 2006; Pantazis et al. 1990; Botto et al. 1992; Colten et al. 1979), increasing the potential for local synthesis of complement components during acute rejection by up to several hundredfold in terms of the content of gene transcript (Farrar et al. 2006; Pratt et al. 2002; Keslar et al. 2008; Li et al. 2010). The production of C3 by tissue-resident cells is essential for transplant rejection in experimental animals, presumably mediated by the effects of C3a and C5a on antigen-presenting cell function and T cell alloreactivity (Keslar et al. 2008; Peng et al. 2008, 2009; Lalli et al. 2008). This is in addition to the destructive capacity of the terminal pathway complement components on renal parenchymal cells.

In contrast to cell-mediated rejection, where complement primarily targets structures in the extravascular tissue space (i.e. the interstitium and tubules, in the case of the kidney), antibody-mediated rejection involves the vascular compartment, where complement is deposited at the site of antibody binding on the donor endothelium (Colvin and Smith 2005). The actions of C5a on endothelium and platelets and effect of C5b-9 on the vessel wall leading to graft thrombosis are well established. Indeed, C4d deposit within the lumen of small capillaries is a hallmark of antibody-mediated classical pathway activation at these sites [to the extent that C4d presence is now formally required for the diagnosis of humoral rejection (Solez et al. 2008)]. Prevention of the cleavage of C3 or C5 or blockade of the actions of C5a and C5b-9 therefore offers therapeutic opportunities to avoid antibody-mediated rejection.

T cell priming and B cell priming by donor antigen takes place in secondary lymphoid organs, i.e. local lymph nodes and spleen. If complement-based therapy is seen as an adjunct to conventional immunosuppression, then the delivery of the therapeutic must take into account the anatomical location of immune priming, as this may require a different approach from the one taken to protect the graft. Nevertheless, since antigen-presenting cells that initiate the immune response originate in the donor organ and migrate out of the graft and into the recipient shortly after transplantation, targeting the graft may offer an opportunity to reduce the potency of donor antigen-presenting cells for allostimulation. At later stages when recipient scavenger cells enter the graft, it may be possible to modify the behaviour of these cells taking up foreign antigen and presenting it to the immune system (Kerekes et al. 1998). The mechanism of complement-enhanced antigen presentation appears to involve C3a-receptor and C5a-receptor signalling (Peng et al. 2006; Li et al. 2010; Lalli et al. 2008; Zhou et al. 2006; Strainic et al. 2008). On the other hand, in the case of antigen uptake and B cell priming, C3b and C3d receptor (CR2, CR3 and CR4) signalling may provide more relevant targets (Fischer et al. 1996; Dempsey et al. 1996).

In summary, candidate targets include C5a and C5b-9 for the prevention of epithelial and endothelial cell injury in the transplant, whereas C3a, C3b, C3d and C5a are eligible targets for more directly preventing alloreactivity. The value of C3 as a therapeutic target is the ability to limit both sets of these reactions.

17.3 The Design of Membrane-Targeted Therapeutic Regulators

Mirococept (also known as APT070) is a derivative of the human complement regulatory protein CR1 (complement receptor 1). Human CR1 (CD35) is a transmembrane protein found on the surface of erythrocytes, neutrophils, monocytes, B cells and T cells (Roozendaal and Carroll 2007). The most common form of CR1 (220 kDa) consists of 30 short consensus repeats, which are cysteine-rich and are a feature of the family of associated regulators (regulators of complement activation) and encoded on chromosome 1. CR1 binds C3b and C4b and enhances their proteolysis and dissociation from the C3 and C5 convertases – the proteolytic enzyme complexes that cleave C3 and C5 and lead to the production of key effectors in the complement cascade. As a potent inhibitor of the C3 and C5 convertases of the classical and alternative pathways, membrane-associated CR1 protects cells from injury caused by the products of complement activation generated in the region of those cells.

The best known therapeutic derivative of CR1 is soluble CR1 (sCR1), a recombinant human protein that lacks the transmembrane portion and is thus secreted by transfected Chinese hamster ovary (CHO) host cells in which it is expressed (Weisman et al. 1990). The soluble protein retains all the inhibitory activities of the native molecule in vitro. Given by intravenous administration, sCR1 serves as an inhibitor of ischaemia-reperfusion injury in a wide variety of models (Rioux 2001). The soluble therapeutic also delays the onset of allograft rejection (Pratt et al. 1996). However, due to its size and systemic route of administration, sCR1 is likely to exert maximal effect in the circulating compartment.

In contrast, Mirococept (23 kDa) contains only a small inhibitory fragment consisting of the three terminal domains of CR1, expressed not in CHO cells but in the more cost-effective *Escherichia coli* host (Dodd et al. 1995). This fragment is sufficient to inhibit the cleavage of C3 and C5 via the alternative and classical pathways. It is linked (by an inserted cysteine residue) to a tail, which consists of a basic peptide component and fatty acid (myristoyl) component. This tail enables the active fragment to attach to cells by a two-step binding and lipid bilayer insertion process (Smith 2002). The cell-bound therapeutic had significantly higher (100-fold) functional inhibitory activity in complement haemolytic assays compared with the non-tailed soluble protein. In effect, the engineered CR1 derivative mimics the function of the native molecule, in the sense of protecting cells decorated with the therapeutic molecule.

17.4 Cytoprotective Effects of Mirococept in Experimental Animals

Mirococept has undergone either systemic or local administration. After intravenous injection in rats, radiolabelled Mirococept was retained selectively by the kidney and was detectable with external imaging (unpublished data). Following infusion into the renal artery of rats, Mirococept was localised throughout the renal capillary bed and on the basolateral surface of the tubule epithelium, as shown by histological staining with monoclonal antibody that detects the therapeutic human fragment (Patel et al. 2006). Rat kidney isografts pretreated with a single intra-arterial injection were resistant to acute kidney injury following ischaemia and reperfusion. Moreover, the treated kidneys showed less chronic vascular damage 20 weeks after the ischaemia-reperfusion insult compared with control-treated organs (Pratt et al. 2003). After prolonged cold storage (16 h), treatment increased the viability of transplanted organs to over 50%, compared with 20% in the control-treated group (Patel et al. 2006). This suggests a way forward both to increase the number of transplant successes using marginal donor organs that might otherwise be unsuitable for transplantation and to improve the functional outcome of standard donor organs.

Given that the cytotopic agent is only detectable for about 24–48 h on the cell surface and is then internalised or shed (Patel et al. 2006), a key question is whether the cytoprotective effect will be sufficient to lower the adaptive immune response. To some extent, this is answered in experiments using limited MHC mismatch between the organ donor and recipient rats (Pratt et al. 2003). Here, treatment of the donor organ with the tailed regulator led to reduced antidonor T cell responses in the recipient spleen and to lower T cell infiltration of the graft. While it is difficult to distinguish the cytoprotective effect of the inhibitor by an action on tissue parenchymal cells from a possible action on passenger antigen-presenting cells transported with the donor organ, it is worth noting that murine antigen-presenting cells treated with Mirococept in vitro have reduced capacity for allostimulation (Peng et al. 2008). This raises the expectation of dual benefit from treating donor organs, both from an anti-inflammatory effect on the graft and from modulation of the immunostimulatory properties of donor antigen-presenting cells.

17.5 Organ Painting in Man

To date, 76 human subjects have received Mirococept, given in single doses ranging from 2 to 100 mg by a variety of routes. Twenty-nine were healthy volunteers (intravenous route), 30 were rheumatoid arthritis patients (intra-articular), 16 were kidney transplant recipients (intrarenal) and one was a liver recipient (intrahepatic). Overall, no drug-related severe adverse reactions occurred. Plasma elimination half-life in healthy subjects was about 3 h, though the dwell time on cells is more important for the envisaged use and appears to be significantly longer than the plasma life. At least for doses below 10 mg, treatment did not result in any significant inhibition of systemic complement activation (Smith et al. 2001). There have been no instances of antibody developed against the therapeutic construct.

A phase 2a study examined the safety and tolerability of Mirococept in renal transplant recipients. A single dose of the therapeutic (10 mg) in carrier solution (Soltran) was given directly into the renal artery of the donor organ using gravity-assisted perfusion ex vivo. This resulted in distribution of tailed therapeutic in a pattern that was similar to the rat kidney transplant experiments, i.e. lining the renal capillaries and arterioles and covering the outer surface of the proximal convoluted tubules of the donor organ (Patel et al. 2006).

The outcome of this pilot study was to show that treatment of the donor kidney was feasible and safe (Smith et al. 2007). Only small amounts of Mirococept were detectable in the recipient serum after transplantation, indicating that the majority of the therapeutic was retained in the graft. Moreover, there was no systemic inhibition of complement measured by C3 and C4 and by complement haemolytic (CH50) assays. Although efficacy was not the primary aim of this small study, there was a difference in early function of the transplant kidneys in the control (n=4) and treatment (n=12) groups, as measured by area-under-the-curve for creatinine between days 1 and 14 post-transplantation. This difference was most obvious in paired donor organs which were removed from the same donor after circulatory arrest and allocated either to treatment (n=2) or control (n=2). This indicated the potential for greater benefit in the presence of severe kidney injury due to ischaemic stress.

The next step is a multicentre phase 2b study to formally test the *e*fficacy of *M*irococept in *p*reventing *i*schaemia-*r*eperfusion *i*njury in *k*idney *al*lografts (EMPIRIKAL). The EMPIRIKAL trial is being supported by the UK Medical Research Council and is expected to begin patient recruitment in 2012. The design involves 280 patients in each arm of a double-blinded randomised controlled trial of Mirococept versus perfusion fluid. The main endpoint is the incidence of delayed graft function (DGF), a manifestation of severe ischaemia-reperfusion injury and defined by requirement for continued dialysis until the function of the transplant kidney resumes. If treatment with the novel therapeutic leads to significant reduction in the incidence of DGF, a surrogate marker for poor long-term outcome of the transplant, then this would be expected to have an impact on clinical outcomes.

17.6 Possible Benefits and Risks of Cytotopic Therapy

A key argument for localised therapy is that retention of the cytotopic agent in the graft offers the opportunity to increase therapeutic efficacy and limit the systemic treatment-related toxicity. This is particularly relevant for transplant patients, whose immune systems are already compromised due to treatment with broadly acting immunosuppressive drugs.

Another vivid illustration currently being explored in rats is the use of a novel cytotopic agent that is designed to prevent coagulation (Melchionna et al. 2010). The therapeutic construct includes a synthetic analogue of hirudin, a thrombin inhibitor found in leech saliva. The synthetic analogue (hirulog) is already licensed for clinical use (Maraganore 1993). When tailed in a manner similar to Mirococept and planted on the cell surface, the antithrombin construct blocks the conversion of fibrinogen to fibrin and inhibits clot formation more effectively than soluble antithrombin. Bound to the vasculature of donor rat kidney, the tailed antithrombin reduces clot formation initiated by high titres of anti-MHC antibodies present in the recipient. Little if any of the bound construct enters the recipient circulation, which means that the transplant surgery can be performed safely without systemic anticoagulation (unpublished observation).

There is a theoretical risk that localised complement inhibition will increase the incidence of infection of the transplanted kidney. Urinary tract infection is common after transplantation and is associated with organisms such as *E. coli* that is frequent in the lower urinary tract. It seems improbable that a single dose of Mirococept, which remains detectable on the cell surface for about 24 h, will have an impact on the incidence of renal tract infection. Moreover, most strains of urinary *E. coli* are complement resistant, and opsonisation by complement appears to facilitate microbial invasion of the urinary tract (Springall et al. 2001). In this context, therapeutic complement regulator blocks the uptake of bacteria by host cells and, paradoxically, lowers the incidence of infection (Springall et al. 2001; Li et al. 2006).

17.7 Alternative Complement-Based Approaches to Prevent Renal Reperfusion Injury

Nature has evolved at least three ways in which cells are protected from complement-mediated injury: (a) membrane-associated regulators (e.g. CR1, DAF) which restrict complement activation on the cell surface, (b) soluble regulators (e.g. C4 binding protein) that limit complement activation in the fluid phase and (c) soluble factors (factor H) capable of acting in the fluid phase and also in the solid phase by virtue of electrostatic binding of factor H to extracellular matrix components. Technology has provided an assortment of therapeutic mimics, such as Mirococept, sCR1, C4-binding protein and factor H (Ricklin and Lambris 2007). In addition, monoclonal antibodies (e.g. anti-C5) and synthetic peptides (e.g. Compstatin, C5a-receptor antagonist) extend the range of therapeutics that may one day be of value in organ transplantation (Lewis et al. 2008; Ricklin and Lambris 2008).

An additional approach builds on the biological phenomenon of RNA interference, in which gene expression can be effectively silenced by specific double-stranded small interfering RNAs (siRNA). Work has led to the development of a set of C3 or C3a receptor- or C5a receptor-specific siRNA molecules (Zheng et al. 2006, 2008). Applied in murine models, these specific siRNA molecules inhibit post-ischaemic acute kidney injury (Zheng et al. 2006, 2008). Translation into clinical benefit would require pretreatment of the donor organ in a timely manner in order to limit the two major peaks of complement gene expression that occur during the first 24–48 h post-transplantation and again during acute rejection (Tang et al. 1999b; Damman et al. 2011b).

17.8 Refining the Targets

More precise definition of the components that trigger complement activation during reperfusion injury and graft rejection may allow more specific approaches, with less disturbance of the physiological role of those components that are not involved. Involvement of factor B and lack of involvement of C4 have indicated that the alternative pathway plays a key role in reperfusion injury of native and transplanted kidneys in mice. On the other hand, recent work in cardiac and intestinal models of post-ischaemic injury in mice has suggested that a large amount of the effect of C3 can be explained by lectin pathway activation (Schwaeble et al. 2011). These components include MASP-2 but not C4, suggesting that a C4-independent route initiated by lectin binding to the target tissue is involved in the cleavage of C3. Confirmed in renal transplant models (unpublished data), this would suggest that the trigger mechanism for complement may involve lectin binding to carbohydrate residues possibly exposed by tubular injury following ischaemic insult. Amplification of the amount of cleaved C3 by the alternative pathway could be secondary to lectin pathway triggering in these circumstances (Selander et al. 2006). Other studies have suggested that classical pathway may be responsible for initiating C3 cleavage in reperfusion injury, though the proposed involvement of natural IgM may be tissue dependent (Weiser et al. 1996) and IgM binding may initiate the lectin pathway (Zhang et al. 2004). These findings offer scope for more selective, less disruptive therapeutic approaches that merit further investigation.

17.9 Concluding Remarks and Future Perspective

A comprehensive review of complement-based therapeutic options goes beyond the range of this article. However, it should be apparent that a number of key principles govern the selection of candidate therapeutic agents earmarked for future development. These include the specific indication (e.g. reperfusion damage, cellular rejection or antibody driven rejection), the biochemical target (complement activation, central or effecter steps) and tissue penetration of the therapeutic to the relevant biological compartment (intra- or extravascular, including the lymphoid compartment where required). Knowledge of this is required for each type of transplant, as anatomical, physiological and biochemical differences are likely to affect the local functions of complement. In an already-burdened patient group, such information will allow maximal selectivity of treatment and will avoid undue impairment of the innate host defence.

References

Botto M, Lissandrini D, Sorio C, Walport MJ (1992) Biosynthesis and secretion of complement component (C3) by activated human polymorphonuclear leukocytes. J Immunol 149:1348–1355

Colten HR, Ooi YM, Edelson PJ (1979) Synthesis and secretion of complement proteins by macrophages. Ann N Y Acad Sci 332:482–490

Colvin RB, Smith RN (2005) Antibody-mediated organ-allograft rejection. Nat Rev Immunol 5:807–817

Damman J, Seelen MA, Moers C, Daha MR, Rahmel A, Leuvenink HG, Paul A, Pirenne J, Ploeg RJ (2011a) Systemic complement activation in deceased donors is associated with acute rejection after renal transplantation in the recipient. Transplantation 92:163–169

Damman J, Nijboer WN, Schuurs TA, Leuvenink HG, Morariu AM, Tullius SG, van Goor H, Ploeg RJ, Seelen MA (2011b) Local renal complement C3 induction by donor brain death is associated with reduced renal allograft function after transplantation. Nephrol Dial Transplant 26:2345–2354

de Vries B, Kohl J, Leclercq WK, Wolfs TG, van Bijnen AA, Heeringa P, Buurman WA (2003) Complement factor C5a mediates renal ischemia-reperfusion injury independent from neutrophils. J Immunol 170:3883–3889

Dempsey PW, Allison ME, Akkaraju S, Goodnow CC, Fearon DT (1996) C3d of complement as a molecular adjuvant: bridging innate and acquired immunity. Science 271:348–350

Dodd I, Mossakowska DE, Camilleri P, Haran M, Hensley P, Lawlor EJ, McBay DL, Pindar W, Smith RA (1995) Overexpression in *Escherichia coli*, folding, purification, and characterization of the first three short consensus repeat modules of human complement receptor type 1. Protein Expr Purif 6:727–736

Farrar CA, Zhou W, Lin T, Sacks SH (2006) Local extravascular pool of C3 is a determinant of postischemic acute renal failure. FASEB J 20:217–226

Fischer MB, Ma M, Goerg S, Zhou X, Xia J, Finco O, Han S, Kelsoe G, Howard RG, Rothstein TL, Kremmer E, Rosen FS, Carroll MC (1996) Regulation of the B cell response to T-dependent antigens by classical pathway complement. J Immunol 157:549–556

Kerekes K, Prechl J, Bajtay Z, Jozsi M, Erdei A (1998) A further link between innate and adaptive immunity: C3 deposition on antigen-presenting cells enhances the proliferation of antigen-specific T cells. Int Immunol 10:1923–1930

Keslar K, Rodriguez ER, Tan CD, Starling RC, Heeger PS (2008) Complement gene expression in human cardiac allograft biopsies as a correlate of histologic grade of injury. Transplantation 86:1319–1321

Lalli PN, Strainic MG, Yang M, Lin F, Medof ME, Heeger PS (2008) Locally produced C5a binds to T cell-expressed C5aR to enhance effector T-cell expansion by limiting antigen-induced apoptosis. Blood 112:1759–1766

Lechler RI, Sykes M, Thomson AW, Turka LA (2005) Organ transplantation – how much of the promise has been realized? Nat Med 11:605–613

Lewis AG, Kohl G, Ma Q, Devarajan P, Kohl J (2008) Pharmacological targeting of C5a receptors during organ preservation improves kidney graft survival. Clin Exp Immunol 153:117–126

Li K, Feito MJ, Sacks SH, Sheerin NS (2006) CD46 (membrane cofactor protein) acts as a human epithelial cell receptor for internalization of opsonized uropathogenic *Escherichia coli*. J Immunol 177:2543–2551

Li K, Sacks SH, Zhou W (2007) The relative importance of local and systemic complement production in ischaemia, transplantation and other pathologies. Mol Immunol 44:3866–3874

Li Q, Peng Q, Xing G, Li K, Wang N, Farrar CA, Meader L, Sacks SH, Zhou W (2010) Deficiency of C5aR prolongs renal allograft survival. J Am Soc Nephrol 21:1344–1353

Maraganore JM (1993) Pre-clinical and clinical studies on Hirulog: a potent and specific direct thrombin inhibitor. Adv Exp Med Biol 340:227–236

Melchionna T, Karegli J, Farrar CA, McVey JH, Dorling A, Sacks SH, Smith RAG (2010) Integrating cytotopic strategy in transplantation: (1) Design of locally-acting coagulation inhibitors for use with anti-complement therapeutics. Mol Immunol 47(13):2203, Ref Type: Abstract

Naesens M, Li L, Ying L, Sansanwal P, Sigdel TK, Hsieh SC, Kambham N, Lerut E, Salvatierra O, Butte AJ, Sarwal MM (2009) Expression of complement components differs between kidney allografts from living and deceased donors. J Am Soc Nephrol 20:1839–1851

Pantazis P, Kalyanaraman VS, Bing DH (1990) Synthesis of the third component of complement (C3) by lectin-activated and HTLV-infected human T-cells. Mol Immunol 27:283–289

Patel H, Smith RA, Sacks SH, Zhou W (2006) Therapeutic strategy with a membrane-localizing complement regulator to increase the number of usable donor organs after prolonged cold storage. J Am Soc Nephrol 17:1102–1111

Peng Q, Li K, Patel H, Sacks SH, Zhou W (2006) Dendritic cell synthesis of C3 is required for full T cell activation and development of a Th1 phenotype. J Immunol 176:3330–3341

Peng Q, Li K, Anderson K, Farrar CA, Lu B, Smith RA, Sacks SH, Zhou W (2008) Local production and activation of complement up-regulates the allostimulatory function of dendritic cells through C3a–C3aR interaction. Blood 111:2452–2461

Peng Q, Li K, Wang N, Li Q, Asgari E, Lu B, Woodruff TM, Sacks SH, Zhou W (2009) Dendritic cell function in allostimulation is modulated by C5aR signaling. J Immunol 183:6058–6068

Pratt JR, Hibbs MJ, Laver AJ, Smith RA, Sacks SH (1996) Effects of complement inhibition with soluble complement receptor-1 on vascular injury and inflammation during renal allograft rejection in the rat. Am J Pathol 149:2055–2066

Pratt JR, Basheer SA, Sacks SH (2002) Local synthesis of complement component C3 regulates acute renal transplant rejection. Nat Med 8:582–587

Pratt JR, Jones ME, Dong J, Zhou W, Chowdhury P, Smith RA, Sacks SH (2003) Nontransgenic hyperexpression of a complement regulator in donor kidney modulates transplant ischemia/reperfusion damage, acute rejection, and chronic nephropathy. Am J Pathol 163:1457–1465

Ricklin D, Lambris JD (2007) Complement-targeted therapeutics. Nat Biotechnol 25:1265–1275

Ricklin D, Lambris JD (2008) Compstatin: a complement inhibitor on its way to clinical application. Adv Exp Med Biol 632:273–292

Rioux P (2001) TP-10 (AVANT Immunotherapeutics). Curr Opin Investig Drugs 2:364–371

Roozendaal R, Carroll MC (2007) Complement receptors CD21 and CD35 in humoral immunity. Immunol Rev 219:157–166

Schwaeble WJ, Lynch NJ, Clark JE, Marber M, Samani NJ, Ali YM, Dudler T, Parent B, Lhotta K, Wallis R, Farrar CA, Sacks S, Lee H, Zhang M, Iwaki D, Takahashi M, Fujita T, Tedford CE, Stover CM (2011) Targeting of mannan-binding lectin-associated serine protease-2 confers protection from myocardial and gastrointestinal ischemia/reperfusion injury. Proc Natl Acad Sci USA 108:7523–7528

Selander B, Martensson U, Weintraub A, Holmstrom E, Matsushita M, Thiel S, Jensenius JC, Truedsson L, Sjoholm AG (2006) Mannan-binding lectin activates C3 and the alternative complement pathway without involvement of C2. J Clin Invest 116:1425–1434

Smith RA (2002) Targeting anticomplement agents. Biochem Soc Trans 30:1037–1041

Smith RA, Dodd I, Oldroyd RG, Harry J, Clark C, Rolan P, Dawes L (2001) Preclinical and clinical progression of a membrane-targeted complement regulator therapeutic. Mol Immunol 38:122, Ref Type: Abstract

Smith RAG, Koffman G, Chowdhury P, Smith KCG, Watson CJ, Nicholson ML, Zhou W, Sacks SH (2007) Membrane-localising complement inhibitors – clinical progress. Mol Immunol 44(16):3915, Ref Type: Abstract

Solez K, Colvin RB, Racusen LC, Haas M, Sis B, Mengel M, Halloran PF, Baldwin W, Banfi G, Collins AB, Cosio F, David DS, Drachenberg C, Einecke G, Fogo AB, Gibson IW, Glotz D, Iskandar SS, Kraus E, Lerut E, Mannon RB, Mihatsch M, Nankivell BJ, Nickeleit V, Papadimitriou JC, Randhawa P, Regele H, Renaudin K, Roberts I, Seron D, Smith RN, Valente M (2008) Banff 07 classification of renal allograft pathology: updates and future directions. Am J Transplant 8:753–760

Springall T, Sheerin NS, Abe K, Holers VM, Wan H, Sacks SH (2001) Epithelial secretion of C3 promotes colonization of the upper urinary tract by *Escherichia coli*. Nat Med 7:801–806

Strainic MG, Liu J, Huang D, An F, Lalli PN, Muqim N, Shapiro VS, Dubyak GR, Heeger PS, Medof ME (2008) Locally produced complement fragments C5a and C3a provide both costimulatory and survival signals to naive CD4+ T cells. Immunity 28:425–435

Tang S, Sheerin NS, Zhou W, Brown Z, Sacks SH (1999a) Apical proteins stimulate complement synthesis by cultured human proximal tubular epithelial cells. J Am Soc Nephrol 10:69–76

Tang S, Zhou W, Sheerin NS, Vaughan RW, Sacks SH (1999b) Contribution of renal secreted complement C3 to the circulating pool in humans. J Immunol 162:4336–4341

Thurman JM, Ljubanovic D, Royer PA, Kraus DM, Molina H, Barry NP, Proctor G, Levi M, Holers VM (2006) Altered renal tubular expression of the complement inhibitor Crry permits complement activation after ischemia/reperfusion. J Clin Invest 116:357–368

Weiser MR, Williams JP, Moore FD Jr, Kobzik L, Ma M, Hechtman HB, Carroll MC (1996) Reperfusion injury of ischemic skeletal muscle is mediated by natural antibody and complement. J Exp Med 183:2343–2348

Weisman HF, Bartow T, Leppo MK, Marsh HC Jr, Carson GR, Concino MF, Boyle MP, Roux KH, Weisfeldt ML, Fearon DT (1990) Soluble human complement receptor type 1: in vivo inhibitor of complement suppressing post-ischemic myocardial inflammation and necrosis. Science 249:146–151

Zhang M, Austen WG Jr, Chiu I, Alicot EM, Hung R, Ma M, Verna N, Xu M, Hechtman HB, Moore FD Jr, and Carroll MC (2004). Identification of a specific self-reactive IgM antibody that initiates intestinal ischemia/reperfusion injury. Proc Natl Acad Sci USA 101:3886

Zheng X, Feng B, Chen G, Zhang X, Li M, Sun H, Liu W, Vladau C, Liu R, Jevnikar AM, Garcia B, Zhong R, Min WP (2006) Preventing renal ischemia-reperfusion injury using small interfering RNA by targeting complement 3 gene. Am J Transplant 6:2099–2108

Zheng X, Zhang X, Feng B, Sun H, Suzuki M, Ichim T, Kubo N, Wong A, Min LR, Budohn ME, Garcia B, Jevnikar AM, Min WP (2008) Gene silencing of complement C5a receptor using siRNA for preventing ischemia/reperfusion injury. Am J Pathol 173:973–980

Zhou W, Farrar CA, Abe K, Pratt JR, Marsh JE, Wang Y, Stahl GL, Sacks SH (2000) Predominant role for C5b-9 in renal ischemia/reperfusion injury. J Clin Invest 105:1363–1371

Zhou W, Patel H, Li K, Peng Q, Villiers MB, Sacks SH (2006) Macrophages from C3-deficient mice have impaired potency to stimulate alloreactive T cells. Blood 107:2461–2469

Chapter 18
Evaluation of the Blood Compatibility of Materials, Cells, and Tissues: Basic Concepts, Test Models, and Practical Guidelines

Kristina N. Ekdahl, Jaan Hong, Osama A. Hamad, Rolf Larsson, and Bo Nilsson

Abstract Medicine today uses a wide range of biomaterials, most of which make contact with blood permanently or transiently upon implantation. Contact between blood and nonbiological materials or cells or tissue of nonhematologic origin initiates activation of the cascade systems (complement, contact activation/coagulation) of the blood, which induces platelet and leukocyte activation.

Although substantial progress regarding biocompatibility has been made, many materials and medical treatment procedures are still associated with severe side effects. Therefore, there is a great need for adequate models and guidelines for evaluating the blood compatibility of biomaterials. Due to the substantial amount of cross talk between the different cascade systems and cell populations in the blood, it is advisable to use an intact system for evaluation.

Here, we describe three such in vitro models for the evaluation of the biocompatibility of materials and therapeutic cells and tissues. The use of different anticoagulants and specific inhibitors in order to be able to dissect interactions between the different cascade systems and cells of the blood is discussed. In addition, we describe two clinically relevant medical treatment modalities, the integration of titanium implants and transplantation of islets of Langerhans to patients with type 1 diabetes, whose mechanisms of action we have addressed using these in vitro models.

K.N. Ekdahl (✉)
Department of Immunology, Genetics and Pathology, Rudbeck Laboratory C5:3, Uppsala University, SE-751 85 Uppsala, Sweden

School of Natural Sciences, Linnæus University, SE-391 82 Kalmar, Sweden
e-mail: Kristina.Nilsson_Ekdahl@igp.uu.se

J. Hong • O.A. Hamad • B. Nilsson
Department of Immunology, Genetics and Pathology, Rudbeck Laboratory C5:3, Uppsala University, SE-751 85 Uppsala, Sweden

R. Larsson
Department of Immunology, Genetics and Pathology, Rudbeck Laboratory C5:3, Uppsala University, SE-751 85 Uppsala, Sweden

Corline Systems AB, PO Box 956, SE-751 09 Uppsala, Sweden

J.D. Lambris et al. (eds.), *Complement Therapeutics*, Advances in Experimental Medicine and Biology 735, DOI 10.1007/978-1-4614-4118-2_18, © Springer Science+Business Media New York 2013

18.1 Background

18.1.1 Hemocompatibility

Modern medicine utilizes a wide range of biomaterials, most of which make contact with blood either permanently or transiently after implantation. The indications for treating patients with various biomedical devices have increased substantially during the last few decades and are expected to increase even further. The emerging generation of biomaterials is being incorporated into new polymeric implants such as artificial skin, implants that are able to regrow tissue, and resorbable cements to repair bone, among others.

Substantial progress has been made in recent years regarding biocompatibility, but many materials and medical treatment procedures are still associated with severe side effects. For example, hemodialysis contributes to inflammation and accelerated arteriosclerosis which may lead to premature death of the patient, and treatment with stents has been associated with thrombosis. These detrimental reactions, initiated by contact between the biomaterial and the various defense systems in the blood, indicate that the enigma of blood compatibility is far from being solved (Nilsson et al. 2007; Ratner 2007; Ekdahl et al. 2011). It is clearly imperative to understand and identify these adverse reactions in order to prevent their occurrence. Consequently, there is a great need for adequate models and guidelines for evaluating the hemocompatibility of biomaterials (Seyfert et al. 2002).

18.1.2 Mechanism of Complement, Coagulation and Contact System Activation on Material Surfaces

18.1.2.1 Adsorption of a Primary Protein Layer

When biomaterials or other nonbiological surfaces are exposed to blood, they momentarily become covered with a layer of plasma proteins, approximating a monolayer, through a process of passive adsorption (Andersson et al. 2005a). Both the composition and the conformation of the proteins within this layer affect the initiation and subsequent activation of the cascade systems of the blood (i.e., the complement and contact activation/coagulation systems) and are therefore of critical importance (Engberg et al. 2011). Proteins that undergo profound conformational changes are often more potent activators than are proteins that bind in a more native conformation. Examples of proteins that undergo conformational changes include IgG, the complement activator of the classical pathway (CP) (Wettero et al. 2001; Tengvall et al. 2001); C3, the complement activator of the alternative pathway (AP) (Andersson et al. 2002); factor XII (FXII), an initiator of contact activation (Schousboe et al. 2008); and fibrinogen, a ligand for GPIIb/IIIa on platelets that triggers their activation (Berglin et al. 2009; Wu et al. 2005).

18.1.2.2 Activation of the Complement System

Recognition molecules within the various complement pathways, i.e., complement C1q of the CP, mannose-binding lectin (MBL) of the lectin pathway (LP), and properdin of the AP, may be bound to or incorporated into the first protein layer and promote complement activation. Their incorporation leads to initial opsonization of the material surface by C4b and C3b and the generation of the anaphylatoxins C3a and the less potent C4a. A powerful amplification by the AP then occurs as a result of the fact that C3b is a subunit of the AP C3 convertase (C3bBb), meaning that each deposited C3b molecule can act as the nucleus of a new convertase complex, and the surface can ultimately become completely covered with C3b and its degradation products iC3b and C3d,g. The subsequently formed C5 convertases then activate C5 to become the potent anaphylatoxin C5a and initiate the formation of the lytic C5b-9 complex, which is the end point of the terminal pathway of complement. It has been demonstrated that the AP is able to contribute to >80% of C5 activation in models where the initial activation is highly specific for the CP, which underscores the potency of its amplification loop (Harboe et al. 2004).

The generated C3a and C5a anaphylatoxins recruit and activate phagocytes, i.e., polymorphonuclear leukocytes (PMNs) and monocytes, thus orchestrating an inflammatory reaction aimed at the surface of the biomaterial or nonhematological therapeutic cell that originally initiated the complement activation.

18.1.2.3 Activation of the Contact Activation/Coagulation System

Activation of the contact system at nonbiological surfaces is initiated by FXII and high molecular weight kininogen (HK), leading to the activation of FXI and the generation of the potent vasoactive peptide bradykinin (BK), respectively. FXIa can initiate activation of the downstream portion of the (intrinsic) coagulation system, FIX and then FX, leading to the generation of thrombin from prothrombin. Under nonpathological conditions in vivo, however, the major initiator of the coagulation system is tissue factor (TF). This molecule is translocated from extravascular sources to the bloodstream (or produced by monocytes), where it is expressed on the activated leukocytes present on material surfaces and also in the fluid phase; thereby, it activates the extrinsic coagulation pathway. TF acts as a receptor for FVIIa, which then activates FIX and FX, ultimately leading to the generation of substantial amounts of thrombin.

Thrombin is a multifunctional enzyme whose two major effects are to activate platelets via their protease-activated receptors (PAR)-1 and -4 and to transform soluble fibrinogen into insoluble fibrin. During the process of fibrin formation, thrombin cleaves both fibrinogen and FXIII, the latter of which is responsible for transforming the initially soluble fibrin into insoluble form.

18.1.2.4 Examples of Cross Talk Between Blood Cascade Systems and Cells

Considerable intercommunication has been identified between the various cascade systems and the cells in the blood. For example, a number of serine proteases within the contact and coagulation systems, e.g., thrombin, kallikrein, and FXIIa, have been shown to be capable of cleaving complement components, in vitro, and thereby generating C3a and C5a independently of the formation of convertases (Lachmann et al. 1982; Ghebrehiwet et al. 1983; Thoman et al. 1984; Huber-Lang et al. 2006; Amara et al. 2010).

An opposite association, i.e., a complement protease that is able to induce activation of coagulation components is MBL-associated serine protease (MASP)-1 which is an enzyme with broad specificity whose substrates include fibrinogen and FXIII, in addition to C3 (Krarup et al. 2008; Gulla et al. 2010; Matsushita et al. 2000).

C3 was implicated to have a role in fibrinolysis already in 1970 (Taylor and Müller-Eberhard 1970). In recent years, this association has been rediscovered when multiple complement components and activation products, including C3, were detected in thrombi from patients with acute myocardial infarction using a proteomic approach (Distelmaier et al. 2009). Furthermore, C3 has been reported to be incorporated into blood clots prepared in vitro, using fibrinogen purified from diabetic patients, and to act as an inhibitor of fibrinolysis (Hess et al. 2011; Howes et al. 2012).

Platelet activation by TRAP in vitro induces activation of both the contact and complement systems. FXIIa and FXIa, which are activated on platelet surfaces, are preferentially regulated by antithrombin (AT) rather than by C1 inhibitor (C1INH), the favored regulator when these enzymes are activated on artificial surfaces (Bäck et al. 2009, 2010). The mechanism by which activated platelets affect complement involves the release of chondroitin sulfate A (CS-A), a potent activator of complement, leading to generation of the anaphylatoxins C3a and C5a and subsequent leukocyte activation, as revealed by the upregulation of CD11b and formation of leukocyte-platelet complexes (Hamad et al. 2008, 2010a, b).

In addition, C5a mediates the upregulation of TF on endothelial cells and PMNs in vitro (Ikeda et al. 1997; Ritis et al. 2006), as well as on PMNs and monocytes from patients undergoing hemodialysis for end-stage renal disease. This upregulation is believed to contribute to the greatly enhanced risk of thrombosis seen in these patients (Kourtzelis et al. 2010).

18.2 In Vitro Evaluation of Hemocompatibility

18.2.1 *Toward a Holistic Approach*

All equipment that is intended to make contact with blood or other tissue(s) must be evaluated with regard to its biocompatibility. In the literature, one can find a plethora of biocompatibility studies utilizing protein solutions, e.g., human or bovine serum albumin, mixtures of a few proteins, diluted plasma or serum (from laboratory animals or humans), or purified cell populations or cell lines in culture. The common aim of most of these studies has been to extrapolate data obtained from simplistic systems to the complex processes taking place in vivo when medical devices containing the tested materials come in contact with the human body. Such studies can yield valuable information but are often hampered by missing components. For example, the behavior of a monocytic cell line in culture, in the presence of heat-inactivated diluted serum (in which complement is nonoperative), will differ considerably from that of monocytes in whole blood, which will be regulated by C5a and other mediators as discussed in Sect. 18.1.2.4 above.

Since the various cascade systems and cell types in the blood are known to interact, the true effect of a biomaterial surface on the blood can only be known when whole blood is used for interaction studies (Nilsson et al. 2007; Gorbet and Sefton 2004). Ideally, a test system should be simple and employ biological material (blood) from humans, because (1) the cascade systems and cells behave differently regarding activation and regulation in individuals from different species, and (2) most commercially available reagents are directed against proteins and cells of human origin.

Real-time studies of platelet adhesion and aggregation can be performed using the Grabowski flow model where recalcified citrate blood or noncitrated blood is perfused in a parallel plate chamber until the chamber becomes occluded (Grabowski 1990; Bogdanov et al. 2003; Grabowski et al. 2012). However, this model only allows measurement for a very limited time. We have therefore taken the approach of investigating the inflammation and thrombus formation elicited by a biomaterial in human whole blood models in order to define the molecular interactions that occur between the biomaterial and plasma proteins and cells, both in the fluid phase and at the material surface. Thereafter, this interaction can be investigated at the molecular level, if necessary, in systems that make use of purified fractionated blood (serum or plasma) or purified proteins and cells.

18.2.2 *Importance of Anticoagulation*

The choice of anticoagulant to be added to the blood is of critical importance for the outcome of such experiments. Not only do substances such as citrate and EDTA inhibit the coagulation cascade by chelating Ca^{2+}, they also inhibit complement because they remove Ca^{2+}, which is needed for initiating the CP, as well as Mg^{2+}, which is necessary for the activity of both convertases. In contrast, the contact activation system is not dependent on the presence of cations, and consequently, its initiation is not affected by the addition of chelators.

Table 18.1 Guidelines for the choice of anticoagulants in blood models for biomaterial testing

	Anticoagulants							
	Chelators		Heparin derivatives			Specific inhibitors		No additive[a]
Additive	EDTA	Citrate	HMW	LMW	Fondaparinux	CTI	Lepirudin	No additive
Complement								
Classical pathway	–	–	+/– [b]	+/– [b]	+/– [b]	+	+	+
Lectin pathway	–	–	+/– [b]	+/– [b]	+/– [b]	+	+	+
Alternative pathway	–	–	+/– [b]	+/– [b]	+/– [b]	+	+	+
Terminal pathway	–	–	+/– [b]	+/– [b]	+/– [b]	+	+	+
Contact/coagulation								
Contact system (intrinsic)	+	+	+	+	+	–	+	+
TF pathway (extrinsic)	–	– (+ [c])	+/– [b]	+/– [b]	+	+	+	+
Common pathway	–	– (+ [c])	+/– [b]	–	–	+	–	+
Platelets	(+)	+	+	+	+	+	+	+

HMW high molecular weight, *LMW* low molecular weight, *fondaparinux* the antithrombin-binding pentasaccharide of heparin, *CTI* corn trypsin inhibitor, *TF* tissue factor, – function is inhibited, + function is active

[a]Requires the use of sampling equipment with low activating surface (heparin coating and/or polymers, e.g., polyvinyl chloride) in combination with adequately designed test systems

[b]Dose dependent: inhibitory at higher concentrations

[c]Recalcified citrate blood or plasma

When heparin is used as an anticoagulant at high doses in studies in vitro, it has a potent complement-dampening effect, whereas at lower doses (e.g., 0.2 IU/mL), it has complement-activating activity; therefore, in order to minimize the inhibition of both systems (still keeping the blood fluidity) to be able to dissect the cross talk between the cascade systems and cellular element of the blood with greater accuracy, it is best to use moderate amounts of heparin, e.g., 1–3 IU/mL (Logue 1977; Keil et al. 1995; Hong et al. 2001, 1999a). Alternatively, specific thrombin inhibitors such as melagatran (Ozmen et al. 2002) and lepirudin (hirudin) or synthetic congeners of lepirudin (Mollnes et al. 2002; Bexborn et al. 2009) may be used. If the blood is collected using heparin-coated equipment in an open system (see Sect. 18.2.3 below), it is possible to avoid using anticoagulants altogether provided that the test system is adequately designed (Andersson et al. 2003; Johnell et al. 2005).

The thrombin inhibitor lepirudin, derived from the leech *Hirudo medicinalis*, binds to and irreversibly inhibits active thrombin (Chang 1983). We have set up experimental systems for selective activation of platelets in whole blood or platelet-rich plasma (PRP, in which the erythrocytes and leukocytes have been removed by centrifugation) using synthetic thrombin receptor-activating peptides (TRAP) such as SFLLRN in conjunction with lepirudin (Bäck et al. 2009, 2010; Hamad et al. 2008, 2010a). This peptide is derived from the sequence of the new amino-terminus of the cleaved PAR-1 and can mimic thrombin receptor activation and act as a full agonist for platelet activation (Shankar et al. 1994), while the presence of lepirudin efficiently inhibits all other thrombin-related events.

It has long been known that heparin is able to bind to various types of blood cells (e.g., monocytes and endothelial cells) and interfere with their expression of TF (Attanasio et al. 1998; Gori et al. 1999; Ettelaie et al. 2011), an effect that is obviously undesirable for models intended for the study of inflammatory reactions in vitro (Bexborn et al. 2009). A comparison of four different anticoagulants (EDTA, citrate, heparin, and lepirudin) showed that lepirudin was the least prone to inhibit LPS-induced expression of TF on monocytes in whole blood (Engstad et al. 1997). A summary of different anticoagulants used in biomaterial testing and their effect on the blood cascade systems and platelets is found in Table 18.1.

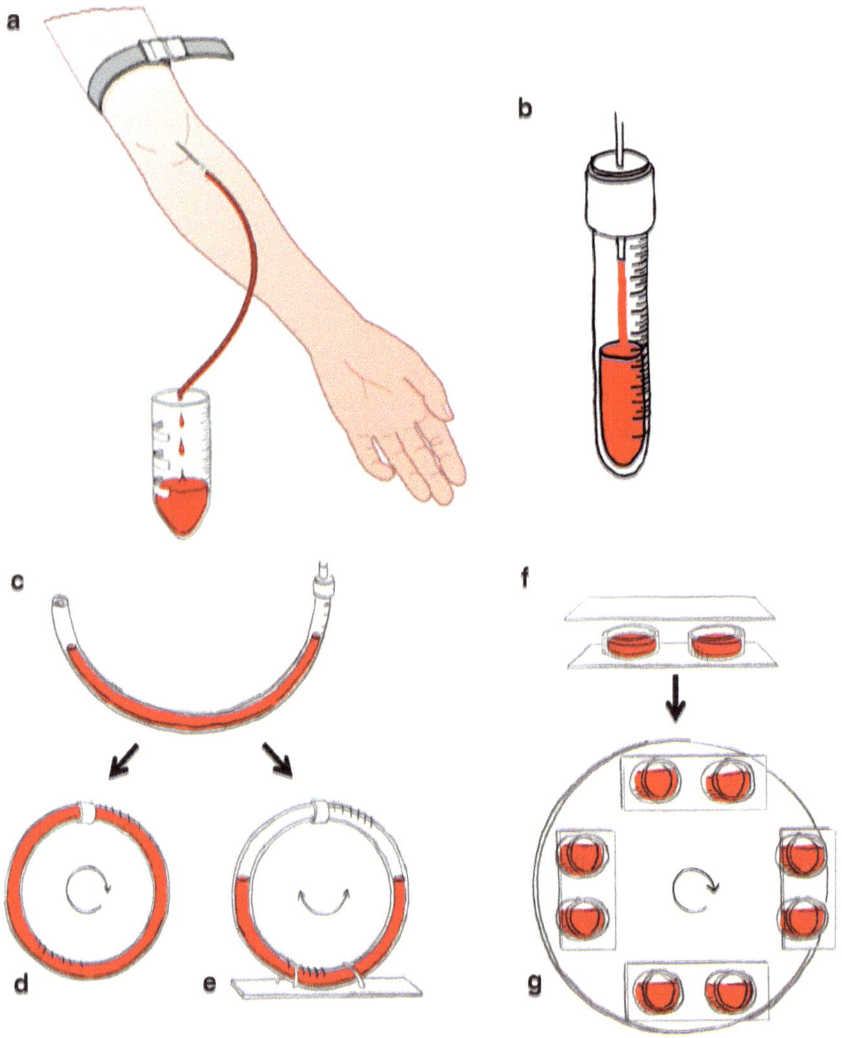

Fig. 18.1 *Summary of alternative procedures for blood collection, the two loop models and the slide chamber model.* Human blood is collected in either an open system (**a**) or evacuated tubes (**b**). It is transferred to plastic tubing (**c**), which is closed with a heparin-coated connector (imperative with adequate design especially when non-anticoagulated blood is used) and either rotated vertically (**d**) or placed on a rocking device (**e**). Alternatively, the blood is transferred to a slide chamber consisting of two polymethylmethacrylate (PMMA) rings fixed to a microscope slide of PMMA, creating two wells (**f**). A second microscope slide (the test surface) is attached, and the device is then rotated vertically (**g**)

18.2.3 Collecting Human Blood for Hemocompatibility Testing

Blood that is to be used for biomaterial evaluation should be collected from healthy blood donors who have not taken any medication for at least 10 days. It is particularly important to exclude individuals who have ingested acetylsalicylic acid, since this drug severely impairs platelet function. The blood can be collected either using a conventional system with evacuated tubes or in an open, more gentle system employing an 18-G (1.2 mm Ø) needle connected to heparin-coated tubing, allowing the blood to flow into heparin-coated plastic tubes by gravity force only (Fig. 18.1a, b).

In order to use the appropriate dose and type of anticoagulant for each individual experiment, it is advisable to add aliquots of stock solutions to the tubes prior to collecting the blood. In the case of the evacuated tubes, this can easily be accomplished using a syringe with a thin needle without compromising the function of the tube. (It should be observed that the concentration of EDTA, citrate, and heparin in commercially available, ready-to-use evacuated tubes (giving final concentrations of 4 mM, 13 mM, and 20 IU/mL, respectively) in most cases is too high to enable studies of the function of the cascade systems and cells in the harvested blood.) The experiment should be initiated as quickly as possible, ideally within 5 min, and the blood should be handled gently using heparin-coated equipment (e.g., pipette tips) in order to minimize the risk of background activation. After incubation, EDTA at a final concentration of 10 mM is added to the samples to stop activation of the complement and coagulation systems, and the samples are transferred to ice, pending centrifugation and further analysis.

It should be emphasized that there are substantial interindividual differences in cascade system and cellular activation. Consequently, it is advisable to use blood from 5 to 10 different donors for evaluation studies.

18.3 In Vitro Whole Blood Models

18.3.1 Tubing Loops for Testing Biomaterials and Soluble Inhibitors

The tubing loop model utilizes plastic tubing, e.g., of polyvinyl chloride, that is filled with fresh blood, plasma, or serum but leaving room for an air bubble (Fig. 18.1c). The tubing is then closed into loops by means of specially designed heparin-coated metal connectors and allowed to rotate vertically at 37°C in a cabinet or water bath (Fig. 18.1d), e.g. Andersson et al. (2003), Johnell et al. (2005), Mollnes et al. (1995), Gong et al. (1996), Nilsson et al. (1998), Lappegård et al. (2004, 2005, 2008), Hussaini et al. (2009). The role of the air bubble is to ensure that the whole volume of blood is rotated in order to avoid clotting due to inadequate flow. If necessary, the incubation can be prolonged up to at least 6 h (Mangsbo et al. 2009). The inner surface of the tubing may be heparin-coated, using different protocols, in order to minimize activation induced by interaction between blood and the plastic surface in experiments designed to study reactions in the fluid phase (e.g., the effect of various inhibitors).

Biomaterials, e.g., different kinds of tubing which can be closed into loops, can be tested directly in this setup, and biomedical devices such as stents, catheters, or lenses may be inserted into heparin-coated loops (Christensen et al. 2001, 2006; Sinn et al. 2011). In this model, it is possible to monitor activation of the cascade systems in the fluid phase (i.e., plasma) and the activation and/or consumption of blood cells. In addition, it is also possible to directly visualize proteins that have adsorbed to the polymer surface. (Further information is given in Table 18.2 and Sect. 18.3.4 below.)

18.3.2 Tubing Loops for Testing Cells and Cell Clusters

A modification of the tubing loop model has been developed to enable studies of interactions between blood and other cells or cell clusters, such as islets of Langerhans (Ozmen et al. 2002; Bennet et al. 1999) and hepatocytes (Gustafson et al. 2011). In this model, the plastic loops are not rotated but are instead placed on a rocking device in a 37°C incubator (Fig. 18.1e). The whole blood loop model can also be further modified so that a section of the heparin-coated tubing is replaced by a segment of a human blood vessel of sufficient diameter. This model allows for the analysis of blood cells and plasma components, as mentioned above, and for immunochemical analysis of the cell clusters or vessel wall in contact with blood (Ozmen et al. 2002; Bennet et al. 2001) (Table 18.2). The blood vessel can be replaced by a segment of material intended for blood contact, e.g., a vascular graft (Fink et al. 2011).

Table 18.2 Examples of available analyses for activation markers of the protein cascade systems and cells in whole blood

Cascade system/cell types	Selected activation markers/analytes in blood/plasma	Selected proteins/cells bound to the material surface
Complement	C3a; C5a; sC5b-9; C1s-CINH	C1q; C4; C3a; C3b; neoC9
Contact activation	FXIIa-AT/CINH; FXIa-AT/CINH; BK	FXII; FXIIa; HK
Coagulation	TAT; soluble fibrin; D-dimer	TF; fibrinogen
Cytokines/chemokines	Multiplex assay	PCR?
Platelets	Count; CD41; CD62; TSP-1; micro particles, Platelet/leukocyte complexes	CD41; ATP
PMNs	Count; oxidative burst; CD11b; MPO Platelet/PMN complexes	Elastase; CD16; CD11b
Monocytes	Count; CD11b; TF Platelet/monocyte complexes	CD14; CD11b; TF
Lymphocytes	CD3; CD4; CD8; DR; CD19	CD3; CD19
Endothelial cells	sVCAM-1; sICAM-1	E-selectin; VCAM-1; ICAM-1

sC5b-9 soluble terminal complement complex, *C1INH* C1 inhibitor, *neoC9* conformationally changed C9 exposing neoepitopes present only in activated C9, *F* (coagulation) factor, *AT* antithrombin, *BK* bradykinin, *HK* high molecular weight kininogen, *TAT* thrombin-antithrombin, *TF* tissue factor, *CD* cluster of differentiation, *ATP* adenosine triphosphate, *TSP-1* thrombospondin-1, *MPO* myeloperoxidase, *DR* HLA-DR, *sVCAM-1* soluble vascular cell adhesion molecule-1, *sICAM-1* soluble intercellular cell adhesion molecule-1

18.3.3 The Slide Chamber Model, Using Whole or Fractionated Blood

We have also developed a slide chamber in vitro model (Hong et al. 1999a) for further dissecting the sequence of events that occur when a biomaterial surface comes into contact with whole blood. In this model, two polymethylmethacrylate (PMMA) rings are fixed to a microscope slide of PMMA, creating two wells. The whole device is coated with heparin to reduce nonspecific activation. The wells are filled with fresh intact or fractionated blood, e.g., PRP, plasma, with or without anticoagulants, or serum. Thereafter, a second microscope slide composed of any (bio-)material is attached, creating two circular chambers, which are then rotated at 37°C in a water bath (Fig. 18.1f, g) (Hong et al. 1999a). Here, it is also possible to conduct immunochemical analyses of adherent cells, including visualization of their activation markers such as TF (Table 18.2) (Bexborn et al. 2009; Nilsson et al. 1998). The test surface is not restricted to transparent materials if fluorescently labeled antibodies are used for detection, i.e., metal surfaces can also be analyzed (Hong et al. 1999b). In a modification of this model, the microscope slide is replaced by a Petri dish coated with cultured cells of interest, such as human endothelial cells. This modification permits analysis of endothelial cell markers in addition to the parameters mentioned above (Table 18.2).

Each of the models is highly flexible, since different parameters can be varied, such as the volume/surface ratio, rotation/rocking velocity (which affects the shear force), incubation time and anticoagulants, and other inhibitors can be varied according to the goal of each experiment.

18.3.4 Techniques to Dissect the Cascade System and Cellular Interactions

After incubation of blood in any of the models, complement, contact system, and coagulation activation markers and activation or decline in the number of cells in the fluid phase (e.g., due to clot formation) can be analyzed using enzyme immunoassays (EIA), fluorescence-activated cell sorting (FACS), cell counts, chemoluminescence, or hemolytic techniques, among others (Table 18.2). When the microscope slide model has been used, it is also possible to perform immunochemical staining of the cells that have bound to the test slide. By using these models, we have shown that various biomaterials behave differently with respect to the activation of complement and coagulation on the solid phase and in the fluid phase, and both phases therefore need to be considered when evaluating activation by biomaterials or therapeutic cells.

Observations made using whole blood models may then be dissected further using different techniques, e.g., surface plasmon resonance (SPR) or quartz crystal balance (QCM) for real-time measurements using purified proteins and/or human serum (Andersson et al. 2002, 2005b; Jokiranta et al. 2001; Arima et al. 2011; Selborn et al. 2003).

18.3.5 Using Specific Inhibitors

In addition to testing the properties of potential or actual biomaterials, the loop and slide chamber models have been used to monitor the effect and reaction mechanisms of a number of soluble pharmaceutical inhibitors. The characterized inhibitors, either alone or in combination, can then be used to dissect the activation mechanisms of the various cascade systems and blood cells induced by materials and therapeutic cells as well as the cross talk between the various systems. For this purpose, complement can be inhibited at different levels: activation of the C1 complex may be inhibited by C1INH (Patston et al. 1991); by specific blocking of monoclonal antibodies (mAb), e.g., against the globular heads of C1q (Hamad et al. 2010a; Wouters et al. 2005); by compstatin, which binds to C3 and blocks its activation by both convertases (Nilsson et al. 1998; Katragadda and Lambris 2006); by eculizumab, an anti-C5 mAb; by C5a receptor antagonists (C5aRA), which block the generation and receptor binding and signaling of C5a, respectively (Lindorfer et al. 2010; Strey et al. 2003); or by soluble complement receptor 1 (Larsson et al. 1997). Clotting can be inhibited at the level of FXII in the contact activation system by corn trypsin inhibitor (CTI) (Mahoney et al. 1984; Hojima et al. 1980); at the level of TF/FVII by active site-inactivated FVIIa (iFVIIa) (Moberg et al. 2002), by a low molecular weight FVIIa inhibitor (Johansson et al. 2005), or by blocking mAbs against TF (Moberg et al. 2002), or by inhibition of thrombin using one of the specific inhibitors melagatran (Ozmen et al. 2002) or lepirudin (Mollnes et al. 2002). Finally, platelet inhibition may be achieved using a number of compounds, including the GPIIb/IIIa inhibitors Ro44 and Reopro (Hong et al. 2001) or various P2Y12 receptor inhibitors (Christensen et al. 2006), or by digesting ADP, which is released by activated platelets and necessary for platelet aggregation with the enzyme apyrase or CD39 (Nilsson et al. 2010).

18.4 Two Problems that Have Been Addressed Using In Vitro Whole Blood Models

18.4.1 The Interaction Between Titanium and Blood

Titanium has long been used for implantation into bone, but the basic mechanisms behind its superior osteointegration have not been well understood. In an initial study, we tested various titanium surfaces in the slide chamber model with whole blood and found that titanium was a potent activator of the intrinsic pathway of coagulation, resulting in substantial platelet activation and the release of a number of growth factors, some of which have been reported to stimulate proliferation of bone cells (Hong et al. 1999b). In a follow-up study, we screened eight different metals commonly used in various medical devices by using the same experimental setup and found that they differed considerably with regard to their thrombogenicity, with titanium being among the most potent, and aluminum among the most inert. In contrast, all surfaces activated complement to a similar degree (Hong et al. 2005).

The purpose of titanium implants is to integrate into the bone as efficiently as possible, and it is therefore beneficial to use implants that induce a potent platelet activation and release of growth factors as possible. The slide chamber model has further been used to screen titanium surfaces used for clinical application with different surface modifications in order to identify surface modification strategies that augment the thrombogenic properties of titanium, with implications for healing into bone (Thor et al. 2007; Hong et al. 2011).

18.4.2 The Instant Blood-Mediated Inflammatory Reaction (IBMIR)

Transplantation of islets of Langerhans to patients with severe type 1 diabetes is a promising treatment. In this procedure, islets of Langerhans are isolated from the pancreas by enzyme digestion followed by gradient centrifugation and then infused into the portal vein of the recipient. However, the procedure is hampered by a severe tissue loss; immediately during transplantation when the nonhematological cells of the islets of Langerhans make contact with the blood, there is substantial damage to the infused cells due to an innate immunity reaction. We have described and dissected these reactions which we have termed the instant blood-mediated inflammatory reaction (IBMIR).

A substantial amount of our work to characterize the IBMIR was performed using the modification of the tubing loop model described above in Sect. 18.3.2, in conjunction with different specific inhibitors as discussed in Sect. 18.3.5. In summary, the IBMIR is an instantaneous thrombotic and inflammatory reaction which consists of powerful activation of the coagulation and complement systems, mediating activation and binding of platelets to the islet surface, followed by leukocyte activation and infiltration into the islets, which ultimately result in the loss of islet integrity. TF, thrombin, anaphylatoxin C5a, as well as numerous cytokines and chemokines, e.g., MCP-1, IL-8, and MIF, have all been demonstrated to contribute to the detrimental effects of the IBMIR (Bennet et al. 1999; Moberg et al. 2002, 2003, 2005; Johansson et al. 2003, 2005; Cabric et al. 2007, 2008; Nilsson et al. 2011).

18.5 Possibilities and Limitations of Using In Vitro Models with Whole Blood

Activation of the cascade systems in the blood is a surface-oriented phenomenon that takes place at both the blood/biomaterial interface and the blood/gas surface (Ekdahl et al. 1992). The activation induced by the biomaterial surfaces of the test system may be attenuated by surface-bound heparin. In contrast, there is constant turnover of the interface between blood and gas, a phenomenon that is not affected by various surface-modification strategies. This means that there will be a background level of activation, in particular of complement, in each experimental system.

In our hands, it is possible to circulate blood in the tubing loop model for at least 6 h with retained pH and oxygen levels (Mangsbo et al. 2009).

Another factor that should be borne in mind is that both the models presented involve a limited quantity of blood that is incubated in a small container. Although there is ample opportunity for cross talk between the various cascade systems and cells in the blood, the models are static in that there can be no turnover of activated components. In vivo, activation products such as the various complexes that are formed between the protease inhibitors C1INH and AT and their substrates (listed in Table 18.2) would be removed from the circulation and replaced by new zymogen molecules. Consequently, it is difficult to monitor low-level activation of any of the cascade systems in vivo, while in the in vitro systems, a low initial activation will be amplified by downstream reactions.

18.6 Conclusion

In conclusion, in vitro techniques such as those described here represent a convenient, biologically relevant (since human blood is being used), economical, and ethically acceptable way of screening materials and therapeutic cells to exclude those that are unsuitable for further in vivo testing in animal models. Insight into the complexity of the molecular-level interactions that take place at the material-blood interface and the cascades stemming from them is necessary for the development of these in vitro techniques and for the development of novel and improved biomaterials.

Acknowledgments This work was supported by grants from the Swedish Research Council (VR) 2009–4675, 2009–4462, the Swedish Research Council, and the Swedish Research Council/SSF/Vinnova contract grant number 60761701 and by faculty grants from the Linnæus University. Reagents for preparing heparin-coated surfaces were obtained from Corline Systems AB. We thank Dr. Deborah McClellan for excellent editorial assistance and Mr. Hans Nilsson for preparing the illustrations in Fig. 18.1.

Conflict of Interest Statement The author Professor R. Larsson is an employee of Corline Systems AB, Uppsala, Sweden. None of the other authors has conflicts of interest.

References

Amara U, Flierl MA, Rittirsch D, Klos A, Chen H, Acker B et al (2010) Molecular intercommunication between the complement and coagulation systems. J Immunol 185(9):5628–5636

Andersson J, Ekdahl KN, Larsson R, Nilsson UR, Nilsson B (2002) C3 adsorbed to a polymer surface can form an initiating alternative pathway convertase. J Immunol 168(11):5786–5791

Andersson J, Sanchez J, Ekdahl KN, Elgue G, Nilsson B, Larsson R (2003) Optimal heparin surface concentration and antithrombin binding capacity as evaluated with human non-anticoagulated blood in vitro. J Biomed Mater Res A 67(2):458–466

Andersson J, Ekdahl KN, Lambris JD, Nilsson B (2005a) Binding of C3 fragments on top of adsorbed plasma proteins during complement activation on a model biomaterial surface. Biomaterials 26(13):1477–1485

Andersson M, Andersson J, Sellborn A, Berglin M, Nilsson B, Elwing H (2005b) Quartz crystal microbalance-with dissipation monitoring (QCM-D) for real time measurements of blood coagulation density and immune complement activation on artificial surfaces. Biosens Bioelectron 21(1):79–86

Arima Y, Toda M, Iwata H (2011) Surface plasmon resonance in monitoring of complement activation on biomaterials. Adv Drug Deliv Rev 63:988–999

Attanasio M, Gori AM, Giusti B, Pepe G, Comeglio P, Brunelli T et al (1998) Cytokine gene expression in human LPS- and IFN gamma-stimulated mononuclear cells is inhibited by heparin. Thromb Haemost 79(5):959–962

Bäck J, Huber-Lang M, Elgue G, Kalbitz M, Sanchez J, Ekdahl KN et al (2009) Distinctive regulation of contact activation by antithrombin and C1-inhibitor on activated platelets and material surfaces. Biomaterials 30(34):6573–6580

Bäck J, Sanchez J, Elgue G, Ekdahl KN, Nilsson B (2010) Activated human platelets induce factor XIIa-mediated contact activation. Biochem Biophys Res Commun 391(1):11–17

Bennet W, Sundberg B, Groth CG, Brendel MD, Brandhorst D, Brandhorst H et al (1999) Incompatibility between human blood and isolated islets of Langerhans: a finding with implications for clinical intraportal islet transplantation? Diabetes 48(10):1907–1914

Bennet W, Sundberg B, Elgue G, Larsson R, Korsgren O, Nilsson B (2001) A new in vitro model for the study of pig-to-human vascular hyperacute rejection. Xenotransplantation 8(3):176–184

Berglin M, Pinori E, Sellborn A, Andersson M, Hulander M, Elwing H (2009) Fibrinogen adsorption and conformational change on model polymers: novel aspects of mutual molecular rearrangement. Langmuir 25(10):5602–5608

Bexborn F, Engberg AE, Sandholm K, Mollnes TE, Hong J, Nilsson Ekdahl K (2009) Hirudin versus heparin for use in whole blood in vitro biocompatibility models. J Biomed Mater Res A 89(4):951–959

Bogdanov V, Balasubramanian V, Hathcock J, Vele O, Lieb M, Nemerson Y (2003) Alternatively spliced human tissue factor: a circulating, soluble, thrombogenic protein. Nat Med 9(4):458–462

Cabric S, Sanchez J, Lundgren T, Foss A, Felldin M, Kallen R et al (2007) Islet surface heparinization prevents the instant blood-mediated inflammatory reaction in islet transplantation. Diabetes 56(8):2008–2015

Cabric S, Eich T, Sanchez J, Nilsson B, Korsgren O, Larsson R (2008) A new method for incorporating functional heparin onto the surface of islets of Langerhans. Tissue Eng C Methods 14(2):141–147

Chang JY (1983) The functional domain of hirudin, a thrombin-specific inhibitor. FEBS Lett 164(2):307–313
Christensen K, Larsson R, Emanuelsson H, Elgue G, Larsson A (2001) Heparin coating of the stent graft—effects on platelets, coagulation and complement activation. Biomaterials 22(4):349–355
Christensen K, Larsson R, Emanuelsson H, Elgue G, Larsson A (2006) Effects on blood compatibility in vitro by combining a direct P2Y12 receptor inhibitor and heparin coating of stents. Platelets 17(5):318–327
Distelmaier K, Adlbrecht C, Jakowitsch J, Winkler S, Dunkler D, Gerner C et al (2009) Local complement activation triggers neutrophil recruitment to the site of thrombus formation in acute myocardial infarction. Thromb Haemost 102(3):564–572
Ekdahl KN, Nilsson B, Pekna M, Nilsson UR (1992) Generation of iC3 at the interface between blood and gas. Scand J Immunol 35(1):85–91
Ekdahl KN, Lambris JD, Elwing H, Ricklin D, Nilsson PH, Teramura Y et al (2011) Innate immunity activation on biomaterial surfaces: a mechanistic model and coping strategies. Adv Drug Deliv Rev 63(12):1042–1050
Engberg AE, Rosengren-Holmberg JP, Chen H, Nilsson B, Lambris JD, Nicholls IA et al (2011) Blood protein-polymer adsorption: implications for understanding complement-mediated hemoincompatibility. J Biomed Mater Res A 94A:74–84
Engstad CS, Gutteberg TJ, Osterud B (1997) Modulation of blood cell activation by four commonly used anticoagulants. Thromb Haemost 77(4):690–696
Ettelaie C, Fountain D, Collier M, Elkeeb A, Xiao Y, Maraveyas A (2011) Low molecular weight heparin downregulates tissue factor expression and activity by modulating growth factor receptor-mediated induction of nuclear factor-κB. Biochim Biophys Acta 1812(12):1591–1600
Fink H, Hong J, Drotz K, Risberg B, Sanchez J, Sellborn A (2011) An in vitro study of blood compatibility of vascular grafts made of bacterial cellulose in comparison with conventionally-used graft materials. J Biomed Mater Res A [Epub ahead of print]
Ghebrehiwet B, Randazzo BP, Dunn JT, Silverberg M, Kaplan AP (1983) Mechanisms of activation of the classical pathway of complement by Hageman factor fragment. J Clin Invest 71(5):1450–1456
Gong J, Larsson R, Ekdahl KN, Mollnes TE, Nilsson U, Nilsson B (1996) Tubing loops as a model for cardiopulmonary bypass circuits: both the biomaterial and the blood-gas phase interfaces induce complement activation in an in vitro model. J Clin Immunol 16(4):222–229
Gorbet MB, Sefton MV (2004) Biomaterial-associated thrombosis: roles of coagulation factors, complement, platelets and leukocytes. Biomaterials 25(26):5681–5703
Gori AM, Pepe G, Attanasio M, Falciani M, Abbate R, Prisco D et al (1999) Tissue factor reduction and tissue factor pathway inhibitor release after heparin administration. Thromb Haemost 81(4):589–593
Grabowski E (1990) Platelet aggregation in flowing blood at a site of injury to an endothelial cell monolayer: quantitation and real-time imaging with the TAB monoclonal antibody. Blood 75(2):390–398
Grabowski E, Curran M, Van Cott E (2012) Assessment of a cohort of primarily pediatric patients with a presumptive diagnosis of type 1 von Willebrand disease with a novel high shear rate, non-citrated blood flow device. Thromb Res [Epub ahead of print]
Gulla K, Gupta K, Krarup A, Gal P, Schwaeble W, Sim R et al (2010) Activation of mannan-binding lectin-associated serine proteases leads to generation of a fibrin clot. Immunology 129(4):482–495
Gustafson EK, Elgue G, Hughes RD, Mitry RR, Sanchez J, Haglund U et al (2011) The instant blood-mediated inflammatory reaction characterized in hepatocyte transplantation. Transplantation 91(6):632–638
Hamad OA, Ekdahl KN, Nilsson PH, Andersson J, Magotti P, Lambris JD et al (2008) Complement activation triggered by chondroitin sulfate released by thrombin receptor-activated platelets. J Thromb Haemost 6(8):1413–1421
Hamad OA, Nilsson PH, Wouters D, Lambris JD, Ekdahl KN, Nilsson B (2010a) Complement component C3 binds to activated normal platelets without preceding proteolytic activation and promotes binding to complement receptor 1. J Immunol 185(5):2686–2692
Hamad O, Nilsson P, Lasaosa M, Ricklin D, Lambris J, Nilsson B et al (2010b) Contribution of chondroitin sulfate A to the binding of complement proteins to activated platelets. PLoS One 5(6):e12889
Harboe M, Ulvund G, Vien L, Fung M, Mollnes TE (2004) The quantitative role of alternative pathway amplification in classical pathway induced terminal complement activation. Clin Exp Immunol 138(3):439–446
Hess K, Alzahran SH, Mathai M, Schroeder V, Carter A, Howell G et al (2011) A novel mechanism for hypofibrinolysis in diabetes: the role of complement C3. Diabetologia [Epub ahead of print]
Hojima Y, Pierce J, Pisano J (1980) Hageman factor fragment inhibitor in corn seeds: purification and characterization. Thromb Res 20:149–162
Hong J, Nilsson Ekdahl K, Reynolds H, Larsson R, Nilsson B (1999a) A new in vitro model to study interaction between whole blood and biomaterials. Studies of platelet and coagulation activation and the effect of aspirin. Biomaterials 20(7):603–611
Hong J, Andersson J, Ekdahl KN, Elgue G, Axen N, Larsson R et al (1999b) Titanium is a highly thrombogenic biomaterial: possible implications for osteogenesis. Thromb Haemost 82(1):58–64

Hong J, Larsson A, Ekdahl KN, Elgue G, Larsson R, Nilsson B (2001) Contact between a polymer and whole blood: sequence of events leading to thrombin generation. J Lab Clin Med 138(2):139–145
Hong J, Azens A, Ekdahl K, Granqvist CG, Nilsson B (2005) Material-specific thrombin generation following contact between metal surfaces and whole blood. Biomaterials 26:1397–1403
Hong J, Kurt S, Thor A (2011) A hydrophilic dental implant surface exhibit thrombogenic properties in vitro. Clin Implant Dent Relat Res [Epub ahead of print]
Howes J, Richardson V, Smith K, Schroeder V, Somani R, Shore A et al (2012) Complement C3 is a novel plasma clot component with anti-fibrinolytic properties. Diab Vasc Dis Res [Epub ahead of print]
Huber-Lang M, Sarma JV, Zetoune FS, Rittirsch D, Neff TA, McGuire SR et al (2006) Generation of C5a in the absence of C3: a new complement activation pathway. Nat Med 12(6):682–687
Hussaini B, Treanor P, Healey N, Tilahun D, Srey R, Lu X et al (2009) Evaluation of blood components exposed to coated arterial filters in extracorporeal circuits. Perfusion 24(5):317–323
Ikeda K, Nagasawa K, Horiuchi T, Tsuru T, Nishizaka H, Niho Y (1997) C5a induces tissue factor activity on endothelial cells. Thromb Haemost 77(2):394–398
Johansson U, Olsson A, Gabrielsson S, Nilsson B, Korsgren O (2003) Inflammatory mediators expressed in human islets of Langerhans: implications for islet transplantation. Biochem Biophys Res Commun 308(3):474–479
Johansson H, Lukinius A, Moberg L, Lundgren T, Berne C, Foss A et al (2005) Tissue factor produced by endocrine cells of the islets of Langerhans is associated with a negative outcome of clinical islet transplantation. Diabetes 54(6):1755–1762
Johnell M, Larsson R, Siegbahn A (2005) The influence of different heparin surface concentrations and antithrombin-binding capacity on inflammation and coagulation. Biomaterials 26(14):1731–1739
Jokiranta TS, Westin J, Nilsson B, Ekdahl KN, Hellwage J, Gordon DL et al (2001) Complement C3b interactions with its ligands measured by Biacore equipment: a new powerful and informative method. Int Immunopharmacol 1:495–506
Katragadda M, Lambris JD (2006) Expression of compstatin in *Escherichia coli*: incorporation of unnatural amino acids enhances its activity. Protein Expr Purif 47(1):289–295
Keil L, Jimenez E, Guma M, Reyes M, Liguori C, DeBari V (1995) Biphasic response of complement to heparin: fluid-phase generation of neoantigens in human serum and in a reconstituted alternative pathway amplification cycle. Am J Hematol 50(4):254–262
Kourtzelis I, Markiewski MM, Doumas M, Rafail S, Kambas K, Mitroulis I et al (2010) Complement anaphylatoxin C5a contributes to hemodialysis-associated thrombosis. Blood 116(4):631–639
Krarup A, Gulla C, Gál P, Hajela K, Sim R (2008) The action of MBL-associated serine protease 1 (MASP1) on factor XIII and fibrinogen. Biochim Biophys Acta 1784:1294–1300
Lachmann PJ, Pangburn MK, Oldroyd RG (1982) Breakdown of C3 after complement activation. Identification of a new fragment C3g, using monoclonal antibodies. J Exp Med 156(1):205–216
Lappegård K, Fung M, Bergseth G, Riesenfeld J, Lambris J, Videm V et al (2004) Effect of complement inhibition and heparin coating on artificial surface-induced leukocyte and platelet activation. Ann Thorac Surg 77(3):932–941
Lappegård K, Riesenfeld J, Brekke O, Bergseth G, Lambris J, Mollnes T (2005) Differential effect of heparin coating and complement inhibition on artificial surface-induced eicosanoid production. Ann Thorac Surg 79(3):917–923
Lappegård KT, Bergseth G, Riesenfeld J, Pharo A, Magotti P, Lambris JD et al (2008) The artificial surface-induced whole blood inflammatory reaction revealed by increases in a series of chemokines and growth factors is largely complement dependent. J Biomed Mater Res A 87(1):129–135
Larsson R, Elgue G, Larsson A, Ekdahl KN, Nilsson UR, Nilsson B (1997) Inhibition of complement activation by soluble recombinant CR1 under conditions resembling those in a cardiopulmonary circuit: reduced up-regulation of CD11b and complete abrogation of binding of PMNs to the biomaterial surface. Immunopharmacology 38(1–2):119–127
Lindorfer M, Pawluczkowycz A, Peek E, Hickman K, Taylor R, Parker CJ (2010) A novel approach to preventing the hemolysis of paroxysmal nocturnal hemoglobinuria: both complement-mediated cytolysis and C3 deposition are blocked by a monoclonal antibody specific for the alternative pathway of complement. Blood 115(11):2283–2291
Logue GL (1977) Effect of heparin on complement activation and lysis of paroxysmal nocturnal hemoglobinuria (PNH) red cells. Blood 50(2):239–247
Mahoney W, Hermodson M, Jones B, Powers D, Corfman R, Reeck G (1984) Amino acid sequence and secondary structural analysis of the corn inhibitor of trypsin and activated Hageman Factor. J Biol Chem 259(13):8412–8416
Mangsbo SM, Sanchez J, Anger K, Lambris JD, Ekdahl KN, Loskog AS et al (2009) Complement activation by CpG in a human whole blood loop system: mechanisms and immunomodulatory effects. J Immunol 183(10):6724–6732
Matsushita M, Thiel S, Jensenius J, Terai I, Fujita T (2000) Proteolytic activities of two types of mannose-binding lectin-associated serine protease. J Immunol 165:2637–2642
Moberg L, Johansson H, Luknius A, Berne C, Foss A, Källen R et al (2002) Production of tissue factor by pancreatic islet cells as a trigger of detrimental thrombotic reactions in clinical islet transplantation. Lancet 360:2039–2045

Moberg L, Olsson A, Berne C, Felldin M, Foss A, Kallen R et al (2003) Nicotinamide inhibits tissue factor expression in isolated human pancreatic islets: implications for clinical islet transplantation. Transplantation 76(9):1285–1288

Moberg L, Korsgren O, Nilsson B (2005) Neutrophilic granulocytes are the predominant cell type infiltrating pancreatic islets in contact with ABO-compatible blood. Clin Exp Immunol 142(1):125–131

Mollnes TE, Riesenfeld J, Garred P, Nordstrom E, Hogasen K, Fosse E et al (1995) A new model for evaluation of biocompatibility: combined determination of neoepitopes in blood and on artificial surfaces demonstrates reduced complement activation by immobilization of heparin. Artif Organs 19(9):909–917

Mollnes TE, Brekke OL, Fung M, Fure H, Christiansen D, Bergseth G et al (2002) Essential role of the C5a receptor in *E. coli*-induced oxidative burst and phagocytosis revealed by a novel lepirudin-based human whole blood model of inflammation. Blood 100(5):1869–1877

Nilsson B, Larsson R, Hong J, Elgue G, Ekdahl KN, Sahu A et al (1998) Compstatin inhibits complement and cellular activation in whole blood in two models of extracorporeal circulation. Blood 92(5):1661–1667

Nilsson B, Ekdahl KN, Mollnes TE, Lambris JD (2007) The role of complement in biomaterial-induced inflammation. Mol Immunol 44(1–3):82–94

Nilsson P, Engberg A, Bäck J, Faxälv L, Lindahl T, Nilsson B et al (2010) The creation of an antithrombotic surface by apyrase immobilization. Biomaterials 31(16):4484–4491

Nilsson B, Nilsson Ekdahl K, Korsgren O (2011) Control of IBMIR (Instant Blood-Mediated Inflammatory Reaction) to improve islets of Langerhans engraftment. Curr Opin Organ Transplant 16(6):620–626

Ozmen L, Ekdahl KN, Elgue G, Larsson R, Korsgren O, Nilsson B (2002) Inhibition of thrombin abrogates the instant blood-mediated inflammatory reaction triggered by isolated human islets: possible application of the thrombin inhibitor melagatran in clinical islet transplantation. Diabetes 51(6):1779–1784

Patston P, Gettins P, Beechem J, Schapira M (1991) Mechanism of serpin action: evidence that C1 inhibitor functions as a suicide substrate. Biochemistry 30(36):8876–8882

Ratner BD (2007) The catastrophe revisited: blood compatibility in the 21st century. Biomaterials 28(34):5144–5147

Ritis K, Doumas M, Mastellos D, Micheli A, Giaglis S, Magotti P et al (2006) A novel C5a receptor-tissue factor cross-talk in neutrophils links innate immunity to coagulation pathways. J Immunol 177(7):4794–4802

Schousboe I, Nystrøm BT, Hansen GH (2008) Differential binding of factor XII and activated factor XII to soluble and immobilized fibronectin–localization of the Hep-1/Fib-1 binding site for activated factor XII. FEBS J 275(20):5161–5172

Selborn A, Andersson M, Fant C, Gretzer C, Elwing H (2003) Methods for research on immune complement activation on modified sensor surfaces. Colloids Surf B: Biointerfaces 27:295–301

Seyfert UT, Biehl V, Schenk J (2002) In vitro hemocompatibility testing of biomaterials according to the ISO 10993–4. Biomol Eng 19(2–6):91–96

Shankar R, de la Motte CA, Poptic EJ, DiCorleto PE (1994) Thrombin receptor-activating peptides differentially stimulate platelet-derived growth factor production, monocytic cell adhesion, and E-selectin expression in human umbilical vein endothelial cells. J Biol Chem 269(19):13936–13941

Sinn S, Scheuermann T, Deichelbohrer S, Ziemer G, Wendel HP (2011) A novel in vitro model for preclinical testing of the hemocompatibility of intravascular stents according to ISO 10993–4. J Mater Sci Mater Med 22(6):1521–1528

Strey CW, Markiewski M, Mastellos D, Tudoran R, Spruce LA, Greenbaum LE et al (2003) The proinflammatory mediators C3a and C5a are essential for liver regeneration. J Exp Med 198(6):913–923

Taylor F, Müller-Eberhard H (1970) Qualitative description of factors involved in the retraction and lysis of dilute whole blood clots and in the aggregation and retraction of platelets. J Clin Invest 49(1):2068–2085

Tengvall P, Askendal A, Lundstrom I (2001) Ellipsometric in vitro studies on the activation of complement by human immunoglobulins M and G after adsorption to methylated silicon. Colloids Surf B: Biointerfaces 20(1):51–62

Thoman ML, Meuth JL, Morgan EL, Weigle WO, Hugli TE (1984) C3d-K, a kallikrein cleavage fragment of iC3b is a potent inhibitor of cellular proliferation. J Immunol 133(5):2629–2633

Thor A, Rasmusson L, Wennerberg A, Thomsen P, Hirsch JM, Nilsson B et al (2007) The role of whole blood in thrombin generation in contact with various titanium surfaces. Biomaterials 28(6):966–974

Wettero J, Bengtsson T, Tengvall P (2001) C1q-independent activation of neutrophils by immunoglobulin M-coated surfaces. J Biomed Mater Res 57(4):550–558

Wouters D, Wiessenberg H, Hart M, Bruins P, Voskuyl A, Daha M et al (2005) Complexes between C1q and C3 or C4: novel and specific markers for classical complement pathway activation. J Immunol Methods 298:35–45

Wu Y, Simonovsky FI, Ratner BD, Horbett TA (2005) The role of adsorbed fibrinogen in platelet adhesion to polyurethane surfaces: a comparison of surface hydrophobicity, protein adsorption, monoclonal antibody binding, and platelet adhesion. J Biomed Mater Res A 74(4):722–738

Chapter 19
Noninvasive Detection of Complement Activation Through Radiologic Imaging

Joshua M. Thurman and Bärbel Rohrer

Abstract A wealth of experimental and clinical data demonstrates that the complement system is involved in the pathogenesis of numerous inflammatory diseases. Complement activation contributes to injury in disorders that involve nearly every tissue in the body. Concerted effort has been expended in recent years to develop therapeutic complement inhibitors. Eculizumab, an inhibitory antibody to C5, was recently approved for the treatment of several diseases, and many other complement inhibitors are in clinical development. As these drugs are developed, the need for improved methods of detecting and monitoring complement activation within particular tissues will be increasingly important. We have developed a magnetic resonance imaging (MRI)-based method for noninvasive detection of complement activation. This method utilizes iron-oxide nanoparticles that are targeted to sites of complement activation with a recombinant protein that contains the C3d-binding region of complement receptor (CR) 2. Iron-oxide nanoparticles darken (negatively enhance) images obtained by T2-weighted MRI. We have demonstrated that the CR2-targeted nanoparticles bind within the kidneys of mice with lupus-like kidney disease (MRL/lpr mice), causing a decrease in the T2 signal within the kidneys. This method discriminates diseased kidneys from healthy controls, and the magnitude of the negative enhancement in the cortex of MRL/lpr mice correlates with their disease severity. This method may be useful for identifying those patients most likely to benefit from complement inhibitors and for monitoring the response of these patients to treatment. These results may open up new avenues to develop tools for the monitoring of disease progression in complement-dependent diseases.

19.1 C3: At the Center of the Complement System

The complement system is an important arm of the body's innate immune defense against invasive pathogens. This system is also necessary for the efficient removal of injured cells and immune complexes (Ricklin et al. 2010). The complement system is a protein cascade comprised of more than 30 components and receptors. It is activated through three distinct initiating pathways, each of which

J.M. Thurman (✉)
Department of Medicine, University of Colorado Denver School of Medicine, Denver, CO 80045, USA
e-mail: joshua.thurman@ucdenver.edu

B. Rohrer
Departments of Ophthalmology and Neurosciences, Division of Research, Medical University of South Carolina, Charleston, SC 29425, USA
e-mail: rohrer@musc.edu

J.D. Lambris et al. (eds.), *Complement Therapeutics*, Advances in Experimental Medicine and Biology 735,
DOI 10.1007/978-1-4614-4118-2_19,

converges on the protein C3. C3 has been termed the central protein of the complement system. During activation, the C3 protein is cleaved into the biologically active fragments C3a and C3b. One of these fragments, C3b, is a component of the C3 and C5 convertases and thus catalyzes the cleavage of even more C3. During the conversion of C3 to C3b, an internal thioester in the C3 molecule is exposed and can form amide and ester bonds with amino and hydroxyl groups on nearby surfaces (Dodds et al. 1996; Janssen et al. 2006).

Fixation of C3b to tissues during complement activation is critical to the effective solid phase activation of the complement system. Hence, tissue-bound C3b (and other C3 activation fragments) can be documented in affected tissues by different techniques, including immunofluorescence microscopy. Indeed, examination of renal tissue for C3 fragment deposition within the glomeruli is a routine part of the evaluation of renal biopsies, and tissue C3 deposits are interpreted as evidence of complement activation. The ability to noninvasively and repeatedly quantify tissue C3 deposits would enable clinicians to monitor inflammatory renal diseases, such as lupus nephritis, and to tailor a patient's treatment based on their response. In this chapter, we will review the evidence that tissue-bound C3 fragments are useful biomarkers of disease, and we will describe a molecular imaging method that has been developed to detect tissue-bound C3 fragments by MRI.

19.2 The Role of Complement in Autoimmune and Inflammatory Diseases

Immune complexes are potent activators of the classical pathway, so the complement system is engaged in immune-complex diseases and in diseases characterized by autoantibodies to self antigens (Walport 2001). The complement system is also activated on damaged and apoptotic cells. Complement activation on the surface of "altered"-self cells is facilitated by recognition molecules, such as C-reactive protein (Griselli et al. 1999), that bind to molecular motifs expressed on the cell surface and then engage the complement system. Apoptotic or stressed cells may also reduce the surface expression of complement regulatory proteins, thus creating a surface favorable to greater complement activation (Elward et al. 2005; Thurman et al. 2009). Other local conditions may also favor complement activation, including a low pH (Fishelson et al. 1987). Finally, it has become clear that congenital deficiency of complement regulatory proteins predisposes to certain diseases (de Cordoba and de Jorge 2008).

As outlined above, complement activation is central to a wide variety of different diseases. But while C3 deposition in tissues may be a sensitive indicator of early inflammation, it is not specific for any disease. The utility of noninvasive detection of complement activation within a given tissue, therefore, might be for monitoring the progression of disease after it is diagnosed and the efficacy of the treatment, rather than serving as a diagnostic tool.

19.3 The Emergence of Complement Therapeutics

The pathogenic role of the complement system in disease has been appreciated for decades. Consequently, concerted effort has been expended to develop therapeutic complement inhibitors (Ricklin and Lambris 2007). Eculizumab, an inhibitory antibody to C5, was originally approved for treatment of paroxysmal nocturnal hemoglobinuria and was also recently approved for atypical hemolytic uremic syndrome (Rother et al. 2007). Eculizumab has proven remarkably effective at ameliorating these previously untreatable diseases, and many new indications are likely to be found in the coming years.

Hand in hand with the discovery of new indications for eculizumab, it is likely that new agents will be developed that modulate the complement system at different levels or that block specific components of this system. The complexity of the complement system—the large number of constituent proteins and their various biologic functions—is also an opportunity for the development of therapeutic agents with narrower functions and potentially fewer side effects. Another important advance has been the development of targeted complement inhibitors (Atkinson et al. 2005, 2008; He et al. 2005; Huang et al. 2008; Song et al. 2003; Rohrer et al. 2009). These agents deliver therapeutic complement regulatory proteins to the anatomic site of complement activation. The ability to target these agents gives them greater therapeutic potency while reducing systemic side effects such as susceptibility to infection (Atkinson et al. 2005).

With the expanding use of complement inhibitory drugs, there is a greater need for methods to detect and monitor complement activation in tissues. A tool that can report on whether the complement system is activated in a given patient and in a particular tissue would be invaluable for guiding the use of many immunomodulatory drugs, but it would be particularly useful for guiding the use of complement inhibitory drugs. The MRI-based method for detecting complement activation described below utilizes a recombinant form of complement receptor (CR) 2 to target tissue-bound C3d. This is the same approach employed by many of the targeted complement inhibitors developed by Tomlinson's group (Song et al. 2003). In addition, by using the same targeting vector, our imaging probe is well suited to identify patients in whom the CR2-targeted inhibitors are likely to be most effective. It may also be useful to combine the therapeutic and diagnostic functions of these targeted agents, a concept referred to as "theragnostics" (Shubayev et al. 2009; Lucignani 2009).

19.4 Current Biomarkers of Complement Activation

Detection of complement activation can be accomplished by three different means: measurement of the levels of complement components or measurement of complement activation fragments in patient plasma and detection of complement deposition on biopsy. Tissue biopsy may be the "gold standard" for detecting complement activation within a tissue, but of course a biopsy cannot be performed on all tissues and serial biopsies are often not feasible. The detection of autoantibodies against complement components (such as C3 nephritic factors or inhibitory antibodies to factor H) or genetic defects in complement regulatory proteins may implicate the complement system in a given patient's disease, but these findings do not inform clinicians as to the activation state of the complement system at a particular time.

The measurement of plasma C3 and C4 has traditionally been performed by physicians to diagnose and monitor autoimmune diseases such as lupus (Ramos-Casals et al. 2004; Ricker et al. 1991; Swaak et al. 1986). The plasma levels of C3 and C4 often fall in patients with active disease, presumably due to consumption. Immune-complex-mediated diseases, such as lupus nephritis, tend to decrease the plasma levels of both C3 and C4 (Hebert et al. 2001), consistent with complement activation through the classical pathway. A decreased plasma C3 level is a fairly sensitive marker of active lupus nephritis (Ramos-Casals et al. 2004; Swaak et al. 1986). In other immune-complex deposition diseases, such as membranoproliferative glomerulonephritis type I, a much lower percentage of patients may have low plasma C3 levels (Cameron et al. 1983). In glomerular diseases that are associated with decreased C3 levels, the sensitivity of this test ranges from 50% to 90%. However, since low complement levels can also be caused by consumption of these proteins or due to underproduction, low C3 or C4 are not specific markers of autoimmunity. Sepsis, atheroembolic disease, pancreatitis, and HIV infection, for example, all can cause low C3 and C4 levels (Hebert et al. 1991). Underproduction of these proteins may be caused by severe malnutrition or liver disease, and production of these proteins may be increased during pregnancy or acute phase response.

Another drawback of measuring intact plasma C3 and C4 is that some diseases associated with complement-mediated tissue injury are not associated with decreased levels of C3. In membranous disease, for example, complement fragments are frequently seen in biopsy tissue and in the urine of patients with active disease as well as in animal models, indicating that the complement system is central to the development of disease, yet the levels of circulating C3 and C4 are usually normal (Morita et al. 2000; West 1998).

The measurement of complement split products has also shown some promise for monitoring disease activity (Manzi et al. 1996; Porcel et al. 1995). These tests have not gained widespread use, however, possibly because their accuracy is affected by sample handling. Furthermore, complement measurements performed on plasma, and even urine, do not localize the activation to a particular organ. Monitoring individual tissues is important because in a multisystem disease such as lupus, the course of treatment may be different depending upon the organ involved.

19.5 Tissue-Bound C3 Fragments as Biomarkers of Renal Disease

C3 is virtually always present in the biopsies of tissues with immune-complex deposition (di Belgiojoso et al. 1975). During activation of C3, the molecule undergoes four sequential cleavages. It is converted first to C3b, then iC3b, C3dg, and finally C3d. Fragments of the C3 molecule are released into the circulation during this process, but millions of C3 fragment molecules are also deposited on affected tissues during activation. The C3 fragments, starting with C3b covalently bind to amino and hydroxyl groups on nearby tissues (Dodds et al. 1996). The bound C3 fragments are thus abundant and durable markers of this process. Targeting molecules that recognize epitopes present on intact C3 will likely bind to the intact protein in the circulation and are less likely to bind well to the tissue deposits. During activation, though, C3 undergoes conformational changes that reveal new epitopes (Janssen et al. 2006). These exposed epitopes enable agents to discriminate between intact and cleaved forms of C3.

Immunofluorescence microscopy of tissue biopsies usually includes staining for C3 deposition, and the detection of C3 within the glomerulus is part of the histologic diagnosis of many types of glomerulonephritis. Many of the characteristics that make the complement system a useful biomarker in tissue biopsies also make it advantageous for molecular imaging. It is a rapidly activated system that is self-amplifying. Although the multiplicity of forms of C3 creates challenges for the development of targeting molecules specific to one particular fragment, agents that are specific to a single C3 fragment species also offer opportunities for extracting greater information about the activity and duration of disease. For example, studies in rat models of glomerular disease demonstrated that C3c antigens (presumably C3b and iC3b) decrease within 24 h of cessation of complement activity (Schulze et al. 1993). C3d was much longer lived, persisting several weeks within the glomeruli after complement activity was terminated. This indicates that C3b and iC3b are likely better biomarkers of "active" disease, but that C3d may be a longer-lived marker.

On the other hand, in a mouse model of lupus nephritis, we found that C3d levels decrease in the most severe stages of disease (Sargsyan et al. 2011). The cleavage of iC3b to C3dg and C3d requires CR1 as cofactor for factor I. The simultaneous accumulation of C3b and iC3b (whose generation also requires factor I) with a decrease in the levels of C3d suggests that there is a reduction in the local availability of CR1 to mediate this final cleavage. Indeed, biopsies of patients with lupus nephritis demonstrate that there is a reduction in the surface expression of CR1 on glomerular podocytes in the setting of lupus nephritis and other types of glomerular disease (Moll et al. 2001). If the local conversion of C3b and iC3b to C3d is reduced in late stages of this disease due to glomerular podocytes injury, then the abundance of C3b and iC3b may become a function of both disease activity (which increases C3b deposition) and of glomerular damage (which decreases iC3b degradation). Consequently, C3d abundance may increase in the early stages of disease but then begin to decline in late stages of disease, when structural damage of the tissues reduces the local availability of CR1.

The decline in C3d in terminally injured tissues may actually be a desirable characteristic as a biomarker. The treatment of autoimmune diseases, such as lupus nephritis, is a balance between the

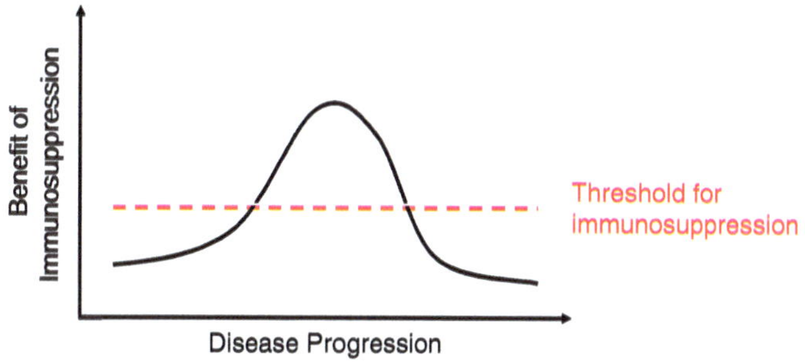

Fig. 19.1 *Relation of inflammation to disease progression in autoimmune disease.* As autoimmune disease progresses, the magnitude of tissue inflammation increases, favoring the use of immunomodulatory therapies. However, as the disease progresses, the inflammation may eventually cause fibrosis and irreversible tissue injury. Because immunosuppressive drugs, themselves, have many adverse side effects; the optimal time to treat is when the tissue inflammation has reached a threshold of severity but before too much irreversible injury has occurred

risks of the disease and the side effects of the therapies used to treat it. In addition to identifying which patients have disease severe enough to warrant the toxicities of treatment, it is equally important to identify those patients whose disease has caused irreversible tissue injury such that they are unlikely to derive much further benefit from treatment (Fig. 19.1). Our results indicate that the abundance of C3d in the kidneys of MRL/lpr mice may increase during the early stages of disease but decline in the late stages of disease. If this dynamic holds true for other renal diseases, then C3d will be a very useful biomarker for identifying patients most likely to benefit from immunosuppression and for guiding the treatment of patients with these diseases.

19.6 Molecular Imaging

The term molecular imaging is used to describe methods of detecting the abundance and location of specific molecules of interest. This typically involves the use of standard radiologic methods in conjunction with a probe or contrast agent that reports on a molecule of interest. Probes targeted to specific inflammation-associated molecules have been used in conjunction with magnetic resonance imaging (MRI) (McAteer et al. 2007), ultrasound (Weller et al. 2003), single photon emission computed tomography (SPECT) (Gratz et al. 2001; Pons et al. 1996), and positron emission tomography (PET) (Roivainen et al. 2003). These imaging modalities differ in their cost, risk to the patient, tissue resolution, and sensitivity (Jaffer and Weissleder 2005). The choice of imaging modality for a particular application usually involves a trade-off of these factors and depends upon which information is most critical for the disease in question. Even within a particular disease, there are settings in which one approach may be more desirable than another. To detect cancer recurrence, for example, the high sensitivity of PET scanning may be desirable. If surgical resection of a tumor is being planned, however, precise anatomic localization may become more important, favoring the use of modalities with greater anatomic resolution, such as PET/CT or MRI.

In addition to the choices among different imaging modalities, various types of imaging probes have been used for molecular imaging. In general, the probes are comprised of a targeting moiety complexed with a reporter element. The targeting moiety is often an antibody to a marker of interest (Barrera et al. 2003), a recombinant protein of interest (Signore et al. 2003), or another molecule known to bind to disease biomarkers (Banati et al. 2000). The utility of each probe is a function of its ability to bind the molecular target with high sensitivity and specificity, as well as the accuracy of the targeted molecule as a biomarker of disease.

19.7 Noninvasive Detection of Complement Activation

To develop a contrast agent capable of reporting on complement activation within specific tissues, we developed an MRI contrast agent that is targeted to tissue-bound C3d. In order to direct the agent to C3d, we employed a recombinant form of complement receptor (CR) 2. The great virtue of CR2 as a targeting moiety is that it binds to C3d but not to intact C3 (Weis et al. 1984). This enables CR2-functionalized contrast agents to bind the tissue-bound C3d without binding C3 in the plasma, increasing the signal to noise ratio. The molecule that we used is a recombinant protein that incorporates two of the C3d-binding regions of CR2 with the Fc region of mouse IgG_1 [CR2-Fc (Gilbert et al. 2006)]. This divalent recombinant protein is approximately 106,600 Da.

As the reporter element of our molecular imaging agent, we have also used iron-oxide (magnetite; Fe_3O_4) superparamagnetic nanoparticles. These particles disrupt the magnetic field in their vicinity and affect the relaxation rate of nearby protons, thus decreasing the local signal generated by a T2-weighted MRI pulse sequence (Josephson et al. 1988). Ten-nanometer crystals were synthesized, and ~75-nm aggregates of the crystals were encapsulated in amine-functionalized phospholipid. The recombinant CR2-Fc was conjugated to the surface of the nanoparticles with the amine groups on the surface of the nanoparticles. In vitro assays using C3-opsonized Chinese hamster ovary (CHO) cells confirmed that the surface CR2-Fc retained its ability to bind the cell surface C3 fragments (Serkova et al. 2010).

When injected systemically, the CR2 protein conjugated to the surface of the nanoparticles adheres to tissue-bound C3d, and the targeted nanoparticles accumulate in inflamed tissues (Fig. 19.2). Because of the effect of the targeted particles on T2-weighted MRI signals, the particles darken the images of the target tissue. We imaged the animals with a 4.7 T MRI scanner and used a series of multislice multiecho (MSME) T2-weighted pulses with 16 various echo times. The T2-relaxation times (in milliseconds, ms) of kidney cortex, inner and outer medulla, as well as muscle (as a control tissue) were then calculated. The baseline signal of a tissue may change with injury due to the altered structure and water content of the tissue, so comparison of the tissues is not based upon an absolute level of T2 signal, but rather it is based upon the change that occurs after the particles bind within

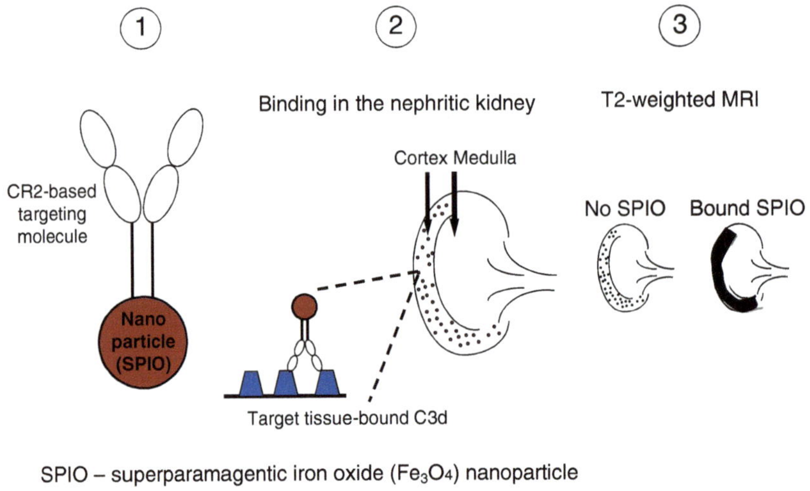

Fig. 19.2 *Method for detecting glomerular C3d deposition by magnetic resonance imaging (MRI).* (**a**) Iron-oxide superparamagnetic nanoparticles are used as an MRI contrast agent. A recombinant protein that contains two copies of the C3d-binding region of complement receptor type 2 (CR2) linked to the Fc region of mouse IgG_1 was conjugated to the nanoparticle surface. (**b**) When injected systemically, the particles bind to tissue deposits of C3d and accumulate in the glomeruli of diseased mice. (**c**) The accumulation of the nanoparticles darkens (negatively enhances) the kidneys of diseased mice when imaged by T2-weighted MRI

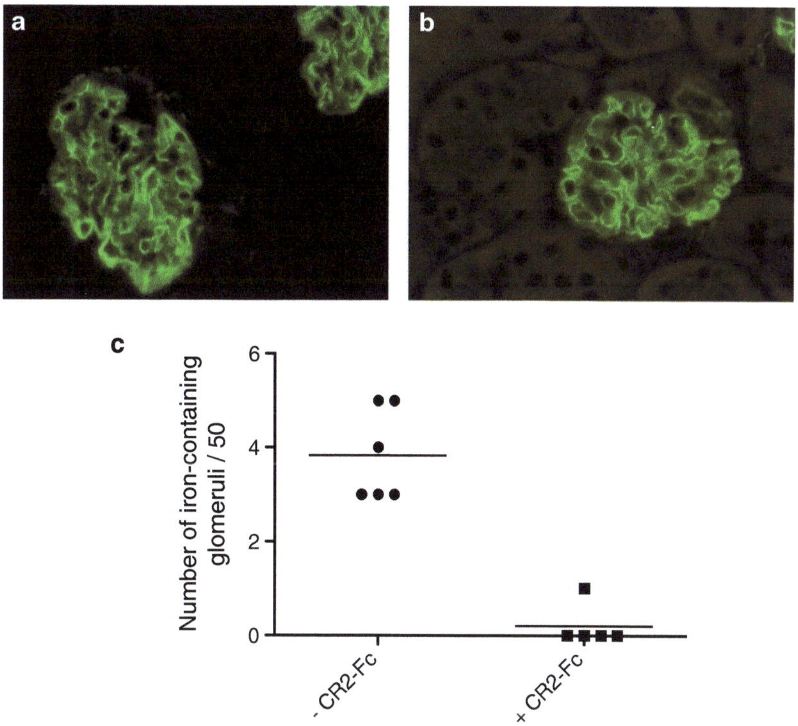

Fig. 19.3 *In vivo specificity of CR2-targeted SPIO.* To confirm that accumulation of CR2-targeted SPIO in nephritic kidneys is mediated by the CR2 moiety, we used unlabeled CR2-Fc to block the binding of CR2-targeted SPIO. For this experiment, $fH^{-/-}$ mice were used. The $fH^{-/-}$ mice have abundant glomerular C3 fragments but do not have glomerular IgG. After preinjecting some of the mice with CR2-Fc, $fH^{-/-}$ mice were injected with CR2-targeted SPIO and harvested after 48 h. (**a**) Representative images of kidneys of $fH^{-/-}$ mice stained for C3 fragments, confirming abundant C3 deposition in the glomeruli of these mice. (**b**) Representative image of the glomerulus of a $fH^{-/-}$ mouse preinjected with CR2-Fc stained for murine IgG in order to detect the CR2-Fc protein. In the mice preinjected with CR2-Fc, diffuse deposition of the CR2-Fc was seen throughout the glomerulus, similar to the pattern of C3 deposition. (**c**) Quantification of iron-containing glomeruli of $fH^{-/-}$ mice 48 h after intravenous CR2-targeted SPIO injection without (–CR2-Fc) or with (+CR2-Fc) preinjection of CR2-Fc, that is, without or with blocking of C3b/iC3b/C3d binding sites of CR2-targeted SPIO. The left and right kidneys of each mouse were examined; $n=3$ mice per condition. $p<0.0001$, Student's *t* test. Kidney Int (Sargsyan et al. 2011)

the tissue. This change in signal, or ΔT2, is proportional to the abundance of the C3d target. We found that the reduction in T2 signal in the kidney lasted 48–72 h after injection of the particles (Serkova et al. 2010).

We tested the particles in MRL/lpr mice, a model of lupus-like renal disease (Serkova et al. 2010). This is a genetically determined disease that progresses as the mice age. We found that the negative enhancement of 16-week MRL/lpr mice could be used to distinguish the diseased animals from age-matched wild-type controls. We examined muscle as a control tissue, and no ΔT2 was seen in muscle after injection of the targeted nanoparticles. Furthermore, untargeted nanoparticles did not cause negative enhancement of the kidneys in MRL/lpr mice, demonstrating the specificity of the method. Experiments using factor H-deficient mice ($fH^{-/-}$ mice), which develop renal disease associated with prominent glomerular C3d deposits but scant glomerular immune complexes, were performed to further confirm the specificity of binding of the CR2-targeted nanoparticles to C3d (Pickering et al. 2002). When unlabeled, CR2-Fc protein was injected into these mice to block the glomerular binding sites for the CR2-targeted nanoparticles; less of the CR2-targeted nanoparticles accumulated in the glomeruli as assessed by iron staining [Fig. 19.3 (Sargsyan et al. 2011)].

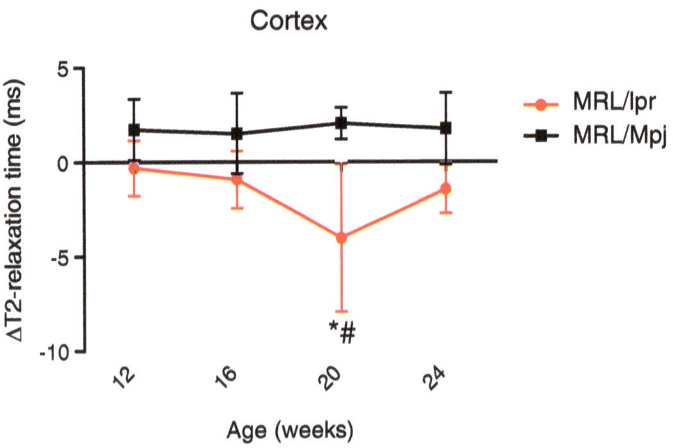

Fig. 19.4 *Noninvasive detection of glomerulonephritis progression.* T2 relaxation times were measured before complement receptor type 2 (CR2)-targeted superparamagnetic iron-oxide injection and 48 h following injection. Changes in T2 relaxation times from preinjected values [ΔT2 relaxation times in milliseconds (ms)] for left kidney cortex of MRL/lpr and control MRL/Mpj mice at 12, 16, 20, and 24 weeks are presented. When ΔT2 relaxation times were compared between the two strains, the MRL/lpr mouse kidneys showed significantly reduced T2 relaxation times at 20 weeks compared to those of age-matched MRL/Mpj mouse kidneys (*$P<0.05$). When ΔT2 relaxation times were compared between the ages within the same strain, the MRL/lpr mouse kidneys showed a significant decrease in ΔT2 relaxation time at 20 weeks for cortex (#$P<0.05$) when compared with earlier ages. [Reproduced from Sargsyan et al. (2011). Detection of glomerular complement C3 fragments by magnetic resonance imaging in murine lupus nephritis. Kidney Int. (Sargsyan et al. 2011)]

As stated above, we have found that the abundance of glomerular C3 deposition increases as MRL/lpr mice age, resulting in peak levels by 20 weeks of age followed by a subsequent decline. The magnitude of ΔT2 in the cortex of MRL/lpr mice injected with CR2-targeted nanoparticles increased as the mice aged from 12 to 20 weeks and then declined at 24 weeks of age (Fig. 19.4). Thus, the ΔT2 mirrored the abundance of glomerular C3d. It also mirrored the severity of the renal disease up to 20 weeks and then declined during the terminal stages of the disease.

19.8 Detection of Complement Activation in Other Organs

The complement system is involved in inflammatory and autoimmune diseases of nearly every organ. We have initially focused on the kidney because of the well-established understanding of complement activation in a wide range of renal diseases, as well as the routine staining of renal biopsies for C3 deposition. However, C3 deposition may be useful as a biomarker of many other diseases, including age-related macular degeneration (AMD) (Anderson et al. 2010), degenerative diseases of the central nervous system (Zhou et al. 2008), and vascular disease (Vlaicu et al. 1985). Because these other organs are not readily accessible to biopsy, the noninvasive detection of complement activation within the tissue offers the promise of revealing early stage tissue inflammation as well as monitoring disease progression. Such information could allow clinicians to titrate the intensity and duration of treatment. The ability to monitor tissue complement activation could also provide a useful surrogate for evaluating drugs in chronic inflammatory diseases.

For each new disease studied, however, the relation of tissue C3 deposits with disease activity will need to be established. The utility of C3d as a disease biomarker will depend upon the kinetics of its accumulation early in the course of the disease. As was seen with glomerular disease, the conversion of C3b/iC3b to C3d may diverge from the overall deposition of the earlier C3 fragments. Each tissue will also differ in its baseline T2 intensity. This is important because the targeted iron-oxide nanoparticles cause a reduction in the T2 signal, creating an effective T2 void. For tissues that are relatively dark at baseline, there will be a reduced dynamic range with which to detect the accumulated particles. The contrast agent will also need to reach the site of inflammation, and the accumulation of a given agent may vary from tissue to tissue. In sites protected by a barrier, such as in the eye and the CNS, the particles may not gain access to the target C3d. On the other hand, inflammation increases vascular permeability, which may increase the access of the nanoparticles to the target site. The requirement of inflammatory changes to permit tissue penetration could, theoretically, add specificity of the any observed contrast effect.

19.9 Detection of Complement Activation Using Other Imaging Modalities

Other imaging modalities may offer greater sensitivity than MRI for detection of complement activation. Because MRI is relatively time consuming, these other modalities may also be more conducive to whole-body imaging. A recombinant form of CR2 has been radiolabeled with (99m) Tc for use as a potential imaging agent (Badar et al. 2011). Radiolabeled targeting proteins can be injected systemically and then detected by single photon emission computed tomography (SPECT) or positron emission tomography (PET). Antibodies to the C3 fragments could also be used for this purpose, provided that they preferentially bind to the activation fragments and not intact C3.

Optical imaging may not be useful for detecting complement proteins in most organs, but this method may be particularly well suited for detecting complement activation in retinal diseases, such as AMD. Uncontrolled activation of the complement cascade is thought to be associated with AMD, and C3 fragments are present in the fundus lesions of these patients (Anderson et al. 2010). Currently, AMD is diagnosed clinically based upon characteristic changes observed during examination of the fundus and the choroidal vasculature. Given the accessibility of this site to visual examination, however, fluorescently labeled proteins targeted to C3 fragments might be detectable by optical imaging of the back of the eye. This could provide a powerful method of detecting AMD at an earlier stage. In addition, this procedure could also be performed serially in order to monitor the progression of disease. There is significant interest in developing complement inhibitors as therapeutics for AMD (Rohrer et al. 2010), and an optical imaging method for monitoring and quantifying complement activation would be of particular use.

19.10 Conclusions

Tissue-bound fragments of C3, such as C3d, are useful biomarkers of inflammatory diseases. The importance of detecting complement activation in real time is even greater now that complement inhibitory agents have entered the clinic. Our experience using CR2-targeted iron-oxide nanoparticles for detection of intrarenal complement activation by MRI demonstrates the feasibility of developing methods to noninvasively monitor complement activation in tissues. Given the frequent involvement of the complement system in a wide range of diseases and the rapid development of therapeutic complement inhibitors, the ability to track the status of the complement system in a given patient will become even more important in the coming years. It is likely that new contrast agents will enable detection of tissue-bound C3d by different imaging modalities. The physicochemical properties of each new molecular imaging agent will determine its ability to report on inflammation in different anatomic locations. As each new agent is developed, it will be necessary to identify the tissues and diseases for which it is best suited.

Acknowledgment This work was supported by a research grant from the Lupus Research Institute (JMT).

References

Anderson DH, Radeke MJ, Gallo NB, Chapin EA, Johnson PT, Curletti CR, Hancox LS, Hu J, Ebright JN, Malek G et al (2010) The pivotal role of the complement system in aging and age-related macular degeneration: hypothesis re-visited. Prog Retin Eye Res 29:95–112

Atkinson C, Song H, Lu B, Qiao F, Burns TA, Holers VM, Tsokos GC, Tomlinson S (2005) Targeted complement inhibition by C3d recognition ameliorates tissue injury without apparent increase in susceptibility to infection. J Clin Invest 115:2444–2453

Atkinson C, Qiao F, Song H, Gilkeson GS, Tomlinson S (2008) Low-dose targeted complement inhibition protects against renal disease and other manifestations of autoimmune disease in MRL/lpr mice. J Immunol 180: 1231–1238

Badar A, DeFreitas S, McDonnell JM, Yahya N, Thakor D, Razavi R, Smith R, Sacks S, Mullen GE (2011) Recombinant complement receptor 2 radiolabeled with [99mTc(CO)3]+: a potential new radiopharmaceutical for imaging activated complement. PLoS One 6:e18275

Banati RB, Newcombe J, Gunn RN, Cagnin A, Turkheimer F, Heppner F, Price G, Wegner F, Giovannoni G, Miller DH et al (2000) The peripheral benzodiazepine binding site in the brain in multiple sclerosis: quantitative in vivo imaging of microglia as a measure of disease activity. Brain 123(Pt 11):2321–2337

Barrera P, Oyen WJ, Boerman OC, van Riel PL (2003) Scintigraphic detection of tumour necrosis factor in patients with rheumatoid arthritis. Ann Rheum Dis 62:825–828

Cameron JS, Turner DR, Heaton J, Williams DG, Ogg CS, Chantler C, Haycock GB, Hicks J (1983) Idiopathic mesangiocapillary glomerulonephritis. Comparison of types I and II in children and adults and long-term prognosis. Am J Med 74:175–192

de Cordoba SR, de Jorge EG (2008) Translational mini-review series on complement factor H: genetics and disease associations of human complement factor H. Clin Exp Immunol 151:1–13

di Belgiojoso GB, Tarantino A, Durante A, Guerra L (1975) Complement deposition in glomerular diseases. Proc Eur Dial Transplant Assoc 11:515–521

Dodds AW, Ren XD, Willis AC, Law SK (1996) The reaction mechanism of the internal thioester in the human complement component C4. Nature 379:177–179

Elward K, Griffiths M, Mizuno M, Harris CL, Neal JW, Morgan BP, Gasque P (2005) CD46 plays a key role in tailoring innate immune recognition of apoptotic and necrotic cells. J Biol Chem 280:36342–36354

Fishelson Z, Horstmann RD, Muller-Eberhard HJ (1987) Regulation of the alternative pathway of complement by pH. J Immunol 138:3392–3395

Gilbert HE, Aslam M, Guthridge JM, Holers VM, Perkins SJ (2006) Extended flexible linker structures in the complement chimaeric conjugate CR2-Ig by scattering, analytical ultracentrifugation and constrained modelling: implications for function and therapy. J Mol Biol 356:397–412

Gratz S, Rennen HJ, Boerman OC, Oyen WJ, Corstens FH (2001) Rapid imaging of experimental colitis with (99m) Tc-interleukin-8 in rabbits. J Nucl Med 42:917–923

Griselli M, Herbert J, Hutchinson WL, Taylor KM, Sohail M, Krausz T, Pepys MB (1999) C-reactive protein and complement are important mediators of tissue damage in acute myocardial infarction. J Exp Med 190:1733–1740

He C, Imai M, Song H, Quigg RJ, Tomlinson S (2005) Complement inhibitors targeted to the proximal tubule prevent injury in experimental nephrotic syndrome and demonstrate a key role for c5b-9. J Immunol 174:5750–5757

Hebert LA, Cosio FG, Neff JC (1991) Diagnostic significance of hypocomplementemia. Kidney Int 39:811–821

Hebert LA, Cosio FG, Birmingham DJ (2001) Complement and complement regulatory proteins in renal disease. In: Neilson EG, Couser WG (eds) Immunologic renal diseases. Lippincott Williams & Wilkins, Philadelphia, pp 367–393

Huang Y, Qiao F, Atkinson C, Holers VM, Tomlinson S (2008) A novel targeted inhibitor of the alternative pathway of complement and its therapeutic application in ischemia/reperfusion injury. J Immunol 181:8068–8076

Jaffer FA, Weissleder R (2005) Molecular imaging in the clinical arena. JAMA 293:855–862

Janssen BJ, Christodoulidou A, McCarthy A, Lambris JD, Gros P (2006) Structure of C3b reveals conformational changes that underlie complement activity. Nature 444:213–216

Josephson L, Lewis J, Jacobs P, Hahn PF, Stark DD (1988) The effects of iron oxides on proton relaxivity. Magn Reson Imaging 6:647–653

Lucignani G (2009) Nanoparticles for concurrent multimodality imaging and therapy: the dawn of new theragnostic synergies. Eur J Nucl Med Mol Imaging 36:869–874

Manzi S, Rairie JE, Carpenter AB, Kelly RH, Jagarlapudi SP, Sereika SM, Medsger TA Jr, Ramsey-Goldman R (1996) Sensitivity and specificity of plasma and urine complement split products as indicators of lupus disease activity. Arthritis Rheum 39:1178–1188

McAteer MA, Sibson NR, von Zur Muhlen C, Schneider JE, Lowe AS, Warrick N, Channon KM, Anthony DC, Choudhury RP (2007) In vivo magnetic resonance imaging of acute brain inflammation using microparticles of iron oxide. Nat Med 13:1253–1258

Moll S, Miot S, Sadallah S, Gudat F, Mihatsch MJ, Schifferli JA (2001) No complement receptor 1 stumps on podocytes in human glomerulopathies. Kidney Int 59:160–168

Morita Y, Ikeguchi H, Nakamura J, Hotta N, Yuzawa Y, Matsuo S (2000) Complement activation products in the urine from proteinuric patients. J Am Soc Nephrol 11:700–707

Pickering MC, Cook HT, Warren J, Bygrave AE, Moss J, Walport MJ, Botto M (2002) Uncontrolled C3 activation causes membranoproliferative glomerulonephritis in mice deficient in complement factor H. Nat Genet 31:424–428

Pons F, Sanmarti R, Herranz R, Collado A, Piera C, Vidal-Sicart S, Munoz-Gomez J, Setoain J (1996) Scintigraphic evaluation of the severity of inflammation of the joints with 99TCm-HIG in rheumatoid arthritis. Nucl Med Commun 17:523–528

Porcel JM, Ordi J, Castro-Salomo A, Vilardell M, Rodrigo MJ, Gene T, Warburton F, Kraus M, Vergani D (1995) The value of complement activation products in the assessment of systemic lupus erythematosus flares. Clin Immunol Immunopathol 74:283–288

Ramos-Casals M, Campoamor MT, Chamorro A, Salvador G, Segura S, Botero JC, Yague J, Cervera R, Ingelmo M, Font J (2004) Hypocomplementemia in systemic lupus erythematosus and primary antiphospholipid syndrome: prevalence and clinical significance in 667 patients. Lupus 13:777–783

Ricker DM, Hebert LA, Rohde R, Sedmak DD, Lewis EJ, Clough JD (1991) Serum C3 levels are diagnostically more sensitive and specific for systemic lupus erythematosus activity than are serum C4 levels. The Lupus Nephritis Collaborative Study Group. Am J Kidney Dis 18:678–685

Ricklin D, Lambris JD (2007) Complement-targeted therapeutics. Nat Biotechnol 25:1265–1275

Ricklin D, Hajishengallis G, Yang K, Lambris JD (2010) Complement: a key system for immune surveillance and homeostasis. Nature Immunol 11:785–797

Rohrer B, Long Q, Coughlin B, Wilson RB, Huang Y, Qiao F, Tang PH, Kunchithapautham K, Gilkeson GS, Tomlinson S (2009) A targeted inhibitor of the alternative complement pathway reduces angiogenesis in a mouse model of age-related macular degeneration. Invest Ophthalmol Vis Sci 50:3056–3064

Rohrer B, Long Q, Coughlin B, Renner B, Huang Y, Kunchithapautham K, Ferreira VP, Pangburn MK, Gilkeson GS, Thurman JM et al (2010) A targeted inhibitor of the complement alternative pathway reduces RPE injury and angiogenesis in models of age-related macular degeneration. Adv Exp Med Biol 703:137–149

Roivainen A, Parkkola R, Yli-Kerttula T, Lehikoinen P, Viljanen T, Mottonen T, Nuutila P, Minn H (2003) Use of positron emission tomography with methyl-11C-choline and 2-18F-fluoro-2-deoxy-D-glucose in comparison with magnetic resonance imaging for the assessment of inflammatory proliferation of synovium. Arthritis Rheum 48:3077–3084

Rother RP, Rollins SA, Mojcik CF, Brodsky RA, Bell L (2007) Discovery and development of the complement inhibitor eculizumab for the treatment of paroxysmal nocturnal hemoglobinuria. Nat Biotechnol 25:1256–1264

Sargsyan SA, Serkova NJ, Renner B, Hasebroock KM, Larsen B, Stoldt C, McFann K, Pickering MC, Thurman JM (2011) Detection of glomerular complement C3 fragments by magnetic resonance imaging in murine lupus nephritis. Kidney Int, 81:152–159

Sargsyan SA, Serkova NJ, Renner B, Hasebroock KM, Larsen B, Stoldt C, McFann K, Pickering MC, Thurman JM (2012) Detection of glomerular complement C3 fragments by magnetic resonance imaging in murine lupus nephritis. Kidney Int, 81:152–159

Schulze M, Pruchno CJ, Burns M, Baker PJ, Johnson RJ, Couser WG (1993) Glomerular C3c localization indicates ongoing immune deposit formation and complement activation in experimental glomerulonephritis. Am J Pathol 142:179–187

Serkova NJ, Renner B, Larsen BA, Stoldt CR, Hasebroock KM, Bradshaw-Pierce EL, Holers VM, Thurman JM (2010) Renal inflammation: targeted iron oxide nanoparticles for molecular MR imaging in mice. Radiology 255:517–526

Shubayev VI, Pisanic TR II, Jin S (2009) Magnetic nanoparticles for theragnostics. Adv Drug Deliv Rev 61:467–477

Signore A, Picarelli A, Annovazzi A, Britton KE, Grossman AB, Bonanno E, Maras B, Barra D, Pozzilli P (2003) 123I-Interleukin-2: biochemical characterization and in vivo use for imaging autoimmune diseases. Nucl Med Commun 24:305–316

Song H, He C, Knaak C, Guthridge JM, Holers VM, Tomlinson S (2003) Complement receptor 2-mediated targeting of complement inhibitors to sites of complement activation. J Clin Invest 111:1875–1885
Swaak AJ, Groenwold J, Bronsveld W (1986) Predictive value of complement profiles and anti-dsDNA in systemic lupus erythematosus. Ann Rheum Dis 45:359–366
Thurman JM, Renner B, Kunchithapautham K, Ferreira VP, Pangburn MK, Ablonczy Z, Tomlinson S, Holers VM, Rohrer B (2009) Oxidative stress renders retinal pigment epithelial cells susceptible to complement-mediated injury. J Biol Chem 284:16939–16947
Vlaicu R, Rus HG, Niculescu F, Cristea A (1985) Immunoglobulins and complement components in human aortic atherosclerotic intima. Atherosclerosis 55:35–50
Walport MJ (2001) Complement. Second of two parts. N Engl J Med 344:1140–1144
Weis JJ, Tedder TF, Fearon DT (1984) Identification of a 145,000 Mr membrane protein as the C3d receptor (CR2) of human B lymphocytes. Proc Natl Acad Sci USA 81:881–885
Weller GE, Lu E, Csikari MM, Klibanov AL, Fischer D, Wagner WR, Villanueva FS (2003) Ultrasound imaging of acute cardiac transplant rejection with microbubbles targeted to intercellular adhesion molecule-1. Circulation 108:218–224
West C (1998) Complement and glomerular diseases. In: Volanakis JE, Frank MM (eds) The human complement system in health and disease. Marcel Dekker, Inc., New York, pp 571–596
Zhou J, Fonseca MI, Pisalyaput K, Tenner AJ (2008) Complement C3 and C4 expression in C1q sufficient and deficient mouse models of Alzheimer's disease. J Neurochem 106:2080–2092

Chapter 20
Highly Multiplexed Proteomic Platform for Biomarker Discovery, Diagnostics, and Therapeutics

Michael R. Mehan, Rachel Ostroff, Sheri K. Wilcox, Fintan Steele, Daniel Schneider, Thale C. Jarvis, Geoffrey S. Baird, Larry Gold, and Nebojsa Janjic

Abstract Progression from health to disease is accompanied by complex changes in protein expression in both the circulation and affected tissues. Large-scale comparative interrogation of the human proteome can offer insights into disease biology as well as lead to the discovery of new biomarkers for diagnostics, new targets for therapeutics, and can identify patients most likely to benefit from treatment. Although genomic studies provide an increasingly sharper understanding of basic biological and pathobiological processes, they ultimately only offer a prediction of relative disease risk, whereas proteins offer an immediate assessment of "real-time" health and disease status.

We have recently developed a new proteomic technology, based on modified aptamers, for biomarker discovery that is capable of simultaneously measuring more than a thousand proteins from small volumes of biological samples such as plasma, tissues, or cells. Our technology is enabled by SOMAmers (*S*low *O*ff-rate *M*odified *Aptamers*), a new class of protein binding reagents that contain chemically modified nucleotides that greatly expand the physicochemical diversity of nucleic acid-based ligands. Such modifications introduce functional groups that are absent in natural nucleic acids but are often found in protein–protein, small molecule–protein, and antibody–antigen interactions. The use of these modifications expands the range of possible targets for SELEX (*S*ystematic *E*volution of *L*igands by *EX*ponential Enrichment), results in improved binding properties, and facilitates selection of SOMAmers with slow dissociation rates.

Our assay works by transforming protein concentrations in a mixture into a corresponding DNA signature, which is then quantified on current commercial DNA microarray platforms. In essence, we take advantage of the dual nature of SOMAmers as both folded binding entities with defined shapes and unique nucleic acid sequences recognizable by specific hybridization probes. Currently, our assay is capable of simultaneously measuring 1,030 proteins, extending to sub-pM detection limits, an average dynamic range of each analyte in the assay of >3 logs, an overall dynamic range of at least 7 logs, and a throughput of one million analytes per week. Our collection includes SOMAmers that specifically recognize most of the complement cascade proteins. We have used this assay to identify potential

M.R. Mehan • R. Ostroff • S.K. Wilcox • F. Steele • D. Schneider • T.C. Jarvis • N. Janjic (✉)
SomaLogic, Inc., 2945 Wilderness Place, Boulder, CO 80301, USA
e-mail: njanjic@somalogic.com

G.S. Baird
Department of Laboratory Medicine, University of Washington, Seattle, WA 98104, USA

L. Gold
SomaLogic, Inc., 2945 Wilderness Place, Boulder, CO 80301, USA

Department of Molecular, Cellular and Developmental Biology, University of Colorado, Boulder, CO 80309, USA

J.D. Lambris et al. (eds.), *Complement Therapeutics*, Advances in Experimental Medicine and Biology 735, DOI 10.1007/978-1-4614-4118-2_20, © Springer Science+Business Media New York 2013

biomarkers in a range of diseases such as malignancies, cardiovascular disorders, and inflammatory conditions. In this chapter, we describe the application of our technology to discovering large-scale protein expression changes associated with chronic kidney disease and non-small cell lung cancer. With this new proteomics technology—which is fast, economical, highly scalable, and flexible—we now have a powerful tool that enables whole-proteome proteomics, biomarker discovery, and advancing the next generation of evidence-based, "personalized" diagnostics and therapeutics.

20.1 Introduction

Development of any new therapeutic requires the identification of a suitable target for therapeutic intervention, discovery of the molecule capable of modulating its activity, completion of preclinical and clinical studies required for approval, and finally—if things go well—regulatory approval. Although this chapter does not discuss specific complement therapeutics, we describe a novel and versatile high-content proteomics assay that can be used to discover new biomarkers or new targets for therapeutic intervention in a hypothesis-free manner across many different diseases and conditions, including those with complement involvement. This powerful tool can also be useful throughout the drug development pipeline, from basic research, such as target validation or mechanism-of-action studies, to later-stage clinical studies, such as monitoring drug safety or patient enrollment. Our assay is based on a large collection (>1,000) of unique affinity reagents called SOMAmers. We illustrate the wide range of applications of our assay through several clinical studies. Because each SOMAmer is a well-defined molecule capable of binding to a specific protein target, we also address the potential use of SOMAmers in diagnostics and therapeutics.

20.2 SELEX Then and Now

Since the first description of SELEX more than 20 years ago (Tuerk and Gold 1990; Ellington and Szostak 1990), aptamers to a wide variety of molecular targets have been described, including proteins, peptides, and small molecules. Aptamers are short single-stranded oligonucleotides that fold into diverse and intricate molecular structures that allow them to bind with high affinity and specificity to their targets. The power of SELEX comes from the extremely large libraries of randomized nucleic acid sequences that can be efficiently screened for molecules with functional properties defined by the selection conditions. A SELEX library with 40 random sequence positions has 4^{40} (~10^{24}) possible combinations and a typical selection screens 10^{14}–10^{15} unique molecules. This is on the order of 10^5 times larger than standard peptide or protein combinatorial molecular libraries (Binz et al. 2005).

These large numbers, which translate to an enormous shape repertoire, make up to a large degree for an imbalance in diversity: with 4 bases compared to 20 amino acids, the chemical diversity of nucleic acid libraries is considerably lower than that of proteins. Other than just the numerical advantage, the physicochemical diversity of amino acids is also considerably greater than that of the four nucleic acid bases.

Until recently, most SELEX experiments have been performed with natural RNA and DNA or with nucleic acid libraries modified at the 2′-position of ribose (such as 2′-fluoro, 2′-amino, or 2′-O-methyl substitutions) for enhanced nuclease resistance. These modifications have been very useful for generating aptamers with improved stability in vitro and longer residence times in vivo. However, their contribution to chemical diversity, and therefore the overall success rate of SELEX, has been modest. Fortunately, there are other positions in nucleic acids that can be modified in a manner compatible with the enzymatic steps of SELEX, and additional diversity has recently been introduced via modifications at the 5′-position of uridine (Vaught et al. 2010). Substantial additional chemical diversity can be

Table 20.1 SELEX library affinities (K_d, nM) with unmodified and modified nucleotides

Target protein	Unmodified DNA	(Bn)dUTP	(iBu)dUTP	(Trp)dUTP
	CH_3	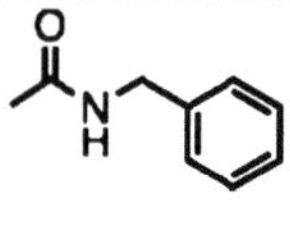	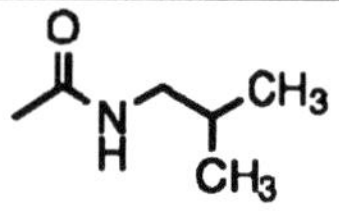	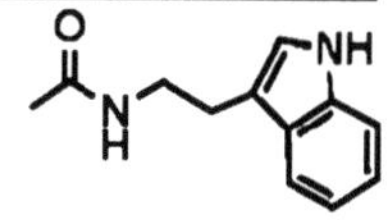
GA733-1 protein[a]	9	3	5	0.5
Osteoprotegerin[a]	40	5	9	0.2
4-1BB[a]	>100[b]	6	>100	4
B7[a]	>100	10	>100	7
B7-2[a]	>100	>100	>100	6
CTLA-4[a]	>100	>100	>100	1
sE-selectin[a]	>100	>100	>100	2
Fractalkine	>100	>100	>100	0.05
gp130, soluble[a]	>100	6	20	1
HMG-1	>100	>100	20	5
IR	>100	2	10	0.2
PAI-1	>100	0.4	0.9	0.2
P-cadherin[a]	>100	4	5	3
sLeptin R[a]	>100	2	>100	0.5

[a]The protein used was expressed as a fusion to the Fc of human IgG_1. No detectable binding of the active library to an alternate Fc fusion protein was observed
[b]No detectable binding up to target protein concentration of 100 nM was observed

introduced with this type of chemistry, permitting the incorporation of functional groups often found in protein–protein interactions, antigen–antibody interactions, and small molecule drugs. Importantly, the use of these modifications has greatly expanded the range of possible targets for SELEX.

When tested against a panel of "difficult" human proteins for which SELEX with unmodified DNA repeatedly failed, in side-by-side experiments, three modified DNA libraries showed considerably better success, as shown in Table 20.1. As an internal control, we included an "easy" target (GA733-1 protein) for which unmodified DNA SELEX results in reproducible affinity enrichment. With these difficult SELEX targets, only modified nucleotide libraries yielded notable affinity enrichment. In this example, depending on the target, certain modifications worked better than others, demonstrating the benefit of applying multiple modifications against the same target to ensure a high probability of success. In light of these results, which have since been repeated in numerous additional SELEX experiments, we have adopted modified nucleotide SELEX as a new standard in all of our selections.

To describe their unique composition and to reflect the deliberate selection for slow kinetics of dissociation from their targets, we have named this new class of binding reagents SOMAmers. The higher success rate of SELEX has allowed us to build a portfolio of SOMAmers that currently contains over 1,000 high-affinity reagents. SOMAmers have been derived for targets that cover a diverse set of molecular functions and several known disease or physiology associations (Fig. 20.1a). This diverse and expanding set of targets provides extensive coverage of major gene families including receptors, kinases, growth factors and hormones, and span a diverse collection of secreted, intracellular and extracellular proteins or domains (Fig. 20.1b, c). Most SOMAmers to protein targets have dissociation constant (K_d) values of less than 1 nM (Gold et al. 2010).

We have recently solved several co-crystal structures of SOMAmers bound to their protein targets (manuscript in preparation). These structures have given us a window into the astonishing diversity and novel structural themes that can be achieved with SOMAmers, including the utilization of hydrophobic moieties unique to SOMAmers for intramolecular motifs and SOMAmer–protein contacts. It is clear that apparently subtle modifications to standard nucleotide chemistry, and at only one of the four bases, open up a new world of possible structures. In addition to known shape complementarity, polar contacts,

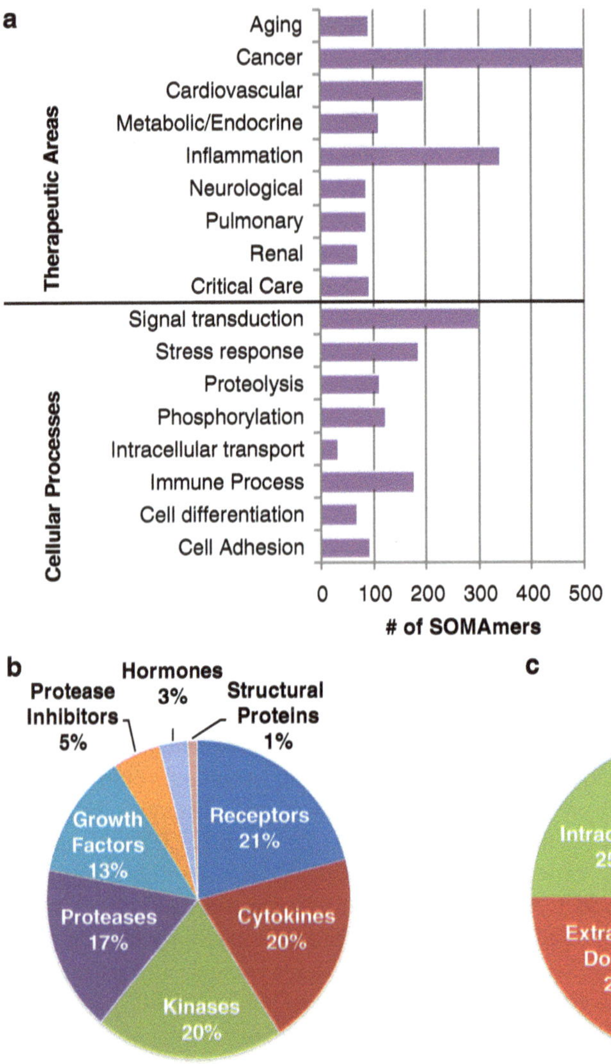

Fig. 20.1 (**a**) SOMAmers bind proteins in numerous disease areas and physiological processes. (**b**) SOMAmers have been generated to various classes of proteins such as kinases, receptors, and even hormones. (**c**) SOMAmers target secreted, intracellular, and extracellular domains of proteins

hydrogen-bonding interactions, and charge–charge interactions observed in aptamers, SOMAmer–protein interactions include hydrophobic contacts, which are known to play a key role in protein–protein interactions.

20.3 Revisiting Shapes and Tapes: High-Content Proteomics

So what can we do with such a unique collection of binding reagents? There are three primary areas that we are pursuing: highly multiplexed proteomic analyses, diagnostics, and therapeutics.

For high-content proteomic analyses, we have developed an assay that takes advantage of the dual nature of SOMAmers as high-affinity binding reagents (shapes) that can also be hybridized to specific

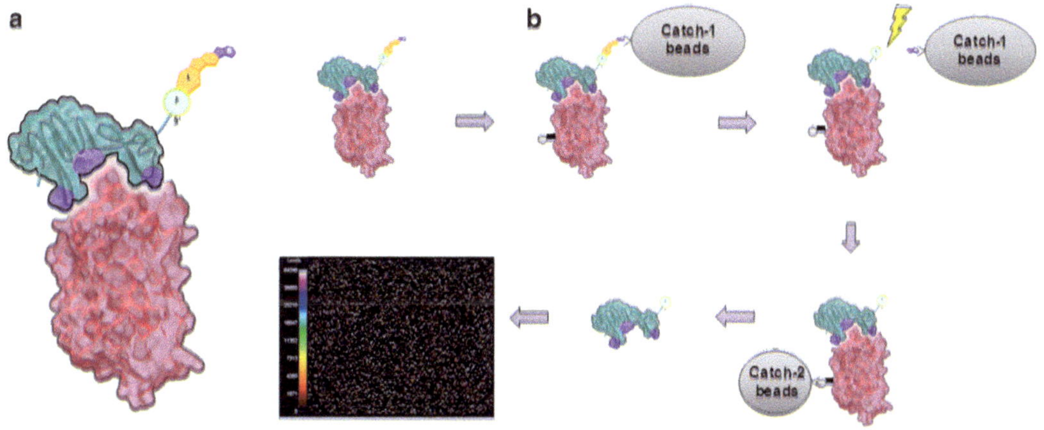

Fig. 20.2 (**a**) Schematic representation of an interaction between a SOMAmer (*green*) and its protein target (*red*). Modified residues at the interface are shown in *purple*. For the assay, the SOMAmer contains a biotin (*B*), a photocleavable linker (*L*), and a fluorophore (*F*) at the 5′ end. (**b**) Schematic sequence of assay steps leading to quantitative readout of target proteins (clockwise from top left): SOMAmer–protein binding, capture onto streptavidin-coated beads ("catch-1"), biotin-labeling of proteins, photocleavage followed by kinetic challenge, second biotin capture ("catch-2"), release from complexes into solution at high pH, and hybridization to microarray. SOMAmers are detected by fluorescent tags when the array is scanned

nucleic acid probes on existing nucleic acid array platforms (tapes). The modifications at the 5′-position of uridine do not interfere with canonical base pairing (as is required for SELEX), and therefore SOMAmers hybridize effectively to capture probes on the oligonucleotides array. Binding of a collection of SOMAmers to their protein targets in a complex mixture (such as plasma, serum, homogenized tissue, cell extracts, and the like) is carried out in solution rather than on a surface to take advantage of more favorable kinetics of binding and dissociation.

In our assay, the sample is incubated with a mixture of SOMAmers, each containing a biotin, a photocleavable group, and a fluorescent tag. SOMAmer–protein complexes are then captured on streptavidin beads in a step we refer to as "catch-1" (Fig. 20.2). The beads are then washed to remove unbound proteins, followed by biotin-labeling of bead-associated proteins. The complexes are then released from the beads back into solution by photocleavage with UV light (biotin that was originally part of the SOMAmer now remains on beads), diluted and challenged with a high concentration of a polyanionic competitor dextran sulfate. It is in this step that a second dimension of specificity (equivalent to a second antibody in a sandwich binding assay) is achieved. Because dissociation rates of cognate SOMAmer–protein interactions are much slower than those of nonspecific interactions, a polyanionic competitor selectively disrupts non-cognate complexes. It is worth noting that one of the reasons such a high degree of multiplexing is possible in our assay is the fact that, since all aptamers are polyanions, there exists a common non-denaturing competitor capable of disrupting virtually all SOMAmer–protein interactions. Such a general competitor is not known for protein–protein or antibody–antigen interactions.

Following this kinetic challenge step, the complexes are recaptured on a second set of beads via the biotinylated proteins ("catch-2"), followed by additional washing to remove unbound SOMAmers. Finally, the SOMAmers that remain associated with beads through their cognate proteins are denatured under basic conditions (DNA is stable to base) and eluted and hybridized to complementary probes printed on a standard DNA microarray. In this manner, protein concentrations in the original sample are quantitatively converted into addressable fluorescence signals on the array.

Based on pull-down experiments that mimic conditions of our assay, we have observed a high degree of specificity with fluorescence signals reflecting cognate SOMAmer–protein interactions,

Table 20.2 CKD population demographics

	Early-stage CKD[a]	Late-stage CKD[b]
N (total=42)	11	31
Gender %F (F/M)	33% (4/11)	45% (14/31)
Ethnicity (A/B/C/H)[c]	1/5/1/4	2/9/0/20
Age (avg years)	62 [51–68]	67 [57–77]
Wt. (avg kg)	89 [73–98]	88 [75–104]
BMI (avg)	30.5 [26.6–36.5]	31.8 [27.1–36.6]
eGFR (median)[d]	70 [62–97]	25 [7–49]

[a]Stages 1–2 based on eGFR
[b]Stages 3–5 based on eGFR
[c]A, Asian; B, African American; C, Caucasian; H, Hispanic
[d]eGFR determined from creatinine clearance (MDRD formula) mL/min/m^2

even at a very high multiplicity. Aside from the intrinsic specificity of SOMAmers and the off-rate differential between specific and nonspecific interactions, the use of two separate bead capture steps clearly contributes to overall specificity. This is because the nature of nonspecific binding interactions between proteins and SOMAmers bound to beads (catch-1) is different from that of SOMAmers binding to proteins immobilized on beads (catch-2). Nonspecific binding of proteins to surfaces is often characterized by high avidity, multivalent interactions, which can have long off-rates. Reversing the polarity of attachment of the complexes on beads (i.e., from SOMAmer-bound to protein-bound) ensures that multivalent interactions on catch-1 beads (e.g., involving a protein with multiple binding sites for nucleic acids) are converted to monovalent interactions on catch-2 beads (where each SOMAmer may interact as a monomer with all immobilized binding sites of that protein). In this manner, differences in off-rates between specific and nonspecific interactions can be fully exploited.

The current version of our assay, which we call SOMAscan, measures 1,030 proteins, a number that we continue to increase regularly. Despite its complexity, the assay has a large overall dynamic range of 7 logs (from 100 fM to 1 μM), low limits of detection (0.3 pM median), and excellent reproducibility (5% median intra-run and inter-run coefficient of variance) (Gold et al. 2010).

20.4 Biomarkers of Chronic Kidney Disease

To establish the utility of SOMAscan with real clinical samples, we obtained and analyzed serum samples from subjects with chronic kidney disease (CKD). The progressive decline in kidney function characteristic of CKD is a substantial global healthcare problem that affects about one-tenth of the population worldwide (Levey et al. 2007). Early detection often translates to better treatment options, and the discovery of new disease-related biomarkers could result in more timely diagnosis, more effective disease monitoring, and improved clinical outcomes. Glomerular filtration rate (GFR) of serum proteins depends primarily on their size and charge (Venturoli and Rippe 2005), and a reduction in GFR that accompanies CKD could be expected to lead to an increase in the concentration of small (<45 kDa), basic proteins.

Of the 42 serum samples obtained from CKD patients at the Rogosin Institute (Cornell University, Ithaca, NY), 11 subjects had early-stage and 31 subjects had late-stage CKD (Table 20.2) based on estimated GFR (eGFR) (Stevens et al. 2006). In this study, we simultaneously measured 614 human proteins, which represented the array size at the time analyses were conducted. Comparison of relative protein amounts for all measured analytes in serum for early-stage versus late-stage CKD was analyzed for significance using the Mann–Whitney test (Fig. 20.3a). Sixty analytes varied significantly between the two groups with a q-value (p-value corrected for false discovery rate) of $\leq 4.2\times10^{-4}$

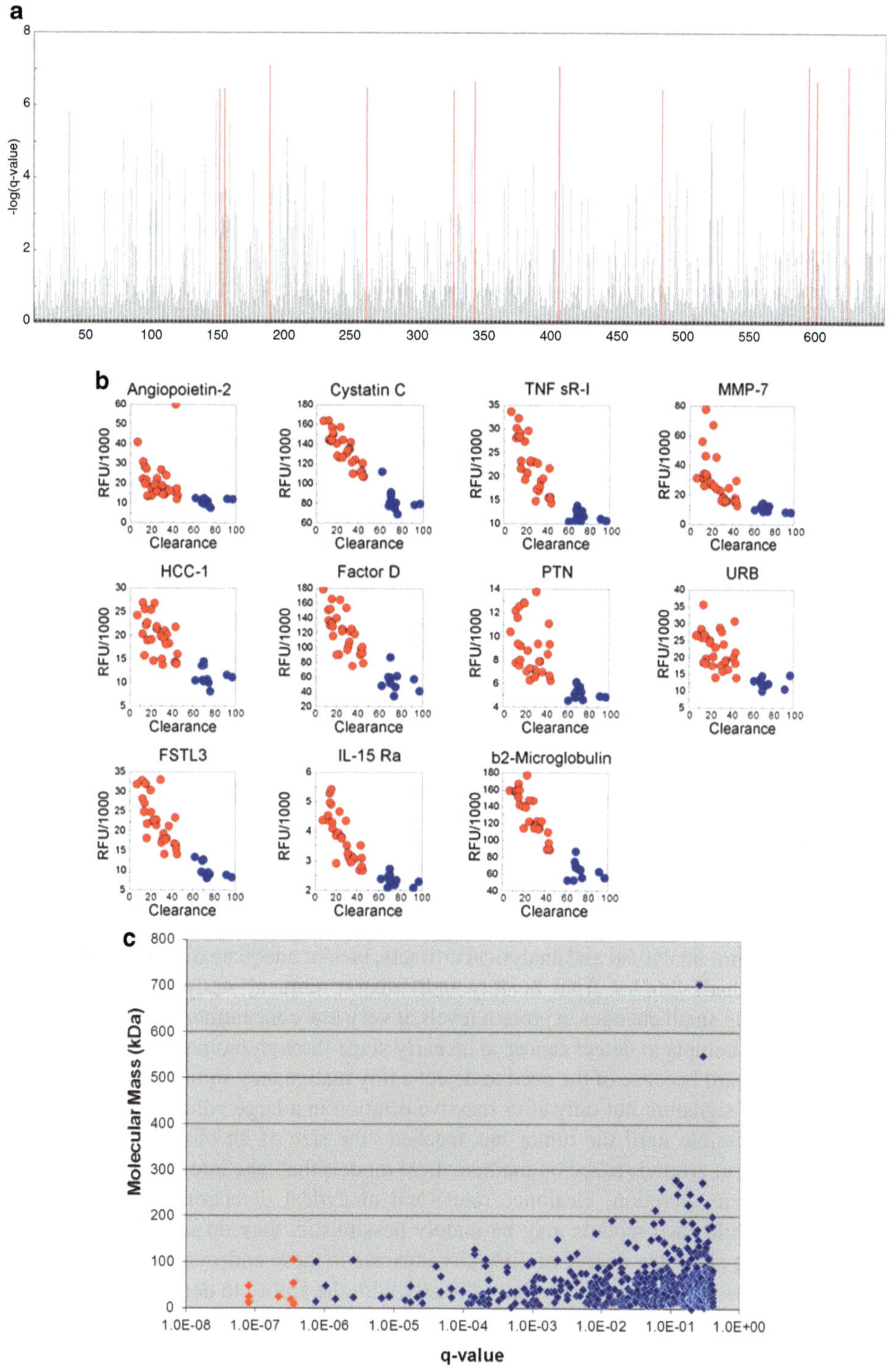

Fig. 20.3 (**a**) CKD biomarker discovery using SOMAscan. Mann–Whitney q-values (p-values corrected for false discovery rate) comparing late-stage versus early-stage CKD. Eleven analytes with the smallest q-values ($\leq 3.4 \times 10^{-7}$) are shown in *red* bars. (**b**) Protein concentrations (expressed as RFU values) as a function of renal clearance (eGFR) for the eleven best biomarkers of late-stage (*red* symbols) versus early-stage (*blue* symbols) CKD. (**c**) Plot of molecular mass of a protein and the probability that it is a CKD biomarker (q-value) (Reproduced from Gold et al. (2010))

(Gold et al. 2010). Eleven analytes with the lowest q-values ($\leq 3.4 \times 10^{-7}$) are shown in Fig. 20.3b. Nine out of eleven are relatively small proteins (<25 kDa), and among this group are two important known markers of CKD, cystatin C ($M_r = 13.3$ kDa) and β_2 microglobulin ($M_r = 11.7$ kDa) (Fassett et al. 2011). Complement factor D ($M_r = 24.4$ kDa) and the soluble form of TNF receptor-1 ($M_r = 21.2$ kDa) have also been reported to be elevated in CKD (Pascual et al. 1988; Riemsdijk-van Overbeeke et al. 2000). Rediscovery of previously known biomarkers is an important validation for a new platform which also lends added credibility to the newly discovered markers. Another observation that supports the notion that these 11 proteins are biomarkers for CKD progression is the inverse correlation between eGFR and protein concentration (Fig. 20.3b).

Our observation that small proteins accumulate in blood as a result of reduced GFR is hardly surprising. However, the fact that many small proteins either change substantially less or not at all with disease progression is less intuitive (Fig. 20.3c) and shows that renal tissue injury and secondary systemic effects resulting from reduced renal function clearly also affect the proteome (protein charge appears to have no obvious influence on the quality of biomarkers). It is also possible that some small proteins may be associated with other proteins in blood so that their larger effective circulating molecular mass results in reduced renal filtration rate. Another possibility is that the epitopes recognized by SOMAmers may not be fully available, for a variety of reasons (e.g., being occupied with another protein), thus interfering with accurate measurement in our assay. In any event, our data clearly show that only a subset of small proteins appear to accumulate in CKD. This finding is consistent with the notion that increased concentration in plasma of such low molecular weight proteins, sometimes referred to as “middle molecules,” contributes to the pathology of kidney disease (Vanholder et al. 2008). High-content proteomic analysis provides a means of unbiased discovery of such proteins and their relationship to disease progression.

20.5 Protein Signatures at the Source and in the Blood

Despite a surge in interest in biomarkers as reflected by the number of publications in the field, the discovery and—more critically—the validation of new biomarkers with demonstrable diagnostic or clinical utility remain a considerable challenge (Diamandis 2010). The reasons for this gap are well known and include pre-analytical and analytical artifacts, lack of adequate disease controls for certain diseases (i.e., matched samples from healthy individuals), clinical study design issues, and the difficulty of detecting small changes in protein levels at very low concentrations. These challenges are most evident with attempts to detect cancer at an early stage through biomarkers in circulation. This effort is especially hard because of the need to detect a tiny malignancy somewhere in the body based on a unique protein signature but only after massive dilution in a large volume of blood. The notion that this is even possible until the tumor has reached “the size of an olive” has been questioned recently on theoretical grounds based on mathematical models that take into account parameters such as kinetics of protein secretion, clearance rates, and analytical detection sensitivities (Hori and Gambhir 2011). While such models may be unduly pessimistic, they do serve a useful purpose of providing a warning about the degree of difficulty intrinsic to these endeavors and also make it clear that an increase in assay sensitivity correlates directly with the ability to detect smaller cancers at an earlier stage (Lutz et al. 2008; Hori et al. 2011). Aside from boosting sensitivity, which has been a central focus of our attention for some time, one way to improve the chances of discovering true cancer biomarkers is to measure protein concentrations at both the source, such as in tumor tissue, as well as in the circulation. Such combined results can support the validity of potential biomarkers and minimize experimental artifacts.

We have recently performed such studies in non-small cell lung cancer (NSCLC). NSCLC is still the leading cause of cancer deaths worldwide largely because the vast majority of cases (84%) are diagnosed at an advanced stage, when a 5-year survival rate is less than 15% (Okada et al. 2005; Kassis et al. 2009; Jemal et al. 2010). In sharp contrast, when surgically removed at an early stage, 5-year survival increases to 86% (Okada et al. 2005; Kassis et al. 2009).

For the NSCLC serum biomarker study, we analyzed 1,326 samples from four independent clinical study centers (Ostroff et al. 2010a). The study included patients diagnosed with clinical stage I–III NSCLC and a control population with a history of long-term tobacco use, including active smokers and ex-smokers with at least 10 pack years of cigarette smoking. At the time this study was conducted, our array size measured 813 human proteins in each sample. Taking extensive precautions to account for pre-analytic variables, we identified 44 candidate biomarkers (Table 20.3) and developed a 12-protein panel that distinguished NSCLC from controls with 91% sensitivity and 84% specificity in a training set and 89% sensitivity and 83% specificity in a blinded, independent verification set (Ostroff et al. 2010a).

Table 20.3 List of potential NSCLC biomarkers identified from serum and tissue samples

Protein name	NSCLC serum	NSCLC tissue
Activin A		Up
Adiponectin		Down
AMPM2	Up	
BGN		Down
BLC	Up	Up
BMP-1	Down	
C1s	Up	
C9	Up	
Cadherin E	Up	Up
Calpain I	Up	
Carbonic anhydrase III		Down
Caspase 3		Up
Catalase	Up	Down
CD30 ligand	Up	
CD36 antigen		Down
CDK5/p35	Up	
CK-MB	Down	
Contactin-5	Down	
CXCL16, soluble		Down
Endostatin	Up	Down
ERBB1	Down	
ESAM		Down
FGF-17	Up	
Fibronectin		Up
FYN	Up	
HSP 90a	Up	
HSP 90b	Up	
IDE		Up
IGFBP-2	Up	Up
IGFBP-5		Up
IGFBP-7		Up
IL-15 Ra	Up	
IL-17B	UP	
IL-8		Up
IMB1	Up	
Kallikrein 7	Down	

(continued)

Table 20.3 (continued)

Protein name	NSCLC serum	NSCLC tissue
KPCI	Up	
LDH-H 1	Up	
LGMN	Up	
LRIG3	Down	
Macrophage mannose receptor	Up	Down
MEK1	Up	
Midkine	Up	
MIP-5	Up	
MK13	Up	Up
MMP-12	Up	Up
MMP-7	Up	Up
NACA	Up	
NAGK	Up	Up
NAP-2		Down
PARC	Up	
P-selectin		Down
PTN	Up	
Renin	Up	
RGM-C	Down	
SCF sR	Down	
SLPI		Down
sL-selectin	Down	
sRAGE		Down
Thrombospondin-1		Up
TPSB2		Down
TrATPase		Down
TSP2		Up
Ubiquitin+1	Up	
uPA		Up
URB		Up
VEGF	Up	Up
vWF		Down
YES	Up	Up

Proteins identified in both studies are shown in boldface font

For the NSCLC tissue biomarker study, we analyzed homogenized lung tissue samples from surgical resections obtained from eight NSCLC patients ranging in age from 47 to 75 years old and diagnosed with pathology-confirmed NSCLC stages IA through IIIB (Mehan et al. submitted). We obtained three samples from each resection: tumor tissue sample, adjacent non-tumor tissue (within 1 cm of the tumor), and distant uninvolved lung tissue (furthest edge of the resection from the tumor). We prepared the samples for analysis by rapidly freezing and homogenizing the tissue after resection and normalizing total protein concentration in each homogenate. In this case, our biomarker discovery array measured 820 human proteins, a nearly identical array size and composition to that used in the NSCLC serum study. The signals generated by most analytes were similar in healthy adjacent and distant tissue. In contrast, comparison of tumor tissues with non-tumor tissue (adjacent or distant) revealed much larger changes in protein expression (Fig. 20.4). Thirty-six proteins with the largest mean fold-change in protein expression between tumor and non-tumor tissue samples are listed in Table 20.3 (alongside the biomarkers identified in the serum NSCLC study). We tested the significance

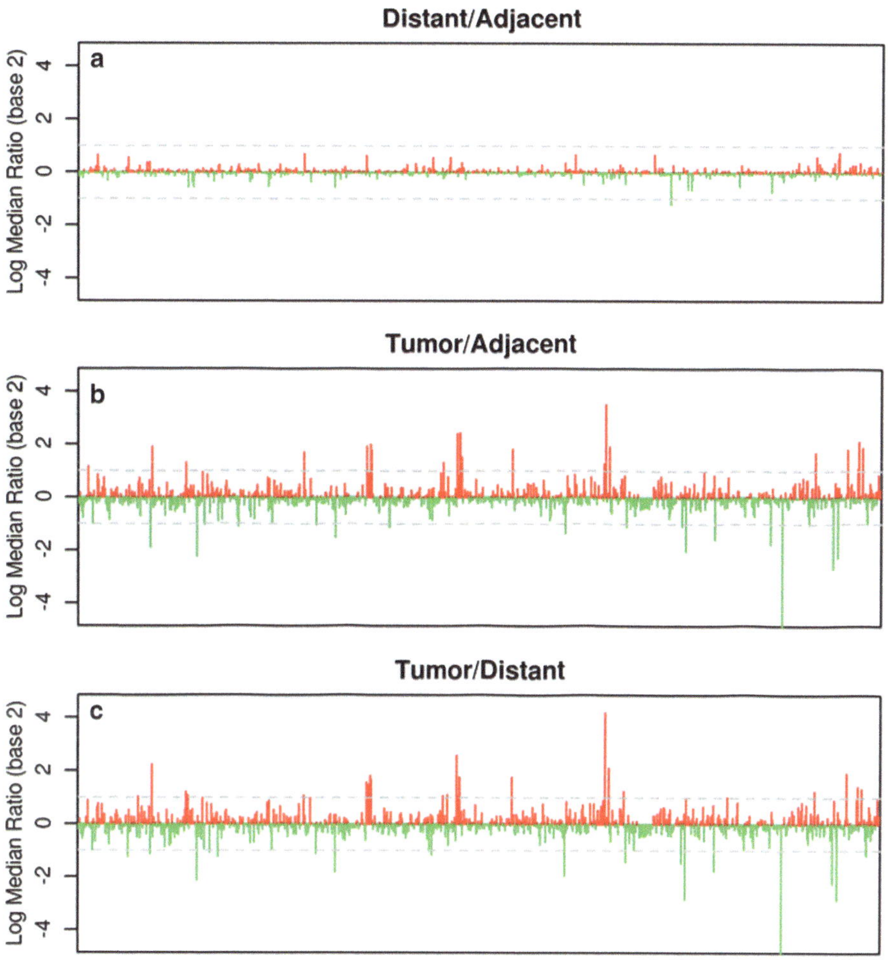

Fig. 20.4 (**a**) Relative changes in protein expression for 813 proteins from eight NSCLC resection samples between adjacent and distant tissue, (**b**) tumor and adjacent tissue, and (**c**) tumor and distant tissue, expressed as log2 median ratios. The *dotted line* represents twofold change

of these changes with the Mann–Whitney test and required a p-value of ≤0.05 after correcting for multiple tests (false discovery rate cutoff of $q<0.05$). We were able to obtain statistically significant results with a relatively small number of samples in part because we had paired tumor and non-tumor tissue samples from each individual, which eliminated population variance associated with cross-sectional study designs. Appropriately chosen reference samples are crucially important in biomarker discovery research (Diamandis 2010; Ioannidis and Panagiotou 2011; Bossuyt 2011). Some of the potential biomarkers identified in this study can be detected in frozen tissue sections with the same SOMAmers used to discover them. We have previously demonstrated that fluorophore-labeled SOMAmers can be used for rapid and selective detection of analytes in tissues (Gupta et al. 2011). For example, the higher level of expression of thrombospondin-2 (TSP-2) in tumor tissues compared to surrounding healthy tissue was confirmed with staining of frozen sections with Cy3-labeled TSP-2 SOMAmer, strikingly revealing its localization to the fibrous tumor-associated stroma (Fig. 20.5).

It is worth keeping in mind that tumor-associated protein is not the same as tumor-specific marker. Inflammation, extracellular matrix remodeling, hypoxia, and tissue necrosis accompany tumor

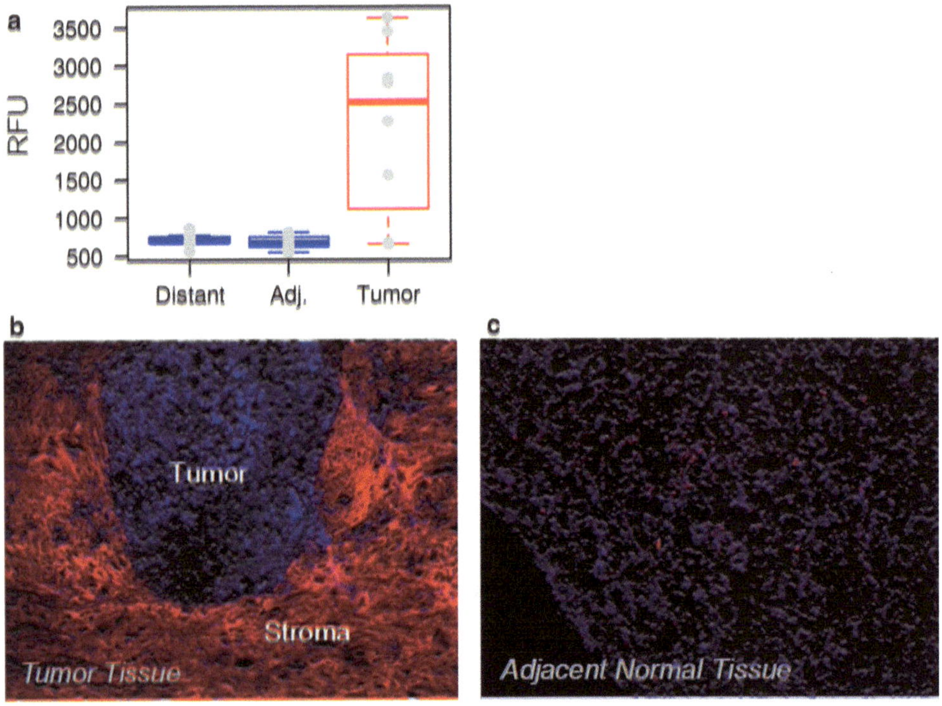

Fig. 20.5 (**a**) TSP-2 expression in tumor tissue (in relative fluorescence units, RFU) compared with adjacent or distal tissue from eight NSCLC samples used in this study. Data points are plotted in box plots showing the median (*line*) and 25% and 75% range of the data (*box outline*), (**b**) Cy3-labeled TSP-2 SOMAmer staining (*red*) of the fibrocollagenous matrix surrounding a tumor nest, and (**c**) corresponding normal lung specimen stained with Cy3-labeled TSP-2 SOMAmer

progression but also many other nonmalignant conditions such as injury, wound healing, or infection. In addition, a difference in the ratio of cell types that constitute a tumor sample compared to that of the normal lung tissue could account, at least in part, for the observed differences in protein expression. Tumor tissue typically consists of cancer cell overgrowth, and may be more or less vascularized than the surrounding normal tissue, leading to an imbalance of cell types between tumor tissue and normal tissue samples.

These two NSCLC studies allow us to compare differential expression of proteins in serum with those in tissues using the same proteomic platform with an essentially identical set of analytes (Table 20.3 and Fig. 20.6). The most obvious difference in these data sets is the much larger magnitude of relative changes in protein expression in tissues compared to serum (Fig. 20.6). Because tumor tissue is the presumed source of the changes in protein expression, this result is not surprising. Twelve of the analytes altered in tumor tissue were also differentially expressed in sera from NSCLC patients (BCA-1, cadherin-1, catalase, endostatin, IGFBP-2, macrophage mannose receptor, MK13, MMP-7, MMP-12, NAGK, VEGF, and YES) (Table 20.3). The direction of change is the same for most but not all biomarkers. Inverse correlations are especially interesting and warrant further studies since they could provide clues regarding the redistribution of certain biomarkers in diseased versus normal tissues. Overall, the availability of high-content proteomic data from both tissue and blood help us better understand complex changes associated with NSCLC and allow us to select the best candidate biomarkers for its early detection.

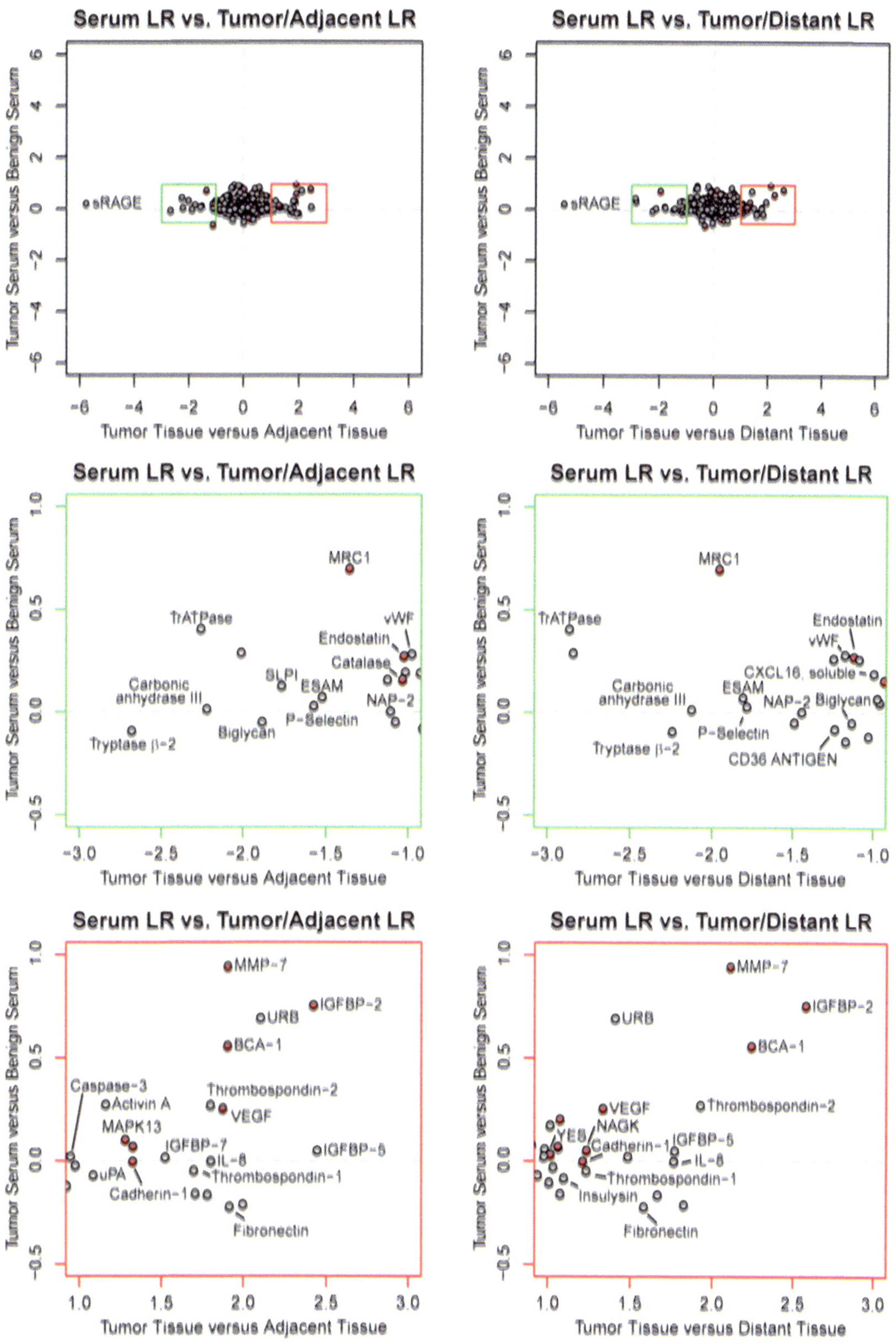

Fig. 20.6 Changes in protein expression in NSCLC tissue compared to serum. The *top two panels* show the log2 ratio (LR) derived from serum samples versus log ratios derived from adjacent tissue and distant tissue, respectively. The *bottom four panels* feature zoomed portions of plots above, indicated by the color of the plot (*green* for decreased and *red* for increased expression compared to non-tumor tissue). Tissue analytes shown in Table 20.3 have been labeled and serum analytes are shown in filled *red* symbols

20.6 Sample Handling Bias and the Role of Complement

As previously mentioned, our current assay measures 1,030 proteins. Represented on the menu are many of the complement proteins (Table 20.4). The role of complement in various disease states is well appreciated, especially among the authors and readers of this book. In our studies, we often identify certain components of the complement pathways as potential disease-related biomarkers, including both examples reviewed above: factor D in CKD and C9 in NSCLC. It is critical that we be able to distinguish between the appearance of complement component as indicators of sample handling artifacts and the possibility that they may be true biomarkers of a disease state.

The discovery and clinical utility of biomarkers in blood can be heavily biased by sample collection. Blood contains powerful cellular and humoral systems that react to injury or foreign and infectious agents. Small challenges can induce the innate immune system (including complement system and cells such as macrophages) to release powerful signals and enzymes and lead to activation of platelets and trigger the coagulation of the blood. These signals are of interest because they are generally related to the processes inside the body and because they can be directly involved in defense and repair systems and serve as markers for disease. However, such process signals are also responsive to the effects of blood sample preparation. Merely drawing blood from a vessel through a needle or exposing blood to air can result in unintended activation of these mechanisms. For example, altering the time, centrifuge speed, or temperature of sample processing steps can alter the apparent protein composition of serum or plasma such that physiologic information is masked by the pre-analytic variability imparted on the sample during collection and processing. The strong susceptibility of these processes and proteins to subtle alterations in sample handling of the proteins can compromise their use as biomarkers due to the concomitant lack of robustness.

Blood samples collected in tubes without anticoagulant are susceptible to profound in vitro changes associated with the coagulation and complement pathways. Both of these catalytic calcium-dependent pathways continue to change as the blood tube is held at room temperature for clotting in order to separate serum from the cellular and fibrinolytic components. During this time, the complement pathways are activated, and we have observed time- and temperature-dependent changes in many complement components, including C3 and C4 (Ostroff et al. 2010a, b; Gold et al. 2010).

Figure 20.7 demonstrates the pre-analytic variability in C3 measured in serum. This marker alone perfectly separates cases and controls in serum samples from a commercial biorepository, yet the separation is due to differences in serum processing, not disease state. Pre-analytic artifacts such as this one may cloud the discovery of true disease markers and may lead investigators down the

Table 20.4 Complement proteins represented on the current version of the SOMAscan menu

Complement proteins	
C1-Esterase inhibitor	C5
C1q	C5a
C1qBP	C6
C1r	C7
C1s	C8
C2	C9
C3	CFHR5
C3a	CRP
C3a desArg	Factor B
C3b	Factor D
iC3b	Factor H
C3d	Factor I
C4	MASP3
C4b	Properdin

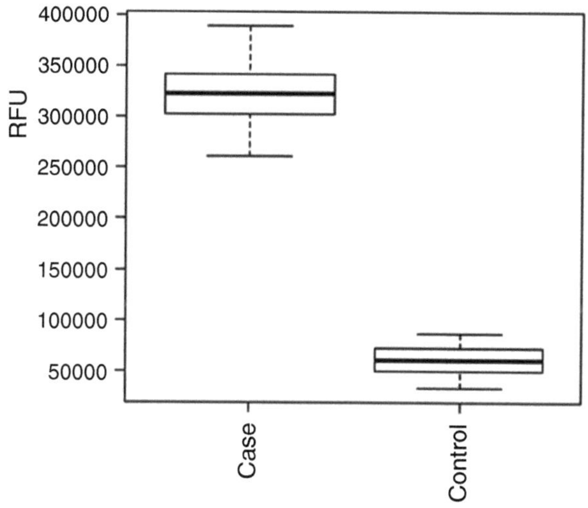

Fig. 20.7 Box plot of complement C3 serum concentration (expressed as relative fluorescence units, RFU) measured in the SOMAmer proteomics assay. NSCLC cases ($n=20$) and control long-term smokers ($n=28$) are clearly separated based on C3 measurements resulting from sample handling differences

wrong path, leading to failed biomarker validation studies. We have expanded our characterization of the pre-analytic variability inherent in poor compliance with sample processing protocols and routinely screen blood sample collections for uniform sample quality prior to embarking on disease biomarker studies.

20.7 And What About Complement (or Any Other) Therapeutics?

The potential of aptamers as therapeutics has been a subject of much interest since the time SELEX was invented. Several comprehensive recent reviews cover the development of aptamer-based therapeutics over the last two decades in considerable detail (Bouchard et al. 2010; Keefe et al. 2010). The net outcome at this time is one drug, numerous drug candidates in development, and high potential yet to be realized. Of course, it is worth remembering that it took quite a few years for antibodies to be embraced as therapeutics following the advent of the hybridoma technology in 1975 (Köhler and Milstein 1975). Nevertheless, the current status and remaining challenges are worth addressing.

The range of indications for which aptamers can be considered, and therefore the overall potential of aptamers as therapeutics, is determined in large part by their pharmacokinetic (PK) and biodistribution properties. In this context, it is useful to classify indications into three categories: conditions where the drug can be administered locally, acute conditions requiring drug treatment for a short period of time, and chronic conditions where long-term treatment is needed. Where feasible, local administration has considerable appeal since it allows systemic PK considerations to be largely bypassed (however, with its own specific demands, such as the need for suitable topical formulations, local biodistribution, and the like). Treatment of acute conditions generally does not require long residence times of the drug, and ensuring adequate exposure is feasible without extensive optimization. It is in the treatment of chronic conditions that long residence times are most needed since infrequent dosing is generally desirable to manage cost and compliance, and for this class of indications further work is required. Because aptamers are macromolecules, oral bioavailability is low, so only local and parenteral administration is practical.

The successful development of Macugen, which remains the only aptamer drug to have received FDA approval to date, was largely the result of the unique advantages of local delivery, coupled with favorable local PK, favorable local safety profile, low systemic exposure, and the clear role of the target (vascular endothelial growth factor, VEGF) in the pathology of age-related macular degeneration (AMD). Following in this track, there are two additional aptamer drug candidates in development for the treatment of AMD, both also delivered by intravitreal injection: E10030, an inhibitor of PDGF B-chain currently in phase 2 testing for AMD, and ARC1905, an inhibitor of C5 currently in phase 1 testing for AMD. Both agents are intended for use together with a VEGF inhibitor. Illustrating the long time frame for development of therapeutics, especially when coupled with often discontinuous nature of drug development in biotechnology companies, both E10030 and ARC1905 in their current form were identified in the late 1990s (Green et al. 1996; Floege et al. 1999; Biesecker et al. 1999). AMD and other ophthalmology indications remain an interesting area for aptamer-based therapeutics.

For systemically acting drugs, the therapeutic agent needs to be available for a sufficient period of time at the body site where its action is needed. Because blood is the reservoir from which drug is delivered to other parts of the body, residence time of a drug in plasma is often used as a conveniently measured surrogate for its availability in tissues. For aptamers (and other nucleic acid-based drug candidates), plasma residence has been primarily determined by nuclease-mediated metabolism and size-mediated clearance. For most aptamer drug candidates in development (as well as for Macugen), substantial nuclease resistance has been achieved primarily with modifications at the 2′-position of ribose (2′-O-methyl and 2′-fluoro substitutions) and, to a lesser degree, phosphorothioate backbone modifications. Increase in size has to date been almost universally achieved with conjugation to 40-kDa polyethylene glycol (PEG). For such constructs, plasma half-lives in primates ranging from 9 to 75 h have been reported (Tucker et al. 1999; Bouchard et al. 2010). Other than PEG, other means by which aptamers can be kept in circulation remain largely unexplored at this time and represent a potentially useful future research direction. Understanding of physicochemical and pharmacological properties of SOMAmers compared to conventional aptamers is another research direction with considerable potential.

20.8 Concluding Remarks

The promise of personalized medicine will only be realized when it is possible to accurately measure an individual's state of wellness in real time, quickly and economically. In addition, the early detection of disease—when it is usually most treatable—is a critical element of future clinical practice. Despite its explosive progress, genomics is only one element of realizing this medical future, as it deals largely with potential risk of disease rather than its actual presence in real time.

We set out over a decade ago to develop a new technology that makes the detection and measurement of even the most subtle protein changes as straightforward as measuring genetic alterations, overcoming the limitations inherent in mass spectrometry-based or antibody-based approaches. Our SOMAmer-based technology now allows us to measure over 1,000 proteins (to date) in a small sample of blood or other tissues, thereby unlocking protein biomarker discovery and promising to revolutionize clinical diagnostics and therapeutics. We have applied our technology successfully to a range of diseases, and we continue to build both our SOMAmer library and to pursue its multiple potential applications.

References

Biesecker G, Dihel L, Enney K, Bendele RA (1999) Derivation of RNA aptamer inhibitors of human complement C5. Immunopharmacology 42:219–230

Binz HK, Amstutz P, Pluckthun A (2005) Engineering novel binding proteins from nonimmunoglobulin domains. Nat Biotechnol 23:1257–1268

Bossuyt PMM (2011) The thin line between hope and hype in biomarker research. J Am Med Assoc 305:2229–2230

Bouchard PR, Hutabarat RM, Thompson KM (2010) Discovery and development of therapeutic aptamers. Annu Rev Pharmacol Toxicol 50:237–257
Diamandis EP (2010) Cancer biomarkers: can we turn recent failures into success? J Natl Cancer Inst 102:1462–1467
Ellington AD, Szostak JW (1990) In vitro selection of RNA molecules that bind specific ligands. Nature 346:818–822
Fassett RG, Venuthurupalli SK, Gobe GC, Coombes JS, Cooper MA, Hoy WE (2011) Biomarkers in chronic kidney disease: a review. Kidney Int 80:806–821
Floege J, Ostendorf T, Janssen U, Burg M, Radeke HH, Vargeese C, Gill SC, Green LS, Janjic N (1999) Novel approach to specific growth factor inhibition in vivo: antagonism of platelet-derived growth factor in glomerulonephritis by aptamers. Am J Pathol 154:169–179
Gold L, Ayers D, Bertino J, Bock C, Bock A, Brody EN, Carter J, Dalby AB, Eaton BE, Fitzwater T, Flather D, Forbes A, Foreman T, Fowler C, Gawande B, Goss M, Gunn M, Gupta S, Halladay D, Heil J, Heilig J, Hicke B, Husar G, Janjic N, Jarvis T, Jennings S, Katilius E, Keeney TR, Kim N, Koch TH, Kraemer S, Kroiss L, Le N, Levine D, Lindsey W, Lollo B, Mayfield W, Mehan M, Mehler R, Nelson SK, Nelson M, Nieuwlandt D, Nikrad M, Ochsner U, Ostroff RM, Otis M, Parker T, Pietrasiewicz S, Resnicow DI, Rohloff J, Sanders G, Sattin S, Schneider D, Singer B, Stanton M, Sterkel A, Stewart A, Stratford S, Vaught JD, Vrkljan M, Walker JJ, Watrobka M, Waugh S, Weiss A, Wilcox SK, Wolfson A, Wolk SK, Zhang C, Zichi D (2010) Aptamer-based multiplexed proteomic technology for biomarker discovery. PLoS One 5:e15004
Green LS, Jellinek D, Jenison R, Östman A, Heldin CH, Janjic N (1996) Inhibitory DNA ligands to platelet-derived growth factor B-chain. Biochemistry 35:14413–14424
Gupta S, Thirstrup D, Jarvis TC, Schneider DJ, Wilcox SK, Carter J, Zhang C, Gelinas A, Weiss A, Janjic N, Baird GS (2011) Rapid histochemistry using slow off-rate modified aptamers with anionic competition. Appl Immunohistochem Mol Morphol 19:273–278
Hori S, Gambhir SS (2011) Mathematical model identifies biomarker-based early cancer detection strategies and limitations. Sci Transl Med 3:109ra116
Hori et al. (2011) Mathematical model identifies biomarker-based early cancer detection strategies and limitations. *Sci Transl Med* 3:109ra116
Ioannidis JPA, Panagiotou OA (2011) Comparison of effect sizes associated with biomarkers reported in highly cited individual articles and in subsequent meta-analyses. J Am Med Assoc 305:2200–2210
Jemal A, Siegel R, Xu J, Ward E (2010) Cancer statistics 2010. CA Cancer J Clin 58:71–96
Kassis ES, Vaporciyan AA, Swisher SG, Correa AM, Bekele BN, Erasmus JJ, Hofstetter WL, Komaki R, Mehran RJ, Moran CA, Pisters KM, Rice DC, Walsh GL, Roth JA (2009) Application of the revised lung cancer staging system (IASLC Staging Project) to a cancer center population. J Thorac Cardiovasc Surg 138:412–418
Keefe AD, Pai S, Ellington A (2010) Aptamers as therapeutics. Nat Rev Drug Discov 9:537–550
Köhler G, Milstein C (1975) Continuous cultures of fused cells secreting antibody of predefined specificity. Nature 256:495–497
Levey AS, Atkins R, Coresh J, Cohen EP, Collins AJ, Eckardt KU, Nahas ME, Jaber BL, Jadoul M, Levin A, Powe NR, Rossert J, Wheeler DC, Lameire N, Eknoyan G (2007) Chronic kidney disease as a global public health problem: approaches and initiatives—a position statement from kidney disease improving global outcomes. Kidney Int 72:247–259
Lutz AM, Willmann JK, Cochran FV, Ray P, Gambhir SS (2008) Cancer screening: a mathematical model relating secreted blood biomarker levels to tumor sizes. PLoS Med 5:e170
Mehan MR, Ayers D, Thirstrup D, Xiong W, Zichi D, Ostroff RM, Brody EN, Walker, JJ, Gold L, Jarvis TC, Janjic N, Baird GS, Wilcox SK (2012) Protein signature of lung cancer tissues. PLoS ONE 7:e35157
Okada M, Nishio W, Sakamoto T, Uchino K, Yuki T, Nakagawa A, Tsubota N (2005) Effect of tumor size on prognosis in patients with non-small cell lung cancer: the role of segmentectomy as a type of lesser resection. J Thorac Cardiovasc Surg 129:87–93
Ostroff RM, Bigbee WL, Franklin W, Gold L, Mehan M, Miller YE, Pass HI, Rom WN, Siegfried JM, Stewart A, Walker JJ, Weissfeld JL, Williams S, Zichi D, Brody EN (2010a) Unlocking biomarker discovery: large scale application of aptamer proteomic technology for early detection of lung cancer. PLoS One 5:e15003
Ostroff R, Foreman T, Keeney TR, Stratford S, Walker JJ, Zichi D (2010b) The stability of the circulating human proteome to variations in sample collection and handling procedures measured with an aptamer-based proteomics array. J Proteomics 73:649–666
Pascual M, Steiger G, Estreicher J, Macon K, Volanakis JE, Schifferli JA (1988) Metabolism of complement factor D in renal failure. Kidney Int 34:529–536
Riemsdijk-van Overbeeke IC, Baan CC, Hesse CJ, Loonen EH, Niesters HG, Zietse R, Weimar W (2000) TNF-alpha: mRNA, plasma protein levels and soluble receptors in patients on chronic hemodialysis, on CAPD and with end-stage renal failure. Clin Nephrol 53:115–123
Stevens LA, Coresh J, Greene T, Levey AS (2006) Assessing kidney function—measured and estimated glomerular filtration rate. N Engl J Med 354:2473–2483

Tucker CE, Chen L-S, Judkins MB, Farmer JA, Gill SC, Drotel DW (1999) Detection and plasma pharmacokinetics of an anti-vascular endothelial growth factor oligonucleotide-aptamer (NX1838) in rhesus monkeys. J Chromatogr B 732:203–212

Tuerk C, Gold L (1990) Systematic evolution of ligands by exponential enrichment: RNA ligands to bacteriophage T4 DNA polymerase. Science 249:505–510

Vanholder R, Van Laecke S, Glorieux G (2008) The middle-molecule hypothesis 30 years after: lost and rediscovered in the universe of uremic toxicity? J Nephrol 21:146–160

Vaught JD, Bock C, Carter J, Fitzwater T, Otis M, Schneider D, Rolando J, Waugh S, Wilcox S, Eaton BE (2010) Expanding the chemistry of DNA for in vitro selection. J Am Chem Soc 132:4141–4151

Venturoli D, Rippe B (2005) Ficoll and dextran vs. globular proteins as probes for testing glomerular permselectivity: effects of molecular size, shape, charge, and deformability. Am J Physiol Renal Physiol 288:F605–F613

Chapter 21
Complement in Action: An Analysis of Patent Trends from 1976 Through 2011

Kun Yang, Robert A. DeAngelis, Janet E. Reed, Daniel Ricklin, and John D. Lambris

Abstract Complement is an essential part of the innate immune response. It interacts with diverse endogenous pathways and contributes to the maintenance of homeostasis, the modulation of adaptive immune responses, and the development of various pathologies. The potential usefulness, in both research and clinical settings, of compounds that detect or modulate complement activity has resulted in thousands of publications on complement-related innovations in fields such as drug discovery, disease diagnosis and treatment, and immunoassays, among others. This study highlights the distribution and publication trends of patents related to the complement system that were granted by the United States Patent and Trademark Office from 1976 to the present day. A comparison to complement-related documents published by the World Intellectual Property Organization is also included. Statistical analyses revealed increasing diversity in complement-related research interests over time. More than half of the patents were found to focus on the discovery of inhibitors; interest in various inhibitor classes exhibited a remarkable transformation from chemical compounds early on to proteins and antibodies in more recent years. Among clinical applications, complement proteins and their modulators have been extensively patented for the diagnosis and treatment of eye diseases (especially age-related macular degeneration), graft rejection, cancer, sepsis, and a variety of other inflammatory and immune diseases. All of the patents discussed in this chapter, as well as those from other databases, are available from our newly constructed complement patent database: www.innateimmunity.us/patent.

21.1 Introduction

21.1.1 Complement: A Powerful Network for Immune Surveillance

Complement has traditionally been viewed as a major early defense system to protect the host from pathogenic organisms, and it is known to be intimately involved in regulation of immune responses. The three major pathways of complement activation are the classical, lectin, and alternative pathways (Fig. 21.1), though in certain circumstances, complement can be activated downstream of these

K. Yang • R.A. DeAngelis • D. Ricklin • J.D. Lambris (✉)
Department of Pathology and Laboratory Medicine, University of Pennsylvania, 401A Stellar-Chance Building, 422 Curie Blvd, Philadelphia, PA 19104, USA
e-mail: lambris@upenn.edu

J.E. Reed
Potter Anderson & Corroon LLP, Wilmington, DE, USA

J.D. Lambris et al. (eds.), *Complement Therapeutics*, Advances in Experimental Medicine and Biology 735, DOI 10.1007/978-1-4614-4118-2_21, © Springer Science+Business Media New York 2013

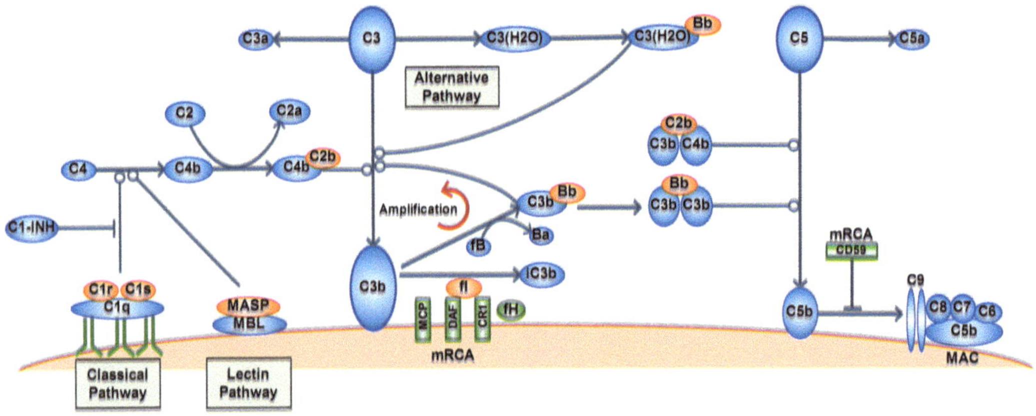

Fig. 21.1 *Activation and regulation of the complement system.* Complement activation has traditionally been described as occurring through three initiation pathways (classical, lectin, and alternative). External proteases are also known to activate complement, independent of these pathways (at the level of C3 and C5), but for simplicity, they are not shown here. The classical and lectin pathways share common mechanisms and differ mainly in their initial activating components. The classical pathway is initiated by the binding of C1q to antigen-antibody complexes, while the lectin pathway is triggered by the binding of MBL or ficolins to pathogen surfaces. This binding leads in both cases to the cleavage of C2 and C4 to form the C4b2b complex (C2b refers to the protease fragment of C2, which has also been referred to as C2a in previous literature), though this activity can be blocked by the C1 inhibitor (*C1-INH*); C4b2b then promotes the cleavage of the central complement component C3 and generates both the anaphylatoxin C3a and the opsonin C3b. The production of C3b can also be promoted by the interaction of C3(H_2O) with factor B (*fB*) and factor D (*fD*) to form an initial alternative pathway C3 convertase. C3b then combines with fB and fD to form the major alternative pathway C3 convertase (*C3bBb*), which leads to an amplification loop for further cleavage of C3. The combination of C3b with the C4b2b or C3bBb complexes leads to the formation of C5 convertases, which cleave C5 into the C5a anaphylatoxin and C5b. C5b then binds to C6, C7, C8, and C9 to form the membrane attack complex (*MAC*). To control the excessive activation of complement, soluble fH and membrane-associated regulators of complement activity (*mRCAs*) such as CR1, DAF, and MCP act as factor I (*fI*) cofactors and degrade C3b to iC3b. CD59, on the other hand, inhibits the assembly of the MAC by binding to C8 and/or C9

initiation events through the actions of external proteases. Though various factors can initiate complement activation, the three pathways all converge at the cleavage of C3 by C3 convertases, resulting in the production of C3a and C3b. C3a is an anaphylatoxin with chemoattractant properties and cell-signaling abilities through the C3a receptor, while C3b and its cleavage product iC3b can act as opsonins, being deposited on pathogenic cells and targeting them for phagocytosis or clearance, as well as the activation of immune cells through an interaction with complement receptors (Markiewski and Lambris 2007). C3b also initiates a self-amplification loop for further complement activation, as its deposition on cells is the basis for the formation of alternative pathway C3 convertase complexes, as well as C5 convertases that cleave C5 to continue downstream activation of the cascade. The cleavage of C5 produces the powerful anaphylatoxin C5a; like C3a, C5a (though signaling via the C5a receptor [C5aR/CD88]) can contribute to immune cell accumulation and activation, generally promoting inflammatory responses. C5b, which is also produced by the cleavage of C5, triggers the formation of a downstream membrane attack complex (MAC); this terminal component of complement activation can be involved in signaling at sublytic doses or can function to kill certain strains of bacteria through insertion in the membrane and subsequent cell lysis (Ricklin et al. 2010).

21.1.2 Complement in Infection and Disease

Because of the essential role of complement in the defense of the host against intruders, decreased activation of the complement system or deficiency in complement components can increase the risk of infection and cause various pathological conditions (Skattum et al. 2011). For example, alcoholic cirrhosis patients with low serum C3 concentrations and decreased hemolytic complement activity have been reported to have an increased risk of infections (Homann et al. 1997). Deficiencies of components of the classical pathway such as C1q, C2, and C4 have been found to be strongly associated with systemic lupus erythematosus (SLE) (Pickering et al. 2000; Pettigrew et al. 2009). On the other hand, uncontrolled, inappropriate, or excessive complement activity can cause damage to host cells and give rise to many diseases ranging from autoimmune to inflammatory pathologies (Markiewski and Lambris 2007; Holers 2003; Ricklin and Lambris 2007; Lambris and Sahu 2001). Indeed, the dual role of complement is illustrated by several pathological conditions. For example, the early increase of complement activation during sepsis may relate to the beneficial opsonization of bacteria. However, complement activation during subsequent phases of sepsis can amplify the initial insult, leading to inflammatory activity, tissue injury, and finally to multiple organ failure and death in many cases (Markiewski et al. 2008a).

Although complement has traditionally been seen as a defense mechanism against pathogens, it has been shown in recent years to also play a role in general immune surveillance, as well as a host of other immune-related functions and inflammatory diseases. Tissue regeneration, lipid metabolism, transplant rejection, age-related macular degeneration (AMD), rheumatoid arthritis, hemodialysis-associated thrombosis, and cancer are just a few of the increasing number of physiological and pathological processes in which complement activity has been implicated (Ricklin et al. 2010). As with pathogen defense, complement can act as both a protective and damaging factor in many of these conditions. For example, complement inhibits tumor growth through antibody-dependent cell-mediated cytotoxicity (ADCC) and complement-dependent cytotoxicity (CDC) (Markiewski and Lambris 2009). However, it has been found that tumors express high amounts of membrane-associated regulators of complement activity (mRCAs) and secrete fluid-phase regulators, which inhibit the activation of complement and contribute to tumor growth (Markiewski and Lambris 2009). In addition, a recent study has shown that complement can promote the progression of tumors in a mouse model of cervical cancer through the generation of C5a in the tumor microenvironment, which enhances the activity of immune-suppressing cells (Markiewski et al. 2008b). Thus, complement activity is critical for certain homeostatic or immune processes, but in some instances, such as over-activation or improper timing or length of activity, it can become detrimental to the overall health of the host.

21.1.3 Protecting Intellectual Property in Complement Research: An Analysis

The multifaceted role of complement in immune defense and disease makes it an attractive biomarker for diagnostic purposes and an obvious target for intervention by complement therapeutics. Thus, it is not surprising to see research and development in complement diagnostics and therapeutics reflected in a large number of granted patents and an increasing number of new patent applications being filed every year. Here, we investigate and analyze the patent landscape as it applies to complement, focusing on granted patents and applications as published by the United States Patent and Trademark Office (USPTO) and, in part, the World Intellectual Property Organization (WIPO). We highlight the history and emerging trends in patent publications on the complement system in different fields. We performed statistical analyses on (1) the specific complement components being targeted by innovations, (2) the areas in which complement-related patents have been focused, (3) the different complement inhibitor forms developed for experimental and potential therapeutic use, and (4) the clinical areas upon which relevant patents were centered. To further assist researchers in the complement field, a complement patent database has been constructed, and all the documents used in the statistical analyses presented here (from the USPTO and WIPO), as well as those retrieved from the European Patent Office (EPO) and Japanese Patent Office (JPO), are available from the database website:

www.innateimmunity.us/patent. Patents stored in the database have full-text links and are classified in terms of target, disease, application, and entity type. An advanced search function is available for finding patents based on specific interests. This database was intended to be thorough; however, there is no way to ensure that all complement-related patent publications are included. Accordingly, this database and the analysis of it as described in this chapter should be taken as reflecting trends only; they are not intended to replace a firsthand search of the USPTO and WIPO databases.

21.2 Study Design and Methodology

21.2.1 Terminology

It should be noted that in the text and figures here, the term "documents" is used to refer to all records retrieved from a database (both applications and granted patents). When a distinction is required between "applications" and "granted patents," these specific terms are used (with "US-granted patents" or "US applications" referring to those obtained from the USPTO). The term "publications" is used in a general sense to refer to data currently being discussed (in most cases, US-granted patents).

21.2.2 Data Retrieval and Analysis

To identify complement-related patent data as presented in this study, the databases of the USPTO and the WIPO were searched, retrieving all documents (applications and granted patents) whose title or abstract contained the keyword "complement." It should be noted that the WIPO searches were confined to publications that were part of the Patent Cooperation Treaty (PCT), which at the time consisted of 144 states. Dates of coverage for the searches were from the earliest dates for which data was available through December 31, 2011. For the USPTO database, the earliest date for records was January 1, 1976, for granted patents and March 15, 2001, for applications (no application records were available prior to this date), while for the WIPO/PCT, which did not distinguish between applications and granted patents, records started on January 1, 1978.

Once all complement-related documents were retrieved, the Python programming language was used to organize and process the resulting enormous amount of data. Documents that were not related to the complement system, as determined by using rules to exclude those containing keywords unrelated to biological systems (e.g., "automobile," "logic calculator," etc.), were discarded from the data pool; the remaining documents were individually examined to verify their association with the complement system.

Although both the USPTO and WIPO/PCT databases were searched, in most cases, only the USPTO-granted patents were used for the comprehensive data analysis because of the difficulty in ensuring that all duplications could be eliminated between these two databases and between US applications and granted patents, which would otherwise skew results when trends were being examined. Also, the USPTO database contained older data, and it was found that in most cases, inventors wanted to protect their intellectual property in the USA as well as abroad, and thus they filed for patents from both offices. Therefore, the USPTO results, specifically the granted patents, were seen to adequately represent general patent trends. However, a comparison between US patents and applications, as well as the US and WIPO/PCT database results, is presented in Fig. 21.2.

Once patents were verified as relating to the complement system, they were classified into various groups based on target, application, or disease, as presented in the results below.

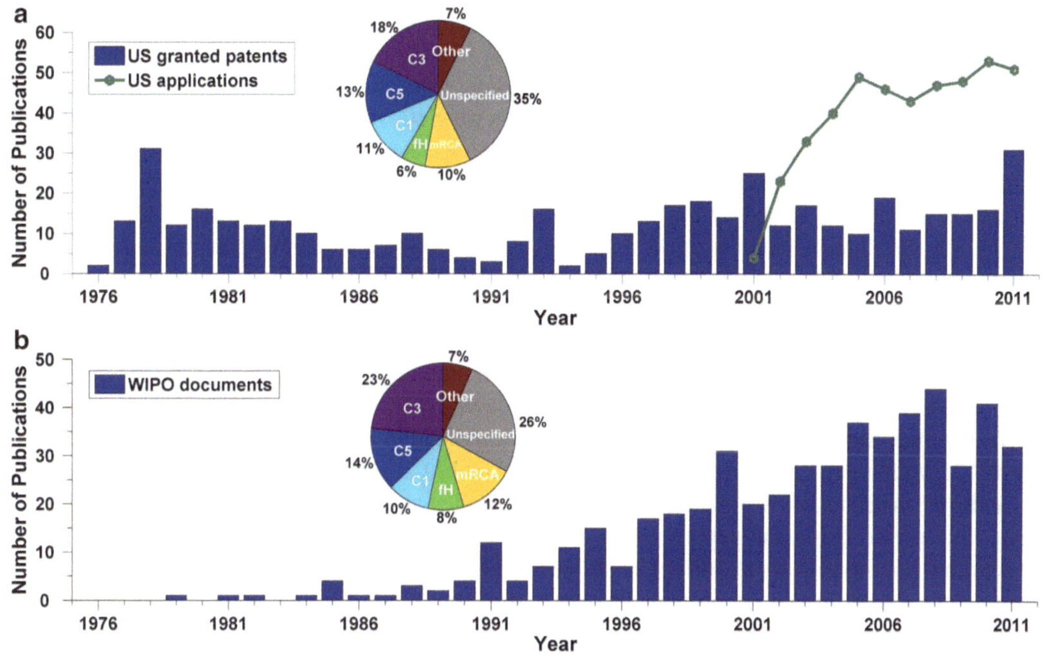

Fig. 21.2 *General publication trend for documents targeting complement components*. Granted patents and applications involving the complement system were collected, covering all available US patents and US applications (**a**) and WIPO/PCT documents (**b**) from 1976 to the end of 2011. Yearly trends for all complement-related publications are shown as column or line graphs. Publications were also classified by their complement targets, and the distribution was analyzed and displayed as a pie chart showing the top five targets (*C1*, *C3*, *C5*, *fH*, and *mRCAs*) from each database; the remaining complement components are grouped under "Other." The "Unspecified" category refers to publications that did not specify a specific complement target. For the analysis displayed in the pie chart in (**a**), duplications between US-granted patents and applications (for the years 2001 onward), in which two or more documents were found to have the same title, abstract, and claims, were removed for accuracy (only one document was included in the analysis). Most likely, these duplicates would represent an application and the resulting granted patent(s) and thus refer to the same innovation. However, if two or more documents were found to have the same title and abstract but differences in their claims, then these were considered two separate documents (as they could represent continuations-in-part), and both were included in the data shown

21.2.3 Patent Database

The complement patent database was created using MySQL. The web interface was implemented using PHP language. The database is running on a Windows 2003 server with IIS6 as the http server. To ensure comprehensive coverage, results were included from not only the USPTO and WIPO/PCT but also the EPO and JPO.

21.3 Results and Discussion

21.3.1 Patent Trends Reflect an Increased Interest in Complement Research

More than 1,000 patent documents related to the complement system have been published by the USPTO and WIPO/PCT since 1976. The publications have followed a general trend of yearly growth

(Fig. 21.2a, b, graphs). Specifically, the number of US applications has increased rapidly over the last 10 years, averaging about 48 per year since 2005 (Fig. 21.2a, graph). Compared to US applications, the number of US-granted patents has remained steadier. When analyzed by decade, the average numbers of granted patents in the 1980s, 1990s, and 2000s were 10, 10, and 16, respectively. Interestingly, over the last decade, the only year for which application data are available, the number of granted patents was less than half the number of applications. As is true for the trend in US applications, the number of documents issued by the WIPO/PCT has generally increased over the years (Fig. 21.2b, graph). The distribution patterns of US documents (applications and granted patents) and WIPO documents targeting different complement components were quite similar (Fig. 21.2a, b, pie charts), with C3, C5, C1, factor H (fH), and mRCAs (defined in this study as including MCP/CD46, DAF/CD55, CR1/CD35, and CD59) as the top five targets. Documents targeting these five complement components accounted for more than half of all publications. Correspondingly, the number of documents targeting other complement components was quite low, consisting of only 7% of all publications. Taken together, the data indicate that interest in the complement system has grown dramatically since the 1970s, with a particularly striking increase over the last 20 years. This is not entirely surprising, since there has been a corresponding increase in knowledge about the complement system during this same time period. Also, the perception of complement has shifted from that of an innate immune system that is primarily important for host defense against pathogens to a much more complex and cross-interactive pathway of proteins involved in a multitude of pathological and homeostatic responses (Ricklin et al. 2010). Thus, interest in and innovations related to complement have grown along with this gain in knowledge. It is also worth noting that while the number of complement-related US-granted patents has remained relatively steady during the past two decades, there has been a blossoming of international publications. This growth may reflect not only an increasing presence of complement research outside of the USA but also the general globalization trends leading US researchers to seek patents both at home and abroad; when the number of applications is taken into account, it is obvious that complement-related research remains prolific in the USA.

21.3.2 Target Analysis Reveals a Shift in the Focus of Complement Patents

We performed further analysis only on US-granted patents (see Methodology for the rationale behind excluding US applications and WIPO/PCT documents) in order to ascertain publication trends in the application of complement components and their modulators to various fields. As with the analysis of all USPTO and WIPO/PCT documents, C3, C5, mRCAs, C1, and fH were found to be the five most popular targets for US-granted patents, with an overall trend for yearly growth since the early 1980s (Fig. 21.3, graphs). It should be noted though that most patents granted in the 1970s and 1980s did not specify their targets. Instead, a large majority of those patents concerned the development of general chemical inhibitors of the complement system, for which the mechanism of action was likely not known, although several also involved the development of assays to detect complement activity. Before 1990, C1 was the most popular of the specified targets, with few patents relating to fH and mRCAs, the only other complement components specifically targeted during this time period (Fig. 21.3, pie charts). C1 was an early popular target for various types of inhibitors. Specific patents were published for amidine compounds with a C1 esterase inhibitory action useful for treating pancreatitis, hemorrhagic diseases, and thrombosis (Fujii et al. 1986, 1987). C1 was also utilized in detection assays, including one to reveal rheumatoid factors in blood (O'connell et al. 1987; Hallgren and Wide 1980; Taguchi et al. 1991), and another early patent described a method for its purification using chromatography (Bing 1983). From 1990 onward, as knowledge about the complement system grew, research interests became more diverse, not only in terms of initiators of complement activity, but also those components involved in its amplification, regulation, and downstream signaling. C3 replaced C1 as the dominant factor of interest and was the subject of 19% of all publications, while interest in

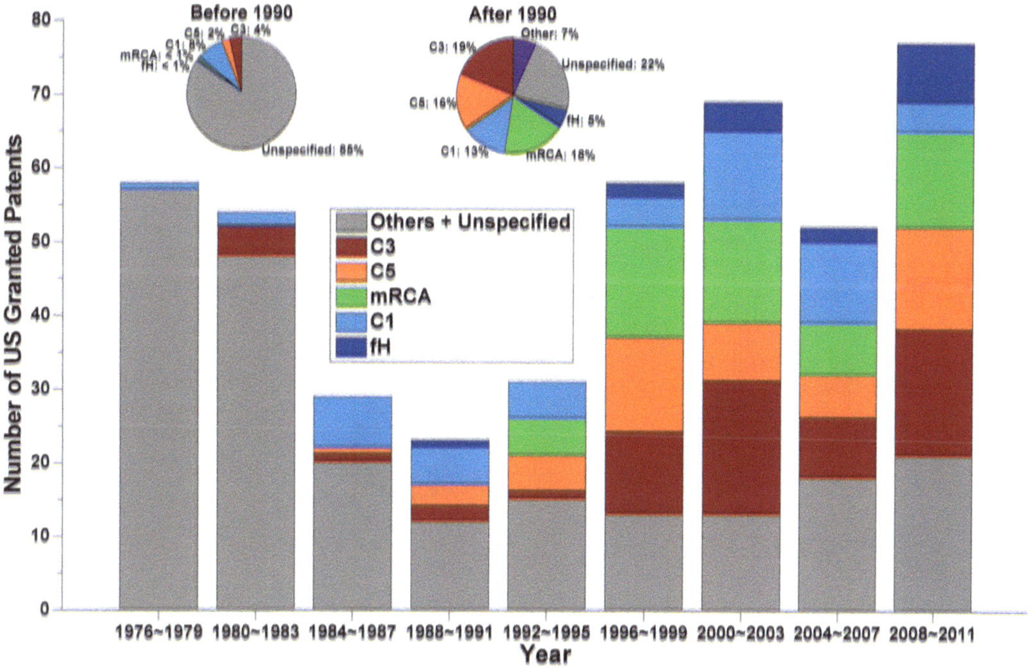

Fig. 21.3 *Patent publication trends for the top five complement targets.* The annual publication of US-granted patents related to one of the top five complement targets (*C1, C3, C5, fH,* and *mRCAs*) is plotted in 3-year periods. The distribution of patents for different targets before 1990 and from 1990 onward was also analyzed and charted, again showing specific distributions for the top five targets. "Other" refers to the remaining complement factors, such as C4, CR2, fB, fD, fI, and MBL-associated serine proteases (MASP), while "Unspecified" refers to publications that did not specify a specific complement target. Note that the "Other" category is not present in the pie chart showing patents prior to 1990 since these proteins were not the subject of any patents in that time period

C5, fH, and mRCAs also grew rapidly during this time. It is not entirely surprising that C3 became such a strong focus for innovation. As the central component of the complement pathway, the regulation of C3 is an obvious choice for nearly complete control of complement activity. Accordingly, it is an attractive target for the development of complement inhibitors, such as the peptide inhibitor compstatin (Qu et al. 2011). In addition, the detection of C3 cleavage is a useful tool for determining the activation of the complement system, and thus innovations for assays to detect this event are highly applicable to complement research in general. Finally, C3 and/or its cleavage products have been linked to a variety of diseases and homeostatic and pathological conditions (Ricklin et al. 2010), further enhancing its appeal as a target for therapeutic regulation. Increased knowledge of the workings of the complement system since 1990 is also evident in the fact that most patents granted since that time have clearly described the mechanisms of their innovations. In fact, only 22% did not specify targets, and the ratio of patents with unspecified targets to total patents generally decreased yearly (Fig. 21.3, gray bar height compared to total bar height).

21.3.3 Publication Trends Demonstrate Varied Applications of Complement Research and the Expansion of Inhibitor Forms

There are many different complement-related innovations described in the patent literature, such as discovering inhibitors against complement activation, developing assays, detecting complement factors as biomarkers to diagnose diseases, and designing new procedures to produce and purify complement

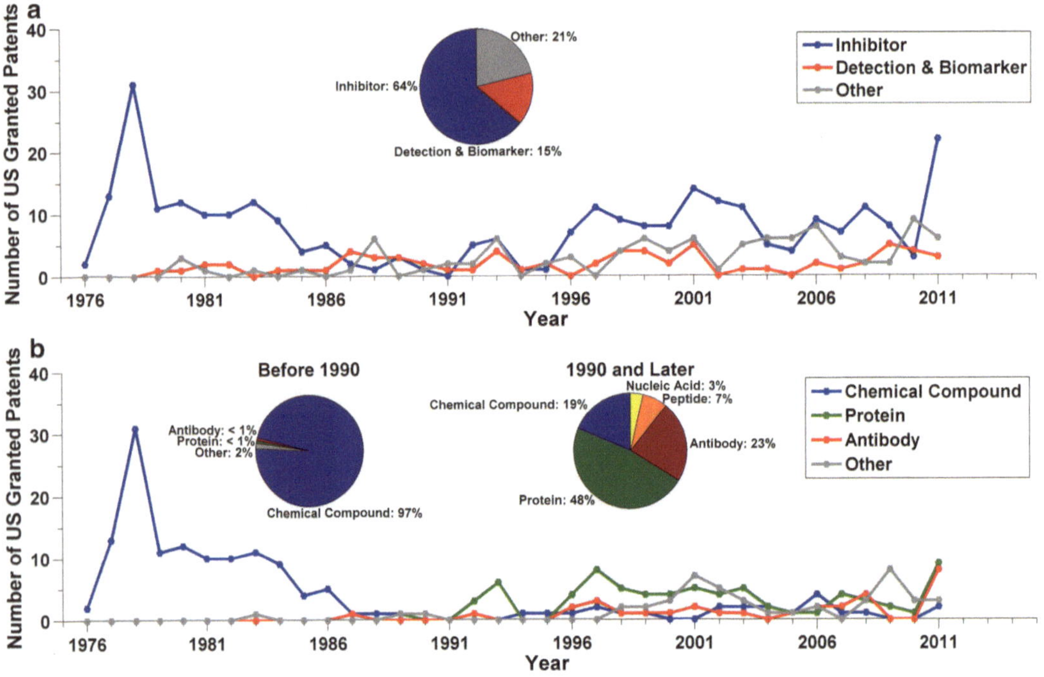

Fig. 21.4 *Publication trends for complement-related patents based on applications or inhibitor forms.* (**a**) The number of complement-related US-granted patents concerning inhibitor discovery, detection of complement components (including as biomarkers), or other applications (e.g., discovering novel complement-related proteins, producing animal models, describing crystal structures of complement factors, or developing absorbents to remove or immobilize complement components) is plotted to show annual trends. The overall distribution percentages are also charted. (**b**) The number of complement-related US-granted patents concerning inhibitors, as shown in (**a**), was further analyzed to determine what types of inhibitors were being proposed and developed, and the major forms are plotted to show annual trends. The data were also examined and charted to display distribution patterns before 1990 and from 1990 onward. "Other" in the graph refers to nucleic acids and peptides

components, among others. Of all these applications, the discovery of complement inhibitors has been the most prevalent, accounting for 64% of all publications (Fig. 21.4a, pie chart). Based on the yearly trends, the development of inhibitors was by far the major focus of early complement-related patents, and there has been another surge in interest over the past 10–15 years (Fig. 21.4a, graph). The granting of US patents for the detection of complement components, including biomarkers for disease diagnosis, and for other innovations has remained somewhat steady over the years, with both categories accounting for similar percentages of all publications (Fig. 21.4a, pie chart). A closer examination of the types of inhibitors being patented shows the overwhelming majority to be chemical compounds, proteins, and antibodies, which were the focus of at least 90% of all publications (Fig. 21.4b, pie charts). Before 1990, almost all complement inhibitors were chemical compounds, which were extensively synthesized by a broad range of pharmaceutical and chemical companies. Very frequently, the corresponding patents did not specify the exact target but instead described general complement inhibitors or inhibitory activity. For example, the American Cyanamid Company has patented more than 100 chemical compounds that can suppress complement activity in body fluids (Conrow and Bernstein 1978; Conrow et al. 1978; Lewis and Bernstein 1981). From 1990 onward, the interest in chemical compounds gradually shifted to antibodies and, largely, proteins. Only 19% of patents granted after 1989 claimed small synthetic compounds as complement inhibitors, while 71% of documents involved protein inhibitors and antibodies. Interestingly, a similar shift from small molecules toward biologics has been observed with FDA-approved drugs over the last 15 years (Mullard 2012). As revealed by yearly trends, the change in proportion for complement inhibitor-related patents seems to be due more to an overall decrease in the number of patents concerning inhibitors (Fig. 21.4a, graph), especially chemical compounds, rather than a large increase in the publication of other forms of inhibitors (Fig. 21.4b, graph). However, the increasing variety of and knowledge about compounds available for use as inhibitors in general since 1990 may explain the corresponding reemergence of patents related to complement inhibitors (Fig. 21.4a, graph), as the existence of new research tools likely spurred an interest in applying those tools to complement regulation.

Despite an overall strong interest in developing complement inhibitors, to date only two complement therapeutics have been approved by the FDA for use in humans (Ricklin and Lambris 2007). One is the purified glycoprotein C1 esterase inhibitor (C1-INH; Cinryze/ViroPharma, Cetor/Sanquin, Berinert/CSL Behring, Lev Pharma), and the other is a C5 antibody (Eculizumab; Soliris/Alexion Pharmaceuticals). Thus, it is not surprising that more and more attention has been paid to proteins and antibodies in recent years. Apart from chemical compounds, proteins, and antibodies, peptides and nucleic acids are the focus of the remaining 7% and 3% of all publications, respectively. Although they represent only a small fraction of all inhibitor-related publications, peptides and nucleic acids have been proposed as experimental therapeutics for several pathologies, and their potential future in disease treatment should not be neglected. For example, the peptidic C3 inhibitor compstatin and its analogs (Lambris and Sahu 2001; Lambris and Katragadda 2011) have been utilized and/or proposed for the treatment of eye disorders (Deschatelets et al. 2007), sepsis (Fung and Mollnes 2007), acute respiratory distress syndrome (ARDS) (Lambris and Ritis 2011), trauma (Francois et al. 2011a), Alzheimer's disease (Dinu 2007), pain (Woolf et al. 2006), and nerve regeneration (Baas and Ramaglia 2010), as well as other pathophysiological conditions.

21.3.4 Emerging Disease Areas Drive New Patent Applications

As mentioned, complement components are involved in many diseases, and they have been used as biomarkers for disease diagnosis, while complement inhibitors have been administered as therapeutics. Our analysis of US-granted patents has revealed eye disorders, transplants, cancer, and sepsis to be the major clinical conditions for which complement and its modulators have been utilized in a diagnostic or therapeutic sense (Fig. 21.5, pie chart).

In the case of eye disorders, about one-third of granted patents have claimed to use complement components to diagnose these diseases, while the other two-thirds focus on the application of complement inhibitors for the purpose of treatment. The three most common subjects of eye disorder-related patents have been fH, C3, and fB (Fig. 21.5, graph). For example, polymorphisms in the C3, fH, and fB genes have been found to predict the occurrence of AMD (Thakkinstian et al. 2011; Zipfel et al. 2010) and thus have been used as diagnostic markers (Allikmets et al. 2011; Day et al. 2010). In other instances, a factor D antibody was patented as a treatment for complement-associated eye conditions such as AMD and choroidal neovascularization (CNV) (Hass et al. 2011). Eculizumab was patented by Alexion Pharmaceuticals, Inc. as a C5-specific antibody for therapeutic use in various complement-related diseases (Evans et al. 2002; Bell 2008; Wang and Matis 2007; Bell and Rother 2009; Rother et al. 2010) and is currently being evaluated for the treatment of AMD (Ricklin and Lambris 2007; Yehoshua et al. 2012). In addition, virus proteins such as smallpox inhibitor of complement enzymes (SPICE) and vaccinia virus complement control protein (VCP), both of which inhibit C3 activity, have been administered locally to the eye to treat disorders such as macular degeneration and choroidal neovascularization (Francois et al. 2011b).

In transplantation medicine, complement activity is known to critically contribute to inflammation and the accommodation or rejection of transplanted tissue (Asgari et al. 2010; Hughes and Cohney 2011). Complement inhibitors have been utilized in transplant recipients through various means to

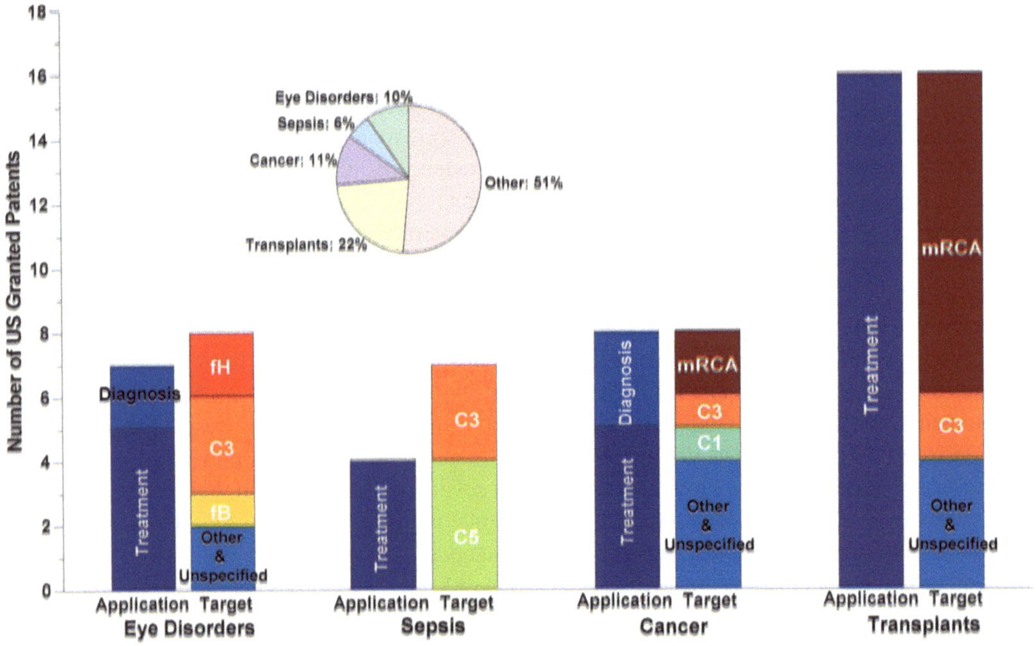

Fig. 21.5 *Distribution of the application of complement and its modulators to various diseases and pathophysiological conditions*. The chart displays the distribution percentages for US-granted patents that pertained to the use of complement and its modulators in the diagnosis and treatment of diseases and pathophysiological conditions, with "Other" referring to atherosclerosis, arthritis, SLE, Alzheimer's disease, spinal cord and neuronal injury, ischemia, infection, and other pathologies. The four main conditions described in these patents are shown in the graph: eye disorders (such as AMD), transplants, cancer, and sepsis. For each condition, the column on the left shows the number of patents related to treatment or diagnosis, and the column on the right indicates the top three complement components targeted, as well as others (in some cases, targets were not specified)

reduce the occurrence of adverse events against transplanted tissues, with mRCAs and C3 (or C3 convertases) being the most common points of intervention (Fig. 21.5, graph). Cells in transplanted tissues have been modified to express mRCAs, or organs have been perfused with membrane-targeted forms of recombinant RCAs in order to suppress complement activation and reduce the chances of complement attack on the tissue (Zhu 2007; Sims and Bothwell 1996; Smith et al. 2010, 2011). Transgenic animals expressing mRCAs have been produced for xenotransplantations (Diamond et al. 2000), and chimeric vaccinia virus proteins against C3b and C4b have been developed to prevent rejection by transplant recipients and improve the function of donor organs and tissues (Rosengard et al. 2000).

Complement and its modulators, especially mRCAs, C3, and C1, have been widely used in the diagnosis and treatment of cancer (Fig. 21.5, graph). One patented method for treating cancer has been the administration of an effective amount of a Coxsackie A-group virus, which recognizes and kills abnormal cells that express the mRCA DAF (Shafren 2008). C3b antibodies, alone or as conjugates with other antibodies, have been used to treat cancer through their binding to C3b on the surface of cancer cells (Taylor et al. 2003). Yet another patent describes analyzing patients' C1qA gene sequence as a means of predicting their response to CD20 antibody therapy (Racila and Weiner 2007).

Inhibitors against C3 and C5 have also been patented for the treatment of sepsis (Fig. 21.5, graph). An immunoadsorber with immobilized antibodies against C3a and/or C5a was developed for blood treatment in sepsis therapy (Heinrich et al. 2005) since excess activity of these anaphylatoxins can contribute to poor outcome in this disease (Bosmann and Ward 2012). Furthermore, complement and its modulators have been used in the diagnosis and treatment of arthritis, diabetes, SLE, ischemia, atherosclerosis, spinal cord and neuronal injury, and other diseases. For example, antibodies that inhibit the cleavage of C5 to C5a and C5b have been used to treat arthritis and prevent excessive downstream complement activation (Wang and Matis 2007). A C3 precursor biopolymer detected through the use of mass spectrometry has been utilized as a biomarker for type II diabetes (Jackowski and Marshall 2006). Finally, the levels of C4d and/or C3d on the surface of T lymphocytes, B lymphocytes, or monocytes in blood samples have been used to diagnose SLE (Ahearn et al. 2009). These are just a few examples of the many disease-related applications of complement that have been patented in the USA, yet they impressively illustrate the large diversity and creativity of complement-related patents, as well as their potential impact on both biomedical research and public health.

21.4 Conclusions and Outlook

The complement system has long been known to be important for host defense against invading pathogens. This alone would make it an attractive target for potential therapeutic applications to enhance the immune response and fight infections. However, studies over the past few decades have revealed an increasing role for complement in a large variety of both pathological and homeostatic processes (Ricklin et al. 2010). In many instances, complement can be beneficial in one respect but becomes harmful to the host in another, namely, when its activity is not properly regulated. Thus, the appeal of targeting complement in an attempt to control its activation and effects has only grown as its many functions and cross-interactions with other biological systems have been increasingly revealed. Further increasing the attractiveness of targeting the complement system is the fact that this system consists of over 50 proteins, which for the most part act in a hierarchical pathway, with the later steps heavily dependent on actions that occur upstream. This situation presents a multitude of potential targets for managing complement activity. Even so, the regulation of complement has not been the only goal of the many innovations patented over the past four decades. Just as important has been the development of assays to detect complement activity, which have aided both experimental research and diagnostic practices. As a result of these new ideas and discoveries, interest in the complement system has continued to grow in the USA and around the world. In addition, the ability to manipulate the production and activity of many complement factors has led to the development of potentially life-saving diagnostic and therapeutic tools. Thus, it seems that the observed trend toward a constantly increasing wealth of complement-related innovations will continue well into the future.

21.5 Disclosure

J.D.L. holds several patents about the development and clinical application of complement inhibitors, including compstatin. He has previously served as a member on the Scientific Advisory Board of Potentia Pharmaceuticals and is the founder of Amyndas Biopharmaceutics, which perform clinical development of compstatin analogs for various indications.

Acknowledgments The authors wish to thank Deborah McClellan for editorial assistance. This work was supported by National Institutes of Health grants AI030040, AI068730, AI071028, AI072106, AI097805, GM097747, DE021685, and EY020633.

References

Ahearn JM, Manzi SM, Liu C-C (2009) Diagnosing and monitoring inflammatory diseases by measuring complement components on white blood cells. US Patent 7,585,640

Allikmets RL, Hageman GS, Dean MC, Gold AM (2011) Variants in complement regulatory genes predict age-related macular degeneration. US Patent 8,012,683

Asgari E, Zhou W, Sacks S (2010) Complement in organ transplantation. Curr Opin Organ Transplant 15:486–491

Baas F, Ramaglia V (2010) Complement inhibition for improved nerve regeneration. US Application 20100143344

Bell L (2008) Methods for reducing morality associated with acute myocardial infarction. US Patent 7,361,339

Bell L, Rother RP (2009) Treatment of paroxysmal nocturnal hemoglobinuria patients by an inhibitor of Complement. US Application 20090220508

Bing DH (1983) Means and methods for purifying Clq, Clr and Cls. US Patent 4,374,061

Bosmann M, Ward PA (2012) Role of C3, C5 and anaphylatoxin receptors in acute lung injury and in sepsis. Adv Exp Med Biol 946:147–159

Conrow RB, Bernstein S (1978) 1,3,5- or 1,3,6-naphthalenetriyltris(sulfonylimino)aryl acids and salts. US Patent 4,087,613

Conrow RB, Poletto JF, Bernstein S (1978) 5-Carboxy-phenylenebis(carbonylimino)benzene carboxylic and dicarboxylic acids and salts. US Patent 4,089,974

Day AJ, Clark SJ, Bishop PN, Sim RB, Blom AM, Heinegard D (2010) Screening assays for binding antagonists of the AMD-associated variant of factor H (H384). US Patent 7,829,301

Deschatelets P, Olson P, Francois C (2007) Compstatin and analogs thereof for eye disorders. US Application 20070238654

Diamond LE, Logan JS, Byrne GW, Sharma A (2000) Method of producing transgenic animals for xenotransplantation expressing both an enzyme masking or reducing the level of the gal epitope and a complement inhibitor. US Patent 6,166,288

Dinu V (2007) Methods of preventing and treating Alzheimer's disease, age-related macular degeneration and other diseases involving extra-cellular debris through the inhibition of the complement system. US Application 20070196367

Evans MJ, Matis LA, Mueller EE, Nye SH, Rollins S, Rother RP, Springhorn JP, Squinto SP, Thomas TC, Wilkins JA (2002) C5-specific antibodies for the treatment of inflammatory diseases. US Patent 6,355,245

Francois C, Deschatelets P, Olson P (2011) Compositions and methods for treatment of trauma. US Application 20110092446

Francois C, Deschatelets P, Olson P (2011) Viral complement control proteins for eye disorders. US Patent 7,947,267

Fujii S, Yaegashi T, Nakayama T, Sakurai Y, Nunomura S, Okutome T (1986) Amidine compound, process for producing same and anti-complement agent comprising same. US Patent 4,570,006

Fujii S, Nakayama T, Nunomura S, Matsui R, Watanabe S-i, Sudo K, Okutome T, Kurumi M, Sakurai Y, Aoyama T (1987) Novel amidine compound. US Patent 4,634,783

Fung SC, Mollnes TE (2007) Methods and compositions for the prevention and treatment of sepsis. US Application 20070274989

Hallgren HR, Wide LE (1980) Method of indicating rheumatoid factors. US Patent 4,184,847

Hass P, Jianping Y, Katschke Jr K, Steffek M, Wiesmann C, Van Lookeren CM (2011) Prevention and treatment of complement-associated eye conditions. US Patent 8,007,791

Heinrich H-W, Hahn H-J, Meyer U, Kruschke P, Wagner H-J (2005) Immunoadsorber for use in sepsis therapy. US Patent 6,881,408

Holers VM (2003) The complement system as a therapeutic target in autoimmunity. Clin Immunol 107:140–151

Homann C, Varming K, Hogasen K, Mollnes TE, Graudal N, Thomsen AC, Garred P (1997) Acquired C3 deficiency in patients with alcoholic cirrhosis predisposes to infection and increased mortality. Gut 40:544–549

Hughes PD, Cohney SJ (2011) Modifiers of complement activation for prevention of antibody-mediated injury to allografts. Curr Opin Organ Transplant 16:425–433

Jackowski G, Marshall J (2006) Complement C3 precursor biopolymer markers predictive of type II diabetes. US Patent 7,097,989

Lambris JD, Katragadda M (2011) Potent compstatin analogs. US Patent 7,888,323

Lambris JD, Ritis K (2011) Method of treating acute respiratory distress syndrome. US Application 20110142837

Lambris JD, Sahu AK (2001) Peptides which inhibit complement activation. US Patent 6,319,897

Lewis AJ, Bernstein S (1981) C6-Modified cyclodextrin sulfate salts as complement inhibitors. US Patent 4,258,180

Markiewski MM, Lambris JD (2007) The role of complement in inflammatory diseases from behind the scenes into the spotlight. Am J Pathol 171:715–727

Markiewski MM, Lambris JD (2009) Unwelcome complement. Cancer Res 69:6367–6370

Markiewski MM, DeAngelis RA, Lambris JD (2008a) Complexity of complement activation in sepsis. J Cell Mol Med 12:2245–2254
Markiewski MM, DeAngelis RA, Benencia F, Ricklin-Lichtsteiner SK, Koutoulaki A, Gerard C, Coukos G, Lambris JD (2008b) Modulation of the antitumor immune response by complement. Nat Immunol 9:1225–1235
Mullard A (2012) 2011 FDA drug approvals. Nat Rev Drug Discov 11:91–94
O'connell JP, Hoke RA, Mize PD (1987) Immunoassay using complement. US Patent 4,693,970
Pettigrew HD, Teuber SS, Gershwin ME (2009) Clinical significance of complement deficiencies. Ann N Y Acad Sci 1173:108–123
Pickering MC, Botto M, Taylor PR, Lachmann PJ, Walport MJ (2000) Systemic lupus erythematosus, complement deficiency, and apoptosis. Adv Immunol 76:227–324
Qu H, Magotti P, Ricklin D, Wu EL, Kourtzelis I, Wu YQ, Kaznessis YN, Lambris JD (2011) Novel analogues of the therapeutic complement inhibitor compstatin with significantly improved affinity and potency. Mol Immunol 48:481–489
Racila EV, Weiner GJ (2007) Assessing response to anti-CD20 therapy by genotyping C1q components. US Patent 7,723,036
Ricklin D, Lambris JD (2007) Complement-targeted therapeutics. Nat Biotechnol 25:1265–1275
Ricklin D, Hajishengallis G, Yang K, Lambris JD (2010) Complement: a key system for immune surveillance and homeostasis. Nat Immunol 11:785–797
Rosengard AM, Ahearn Jr JM, Sanfilippo AP, Baldwin WM III (2000) Use of chimeric vaccinia virus complement control proteins to inhibit complement. US Patent 6,140,472
Rother RP, Wang H, Zhong Z (2010) Prolongation of survival of an allograft by inhibiting complement activity. US Application 20100135992
Shafren DR (2008) Methods for treating malignancies expressing ICAM-1 using coxsackie a group viruses. US Patent 7,361,354
Sims PJ, Bothwell ALM (1996) Cells expressing high levels of CD59. US Patent 5,573,940
Skattum L, van Deuren M, van der Poll T, Truedsson L (2011) Complement deficiency states and associated infections. Mol Immunol 48:1643–1655
Smith RA, Dodd I, Mossakowkska DEI (2010) Conjugates of soluble peptidic compounds with membrane-binding agents. US Patent 7655617
Smith RAG, Pratt JR, Sacks SH (2011) Method of preparing an organ by perfusion. US Patent 7,888,318
Taguchi F, Mitsui I, Hara K, Hayashi M, Ezawa K, Fukunaga K, Kuranari J (1991) Test method involving substance-conjugated complement component C1q. US Patent 5,035,995
Taylor RP, Nardin A, Sutherland WM, Sokoloff MH, Chung L (2003) Methods for the prevention and treatment of cancer using anti-C3b(i) antibodies. US Patent 6,572,856
Thakkinstian A, McKay GJ, McEvoy M, Chakravarthy U, Chakrabarti S, Silvestri G, Kaur I, Li X, Attia J (2011) Systematic review and meta-analysis of the association between complement component 3 and age-related macular degeneration: a HuGE review and meta-analysis. Am J Epidemiol 173:1365–1379
Wang Y, Matis L (2007) Methods for the treatment of inflammatory joint disease. US Patent 7,279,158
Woolf C, Costigan M, Griffin R (2006) Methods for treatment of pain. US Application 20060241074
Yehoshua Z, Garcia Filho CAA, Gregori G, Li Y, Feuer WJ, Penha FM, Sadda SR, Zhang L, Zhang K, Rosenfeld PJ (2012) Systemic complement inhibition with eculizumab for the treatment of geographic atrophy in AMD patients: the complete study. In: The 2012 ARVO annual meeting
Zhu A (2007) Modified organs and cells for xenotransplantation. US Patent 7,166,278
Zipfel PF, Lauer N, Skerka C (2010) The role of complement in AMD. Adv Exp Med Biol 703:9–24

Index

J.D. Lambris et al. (eds.), *Complement Therapeutics*, Advances in Experimental Medicine and Biology 735, DOI 10.1007/978-1-4614-4118-2, © Springer Science+Business Media New York 2013

MIX
Papier aus verantwortungsvollen Quellen
Paper from responsible sources
FSC® C105338

If you have any concerns about our products,
you can contact us on
ProductSafety@springernature.com

In case Publisher is established outside the EU,
the EU authorized representative is:
Springer Nature Customer Service Center GmbH
Europaplatz 3, 69115 Heidelberg, Germany

Printed by Libri Plureos GmbH
in Hamburg, Germany